Texts in Mathematics

Volume 9

Ordinary Differential Equations
Exercises and Problems

Texts in Mathematics Series Editor
Dov Gabbay   dov.gabbay@kcl.ac.uk

Ordinary Differential Equations

Exercises and Problems

Andrei Bourchtein

Ludmila Bourchtein

ISBN 978-1-84890-464-4

College Publications
Scientific Director: Dov Gabbay
Managing Director: Jane Spurr

http://www.collegepublications.co.uk

Cover designed by Laraine Welch

To Victoria for the highest scores of Lexile and Quantile;
To Valentina for the memorable essay on children's homework;
To Maxim and Natalia for very deep learning and dedication;
To Haim and Maria for the best memories

Preface

This book presents a variety of methods of analytic solution of ordinary differential equations and features a wealth of examples, both solved in detail and proposed. The goal of this work is to give a broad training in techniques of solving equations and problems involving differential equations, which are important both in mathematics and applications.

The book is aimed primarily at undergraduate students of mathematics, physics, chemistry and various engineering courses, who need a solid knowledge of differential equations in their professional activities. The text contains many examples of scientific problems solved by constructing models involving differential equations, which introduce students to a number of interesting aspects of applications.

Another group of readers who may be interested in this book are professors and instructors of differential equation courses, since the text contains an exposition of all the principal analytic techniques usually studied in undergraduate programs, which are illustrated with numerous examples. In this way, teachers/instructors could use this material both to teach classes and to administer exams/tests on the studied subjects.

All the exposition is kept at a level accessible to undergraduate students who have studied Calculus and Linear Algebra. The subject is developed gradually, starting with simple first order equations, proceeding to higher order equations and ending with systems of equations. The diverse types of equations and techniques of their solution considered throughout the text are employed to solve different application problems from various areas of science and engineering.

The book consists of fourteen chapters, which can be grouped into three parts: first order equations, higher order equations (especially second order equations) and systems of equations. The exposition begins with a general introductory section, explaining the basic concepts of equations and their solutions in Chapter 1. Chapters 2 to 5 deal with first order equations, arriving at their geometric and physical applications in Chapter 5. The next six chapters present concepts, results and methods of solution of higher order equations. The largest segment of this part is dedicated to the study of linear equations. Chapter 11 (the last of this part) considers different physical problems whose mathematical modeling leads to higher order differential equations. Finally, the last three chapters are dedicated to the study of systems of equations, culminating in a presentation of some application models, whose mathematical investigation involves the construction of systems of associated differential equations and the analysis of properties of their solutions.

Each chapter is comprised of sections and subsections, which are numbered separately within each chapter. For example, section 2 can be found in each of the fourteen chapters, while subsection 2.1 means the first subsection within the second section. When a section or subsection is referenced outside its own chapter, then the corresponding chapter number is indicated. Figures are numbered sequentially within each chapter regardless of the section in which they are found. For instance, Figure 5.6 indicates the sixth figure in the fifth chapter. Definitions, theorems, propositions, lemmas and examples are numbered independently within each section and subsection.

Numerous standard examples are included in each section for developing and training of essential techniques and also to test the understanding of concepts. Some of the posed problems are more complicated, they serve to illustrate finer points of theory and methods of solution. Different complementary problems are proposed as exercises for the reader during the exposition of the subject and also at the end of many sections and chapters. The answers to most of these exercises are given at the end of the book.

The main features of the text are the following:

1. Completeness: the text contains an exposition of all the methods of solution usually studied in undergraduate courses.

2. Self-sufficiency: all the background topics related to the techniques of solution of ordinary differential equations are covered in the text, and, consequently, this work can be used as both a textbook and a source for self-study.

3. Accessibility: all the subjects are developed gradually and all the exposition is kept at the level accessible to undergraduate students who have studied Calculus and Linear Algebra.

4. Exercises: there are a large number of problems and exercises, solved and proposed, which should make the book suitable for both classroom use and self-study.

Contents

Chapter 1

Initial concepts

1 Initial definitions: equation and its solution

Before introducing the first definitions, let us recall the standard mathematical notations to be used. The first-order derivative of a function $y(x)$ is denoted by $y' = y_x = \frac{dy}{dx}$; the derivative of order n is denoted by $y^{(n)} = y_{nx} = \frac{d^n y}{dx^n}$. The increment of the independent variable x has the notation dx and the differential dy of the function $y(x)$ is, by definition, $dy = y'dx$. Usually, we will work in this text with real functions of one and several real variables, so we will not mention this fact any more. In exceptional cases, when we will use other types of functions, we will state this explicitly.

Definition of an ordinary differential equation of order n. The equation in the form

$$F(x, y, y', \ldots y^{(n)}) = 0,$$

where F is a given function of $n + 2$ arguments and $y(x)$ is an unknown function of independent variable x, to be determined by solving this equation, is called an *ordinary differential equation of order n.*

Roughly speaking, an equation that links the independent variable x, the unknown function $y(x)$ and its derivatives up to order n is called an ordinary differential equation of order n.

Some important remarks about the definition.

Remark 1. It will often be clear from the context that we are considering an ordinary differential equation and, in these cases, we will simply call it an equation. The abbreviation we will often use is ODE.

Remark 2. The order of the equation is always the highest order of the derivative of y actually present in the equation. This means that in an equation of order n, the derivative $y^{(n)}$ cannot be eliminated using equivalence transformations. The analytical verification of this, in addition to elementary transformations, can be done by calculating the partial derivative of F with respect to $y^{(n)}$: if $F_{y^{(n)}} = 0$, then F does not depend on nth derivative. For example, the equation $\cos^2 y'' + \sin^2 y'' + y' = 0$ is not a 2nd order equation but a 1st order equation, since it is equivalent to the equation $y' = -1$. Calculating the partial derivative of F, we get: $F_{y''} = -2\cos y'' \sin y'' + 2\sin y'' \cos y'' = 0$. The equation $\cos^2 y' + \sin^2 y' + y = 0$ is not an ODE, but a simple definition of a constant function. For the partial derivative of F we have: $F_{y'} = -2\cos y' \sin y' + 2\sin y' \cos y' = 0$, that is, F does not depend on y'.

Remark 3. In general, the function y can depend on other variables, but an ODE can have derivatives of y only with respect to x (or any other letter used for notation of the main independent variable). All other variables can enter only as passive parameters, thus representing a family (set) of ODEs. If (partial) derivatives are present in an equation with respect to different variables of the unknown function, then this equation is called a partial differential equation, which is a more complicated type of equation than ODEs and is not part of our study. For example, the equation $y_x + y = x + t$, where $y(x, t)$ is a function of two variables, is an ODE of the 1th order with parameter t. Or, equivalently, a family of ODEs, depending on t, since for each fixed t we get a specific ODE of the 1th order. On the other hand, the equation $y_x + y_t = x + t$ is not an ODE but a partial differential equation, because the equation includes both the derivatives with respect to x and t.

Definition of the explicit/normal form of ODE of order n. The equation in the form

$$y^{(n)} = f(x, y, y', \ldots y^{(n-1)}),$$

where f is a given function of $n + 1$ arguments and $y(x)$ is an unknown function of variable x, is called an ODE of order n in *explicit/normal form*.

Remark. Obviously, the equation in explicit/normal form is the particular case of the general definition of the equation of order n when the function F has the explicit form with respect to the highest derivative.

Definition of the solution (particular solution). Function $y(x)$ is an *solution* (more precisely, *particular solution*) of an ODE if, substituted into that ODE, it transforms the equation into an identity.

Remark 1. As we will see below, there are other types of solutions to the same equation. To distinguish this type from others, it is called a *particular solution*.

Remark 2. A solution is usually considered on a set X of the values of x defined explicitly or implicitly by the form of the function $y(x)$ and the original equation. In most cases we will consider an interval I of any type: open, semi-open or closed, finite or infinite.

Remark 3. The definition of the solution implicitly assumes that the solution must be a function differentiable on X at least as many times as the order of the equation.

Remark 4. The definition itself indicates a simple way of checking whether or not a function is a solution of the given equation on the set X: first, you have to check whether the function is differentiable n times on X and, if so, then substitute it, along with its derivatives, into the original ODE and verify whether or not the latter becomes an identity.

Remark 5. Any analytical form of a function is accepted to define a solution. Remember that there are three analytical forms: explicit, implicit and parametric. The first is the simplest one: $y = f(x)$, for example $y = \sqrt{1 - x^2}$. In the second, the relation between y and x is not solved for y: $H(x, y) = 0$, for example, $y^2 + x^2 = 1, y \geq 0$. The third form is the most general and complicated, where the function y and variable x are linked via an additional parameter. Here we'll use the following representation: $\begin{cases} x = \varphi(t) \\ y = \psi(t) \end{cases}$, for example, $\begin{cases} x = \cos t \\ y = \sin t \end{cases}$, $t \in [0, \pi]$. In the shown examples, all the forms define the same function and the most complex, parametric, can be reduced to explicit. In general, this reduction is not possible and it is necessary to work with the implicit and parametric forms, including to use the respective formulas of differentiation: $y_x = -\frac{H_x}{H_y}$ for the implicit one, and $y_x = \frac{y_t}{x_t}$ for the parametric one.

Remark 6. Obviously, there are equations that do not admit any solution. For example, $y'^2 = -1$.

Definition of the general solution. A function $y(x, C_1, \ldots, C_n)$ of independent variable x and n independent parameters $C_1, \ldots, C_n$ is a *general solution* of an ODE of order n if for any particular choice of $C_1, \ldots, C_n$ it represents a particular solution of the same ODE.

Remark 1. Notice that the number of parameters in a general solution must coincide with the order of the equation.

Remark 2. A general solution may or may not contain all the particular solutions of the corresponding ODE. A particular solution that is not in the form of the general solution is often called special or singular. For example, the equation $(y - x)y' = 0$ has the general solution $y = C$ (this is easy to check by substituting its derivative $y' = 0$ into the equation) and the singular solution $y = x$, which cannot be obtained from the general solution (since the first factor is zero for this function, it is a particular solution).

Remark 3. Just as in the case of a particular solution, the general solution can be defined in any analytical form.

Remark 4. In general, except for special cases, an equation of order n determines its general solution (with n parameters). It turns out that, under mild conditions, the converse is also valid: a family of functions with n parameters generates an ODE of order n for which this family is the general solution. We will not specify the conditions for the validity of the converse here.

2 Exemplification of initial concepts

In order to illustrate the concepts introduced so far and show different situations that can occur, we gather various examples in Table 1 and investigate some of them.

Table 1. ODEs and their solutions. exp=explicit, imp=implicit

ODE	form, order	particular solution	general solution	special solution
1. $y' = 0$	exp, 1	$y = 1$	$y = C$	none
2. $xy' = 2y$	exp, 1	$y = 5x^2$	$y = Cx^2$	none
3. $y'' = 2$	exp, 2	$y = x^2$	$y = x^2 + Ax + B$	none
4. $y'' = y$	exp, 2	$y = e^x$	$y = Ae^x + Be^{-x}$	none
5. $xy' - y = xe^x$	exp, 1	$y = x\left(\int \frac{e^x}{x}dx + 1\right)$	$y = x\left(\int \frac{e^x}{x}dx + C\right)$	none
6. $y' = -\frac{x}{y}$	exp, 1	$x^2 + y^2 = 1$	$x^2 + y^2 = C^2$	none
7. $y' = \frac{2x+y+1}{4x+2y-3}$	exp, 1	$2x + y - 1 = e^{2y-x}$	$2x + y - 1 = Ce^{2y-x}$	none
8. $y'\ln\frac{y'}{4} = 4x$	imp, 1	$\begin{cases} x = t\ln t \\ y = t^2(2\ln t + 1) \end{cases}$	$\begin{cases} x = t\ln t \\ y = t^2(2\ln t+1)+C \end{cases}$	none
9. $y'^2 + e^{y'} = x$	imp, 1	$\begin{cases} x = t^2 + e^t \\ y = \frac{2}{3}t^3 + (t-1)e^t \end{cases}$	$\begin{cases} x = t^2 + e^t \\ y = \frac{2}{3}t^3 + (t-1)e^t + C \end{cases}$	none
10. $e^{y'} = 0$	imp, 1	none	none	none
11. $y' = 1$	exp, 1	$y = x$	$y = x + C$	none
12. $yy' = y$	exp, 1	$y = x$	$y = x + C$	$y = 0$
13. $yy'^2 = y$	imp, 1	$y = x$	$y = \pm x + C$	$y = 0$
14. $yy'^2 = -y$	imp, 1	$y = 0$	none	$y = 0$
15. $y' = \cos(y-x)$	exp, 1	$y = x$	$\cot\frac{y-x}{2} = x + C$	$y = x + 2k\pi,$ $k \in \mathbb{Z}$
16. $y^{(n)} = 1$	exp, n	$y = x$	$y = \frac{x^n}{n!} + C_1 x^{n-1} +$ $\ldots + C_{n-1}x + C_n$	none

Following definitions, we will check several results in Table 1 (except for the existence of special solutions, which requires a knowledge of the methods of solution presented later), leaving others to the reader.

1. $y' = 0$.
Differentiating the function $y = 1$, we get $y' = 0$, which shows that it is a particular solution. Differentiating the function $y = C$, we have $y' = 0$ for any value of the parameter C, which means that $y = C$ is the general solution (the number of parameters coincides with the order of the equation).

2. $xy' = 2y$.
Differentiating the function $y = 5x^2$, we get $y' = 10x$ and substituting it into the equation we obtain the identity: $xy' = x \cdot 10x = 2 \cdot 5x^2 = 2y$, which means that $y = 5x^2$ is a particular solution. Differentiating the function $y = Cx^2$, we get $y' = 2Cx$ and substituting it into the equation we obtain the identity for any C: $xy' = x \cdot 2Cx = 2 \cdot Cx^2 = 2y$. Since the number of parameters is equal to the order of the equation, $y = Cx^2$ is the general solution.

3. $y'' = 2$.
Differentiating $y = x^2 + Ax + B$ twice, we get $y'' = 2$, that is, the equation is satisfied. In particular, for $A = B = 0$ we have the indicated particular solution. Since the number of parameters is equal to the order of the equation, $y = x^2 + Ax + B$ is the general solution.

4. $y'' = y$. Task for the reader.

5. $xy' - y = xe^x$.
Let us consider the proposal for the general solution $y = x \left(\int \frac{e^x}{x} dx + C \right)$, because the particular one is obtained when $C = 1$. Remember that the integral of $\frac{e^x}{x}$ cannot be calculated in elementary functions, so we have to perform the differentiation in the integral form of the solution: $y' = \int \frac{e^x}{x} dx + C + x \cdot \frac{e^x}{x} = \frac{y}{x} + e^x$. Multiplying the last expression by x, we get the original equation. Therefore, the function $y = x \left(\int \frac{e^x}{x} dx + C \right)$ is the solution of the equation for any value of the parameter C, which means that it is the general solution and for $C = 1$ we have the particular one.

6. $y' = -\frac{x}{y}$.
Again, let us check right away the proposed general solution $x^2 + y^2 = C^2$. Differentiating both sides with respect to x (and recalling that y is a function of x), we get $2x + 2yy' = 0$, or, isolating the derivative, $y' = -\frac{x}{y}$, that is, the equation is satisfied for any C. Consequently, $x^2 + y^2 = C^2$ is the general solution and $x^2 + y^2 = 1$ is the particular one. We notice that the implicit

form $x^2 + y^2 = C^2$ defines the two functions $y = \pm\sqrt{C^2 - x^2}$ and it was verified that both are solutions of the original equation. This can also be demonstrated by using the explicit form (this is a task for the reader).

7. $y' = \frac{2x+y+1}{4x+2y-3}$. Task for the reader.

8. $y' \ln\frac{y'}{4} = 4x$.

Since there is no way to transform the parametric form into an implicit or explicit form, we perform the parametric differentiation of the proposed general solution: $y_x = \frac{y_t}{x_t} = \frac{2t(2\ln t+1)+t^2\frac{2}{t}}{\ln t+t\frac{1}{t}} = \frac{4t(\ln t+1)}{\ln t+1} = 4t$. So $y' \ln\frac{y'}{4} = 4t\ln t = 4x$,

that is, the equation is satisfied for any C. Therefore, $\begin{cases} x = t\ln t \\ y = t^2(2\ln t+1)+C \end{cases}$
is the general solution and for $C = 0$ we get the particular one.

9. $y'^2 + e^{y'} = x$. Task for the reader.

10. $e^{y'} = 0$. Task for the reader.

11. $y' = 1$. Task for the reader.

12. $yy' = y$. Task for the reader.

13. $yy'^2 = y$. Task for the reader.

14. $yy'^2 = -y$.

The given equation can be written in the form $y(y'^2+1)=0$, which shows that there are two options: the first $y = 0$ and the second $y'^2 + 1 = 0$. The second does not generate any solution, since the sum of a non-negative expression and a positive number cannot be 0. This leaves the only solution $y = 0$. Since the equation does not possess a general solution, this solution is singular.

15. $y' = \cos(y - x)$.

Let us check the proposed general solution and the singular ones (the particular solution is one of the singulars). Differentiating both sides of the implicit relation $\cot\frac{y-x}{2} = x + C$, we get $-\frac{1}{\sin^2\frac{y-x}{2}} \cdot \frac{y'-1}{2} = 1$ or $y' = 1 - 2\sin^2\frac{y-x}{2}$.
Using the trigonometric formula $\cos 2\alpha = 1 - 2\sin^2\alpha$, we transform the last equation into the form of the original one: $y' = \cos(y - x)$. Thus, we have the general solution of the given equation. To check the singular solutions, we use the trivial result $y' = 1$ and note that $\cos(y-x) = \cos 2k\pi = 1$ for any $k \in \mathbb{Z}$. in this way, we confirm that $y = x + 2k\pi$, $k \in \mathbb{Z}$ are solutions of the original equation (including for $k = 0$ which gives the particular solution).

16. $y^{(n)} = 1$. Task for the reader.

Remark. Although the methods of finding the solutions presented in Table 1 are not of our concern at the moment, we will give some brief comments on how these solutions can be obtained. These brief explanations are summarized in Table 2. In the cases when the calculation of a solution or verification of its non-existence is trivial, a short explanation about the used technique is provided; in other cases, the reference to later sections where the corresponding methods of solution will be studied is pointed out.

Table 2. ODEs of Table 1 and methods of their solution

ODE	reference to the method of solution
1. $y' = 0$	trivial, direct integration of the function on the right-hand side of the equation
2. $xy' = 2y$	Example 3 in section 1.1 of Chapter 3 (separable equations)
3. $y'' = 2$	trivial, direct integration of the function on the right-hand side of the equation
4. $y'' = y$	Example 1 in section 1 of Chapter 9 (linear equations of the second order)
5. $xy' - y = xe^x$	Example 4 in section 4.1 of Chapter 3 (linear equations of the first order)
6. $y' = -\frac{x}{y}$	Example 4 in section 1.1 of Chapter 3 (separable equations)
7. $y' = \frac{2x+y+1}{4x+2y-3}$	Example 4 in section 1.2 of Chapter 3 (equations reducible to separable)
8. $y' \ln \frac{y'}{4} = 4x$	Example 7 in section 3 of Chapter 4 (implicit equations solved for x)
9. $y'^2 + e^{y'} = x$	Example 8 in section 3 of Chapter 4 (implicit equations solved for x)
10. $e^{y'} = 0$	has no solution because the left-hand side is always positive
11. $y' = 1$	trivial, direct integration of the function on the right-hand side of the equation
12. $yy' = y$	reducible to the trivial equation 11 via division by y; consequently, it has the same solutions as 11 plus a particular solution $y = 0$
13. $yy'^2 = y$	reducible to the equation $y'^2 = 1$ via division by y and the last is decoupled into the two trivial equations $y' = \pm 1$; consequently, it has all the solutions $y = x + C$ of equation 11, also the solutions $y = -x + C$ of equation $y' = -1$ and still a particular solution $y = 0$
14. $yy'^2 = -y$	has only one solution $y = 0$, since its simplified form $y'^2 = -1$ (after division by y) has no solution
15. $y' = \cos(y-x)$	Example 2 in section 1.2 of Chapter 3 (equations reducible to separable)
16. $y^{(n)} = 1$	trivial, successive integration of the function on the right side of the equation

3   Differential equation for family of functions

As mentioned before, usually a family of functions with n parameters gives rise to an ODE of order n for which it is the general solution. Let us illustrate this with a few examples.

1. $y = x + C$.

The family contains the only parameter, so we have to look for a first-order equation (without any parameter). Deriving the functions of the family, we get the equation we are looking for: $y' = 1$.

2. $y = Cx$.

Again, we are looking for the first-order equation (without any parameter). Differentiating the functions of the family, we get $y' = C$, which is not the required result, because this equation contains the parameter C (it's easy to see that the general solution of this equation is $y = Cx + B$, where C and B are two independent parameters). If we differentiate it the second time, then the parameter will disappear, but the order of the equation will be raised to the second: $y'' = 0$ (we note that the general solution of this equation is also $y = Cx + B$, where C and B are two arbitrary parameters). So, we have to find another way to eliminate the parameter C without increasing the order of the equation. The answer comes from the original definition of the family, where we find $C = \frac{y}{x}$ and, consequently, $y' = \frac{y}{x}$ is the equation we are looking for.

3. $y = \sin(x + C)$.

Differentiating the functions of the set, we get $y' = \cos(x + C)$. To eliminate C, we use the trigonometric identity: $y'^2 + y^2 = \cos^2(x+C) + \sin^2(x+C) = 1$ and find the required equation.

4. $y = Ax^2 + Bx$.

The family contains two parameters, so we have to find an equation of the second order (without any parameters). Therefore, we necessarily have to calculate the second derivative of y, although this calculation may involve some transformations between y and y' if this simplifies the result. In this particular example, we perform the calculations directly, without modifications: $y' = 2Ax + B$ and $y'' = 2A$. Now we use these two expressions to represent A and B in terms of the derivatives: for A we already have a direct relationship with y'' and by substituting it into the first derivative we get $y' = y''x + Bx$, from which we can express $Bx = y' - y''x$. Substituting the expressions for A and B into the original family, we get the equation of the second order we are looking for: $y = \frac{y''}{2}x^2 + y' - y''x$.

5. $(x - A)^2 + By^2 = 1$.

Differentiating this implicit relationship, we get $2(x - A) + 2Byy' = 0$ or $x - A + Byy' = 0$. The second differentiation eliminates the parameter A: $1 + B(y'^2 + yy'') = 0$. So, we express B in the form $B = -\frac{1}{y'^2 + yy''}$, and then we find $x - A = \frac{yy'}{y'^2 + yy''}$. Bringing these expressions of the parameters in the original family, we find the equation $(\frac{yy'}{y'^2 + yy''})^2 - \frac{1}{y'^2 + yy''}y^2 = 1$.

Exercises for section 3

Find ODE whose general solution is a given family of functions:
1. $y = (x - C)^3$.
2. $y = e^{Cx}$.
3. $x^2 + Cy^2 = 2y$.
4. $y = Ax^2 + Be^x$.
5. $\ln y = Ax + By$.
6. $y = Ax^3 + Bx^2 + Cx$.

Chapter 2

Explicit equations of the first order: initial concepts

1 Basic definitions

Let us briefly review the initial concepts of ordinary differential equations specifying them for the first-order equations.

Definition of an ordinary differential equation of the first order. The equation in the form

$$F(x, y, y') = 0,$$

where F is a given function of 3 arguments and $y(x)$ is an unknown function of independent variable x, to be determined by solving this equation, is called an *ordinary differential equation (ODE) of the first order*.

Definition of the explicit/normal form of ODE of the first order. The equation in the form

$$y' = f(x, y),$$

where f is a given function of 2 variables and $y(x)$ is an unknown function of independent variable x, is called an *ODE of the first order in the explicit/normal form*.

Definition of a particular solution. A specific function $y(x)$ is a *particular solution* of the equation $F(x, y, y') = 0$ if, substituted into this equation, it transforms the equation into an identity.

Definition of a general solution. Function $y(x, C)$ of independent variable x and parameter C is a *general solution* of equation $F(x, y, y') = 0$

if for any specific choice of C it represents a particular solution of the same equation.

2 Geometric interpretation of the equation and its solution

Given a differentiable function $y(x)$, the geometric meaning of its derivative $y'(x_0)$ at a point x_0 is the slope of the tangent line to the graph of the function $y(x)$ at the point $P_0 = (x_0, y(x_0))$. (Henceforth the term "slope of the tangent line" is an abbreviated form of the following exact meaning: "the tangent of the angle of inclination that the tangent line forms with the positive direction of the x-axis"). If we consider the derivatives $y'(x)$ at different points, then we have a set, also called a field, of slopes. Consequently, the equation $y' = f(x, y)$ determines the field of slopes at all points on the Cartesian plane where the function $f(x, y)$ is defined. Therefore, the problem of solving this equation consists, geometrically, of finding all the curves $y(x)$ whose slopes at any of their points coincide with those given by the equation. The set of these curves represents the general solution (and also singular solutions if there are any). A specific curve of this set represents a particular solution that passes through a specific point.

3 Physical interpretation of the equation and its solution

Given a differentiable function $y(x)$, which represents the position y of a point body in a rectilinear motion at any instant x, the physical meaning of its derivative $y'(x_0)$ is the instantaneous velocity at the moment x_0 (in abbreviated form, simply velocity). If we consider velocities $y'(x)$ at different moments in time, then we have a set, also called a field, of velocities. Consequently, the equation $y' = f(x, y)$ determines the velocity field at all times x and positions y where the function $f(x, y)$ is defined. Therefore, the problem of solving this equation consists, physically, in finding all the laws of position $y(x)$ whose derivatives at any instant and point of location coincide with those given by the equation. The set of these laws generates the general solution (and also singular solutions if there are any). A specific law from this set represents a particular solution to the equation.

4 Cauchy problem (initial value problem)

As we have seen in the examples of equations and their solutions, a first-order differential equation usually generates a general solution containing an arbitrary parameter in addition to the main independent variable.

Taking advantage of the physical interpretation, it's natural to conclude that knowing the speed of an object at any given moment, we still can't uniquely determine the position of that object, because the same speed can be observed on any route, whether it is on the road connecting Los Angeles and San Francisco, or on the road from Los Angeles to SanDiego, or on the road from San Francisco to Lake Tahoe, etc. So, physically, we can expect to be able to identify a single position law only if we have an additional information about the object's position, for example, its location at a specific instant.

Similarly, knowing only the slope field, we usually obtain a family of curves with given slopes. To single out a specific curve from this set, we need some additional information, for example, a specific point through which the desired curve passes.

In analytical terms, this leads to setting up an additional condition, along with the differential equation, in the form $y(x_0) = y_0$. This condition, according to the physical interpretation, is called the *initial condition* or the *initial value*.

Definition of the Cauchy problem. The differential equation together with the initial condition form the *Cauchy problem*:

$$\begin{cases} F(x, y, y') = 0 \\ y(x_0) = y_0 \end{cases}.$$

This problem is also called the *initial condition or initial value problem*.

If the Cauchy problem is formulated for an explicit equation, then there is an important result, called the *Cauchy theorem* (also the *Picard theorem* or *Picard-Lindelöf* theorem) that guarantees the existence and uniqueness of a solution of the problem

$$\begin{cases} y' = f(x, y) \\ y(x_0) = y_0 \end{cases}.$$

Cauchy theorem (theorem of existence and uniqueness of solutions). If there exists a neighborhood of the point (x_0, y_0) in $\mathbb{R}^2$ where the function f and its partial derivative f_y are continuous functions, then there exists a neighborhood of x_0 in $\mathbb{R}$ where the solution of the Cauchy problem exists and is unique.

Notice that the conditions of the Cauchy theorem are sufficient, but not necessary. Let us consider some illustrative examples.

Examples.

1. $\begin{cases} y' = 2xy \\ y(0) = 0 \end{cases}$.

The function $f(x, y) = 2xy$ and its partial derivative $f_y(x, y) = 2x$ are continuous on $\mathbb{R}^2$. Therefore, the conditions of the Cauchy theorem are satisfied and, consequently, the solution of this problem exists and is unique. Obviously, it is given by the formula $y \equiv 0$.

2. $\begin{cases} y' = \operatorname{sgn} x \\ y(0) = 0 \end{cases}$.

The function $f(x, y) = \operatorname{sgn} x$ is not continuous at $x = 0$, which means that the conditions of the Cauchy theorem do not met. In this case, the solution of the problem does not exist, because the derivative of any function (when exists) should satisfy the intermediate value property, and consequently, cannot have a jump discontinuity, which is not the case of the function $\operatorname{sgn} x$ in a neighborhood of $x = 0$ (recall Darboux's theorem in Calculus).

3. Peano's example. The Cauchy problem $\begin{cases} y' = 3y^{2/3} \\ y(0) = 0 \end{cases}$.

The function $f(x, y) = 3y^{2/3}$ is continuous, but not differentiable with respect to y in a neighborhood of $y = 0$ and, consequently, the conditions of the Cauchy theorem are not fulfilled. In this case, one can easily check that both $y_1(x) \equiv 0$ and $y_2(x) = x^3$ are solutions of this problem in $\mathbb{R}$. Other solutions to the same problem can be found as a combination of these two functions:

$y(x) = \begin{cases} (x - x_1)^3, x < x_1 \\ 0, x_1 \leq x \leq x_2 \\ (x - x_2)^3, x > x_2 \end{cases}$, where $x_1 \leq 0 \leq x_2$. It can be shown that any

solution of this problem in $\mathbb{R}$ is one of the obtained functions.

If one searches for the solution of the same problem on the interval $[0, +\infty)$, then $y(x) = \begin{cases} 0, x \leq x_0 \\ (x - x_0)^3, x > x_0 \end{cases}$, where $x_0 \geq 0$ is the corresponding family of solutions. Again, all the solutions of the Cauchy problem on $[0, +\infty)$ are contained in this family.

4. $\begin{cases} y' = 3(1 + y^{2/3}) \\ y(0) = 0 \end{cases}$.

The function $f(x, y) = 3(1 + y^{2/3})$ is continuous, but not differentiable with respect to y in a neighborhood of $y = 0$, which means that the conditions of the Cauchy theorem are not fulfilled. Nevertheless, this problem has the unique solution $y^{1/3} - \arctan y^{1/3} = x$. It is easy to verify that this function satisfies both the equation and initial condition. The uniqueness can be shown by direct solution of the equation. Anticipating general exposition of the method of solution of separable equations, which is the type of the equation $y' = 3(1+y^{2/3})$ (see section 1 of Chapter 3 for detailed explanation), we can solve this equation for completeness. To do this, let us rewrite the given equation in the terms of the inverse function $x(y)$: $x_y = \frac{1}{3(1+y^{2/3})}$ (recall that $x_y = \frac{1}{y_x}$). To solve for $x(y)$ it is sufficient to calculate the integral of the right-hand side, that can be made by substitution $y = z^3$:

$$\int \frac{1}{3(1+y^{2/3})}dy = \int \frac{3z^2}{3(1+z^2)}dz = \int \frac{z^2+1-1}{1+z^2}dz$$

$$= z - \arctan z + C = y^{1/3} - \arctan y^{1/3} + C.$$

Hence, $x(y) = y^{1/3} - \arctan y^{1/3} + C$. Applying the initial condition, we obtain the unique solution $x = y^{1/3} - \arctan y^{1/3}$, which is the implicit form of the solution of the original problem.

5. $\begin{cases} y' = f(x) \\ y(0) = 0 \end{cases}$, where $f(x) = \begin{cases} 2x\sin\frac{1}{x} - \cos\frac{1}{x}, x \neq 0 \\ 0, x = 0 \end{cases}$.

In this case, $f(x)$ is not even continuous at the origin ($2x\sin\frac{1}{x}$ approaches 0, while the limit of $\cos\frac{1}{x}$ does not exist), that is, the function $f(x,y)$ violates the conditions of the Cauchy theorem even stronger than in Example 4. Nonetheless, the problem has the unique solution $y(x) = \begin{cases} x^2\sin\frac{1}{x}, x \neq 0 \\ 0, x = 0 \end{cases}$.

Indeed, it is easy to check that this function satisfies both the equation and initial condition. At the same time, the uniqueness of the solution follows from the fact that $y(x)$ is one of the antiderivatives of $f(x)$, which implies that the set of all antiderivatives (indefinite integral) is represented in the form $y(x) + C$, where C is an arbitrary constant, but only one function of this set, namely $y(x)$, satisfies the given initial condition.

Chapter 3

Explicit equations of the first order: methods of solution

First of all, let us establish that the problem of solving ordinary differential equations is understood here from a practical point of view, that is, as finding all the solutions of the given equation or, at least, finding its general solution. If we work with an initial value problem, then we need to find all solutions to that problem (of course, if the conditions of Cauchy's theorem are satisfied, then we look for a single solution).

It should be kept in mind that solution of differential equations, even of the first order and in explicit form, is a far from trivial problem in practice. First, not all equations of this simple type can be solved. This means that there are explicit first-order equations (actually, most of these equations) for which a method of solution is unknown, even when the theory guarantees the existence of solutions. Second, the equations that can be solved practically do not have a universal algorithm for their solution. On the contrary, there are different groups (types) of equations, each of which admits its own specific method of solution. In other words, there is no general theory for solution of explicit first-order equations, but rather different methods for solving some subtypes of these equations. Therefore, in practical solution of equations, one of the important ingredients is determining the specific type of the original equation that indicates its specific way of solution. In accordance with this logic of the relationship between types of equations and their methods of solution, below we will consider separately each important group of equations that allows practical solution and present their specific treatment. Finally, even when there is an algorithm for solution of a specific equation, this usually requires calculation of integrals to represent a solution in the usual form, and this integration phase generates its own additional problems since integration is usually a more laborious operation than differentiation and it cannot always be carried out in terms of elementary functions. Thus, in many cases, a solution algorithm must stop at the integration stage, without expressing a sought solution in the desired analytical form (so called closed form).

1   Separable equations and reducible to them

1.1   Separable equations

Definition. An *equation of separable variables* (or simply *separable equation*) has the form

$$y' = f(x)g(y).$$

Remark. A separable equation can be written in the form $A(x)B(y)dx + C(x)D(y)dy = 0$. Recalling that, by definition of the differential, $dy = y'dx$ and dx is an (arbitrary and non-zero) increment of the independent variable x, solving for y' we transform this equation into the form of the definition: $y' = -\frac{A(x)B(y)}{C(x)D(y)}$ with $f(x) = -\frac{A(x)}{C(x)}$ and $g(y) = \frac{B(y)}{D(y)}$. In this transformation, the points where $C(x) = 0$ and the functions such that $D(y) = 0$ are considered separately and have no influence on the type of the equation. The conversion from $y' = f(x)g(y)$ to the second form is even simpler: just use the formula $dy = y'dx$ and rewrite the equation in the form $f(x)g(y)dx - dy = 0$. Thus, the two representations of a separable equation are equivalent.

Method of solution.
Let us start with a simpler equation in the form $y' = f(x)$ ($g(y) = 1$). Since $f(x)$ is a given function of the variable x, we can simply integrate both sides of the equation to find the general solution: $y = \int f(x)dx = F(x) + C$, where $\int f(x)dx$ is indefinite integral of $f(x)$ and $F(x)$ is an antiderivative of $f(x)$. (Sometimes the symbol $\int f(x)dx$ is convenient to understand as one of the antiderivatives of $f(x)$, in which case we write the general solution in the form $y = \int f(x)dx + C$.)

Consider now the second specific situation when $f(x) = 1$. In this case, we cannot integrate with respect to x, since $g(y)$ is a given function of y, but not of x, because y is an unknown solution. However, we can interchange the roles of independent and dependent variables, and for the function $x(y)$ we have the equation $\frac{1}{x'} = g(y)$ or $x' = \frac{1}{g(y)}$. Integrating now with respect to y, we get the general solution in the implicit form $x = \int \frac{1}{g(y)} dy = G(y) + C$, where $G(y)$ is an antiderivative of $\frac{1}{g(y)}$. (Recall that we have already used this technique to solve Example 4 in section 4 of Chapter 2.)

These two specific situations prepare a ground for a treatment of the general case. Given the equation $y' = f(x)g(y)$, we first separate the variables, leaving $f(x)$ on the right-hand side and shifting $g(y)$ to the left-hand side: $\frac{y'}{g(y)} = f(x)$. Now we integrate both sides with respect to x: $\int \frac{y'}{g(y)} dx = \int f(x) dx$. On the right-hand side we have immediately $\int f(x) dx = F(x) + A$, since $f(x)$ is a known function of x, but on the left-hand side x enter through the unknown function y. However, changing the variable of integration from x to y (recall that in this substitution $dy = y' dx$), we arrive at the integral that can be calculated: $\int \frac{y'}{g(y)} dx = \int \frac{dy}{g(y)} = G(y) + B$. Therefore, we have the general solution in the implicit form

$$\int \frac{dy}{g(y)} = G(y) + B = F(x) + A = \int f(x) dx,$$

or, joining the two constants (which are not independent in this form of solution), we finally obtain $G(y) = F(x) + C$, where $G(y)$ is an antiderivative of $\frac{1}{g(y)}$, $F(x)$ is an antiderivative of $f(x)$ and C is an arbitrary parameter.

Using more formal, algorithmic representation, the procedure of solving the equation $y' = f(x)g(y)$ can be schematized as follows. First, write y' in the form $y' = \frac{dy}{dx}$ and separate the variables, rewriting the equation in the form $\frac{dy}{g(y)} = f(x) dx$. Second, integrate the left-hand side with respect to y and the right-hand side with respect to x: $\int \frac{dy}{g(y)} = \int f(x) dx$. Finally, join the two constants of integration and obtain the general solution in the form $G(y) = F(x) + C$. Notice that this formal algorithm is valid only because it was previously justified.

After finding the general solution using this algorithm, we still have to check whether there were lost or added solutions in this procedure. What can cause a loss or undue addition of solutions is the division by $g(y)$, which is a necessary step in the algorithm of solution of a separable equation. If this division eliminates some functions from consideration, then it may turn out that some of them are solutions of the original equation. In this case, you need to add these solutions to the general solution (or remove undue solutions from the general one).

Remark. Although the method of solution is quite simple from a theoretical point of view, the need to calculate indefinite integrals (or antiderivatives) can generate serious technical problems in the practical procedure. Just recall that, unlike the operation of differentiation, the integration is usually a more intricate procedure that cannot always be carried out in terms of elementary functions: for example, the integration of such "harmless" for differentiation functions as e^{x^2}, $\cos x^2$, $\frac{\sin x}{x}$, etc. cannot be performed in closed form (within the class of elementary functions). Therefore, in practice, the possibility of finding a solution of a separable equation in closed form is limited by the possibility to calculate the integrals of functions involved in the method of solution.

Examples.

To illustrate the technique of solution, we solve some equations below and propose others to the reader.

1. $y' = \cos x$.
This is a separable equation with $f(x) = \cos x$ and $g(y) = 1$. Integrating with respect to x, we find the general solution: $y = \int \cos x\, dx = \sin x + C$. According to the employed procedure, this general solution contain all particular solutions.

2. $y' = y^2$.
This is a separable equation with $f(x) = 1$ and $g(y) = y^2$. Dividing by y^2 and integrating the left-hand side with respect to y and the right-hand side with respect to x, we obtain $\int \frac{1}{y^2} dy = \int 1 dx$ or $-\frac{1}{y} = x + C$. The last relation is the general solution, which can also be expressed in the form $y = -\frac{1}{x+C}$. In this process of finding the general solution we were forced to divide by y, that eliminates the function $y = 0$, which is obviously a particular solution of the original equation. Then, we have one more (special) solution $y = 0$, which is not included in the general solution.

3. $xy' = 2y$ (the second equation in Table 1 of section 2 in Chapter 1). By writing the equation in the form $y' = \frac{2y}{x}$ we can see that this is a separable equation with $f(x) = \frac{1}{x}$ and $g(y) = 2y$. (There is no need to rewrite the equation in this form for classification if the type of equation can already be specified in the original form.) Separating the variables and integrating each side with respect to its variable, we get: $\int \frac{dy}{y} = \int \frac{2dx}{x}$. The results of integration are as follows: $\ln|y| = 2\ln|x| + C$. This is the general solution. In the course of this algorithm, we have to divide by y that eliminates the function $y = 0$ from consideration, which is obviously another solution of the original equation. This solution is not included in the form of the general

solution $\ln |y| = 2 \ln |x| + C$, and so it is special with respect to this general solution. However, we note that the general solution can also be written in the form $\ln |y| = \ln x^2 + \ln B$, $B > 0$ ($C = \ln B$ because the variation of B in $(0, +\infty)$ guarantees that $\ln B$ spans all the values in $(-\infty, +\infty)$). Using the properties of logarithms we then have $|y| = Bx^2$ or, opening the absolute value, $y = \pm Bx^2$, which equals to $y = Ax^2$, $A \neq 0$. This form of the general solution is still equivalent to the logarithmic form. It now remains to see that if we assign 0 to the constant A, then we get the solution $y = 0$ lost in the solution algorithm. Therefore, we can use the general solution in the form $y = Ax^2$, $\forall A$, which includes both the general solution in the logarithmic form and the special solution $y = 0$. This form of the general solution is not only simpler, but also contains all particular solutions and, regarding this form, $y = 0$ is a regular particular solution (not a special one).

In many exercises to follow, we will use this type of transformation (without providing again all the details), which allows us to recover special solutions in the modified form of the general solution.

4. $y' = -\frac{x}{y}$ (the sixth equation in Table 1 of section 2 in Chapter 1). Obviously, this is a separable equation with $f(x) = -x$ and $g(y) = \frac{1}{y}$. Separating the variables and applying the integral to each side with respect to its variable, we get: $\int y dy = -\int x dx$. After integration, we obtain $\frac{y^2}{2} = -\frac{x^2}{2} + C$ or $y^2 + x^2 = C$, $C > 0$ (constant $C < 0$ does not generate any function and for $C = 0$ the function is defined only at $x = 0$, so it is not differentiable). Another way of writing the same general solution is $y^2 + x^2 = A^2$, $A \neq 0$. Although the equation was multiplied by y during the procedure of solution, this did not generate the false solution $y = 0$ (not allowed by the original equation) because $y = 0$ is not included in the obtained general solution. Furthermore, at the points $x = \pm A$, where the function vanishes, its derivative does not exist, and therefore, the solution cannot be considered. So for any $A \neq 0$ the points where $y = 0$ are automatically eliminated.

5. $xyy' = \sqrt{y^2 + 1}$, $y(1) = 0$.
In this case, we have the Cauchy problem. Writing the equation in the form $y' = \frac{\sqrt{y^2+1}}{xy}$, we can see that this is a separable equation with $f(x) = \frac{1}{x}$ and $g(y) = \frac{\sqrt{y^2+1}}{y}$. Separating the variables and integrating each side with respect to its variable, we get: $\int \frac{y}{\sqrt{y^2+1}} dy = \int \frac{dx}{x}$. The results of integration are: $\sqrt{y^2 + 1} = \ln |x| + C$, which gives the general solution. To find this solution it was required to divide by $\sqrt{y^2 + 1}$, but this does not cause a loss of any function. Therefore, the general solution contains all the particular ones.

Substituting the initial condition into the general solution, we have $\sqrt{1} = \ln 1 + C$ or $C = 1$. Thus, the solution to the problem is $\sqrt{y^2 + 1} = \ln |x| + 1$.

6. $y' = 3\sqrt[3]{y^2}$, $y(0) = 1$.

This is the Cauchy problem for a separable equation with $f(x) = 1$ and $g(y) = 3\sqrt[3]{y^2}$. By separating the variables and integrating each side with respect to its variable, we get: $\int \frac{1}{3\sqrt[3]{y^2}} dy = \int dx$. The results of integration are: $\sqrt[3]{y} = x + C$. This is the general solution. To find it, we had to divide by $\sqrt[3]{y^2}$, that eliminated the function $y = 0$ which is a particular solution. However, this has no influence on finding the solution to the problem. In fact, the solution $y = 0$ does not satisfy the initial condition, and so the solution to the problem should be found in the form of the general solution. Substituting the initial condition into it, we specify $C = 1$, that is, the solution to the problem is $\sqrt[3]{y} = x + 1$.

7. $(x + 1)y' + xy = 0$, $y(0) = 1$.

We have the Cauchy problem. By writing the equation in the form $y' = -\frac{xy}{x+1}$, we can see that this is a separable equation with $f(x) = -\frac{x}{x+1}$ and $g(y) = y$. Separating the variables and integrating each side with respect to its variable, we get: $\int \frac{dy}{y} = -\int \frac{x}{x+1} dx$. The results of integration are:

$$\ln |y| = -\int \frac{x + 1 - 1}{x + 1} dx = -\int \left(1 - \frac{1}{x + 1}\right) dx = -x + \ln |x + 1| + C.$$

This is the general solution. To find it, we had to divide by y, but we can recover the solution $y = 0$, representing the general solution in the form $y = C(x + 1)e^{-x}$. Substituting the initial condition, we specify C: $1 = C(0 + 1)e^0$, that is, $C = 1$. Therefore, the solution to the problem is $y = (x + 1)e^{-x}$.

8. $(1 + y^2)dx + xydy = 0$.

Representation of the equation in the form $y' = -\frac{1+y^2}{xy}$ shows that this is a separable equation with $f(x) = -\frac{1}{x}$ and $g(y) = \frac{1+y^2}{y}$. Separating the variables and integrating each side with respect to its variable, we get: $\int \frac{y}{1+y^2} dy = -\int \frac{dx}{x}$. The results of integration are: $\ln 1 + y^2 = -\ln |x| + C$ or $1 + y^2 = \frac{C}{x}$. This is the general solution. No function has been disregarded or added in this procedure.

9. $3x^2ydx + 2\sqrt{4 - x^3}dy = 0$.

Writing the equation in the form $y' = -\frac{3x^2y}{2\sqrt{4-x^3}}$, we can see that this is a separable equation with $f(x) = -\frac{3x^2}{2\sqrt{4-x^3}}$ and $g(y) = y$. Separating the variables and integrating each side with respect to its variable, we get:

$\int \frac{dy}{y} dy = -\int \frac{3x^2}{2\sqrt{4-x^3}} dx$. The results of integration are: $\ln|y| = \sqrt{4-x^3}+C$ or $y = Ce^{\sqrt{4-x^3}}$. We note that the general solution in the logarithmic form requires the addition of one more particular solution $y = 0$ which was lost when we divided the equation by y, but in the exponential form this solution was recovered.

10. $(\sqrt{xy} + \sqrt{x})y' - y = 0$, $y(1) = 1$.
This is the Cauchy problem. By writing the equation in the form $y' = \frac{y}{\sqrt{x}(\sqrt{y}+1)}$ we make sure that this is a separable equation with $f(x) = \frac{1}{\sqrt{x}}$ and $g(y) = \frac{y}{\sqrt{y}+1}$. Separating the variables and integrating each side with respect to its variable, we obtain: $\int \frac{\sqrt{y}+1}{y} dy = \int \frac{dx}{\sqrt{x}}$. Calculating the integral on the left side, we get

$$\int \frac{\sqrt{y}+1}{y} dy = \int \frac{1}{\sqrt{y}} + \frac{1}{y} dy = 2\sqrt{y} + \ln|y| + C.$$

Together with the integral of the right side, we have the general solution $2\sqrt{y} + \ln|y| = 2\sqrt{x} + C$. During this procedure, the particular solution $y = 0$ was disregarded, but it does not satisfy the initial condition. To find the solution to the problem, we apply the initial condition to the general solution: $2\sqrt{1} + \ln 1 = 2\sqrt{1} + C$, whence $C = 0$. Thus, the solution to the problem is $2\sqrt{y} + \ln|y| = 2\sqrt{x}$.

11. $e^{x+3y}dy - xdx = 0$.
Representation of the equation in the form $y' = xe^{-x}e^{-3y}$ makes it clear that we have a separable equation with $f(x) = xe^{-x}$ and $g(y) = e^{-3y}$. By separating the variables and integrating each side with respect to its variable, we get: $\int e^{3y}dy = \int xe^{-x}dx$. The results of integration provide the general solution $\frac{1}{3}e^{3y} = -(x+1)e^{-x} + C$.

12. $\frac{\tan y}{\cos^2 x}dx + \frac{\tan x}{\cos^2 y}dy = 0$.
By writing the equation in the form $y' = -\frac{\tan y \cos^2 y}{\tan x \cos^2 x}$ we can see that this is a separable equation with $f(x) = -\frac{1}{\tan x \cos^2 x}$ and $g(y) = \tan y \cos^2 y$. Separating the variables and integrating, we get: $\int \frac{1}{\tan y \cos^2 y}dy = -\int \frac{1}{\tan x \cos^2 x}dx$. By calculating the integrals

$$\int \frac{1}{\tan y \cos^2 y} dy = \int \frac{d(\tan y)}{\tan y} = \ln|\tan y| + A$$

and

$$\int \frac{1}{\tan x \cos^2 x} dy = \int \frac{d(\tan x)}{\tan x} = \ln|\tan x| + B,$$

we obtain the general solution $\ln|\tan y| = -\ln|\tan x| + C$, or, eliminating the logarithms, $\tan y \cdot \tan x = C$.

13. $y - xy' = 1 + x^2 y'$.

This equation is separable, which can be seen more clearly if we rewrite it in the form $y' = \frac{y-1}{x^2+x}$ with $f(x) = \frac{1}{x^2+x}$ and $g(y) = y - 1$. Separating the variables and integrating, we get: $\int \frac{dy}{y-1} = \int \frac{dx}{x^2+x}$. By calculating the integrals

$$\int \frac{dy}{y-1} = \ln|y-1| + A$$

and

$$\int \frac{dx}{x^2+x} = \int \frac{1}{x} - \frac{1}{x+1}dx = \ln|x| - \ln|x+1| + B = \ln\frac{x}{x+1} + B,$$

we obtain the general solution $\ln|y-1| = \ln\frac{x}{x+1} + C$, or in the form without the logarithms $y - 1 = \frac{Cx}{x+1}$.

14. $(\sin(x+y) + \sin(x-y))dx + \frac{dy}{\cos y} = 0$.

Using the trigonometric formula $\sin(x+y) + \sin(x-y) = 2\sin x \cos y$ and isolating the derivative, we detect that the equation takes the form of separable variables $y' = -2\sin x \cos^2 y$ with $f(x) = -2\sin x$ and $g(y) = \cos^2 y$. Separating the variables and integrating, we get: $\int \frac{dy}{\cos^2 y} = -2\int \sin x dx$. By integrating, we find the general solution $\tan y = 2\cos x + C$.

15. $(\cos(x-2y) + \cos(x+2y))y' = \frac{1}{\cos x}$.

Using the trigonometric formula, we transform the equation into the form $2\cos x \cdot \cos 2y \cdot y' = \frac{1}{\cos x}$ and note that this is a separable equation with $f(x) = \frac{1}{\cos^2 x}$ and $g(y) = \frac{1}{2\cos 2y}$. Separating the variables and integrating, we get: $\int 2\cos 2y dy = \int \frac{dx}{\cos^2 x}$. Calculating the integrals, we find the general solution $\sin 2y = \tan x + C$.

16. $\sin xy' = y\cos x + 2\cos x$.

This equation is separable with $f(x) = \frac{\cos x}{\sin x}$ and $g(y) = y + 2$. Separating the variables and integrating, we get: $\int \frac{dy}{y+2} = \int \frac{d(\sin x)}{\sin x}$. By integrating, we obtain the general solution $\ln|y+2| = \ln|\sin x| + C$ or $y + 2 = C\sin x$.

Exercises for section 1.1

Solve the following separable equations:

1. $\sin x \cdot \tan y dx - \frac{dy}{\sin x} = 0$.
2. $(xy^3 + x)dx + (x^2y^2 - y^2)dy = 0$.
3. $(1 + y^2)dx - (y + yx^2)dy = 0$.

4. $y' = 2xy + x$.
5. $2xyy' = 1 - x^2$.
6. $y' = e^{x^2}x(1 + y^2)$.
7. $y'\cot x + y = 2$.
8. $(1 + e^{3y})xdx = e^{3y}dy$.
9. $y - xy' = 1 + x^2y'$.
10. $2x^2yy' + y^2 = 2$.
11. $y - xy' = 2(1 + x^2y')$.
12. $(1 + e^x)ydy - e^ydx = 0$.
13. $(x + xy^2)dy = (y^2 - y)dx$.
14. $y' = \sin^2 y$.
15. $y' = (y - 1)x$.
16. $x^2y^2y' + 1 = y$.
17. $xy' + y = y^2$.
18. $dy = e^{x+y}dx$.

1.2   Equations reducible to separable

Theoretically, among different types of explicit first-order equations that can be solved analytically (in closed form), only two types are independent (basic): the separable equations and the exact equations (the second type will be studied later). All other types are somehow reduced to one of the two independent types. However, some of the reducible types, due to various reasons, including their importance and frequency in applications, have special names and are usually treated separately from the basic equations. Other equations are not treated independently of those to which they are reduced. Again, this separation is just a matter of traditions, patterns of the use and the importance of different types of equations.

One of the equations that has no special name and is traditionally treated as *reducible to a separable equation* is $y' = f(ax + by + c)$, where a, b, c are arbitrary constants.

Method of solution.
First, we note that if one of the coefficients a or b is zero, then we already have a separable equation. To rule out this uninteresting situation, let us assume from now on that $ab \neq 0$. Substituting the unknown function in the original equation by the formula $z = ax + by + c$, and using the corresponding relationship between the derivatives $z' = a + by'$, we get $\frac{z'-a}{b} = f(z)$, which is a separable equation. In this way, the problem is solved (the method of solution of separable equations was discussed in the preceding section).

Examples.

1. $y' = (2x - 4y + 1)^2$.

First, we verify that this equation is of the studied type: $f(z) = z^2$, $z = 2x - 4y + 1$. Then, we can use the described method of reducing to a separable equation: substituting z for y in the original equation (and using the relation $z' = 2 - 4y'$), we obtain a separable equation: $-\frac{z'-2}{4} = z^2$ or $z' = 2 - 4z^2$. To solve the latter, we need to perform two integrations: $\int \frac{dz}{1-2z^2} = 2\int dx$. The integral on the left side can be calculated using reduction to simple fractions:

$$\int \frac{dz}{1 - 2z^2} = \frac{1}{2}\int \frac{1}{1 - \sqrt{2}z} + \frac{1}{1 + \sqrt{2}z}dz$$

$$= \frac{1}{2\sqrt{2}}\left(-\ln|1 - \sqrt{2}z| + \ln|1 + \sqrt{2}z|\right) + A = \frac{1}{2\sqrt{2}}\ln\left|\frac{1 + \sqrt{2}z}{1 - \sqrt{2}z}\right| + A.$$

Joining with the integral on the right side, we have the general solution of the transformed equation

$$\frac{1}{2\sqrt{2}}\ln\left|\frac{1 + \sqrt{2}z}{1 - \sqrt{2}z}\right| = 2x + B \text{ or } \ln\left|\frac{1 + \sqrt{2}z}{1 - \sqrt{2}z}\right| = 4\sqrt{2}x + D.$$

We can also change this form to one without the logarithm: $\frac{1+\sqrt{2}z}{1-\sqrt{2}z} = Ce^{4\sqrt{2}x}$, but it should be noted that this is not an equivalent transformation, because the last representation of the general solution has one more function (which corresponds to $C = 0$) than the logarithmic form. However, the function $1 + \sqrt{2}z = 0$ is the function disregarded when we divided by $1 - 2z^2$ and it satisfies the equation $z' = 2 - 4z^2$. Therefore, by adding it, we only recover one more solution in the exponential form of the general solution. It remains to see that the function $1 - \sqrt{2}z = 0$, also eliminated when dividing by $1 - 2z^2$, is also a solution to the equation $z' = 2 - 4z^2$. Therefore, along with the general solution in the exponential form, we have the special solution $z = \frac{1}{\sqrt{2}}$. Back to the original unknown, we get the general solution

$$\frac{1 + \sqrt{2}(2x - 4y + 1)}{1 - \sqrt{2}(2x - 4y + 1)4} = Ce^{4\sqrt{2}x}$$

and the special solution $2x - 4y + 1 = \frac{1}{\sqrt{2}}$. This concludes the algorithm of solution. Note that the general solution can still be written in the form that includes the remaining special solution (this is a homework for the reader).

2. $y' = \cos(y - x)$ (equation 15 in Table 1 of section 2 in Chapter 1). It is obvious that this equation is of the studied type: $f(z) = \cos z$, $z =$

$y - x$. Then, we substitute z for y and obtain a separable equation: $z' + 1 = \cos z$. Separating variables, we have $\int \frac{dz}{\cos z - 1} = \int dx$. To perform the integration on the right-hand side, it is simpler to use the trigonometric formula: $\int \frac{dz}{\cos z - 1} = \int \frac{dz}{-2 \sin^2 z} = \cot \frac{z}{2}$. So, the general solution comes in the form $\cot \frac{z}{2} = x + C$. Returning to the original unknown, we get $\cot \frac{y-x}{2} = x + C$ which is the general solution of the original equation. In the process of solution, it was necessary to divide by $\cos z - 1 = \cos(y-x) - 1$. Let us check whether the functions $\cos(y - x) = 1$ are special solutions of the given equation. It is more convenient to rewrite these functions in explicit form, solving the trigonometric equation: $y - x = 2k\pi$, $k \in \mathbb{Z}$. These functions are not included in the general solution ($\cot t$ is not defined at the points $t = k\pi$, $k \in \mathbb{Z}$), but these functions satisfy the original equation, because $y' = 1 = \cos(y - x)$. Therefore, all these functions are special solutions of the original equation.

3. $y' = \frac{1}{x+y-1}$.

First, we identify that this equation is of the type $y' = f(ax + by + c)$ with $a = 1, b = 1, c = -1$, $f(z) = \frac{1}{z}$, $z = x + y - 1$. Therefore, we can reduce it to a separable equation by substituting $z = x + y - 1$ for y in the original equation: $z' - 1 = \frac{1}{z}$ or $z' = \frac{z+1}{z}$. Separating variables and integrating, we get $\int \frac{z}{z+1} dz = \int dx$. The integral on the left-hand side gives

$$\int \frac{z+1-1}{z+1} dz = \int 1 - \frac{1}{z+1} dz = z - \ln|z + 1| + C.$$

Then, we get the general solution $z - \ln|z + 1| = x + C$. For the original equation we have $x + y - 1 - \ln|x + y| = x + C$ or in simplified form $y - \ln|x + y| = C$. In the course of this procedure, the function $z + 1 = 0$ was eliminated from consideration, which implies that a particular solution of the original equation $y = -x$ is not included in the general solution (it is elementary to check this).

4. $(x - 2y - 1)dx + (3x - 6y + 2)dy = 0$, $y(2) = 1$.

This is the Cauchy problem. First, we find all the solutions to the given equation, and then we apply the initial condition. To do this, we rewrite this equation in the form $y' = -\frac{x-2y-1}{3x-6y+2}$ and identify that it is of the type $y' = f(ax+by+c)$ with $a = 1, b = -2, c = -1$, $f(z) = -\frac{z}{3z+5}$, $z = x - 2y - 1$. Therefore, we can transform it into a separable equation by substituting $z = x - 2y - 1$ for y in the original equation: $\frac{1}{2}(1 - z') = -\frac{z}{3z+5}$, or simplifying, $z' = \frac{5z+5}{3z+5}$. Separating variables and integrating, we get $\int \frac{3z+5}{5z+5} dz = \int dx$. The integral on the left-hand side gives

$$\int \frac{3z + 3 + 2}{5z + 5} dz = \int \frac{3}{5} + \frac{2}{5z + 5} dz = \frac{3}{5}z + \frac{2}{5} \ln|5z + 5| + C.$$

Then, we obtain the general solution

$$\frac{3}{5}z + \frac{2}{5}\ln|5z + 5| = x + C \text{ or } 3z + 2\ln|z + 1| = 5x + C.$$

Returning to y, we have $3(x - 2y - 1) + 2\ln|x - 2y| = 5x + C$ or, in a simpler form, $x + 3y - \ln|x - 2y| = C$. Under this procedure, the function $z + 1 = 0$ was disregarded, which means that $x - 2y = 0$ was not taking into consideration, although this is a particular solution of the original equation (the reader can check this easily). This function must be added to the set of solutions, because it is not included in the general solution. As we will see it next, this particular solution is a key to solve the Cauchy problem.

Attempt to substitute the initial condition into the general solution shows that this is impossible (the logarithm argument vanishes), whatever the value of the constant C. Thus, among the solutions that make up the general solution set, there is none that satisfies the initial condition. However, the special solution $x = 2y$ satisfies the initial condition and is therefore the only solution to the Cauchy problem.

5. $y' + y = 2x + 1$.
First, we identify that this equation is of the type $y' = f(ax + by + c)$ with $a = 2, b = -1, c = 1$ and $f(z) = z, z = 2x - y + 1$. Therefore, we can reduce it to a separable equation by replacing the function y with $z = 2x - y + 1$. Substituting in the original equation, we get: $2 - z' = z$. Separating variables and integrating, we obtain $\int \frac{dz}{2-z} = \int dx$, which leads to the general solution $\ln|z - 2| = -x + C$ or $z - 2 = Ce^{-x}$. Returning to y, we have $2x - y - 1 = Ce^{-x}$.

Exercises for section 1.2

Solve the following equations, transforming them into separable:
1. $y' = \cos(x - y - 1)$.
2. $y' + 1 = \frac{1}{(x+y)+(x+y)^2}$.
3. $y' - y = 2x - 3$.
4. $y' = \sqrt{4x + 2y - 1}$.
5. $y' = x + y + 1$.
6. $y' = \sqrt{y - x} + 1$.
7. $(2x + y + 2)dx - (4x + 2y + 9)dy = 0$.
8. $(y - 3x + 2)dx + (3x - y - 1)dy = 0$.

2 Homogeneous equations and reducible to them

2.1 Homogeneous equations

Definition. A homogeneous equation has the form $y' = f\left(\frac{y}{x}\right)$.

    Remark. Another way to define a homogeneous equation, which is also found in textbooks, is as follows: $A(x, y)dx + B(x, y)dy = 0$, where A and B are homogeneous functions of the same degree. Recall that a function $C(x, y)$ is called homogeneous of degree k if $C(ax, ay) = a^k C(x, y)$ for any real parameter a (guaranteed that the point (ax, ay) belongs to the domain of $C(x, y)$) and some number k. Starting from the definition $y' = f\left(\frac{y}{x}\right)$, we can rewrite it in the form $f\left(\frac{y}{x}\right)dx - dy = 0$ and immediately note that the functions $A = f\left(\frac{y}{x}\right)$ and $B = 1$ are homogeneous of degree 1. Conversely, we can transform the equation $A(x, y)dx + B(x, y)dy = 0$ as follows:

$$y' = -\frac{A(x, y)}{B(x, y)} = -\frac{A\left(x \cdot 1, x \cdot \frac{y}{x}\right)}{B\left(x \cdot 1, x \cdot \frac{y}{x}\right)} = -\frac{x^k A\left(1, \frac{y}{x}\right)}{x^k B\left(1, \frac{y}{x}\right)} = -\frac{A\left(1, \frac{y}{x}\right)}{B\left(1, \frac{y}{x}\right)} = f\left(\frac{y}{x}\right),$$

that is, we have arrived at the original definition of the homogeneous equation. Thus, the two representations of homogeneous equation are equivalent.

    Method of solution.

    A homogeneous equation is reducible to that of separable variables. Simply introducing the new unknown function z by the formula $z = \frac{y}{x}$, or $y = xz$ and using the corresponding relation between the derivatives $y' = z + xz'$, we transform the equation $y' = f\left(\frac{y}{x}\right)$ into the following form: $xz' + z = f(z)$ or $z' = \frac{f(z)-z}{x}$, which is a separable equation for z. Thus, the problem is solved, since the method of solution of separable equations is already known.

    Examples.

    1. $xy' = x + y$.
Representation of this equation in the form $y' = 1 + \frac{y}{x}$ shows that it is homogeneous with $f(z) = 1 + z$, $z = \frac{y}{x}$. Substituting the function z for y by the formula $y = xz$, and using the relation $y' = z + xz'$, we arrive at the separable equation $z + xz' = 1 + z$. Simplifying and integrating, we get $z = \int \frac{1}{x}dx = \ln|x| + C$. Returning to the function y, we have the general solution $y = x\ln|x| + Cx$ which, according to the procedure of solution, includes all particular solutions.

    We notice that the solution technique requires dividing by x (in the definition of z and in the separation of variables). Normally, the loss of some points in the domain of the solutions does not worry us, although often these points can be restored by making necessary adjustments to the obtained solutions.

2. $xy' = y + \sqrt{x^2 - y^2}$.

Writing this equation in the form $y' = \frac{y}{x} + \sqrt{1 - \frac{y^2}{x^2}}$ we see that it is homogeneous with $f(z) = z + \sqrt{1 - z^2}$, $z = \frac{y}{x}$. Replacing the function y by $z = \frac{y}{x}$, we get the separable equation $z + xz' = z + \sqrt{1 - z^2}$. Simplifying and integrating, we obtain $\int \frac{1}{\sqrt{1-z^2}} dz = \int \frac{1}{x} dx$. The two integrals are elementary and we find the implicit form of the solution $\arcsin z = \ln |x| + C$. Returning to the function y, we have the general solution $\arcsin \frac{y}{x} = \ln |x| + C$, which, according to the procedure of solution, includes all particular solutions, except, possibly, for $y = \pm x$ (which corresponds to $z = \pm 1$). To check whether these two functions are particular solutions or not, we substitute them into the original equation: $x \cdot (\pm 1) = \pm x + \sqrt{x^2 - (\pm x)^2}$, that is, $\pm x = \pm x$. Since we obtain the identity, the functions $y = \pm x$ are additional particular solutions.

3. $x^2 y' = 4x^2 + 5xy + y^2$.

Representing this equation in the form $y' = \frac{4x^2 + 5xy + y^2}{x^2} = 4 + 5\frac{y}{x} + \frac{y^2}{x^2}$ we see that it is homogeneous with $f(z) = 4 + 5z + z^2$, $z = \frac{y}{x}$. Substituting the function z for y and using $y' = z + xz'$, we get the separable equation $z + xz' = 4 + 5z + z^2$. Simplifying, separating variables and integrating, we obtain $\int \frac{1}{4 + 4z + z^2} dz = \int \frac{1}{x} dx$ or $\int \frac{1}{(z+2)^2} dz = \int \frac{1}{x} dx$. Integration gives the general solution for z: $-\frac{1}{z+2} = \ln |x| + C$. Returning to the function y, we have the general solution $-\frac{1}{\frac{y}{x}+2} = \ln |x| + C$ or $\frac{y}{x} + 2 = -\frac{1}{\ln |Cx|}$.

Notice that in the course of solution it was required to divide by $(z+2)^2$ and, at this step, the function $\frac{y}{x} = -2$ was disregarded and was not recovered later in the form of the general solution. Therefore, we still have to check whether $y = -2x$ is the solution of the equation or not. Substitution in the original equation shows that this is a particular solution, which is not found in the general solution.

4. $xy' = y \ln \frac{y}{x}$.

Representation of this equation in the form $y' = \frac{y}{x} \ln \frac{y}{x}$ shows that it is homogeneous with $f(z) = z \ln z$, $z = \frac{y}{x}$. Replacing the function y by z and using $y' = z + xz'$, we obtain the separable equation $z + xz' = z \ln z$. Separating variables and integrating, we get $\int \frac{1}{z(\ln z - 1)} dz = \int \frac{1}{x} dx$ or $\int \frac{d(\ln z - 1)}{(\ln z - 1)} = \int \frac{1}{x} dx$. Integration gives the general solution for z: $\ln |\ln z - 1| = \ln |x| + C$ or $\ln z - 1 = Cx$. Returning to the function y, we have the general solution $\ln \frac{y}{x} = Cx + 1$. Although this algorithm requires the division by $z(\ln z - 1)$, no solution is lost due to the use of the exponential form for $\ln \frac{y}{x}$ in the general solution (note that $y \neq 0$ due to the form in which the equation is formulated).

5. $xy' = y - xe^{y/x}$.

Writing this equation in the form $y' = \frac{y}{x} - e^{y/x}$, we see that it is homogeneous with $f(z) = z - e^z$, $z = \frac{y}{x}$. By substituting the function z for y, we arrive at the separable equation $z + xz' = z - e^z$. Simplifying, separating variables and integrating, we get $\int e^{-z} dz = -\int \frac{dx}{x}$. Integration gives the general solution for z: $e^{-z} = \ln|x| + C$ or $z = -\ln(\ln|Cx|)$. Returning to the function y, we have the general solution $y = -x \ln(\ln|Cx|)$.

6. $(y^2 - 3x^2)dy + 2xydx = 0$, $y(0) = 1$.

This is the Cauchy problem. Writing the equation in the form $y' = -\frac{2xy}{y^2 - 3x^2} = -\frac{2y/x}{(y/x)^2 - 3}$, we see that it is homogeneous with $f(z) = -\frac{2z}{z^2 - 3}$, $z = \frac{y}{x}$. By substituting the function z for y, we arrive at the separable equation $z + xz' = -\frac{2z}{z^2 - 3}$. Simplifying, separating variables and integrating, we get $\int \frac{z^2 - 3}{z^3 - z} dz = -\int \frac{dx}{x}$. To calculate the first integral we use simple fractions:

$$\int \frac{z^2 - 3}{z^3 - z} dz = \int \frac{3}{z} - \frac{1}{z-1} - \frac{1}{z+1} dz = 3\ln|z| - \ln|z-1| - \ln|z+1| + C.$$

So, the general solution for z is

$$3\ln|z| - \ln|z-1| - \ln|z+1| = -\ln|x| + C \text{ or } \ln\frac{z^3}{z^2 - 1} = -\ln|Cx|.$$

Additionally, eliminating the logarithm, we have $\frac{z^2 - 1}{z^3} = Cx$. Returning to the function y, we have the general solution $\frac{(y/x)^2 - 1}{(y/x)^3} = Cx$ or $y^2 - x^2 = Cy^3$. Applying the initial condition we get $1^2 - 0^2 = C \cdot 1^3$, that is, $C = 1$. So the solution to the problem is $y^2 - x^2 = y^3$. When solving the problem, it was necessary to divide by $z^3 - z$, but this did not cause the problem of losing any solution, because the solutions $y = \pm x$ were recovered when we eliminated the logarithm and passed to the form $y^2 - x^2 = Cy^3$ of the general solution, and the solution $y = 0$, although not found in the general solution, does not satisfy the initial condition.

7. $xy^2 dy = (x^3 + y^3)dx$.

Writing the equation in the form $y' = \frac{x^3 + y^3}{xy^2} = \frac{1 + (y/x)^3}{(y/x)^2}$, we see that it is homogeneous with $f(z) = \frac{1 + z^3}{z^2}$, $z = \frac{y}{x}$. By substituting the function z for y, we arrive at the separable equation $z + xz' = \frac{1 + z^3}{z^2}$. Simplifying, separating variables and integrating, we get $\int z^2 dz = \int \frac{dx}{x}$. So the general solution for z comes in the form $\frac{z^3}{3} = \ln|x| + C = \ln|Cx|$. Returning to the function y, we have the general solution $\frac{1}{3}\frac{y^3}{x^3} = \ln|Cx|$. This solution contains all the particular ones, since no function was disregarded in this procedure.

8. $y - xy' = \frac{x}{\cos y/x}$.

Representing the equation in the form $y' = \frac{y}{x} - \frac{1}{\cos y/x}$, we clarify that it is homogeneous with $f(z) = z - \frac{1}{\cos z}$, $z = \frac{y}{x}$. By replacing the function y by z, we obtain the separable equation $z + xz' = z - \frac{1}{\cos z}$ or $xz' = -\frac{1}{\cos z}$. Separating variables and integrating, we get $\int \cos z dz = -\int \frac{dx}{x}$. Then the general solution for z comes in the form $\sin z = -\ln|x| + C$. Returning to the function y, we have the general solution $\sin \frac{y}{x} = \ln|Cx|$, which contains all particular ones.

9. $(x - y)dx + (x + y)dy = 0$.

This is a homogeneous equation, that can be seen clearly if we solve for y' and divide numerator and denominator by x: $y' = -\frac{x-y}{x+y} = -\frac{1-\frac{y}{x}}{1+\frac{y}{x}}$. Then, the right-hand side depends only on $z = \frac{y}{x}$ in the form $f(z) = -\frac{1-z}{1+z}$. By replacing the function y with z, we get the separable equation $z + xz' = -\frac{1-z}{1+z}$ or $xz' = -\frac{1+z^2}{1+z}$. Separating variables and integrating, we get $\int \frac{1+z}{1+z^2}dz = -\int \frac{dx}{x}$. The integral on the left-hand side gives

$$\int \frac{1+z}{1+z^2}dz = \int \frac{1}{1+z^2}dz + \frac{1}{2}\int \frac{d(1+z^2)}{1+z^2} = \arctan z + \frac{1}{2}\ln(1+z^2) + C.$$

So, the general solution for z comes in the form $\arctan z + \frac{1}{2}\ln(1+z^2) = -\ln|x| + C$. Returning to the function y, we have the general solution

$$\arctan \frac{y}{x} + \frac{1}{2}\ln\left(1 + \frac{y^2}{x^2}\right) = -\ln|x| + C,$$

which contains all particular ones.

10. $xy' - y = (x + y)\ln\frac{x+y}{x}$.

If we solve for the derivative $y' = \frac{y}{x} + \left(1 + \frac{y}{x}\right)\ln\left(1 + \frac{y}{x}\right)$, then it becomes clear that the right-hand side depends only on $z = \frac{y}{x}$ in the form $f(z) = z + (1+z)\ln(1+z)$, that is, the equation is homogeneous. By replacing the function y with z, we get the separable equation $z + xz' = z + (1+z)\ln(1+z)$ or $xz' = (1+z)\ln(1+z)$. Separating variables and integrating, we obtain $\int \frac{1}{(1+z)\ln(1+z)}dz = \int \frac{dx}{x}$. The integral on the left-hand side gives

$$\int \frac{1}{(1+z)\ln(1+z)}dz = \int \frac{d(\ln(1+z))}{\ln(1+z)}dz = \ln|\ln(1+z)| + C.$$

So, the general solution for z comes in the form $\ln|\ln(1+z)| = \ln|x| + C$ or, cutting the logarithms, $\ln(1+z) = Cx$. Returning to the function y, we have the general solution $\ln\left(1+\frac{y}{x}\right) = Cx$, which contains all particular ones.

11. $(y + \sqrt{xy})dx = xdy$.

If we solve for the derivative $y' = \frac{y}{x} + \sqrt{\frac{y}{x}}$, then it becomes clear that the right-hand side depends only on $z = \frac{y}{x}$ in the form $f(z) = z + \sqrt{z}$, that is, the equation is homogeneous. By replacing the function y with z, we get the separable equation $z + xz' = z + \sqrt{z}$ or $xz' = \sqrt{z}$. Separating variables and integrating, we obtain $\int \frac{dz}{\sqrt{z}} = \int \frac{dx}{x}$ and, after integration, $2\sqrt{z} = \ln|x| + C$. Returning to the function y, we have the general solution $2\sqrt{\frac{y}{x}} = \ln|x| + C$ or $y = \frac{x}{4}\ln^2|Cx|$. During the process of solution, we divided by $\sqrt{z}$, and therefore, the function $y = 0$ was eliminated from consideration. Substituting it into the original equation, we see that this is one more solution.

12. $(x - y\cos\frac{y}{x})dx + x\cos\frac{y}{x}dy = 0$, $y(1) = 2$.

As usual, we start the solution of this Cauchy problem by finding all the solutions of the given equation. Representation of the equation in the form $(1 - \frac{y}{x}\cos\frac{y}{x})dx + \cos\frac{y}{x}dy = 0$ makes it clear that all the terms depend only on the quotient $\frac{y}{x}$, that is, the equation is homogeneous. By changing the function by the formula $y = xz$, we get the separable equation

$$(1 - z\cos z)dx + \cos z(xdz + zdx) = 0 \text{ or } dx + x\cos z\,dz = 0.$$

Separating variables and integrating, we get $\int \cos z\,dz = -\int \frac{dx}{x}$, from which $\sin z = C - \ln|x|$. Returning to the function y, we have the general solution $\sin\frac{y}{x} = C - \ln|x|$. Finally, applying the initial condition, we specify the constant C: $\sin\frac{2}{1} = C - \ln 1$, that is, $C = \sin 2$. Therefore, the solution to the Cauchy problem comes in the form $\sin\frac{y}{x} = \sin 2 - \ln|x|$.

13. $(x^2 + 2xy - y^2)dx + (y^2 + 2xy - x^2)dy = 0$.

This is a homogeneous equation that can be written as

$$y' = -\frac{x^2 + 2xy - y^2}{y^2 + 2xy - x^2} = -\frac{1 + 2\frac{y}{x} - \frac{y^2}{x^2}}{\frac{y^2}{x^2} + 2\frac{y}{x} - 1},$$

where the right-hand side depends only on $z = \frac{y}{x}$ in the form $f(z) = -\frac{1+2z-z^2}{z^2+2z-1}$. By replacing the function y with z, we get the separable equation

$$z + xz' = -\frac{1 + 2z - z^2}{z^2 + 2z - 1} \text{ or } xz' = -\frac{z^3 + z^2 + z + 1}{z^2 + 2z - 1}.$$

Separating variables and integrating, we get $\int \frac{z^2+2z-1}{(z+1)(z^2+1)}dz = -\int \frac{dx}{x}$. The integral on the left-hand side can be calculated as follows:

$$\int \frac{z^2 + 2z - 1}{(z+1)(z^2+1)}dz = \int \frac{(-z^2-1)+(2z^2+2z)}{(z+1)(z^2+1)}dz$$

$$= \int -\frac{1}{z+1} + \frac{2z}{z^2+1}dz = -\ln|z+1| + \ln(z^2+1) + C.$$

So, the general solution comes in the form $-\ln|z+1| + \ln(z^2+1) = -\ln|x| + C$ or, eliminating the logarithm, $\frac{z^2+1}{z+1} = \frac{C}{x}$. Returning to the function y, we have the general solution $\frac{y^2+x^2}{y+x} = C$. During the process of solution, we divided by $z+1$ and therefore the function $y = -x$ was disregarded. Substituting it into the original equation, we see that it is one more solution.

14. $xy' = 3y - 2x - 2\sqrt{xy - x^2}$.
This is a homogeneous equation with $f(z) = 3z - 2 - 2\sqrt{z-1}$, $z = \frac{y}{x}$. By replacing the function y with z, we obtain the separable equation

$$z + xz' = 3z - 2 - 2\sqrt{z-1} \text{ or } xz' = 2z - 2 - 2\sqrt{z-1}.$$

Separating variables and integrating, we get $\int \frac{dz}{2(z-1)-2\sqrt{z-1}}dz = \int \frac{dx}{x}$. The integral on the left-hand side can be calculated as follows:

$$\int \frac{dz}{2(z-1) - 2\sqrt{z-1}}dz = \int \frac{dz}{2\sqrt{z-1}(\sqrt{z-1}-1)}dz$$

$$= \int \frac{d(\sqrt{z-1}-1)}{\sqrt{z-1}-1}dz = \ln|\sqrt{z-1}-1| + C.$$

So, the general solution comes in the form $\ln|\sqrt{z-1}-1| = \ln|x| + C$ or, eliminating the logarithm, $\sqrt{z-1} - 1 = Cx$, or still, making z explicit, $z = 1 + (Cx+1)^2$. Returning to the function y, we have the general solution $y = x + x(Cx+1)^2$. During the process of solution, we divided by $\sqrt{z-1}$ and the function $z = 1$, that is, $y = x$ was not recovered in the general solution. Substituting it into the original equation, we see that this is one more solution.

Exercises for section 2.1

Solve the following homogeneous equations:
1. $xy' = \sqrt{x^2 - y^2} + y$.
2. $y = x(y' - \sqrt[x]{e^y})$.
3. $ydx + (2\sqrt{xy} - x)dy = 0$.

4. $xy + y^2 = (2x^2 + xy)y'$.

5. $xy' + y(\ln \frac{y}{x} - 1) = 0$.

6. $(2x - y)dx + (x + y)dy = 0$.

7. $(4x^2 + 3xy + y^2)dx + (4y^2 + 3xy + x^2)dy = 0$.

8. $(2\sqrt{xy} - y)dx + xdy = 0$.

9. $y^2 + x^2y' = xyy'$.

10. $(y^2 - 2xy)dx + x^2dy = 0$.

11. $2x^3y' = 2x^2y - y^3$.

12. $(9x^2 + y^2)y' = 2xy$.

13. $y' = \frac{y}{x} - e^{y/x}$.

14. $xy' - y = x\tan \frac{y}{x}$.

2.2 Equations reducible to homogeneous

One of the equations that has no special name and is traditionally treated as *reducible to homogeneous* is $y' = f\left(\frac{ax+by+c}{\alpha x+\beta y+\gamma}\right)$, where a, b, c and α, β, γ are arbitrary constants. Before presenting the method of solution, let us filter out some uninteresting cases related to the special choices of constants. If $a = 0, b = 0$ or $\alpha = 0, \beta = 0$, then we have the equation of the type $y' = f_1(a_1x + b_1y + c_1)$ that has already been considered (it is reducible to a separable equation). If $a = 0, \alpha = 0$ or $b = 0, \beta = 0$, then we have a separable equation. In what follows, we will disregard all these cases.

Method of solution.

Let us consider the two lines $ax + by + c = 0$ and $\alpha x + \beta y + \gamma = 0$, formed from the numerator and denominator of the function f. If the two lines are parallel, then analytically this means that the determinant Δ of the system $\begin{cases} ax + by + c = 0 \\ \alpha x + \beta y + \gamma = 0 \end{cases}$ is equal to zero (in this case, the system has no solution). Then, we have $\Delta = a\beta - b\alpha = 0$ or $a\beta = b\alpha$ or still $\frac{\alpha}{a} = \frac{\beta}{b} = k$. Therefore, the right-hand side of the equation can be written as $f\left(\frac{ax+by+c}{k(ax+by)+\gamma}\right) = F(ax + by)$, and consequently, we come back to the equation $y = F(ax + by)$ reducible to a separable equation.

Let us now investigate the situation when the lines $ax + by + c = 0$ and $\alpha x + \beta y + \gamma = 0$ are not parallel, in which case we have a new type of equation. Analytically this condition is equivalent to $\Delta = a\beta - b\alpha \neq 0$ and we can find the only solution (x_0, y_0) of the system $\begin{cases} ax + by + c = 0 \\ \alpha x + \beta y + \gamma = 0 \end{cases}$, which represents the point of intersection of the two lines. Finding the coordinates x_0 and y_0, we simultaneously change the independent variable and the unknown function using the formulas $t = x - x_0$ and $z = y - y_0$. Recalculating the derivative and expressions of the equation in terms of t and z, we get:

$$y_x = y_t \cdot t_x = y_t \cdot 1 = (z + y_0)_t = z_t$$

and

$$ax + by + c = a(t + x_0) + b(z + y_0) + c = at + bz + (ax_0 + by_0 + c) = at + bz,$$

$$\alpha(t + x_0) + \beta(z + y_0) + \gamma = \alpha t + \beta z + (\alpha x_0 + \beta y_0 + \gamma) = \alpha t + \beta z.$$

Substituting these results into the original equation, we obtain

$$z_t = f\left(\frac{at + bz}{\alpha t + \beta z}\right) = f\left(\frac{a + b\frac{z}{t}}{\alpha + \beta\frac{z}{t}}\right) = F\left(\frac{z}{t}\right),$$

that is, we arrive at a homogeneous equation for the unknown function z of the variable t.

Examples.

1. $(2x + y + 1)dx = (4x + 2y - 3)dy$.
Writing the equation in the form $y' = \frac{2x+y+1}{4x+2y-3}$ and noting that the lines $2x + y + 1 = 0$ and $4x + 2y - 3 = 0$ are parallel, we decide that the equation is of the type $y' = F(ax + by + c)$. Now we can make any change that replaces $2x + y$ with the new function z. To simplify the numerator, let us use $z = 2x + y + 1$. Then, $z' = 2 + y'$, $4x + 2y - 3 = 2z - 5$ and the equation takes the form of separable variables: $z' - 2 = \frac{z}{2z-5}$ or $z' = \frac{5z-10}{2z-5}$. Integrating, we get $\int \frac{2z-5}{5z-10}dz = \int dx$. The integral on the left side can be calculated as follows:

$$\int \frac{2z-5}{5z-10}dz = \int \frac{2z-4-1}{5z-10} = \frac{2}{5}\int dz - \frac{1}{5}\int \frac{1}{z-2}dz = \frac{2}{5}z - \frac{1}{5}\ln|z-2| + C.$$

Adding the right-hand side, we find $\frac{2}{5}z - \frac{1}{5}\ln|z - 2| = x + C$. Returning to the function y, we have a general solution in the implicit form $\frac{2}{5}(2x + y + 1) - \frac{1}{5}\ln|2x + y - 1| = x + C$. In the course of the solution, we divided by $z - 2$, that is, by $2x + y - 1$. Then, we need to check whether the function $2x + y - 1 = 0$ is a special solution or not. Since $dy = y'dx = -2dx$, substitution into the equation gives the identity

$$((2x + y) + 1)dx = (1 + 1)dx = (2 - 3) \cdot (-2dx) = (2(2x + y) - 3)dy.$$

Therefore, $y = 1 - 2x$ is a special solution (not included in the general solution).

2. $(2x - 4y)dx + (x + y - 3)dy = 0$.

We rewrite the equation in the form $y' = -\frac{2x-4y}{x+y-3}$ and note that the lines $2x - 4y = 0$ and $x + y - 3 = 0$ are not parallel ($\Delta = 2 \cdot 1 - (-4) \cdot 1 \neq 0$). So, we have to found the only solution of the system $\begin{cases} 2x - 4y = 0 \\ x + y - 3 = 0 \end{cases}$, which is $x_0 = 2, y_0 = 1$. Now we replace the independent variable and unknown function by the formulas $t = x - 2$ and $z = y - 1$. The derivatives are related by the formula $y_x = z_t$. Substituting these relations into the original equation, we get the homogeneous equation $z_t = -\frac{2t-4z}{t+z} = \frac{4z-2t}{z+t}$.

Now we follow the method of solution of homogeneous equations. Using the new function $u = \frac{z}{t}$, we arrive at the separable equation $tu_t + u = \frac{4u-2}{u+1}$ or $tu_t = \frac{-u^2+3u-2}{u+1}$. Separating the variables and integrating, we get $\int \frac{u+1}{u^2-3u+2}du = -\int \frac{1}{t}dt$. To calculate the integral on the left side, we note that $u^2 - 3u + 2 = (u - 1)(u - 2)$ and use simple fractions:

$$\int \frac{u + 1}{(u - 1)(u - 2)}du = \int \frac{-2}{u - 1} + \frac{3}{u - 2}du = -2\ln|u - 1| + 3\ln|u - 2| + C.$$

Therefore, the general solution is found in the form $-2\ln|u-1|+3\ln|u-2| = -\ln|t|+C$. Converting to the form without logarithms, we get $\frac{(u-1)^2}{(u-2)^3} = Ct$. The function $u = 1$ was eliminated from consideration when we divided by $u^2 - 3u + 2$, but it was restored in the conversion from the logarithmic form to the form without logarithms. Since this is a solution for the u-equation, this transformation of the form of the general solution recovers this solution. It is obvious that the function $u = 2$ is also a solution to the equation for u, but it is not in the last form of the general solution (neither in the logarithmic form).

Returning to z, we have the general solution $\frac{(z-t)^2}{(z-2t)^3} = C$ and the special one $z = 2t$. Finally, for y and x we get the general solution in the implicit form $\frac{(y-x+1)^2}{(y-2x+3)^3} = C$ and the special one in the explicit form $y - 1 = 2(x - 2)$.

3. $(y' + 1)\ln\frac{y+x}{x+3} = \frac{y+x}{x+3}$.

First, we find the only solution of the system $\begin{cases} y + x = 0 \\ x + 3 = 0 \end{cases}$, which is $x_0 = -3$, $y_0 = 3$. Now we change the independent variable and unknown function by the formulas $t = x + 3$ and $z = y - 3$. The derivatives are related by the formula $y_x = z_t$. Substituting these relationships into the original equation, we get the equation $(z_t + 1)\ln\frac{z+t}{t} = \frac{z+t}{t}$. To better see that this equation is homogeneous, we can write it in the form $(z_t + 1)\ln(\frac{z}{t} + 1) = \frac{z}{t} + 1$.

Now we follow the method of solution of homogeneous equations. Using the new function $u = \frac{z}{t}$, we obtain the separable equation $(tu_t + u + 1)\ln(u+1) = u+1$. To simplify further calculations, we replace the unknown function u with v by the formula $v = u+1$. This leads to the equation $(tv_t + v)\ln v = v$. Separating the variables and integrating, we get $\int \frac{\ln v}{v(1-\ln v)}dv = \int \frac{1}{t}dt$. The integral on the left side can be calculated using the change of the variable $p = \ln v$:

$$\int \frac{\ln v}{v(1 - \ln v)}dv = \int \frac{\ln v d(\ln v)}{1 - \ln v} = \int \frac{pdp}{1-p} = \int \frac{p-1+1}{1-p}dp$$

$$= \int -1 + \frac{1}{1-p}dp = -p - \ln|1-p| + C = -\ln v - \ln|1 - \ln v| + C.$$

So, the general solution for v is found in the form $\ln v + \ln|1 - \ln v| = -\ln|t| + C$ or, eliminating the external logarithm, $v(1 - \ln v) = \frac{C}{t}$. Returning to $u = v - 1$, we get $(u+1)(1 - \ln(u+1)) = \frac{C}{t}$. Recalling that $t = x+3$ and $u = \frac{z}{t} = \frac{y-3}{x+3}$, we find the general solution of the original equation:

$$\left(\frac{y-3}{x+3} + 1\right)\left(1 - \ln\left(\frac{y-3}{x+3} + 1\right)\right) = \frac{C}{x+3}.$$

We can simplify a bit this relationship, by noting that $\frac{y-3}{x+3} + 1 = \frac{y+x}{x+3}$, to the following form: $\frac{y+x}{x+3}\left(1 - \ln\frac{y+x}{x+3}\right) = \frac{C}{x+3}$.

   4. $y' = \frac{x+2y-3}{4x-y-3}$, $y(1) = 1$.
This is the Cauchy problem. We start with finding all the solutions of the differential equation. To do this, we first solve the system $\begin{cases} x + 2y - 3 = 0 \\ 4x - y - 3 = 0 \end{cases}$ whose only solution is $x_0 = 1, y_0 = 1$. Next, changing the independent variable and unknown function by the formulas $t = x - 1$ and $z = y - 1$, we transform the given equation into a homogeneous one $z_t = \frac{t+2z}{4t-z} = \frac{1+2z/t}{4-z/t}$. Third, introducing one more function $u = \frac{z}{t}$, we get the separable equation $tu_t + u = \frac{1+2u}{4-u}$ or $tu_t = \frac{1-2u+u^2}{4-u}$. Now we solve the last equation by separating the variables and integrating: $\int \frac{4-u}{(u-1)^2}du = \int \frac{1}{t}dt$. The integral on the left side can be calculated as follows:

$$\int \frac{3+1-u}{(u-1)^2}du = \int \frac{3}{(u-1)^2} - \frac{1}{u-1}du = -\frac{3}{u-1} - \ln|u-1| + C.$$

So, the general solution for u is found in the form $\frac{3}{u-1} + \ln|u-1| = -\ln|t| + C$. Finally, we return to the original unknown function. Recalling that $u = \frac{y-1}{x-1}$, we get the general solution of the original equation:

$$\frac{3}{\frac{y-1}{x-1}-1}+\ln\left|\frac{y-1}{x-1}-1\right|=-\ln|x-1|+C \text{ or } \frac{3(x-1)}{y-x}+\ln|y-x|=C.$$

In the course of solution, the function $\frac{y-1}{x-1}=1$, whose simplified form is $y=x$, was eliminated from consideration. Substituting it into the original equation, we see that $y=x$ is a particular solution, not included in the general solution.

To find the solution to the Cauchy problem, we substitute the initial condition in all the solutions found, that is, in the general solution and the particular solution. Substitution in the general solution shows that the condition cannot be satisfied under any value of C, because it leads to indeterminate expressions (division by 0 and logarithm of 0). However, the particular solution $y=x$, lost in the form of the general solution, satisfies the initial condition and is therefore the only solution to the Cauchy problem.

5. $(x-y-1)dx+(x+y+3)dy=0$.
To reduce this equation to homogeneous one, we find the solution $x_0=-1$, $y_0=-2$ of the system $\begin{cases} x-y-1=0 \\ x+y+3=0 \end{cases}$. Now, changing the independent variable and unknown function by the formulas $t=x+1$ and $z=y+2$, we transform the given equation into the homogeneous one $z_t=-\frac{t-z}{t+z}=-\frac{1-z/t}{1+z/t}$. Using one more function $u=\frac{z}{t}$, we arrive at the separable equation $tu_t+u=-\frac{1-u}{1+u}$ or $tu_t=-\frac{1+u^2}{1+u}$. Separating the variables and integrating, we get $\int\frac{1+u}{1+u^2}du=-\int\frac{1}{t}dt$. The integral on the left side can be calculated as follows:

$$\int\frac{1+u}{1+u^2}du=\int\frac{1}{1+u^2}du+\frac{1}{2}\int\frac{d(1+u^2)}{1+u^2}du=\arctan u+\frac{1}{2}\ln(1+u^2)+C.$$

So, the general solution for u is $\arctan u+\frac{1}{2}\ln(1+u^2)=-\ln|t|+C$. Recalling that $u=\frac{z}{t}=\frac{y+2}{x+1}$, we obtain the general solution to the original equation:

$$\arctan\frac{y+2}{x+1}+\frac{1}{2}\ln\left(1+\left(\frac{y+2}{x+1}\right)^2\right)=\ln\left|\frac{C}{x+1}\right|.$$

6. $(2x-y+1)dx+(2y-x-1)dy=0$.
To reduce this equation to homogeneous one, we find the solution $x_0=-\frac{1}{3}$, $y_0=\frac{1}{3}$ of the system $\begin{cases} 2x-y+1=0 \\ 2y-x-1=0 \end{cases}$. Then, changing the independent

variable and unknown function by the formulas $t = x + \frac{1}{3}$ and $z = y - \frac{1}{3}$, we transform the given equation into the homogeneous $z_t = -\frac{2t-z}{2z-t} = -\frac{2-z/t}{2z/t-1}$. Using one more function $u = \frac{z}{t}$, we obtain the separable equation $tu_t + u = -\frac{2-u}{2u-1}$ or $tu_t = -\frac{2u^2-2u+2}{2u-1}$. Separating the variables and integrating, we get $\int \frac{2u-1}{u^2-u+1}du = -2\int \frac{dt}{t}$. Calculating the integrals we have

$$\int \frac{d(u^2-u+1)}{u^2-u+1}du = \ln(u^2-u+1) = -2\ln|t| + C$$

or, eliminating the logarithm, $u^2 - u + 1 = \frac{C}{t^2}$. Recalling that $u = \frac{z}{t} = \frac{y-\frac{1}{3}}{x+\frac{1}{3}}$, we get the general solution of the original equation:

$$\left(\frac{y-\frac{1}{3}}{x+\frac{1}{3}}\right)^2 - \frac{y-\frac{1}{3}}{x+\frac{1}{3}} + 1 = \frac{C}{\left(x+\frac{1}{3}\right)^2}.$$

Multiplying both sides by $\left(x+\frac{1}{3}\right)^2$ and simplifying, we can represent the general solution in the form $y^2 - xy + x^2 - y + x = C$.

7. $(2x - 4y + 6)dx + (x + y - 3)dy = 0$.

To reduce this equation to a homogeneous one, we find the solution $x_0 = 1$, $y_0 = 2$ of the system $\begin{cases} 2x - 4y + 6 = 0 \\ x + y - 3 = 0 \end{cases}$. Then, changing the independent variable and unknown function by the formulas $t = x - 1$ and $z = y - 2$, we transform the given equation into a homogeneous one $z_t = -\frac{2t-4z}{t+z} = -\frac{2-4z/t}{t+z/t}$. Using one more function $u = \frac{z}{t}$, we get the separable equation $tu_t + u = -\frac{2-4u}{1+u}$ or $tu_t = -\frac{u^2-3u+2}{u+1}$. Separating the variables and integrating, we get $\int \frac{u+1}{u^2-3u+2}du = -\int \frac{dt}{t}$. Calculating the integrals we have

$$\int \frac{u+1}{u^2-3u+2}du = \int \frac{3}{u-2} - \frac{2}{u-1}du = 3\ln|u-2| - 2\ln|u-1| = -\ln|t| + C,$$

or, eliminating the logarithm, $\frac{(u-2)^3}{(u-1)^2} = \frac{C}{t}$. Recalling that $u = \frac{z}{t} = \frac{y-2}{x-1}$, we get the general solution of the original equation:

$$\frac{\left(\frac{y-2}{x-1} - 2\right)^3}{\left(\frac{y-2}{x-1} - 1\right)^2} = \frac{C}{x-1}.$$

Simplifying, we can represent the general solution in the form $\frac{(y-2x)^3}{(y-x-1)^2} = C$. Note that $y = x + 1$ is another particular solution, lost in the applied algorithm and not included in the general solution.

8. $(x + y - 2)dx + (x - y + 4)dy = 0$.

To reduce this equation to homogeneous one, we find the solution $x_0 = -1$, $y_0 = 3$ of the system $\begin{cases} x + y - 2 = 0 \\ x - y + 4 = 0 \end{cases}$. The change of the independent variable and unknown function $t = x + 1$ and $z = y - 3$ transforms the given equation into a homogeneous one $z_t = -\frac{t+z}{t-z} = -\frac{1+z/t}{1-z/t}$. Using one more function $u = \frac{z}{t}$, we arrive at the separable equation $tu_t + u = -\frac{1+u}{1-u}$ or $tu_t = \frac{u^2 - 2u - 1}{1-u}$. Separating the variables and integrating, we get

$$\int \frac{u-1}{u^2 - 2u - 1}du = \int \frac{1}{2}\frac{d(u^2 - 2u - 1)}{u^2 - 2u - 1} = -\int \frac{1}{t}dt.$$

The result of the integration is

$$\ln|u^2 - 2u - 1| = -2\ln t + C \text{ or } u^2 - 2u - 1 = \frac{C}{t^2}.$$

Recalling that $u = \frac{z}{t} = \frac{y-3}{x+1}$, we get the general solution of the original equation:

$$(y-3)^2 - 2(y-3)(x+1) - (x+1)^2 = C \text{ or } y^2 - 2xy - x^2 - 8y + 4x = C.$$

9. $y' = 2\left(\frac{y+1}{x+y-2}\right)^2$.

To reduce this equation to homogeneous one, we find the solution $x_0 = 3$, $y_0 = -1$ of the system $\begin{cases} y + 1 = 0 \\ x + y - 2 = 0 \end{cases}$. The change of the independent variable and unknown function $t = x - 3$ and $z = y + 1$ transforms the given equation into a homogeneous one $z_t = 2\left(\frac{z}{t+z}\right)^2$. Introducing one more function $u = \frac{z}{t}$, we arrive at the separable equation $tu_t + u = 2\left(\frac{u}{1+u}\right)^2$ or $tu_t = -\frac{u^3 + u}{(u+1)^2}$. Separating the variables and integrating, we get $\int \frac{(u+1)^2}{u^3 + u}du = -\int \frac{dt}{t}$. The integral on the left side can be calculated as follows:

$$\int \frac{u^2 + 1 + 2u}{u(u^2+1)}du = \int \frac{1}{u} + \frac{2}{u^2+1}du = \ln|u| + 2\arctan u + C.$$

So, $\ln|u| + 2\arctan u = -\ln|t| + C$. Recalling that $u = \frac{z}{t} = \frac{y+1}{x-3}$, we obtain the general solution of the original equation:

$$\ln\left|\frac{y+1}{x-3}\right| + 2\arctan\frac{y+1}{x-3} = -\ln|x-3| + C.$$

In this procedure, the function $y = -1$ has been disregarded, and it is easy to see that this is an additional solution of the original equation.

Exercises for section 2.2

Solve the following equations, transforming them into homogeneous or separable:

1. $(x - y)dx + (2y - x + 1)dy = 0.$
2. $(3y - 7x + 7)dx - (3x - 7y - 3)dy = 0.$
3. $(y + 2)dx = (2x + y - 4)dy.$
4. $(x - 2y - 1)dx + (3x - 6y + 2)dy = 0.$
5. $(x + y + 1)dx + (x - y + 3)dy = 0.$
6. $(2x - y - 2)dx + (x + y - 4)dy = 0.$
7. $(2x + y - 3)y' + y + 1 = 0.$
8. $(x + 4y)y' = 2x + 3y - 5.$

3   Exact equations. Integrating factor

3.1   Exact equations

Definition. An equation $A(x, y)dx + B(x, y)dy = 0$ is called an *equation of exact differential* (or simply an *exact equation*) if its left-hand side represents the differential of a function of two variables, that is, if there exists a function $F(x, y)$ such that $A(x, y)dx + B(x, y)dy = dF(x, y)$.

Method of solution.

If the function $F(x, y)$ in the definition is known, then the equation takes the form $dF(x, y) = 0$, and consequently, its general solution in implicit form is $F(x, y) = C$ (recall that a function is constant in a simply connected region if and only if its differential is zero in that region). The main problem is that the function $F(x, y)$ is usually unknown and it is not even known whether the equation $A(x, y)dx + B(x, y)dy = 0$ is exact or not. Therefore, the first step is to check whether the equation really is exact and, if so, in the second step to find the function $F(x, y)$.

To better understand the verification procedure, recall that, by the definition, $dF(x, y) = F_x(x, y)dx + F_y(x, y)dy$, where dx and dy are increments of the independent variables x and y. So, taking into account that these increments are arbitrary, from the relation $dF(x, y) = F_x(x, y)dx + F_y(x, y)dy = A(x, y)dx + B(x, y)dy$ it follows directly that $F_x = A$ and $F_y = B$. Assuming

that F is a smooth function (for example, it has continuous partial derivatives of the second order), we have the equality of the mixed derivatives,
that is, $F_{xy} = (F_x)_y = A_y = B_x = (F_y)_x = F_{yx}$. Therefore, if $Adx + Bdy$ is
an exact differential of $F(x, y)$, then the relation $A_y = B_x$ is satisfied.

It turns out that the converse statement is also true in a simply connected
region, that is, a region satisfying the following two properties: any two
points can be connected by a continuous curve lying entirely within the
region; and any closed curve contains inside it only the points of the region
(roughly speaking, a region that has no disjoint parts and no holes). If
$A_y = B_x$ in a simply connected region, then $Adx + Bdy$ is an exact differential
of a function F in that region. The demonstration of this result is rather
simple and, on the other hand, it offers a general procedure for finding
the function F, so let us consider it. To show that $Adx + Bdy = dF =$
$F_x(x, y)dx + F_y(x, y)dy$, we have to find F such that the two equations
$\begin{cases} F_x = A \\ F_y = B \end{cases}$ hold simultaneously. To show that the equations of the system
are compatible (admit a solution), let us transform the first equation into a
form more similar with the second (the same can be done with the second
equation). To do this, we integrate the first equation with respect to x,
considering y as a parameter: $F = \int Adx + g(y)$, where $g(y)$ is an arbitrary
function of y (for convenience, here we understand $\int Adx$ as one of the
antiderivatives of A for any fixed y). Now, differentiating this relation with
respect to y, we get $F_y = \int A_y dx + g_y(y)$. Due to the condition $A_y = B_x$,
we can rewrite: $F_y = \int B_x dx + g_y(y) = B + h(y) + g_y(y)$, where $h(y)$ is
one more arbitrary function of y. This is another form of the first equation,
which is similar to the second. Comparing the two equations, we have that
$B + h(y) + g_y(y) = B$ or, simplifying, $g_y(y) = -h(y)$. This is a compatibility
condition between two equations, which can always be met: by choosing an
arbitrary function $h(y)$, it is simple to find $g(y)$ (which will no longer be
arbitrary) by integrating $h(y)$. Specifying $g(y)$, we obtain the function F
determined by the formula $F = \int Adx + g(y)$. Thus, we have shown that the
system has a solution, that is, there exists a function F whose differential is
$Adx + Bdy$. Besides, we have given an algorithm for finding F, which can
be used to solve specific equations.

Remark 1. According to the method of solution, the found general
solution contains all solutions of an exact equation.

Remark 2. Since the function $g(y)$ is obtained through integration, it
is determined only up to an additive constant, and the same is true for the
function $F(x, y)$, since its expression involves $g(y)$. This is in line with the
elementary fact that if $F(x, y)$ is a function whose differential is equal to
$Adx + Bdy$, then any function of the family $F(x, y) + C_1$, where C_1 is an

arbitrary constant, has the same property (just recall that the differential of a constant is zero). Actually, we can always choose a specific value of this constant (the one we find most convenient), because the solution of the equation is represented in the form $F(x, y) = C$, where C is also an arbitrary constant.

Remark 3. A separable equation $y' = f(x)g(y)$ becomes exact after separation of variables in the form $\frac{dy}{g(y)} = f(x)dx$. However, separable equations posses their own algorithm of solution (by integration of both sides of the last formula), which is more direct and in many cases faster than the solution of the last equation as the exact one. For this reason, separable equations normally are not reduced to exact and are considered as a basic, irreducible type.

Examples.

1. $e^x \sin y \, dx + (1 + e^x) \cos y \, dy = 0$.
First, we analyze whether this equation is exact or not by checking the condition $A_y = B_x$, where $A = e^x \sin y$ and $B = (1 + e^x) \cos y$. Since $A_y = e^x \cos y = B_x$, the equation is exact. Second, we find the function F such that $dF = Adx + Bdy$, by solving the system $\begin{cases} F_x = A = e^x \sin y \\ F_y = B = (1 + e^x) \cos y \end{cases}$.
The two equations are simple to integrate, so we can start with the first: $F = \int e^x \sin y \, dx = e^x \sin y + g(y)$ and $F_y = e^x \cos y + g_y(y)$. Comparing the last expression for F_y with the second equation of the system, we conclude that $g_y(y) = \cos y$, and consequently, $g(y) = \sin y + C_1$. Substituting this expression for $g(y)$ into the formula for F, we get $F = e^x \sin y + \sin y + C_1$. With the function F found, we have the general solution of the original equation in the form $F = C$, that is, $e^x \sin y + \sin y + C_1 + C_1 = C$, or $(e^x + 1) \sin y = C$. According to Remark 2, the arbitrary constant C_1 that appears in the expression of $F(x, y)$ can be discarded, because the form of the general solution already has a constant that can be joined with C_1.

It is interesting to note that this equation can also be classified as separable. This is easy to see by writing it in the form $y' = -\frac{e^x}{1+e^x} \cdot \frac{\sin y}{\cos y}$. Sometimes the method of separable equations is more straightforward and simpler than the algorithm of exact equations. Let us solve the same equation as separable one and compare the two procedures and the two solutions. First, we separate the variables: $\frac{\cos y}{\sin y} = -\frac{e^x}{1+e^x}$. Integrating the two sides, we get:

$$\int \frac{\cos y}{\sin y} dy = \int \frac{d(\sin y)}{\sin y} = \ln|\sin y| + C_1$$

for the left side and

$$\int \frac{e^x}{1+e^x} dx = \int \frac{d(1+e^x)}{1+e^x} = \ln(1+e^x) + C_2$$

for the right side (without negative sign). Therefore, the general solution appears in the form

$$\ln|\sin y| = -\ln(1+e^x) + B \text{ or } \sin y = \frac{B}{1+e^x}, B \neq 0.$$

Taking into account that the division by $\sin y$ removes from consideration the particular solutions $\sin y = 0$, that is, $y = k\pi$, $k \in \mathbb{Z}$, we can restore these solutions, incorporating them into the general solution, by using the following form: $\sin y = \frac{B}{1+e^x}$, $\forall B$. Comparing the solutions obtained by the two methods, we see that the final result is the same (as it should be). In this case, the two techniques require approximately the same amount of technical work.

2. $x(2x^2 + y^2)dx + y(x^2 + 2y^2)dy = 0$.
First, we analyze whether this equation is exact or not by checking the condition $A_y = B_x$, where $A = x(2x^2 + y^2)$ and $B = y(x^2 + 2y^2)$. Since $A_y = 2xy = B_x$, the equation is exact. Second, we find the function F such that $dF = Adx + Bdy$, by solving the system $\begin{cases} F_x = A = x(2x^2 + y^2) \\ F_y = B = y(x^2 + 2y^2) \end{cases}$.
The two equations are simple to integrate, so we can start with the first: $F = \int 2x^3 + xy^2 dx = \frac{x^4}{2} + \frac{x^2y^2}{2} + g(y)$ and $F_y = x^2y + g_y(y)$. Comparing the last expression for F_y with the second equation of the system, we conclude that $g_y(y) = 2y^3$, and consequently, $g(y) = \frac{y^4}{2} + C_1$. Substituting this expression for $g(y)$ into the formula for F, we get $F = \frac{x^4}{2} + \frac{x^2y^2}{2} + \frac{y^4}{2} + C_1$. With the function F found, we have the general solution of the original equation in the form $F = C$, that is,

$$\frac{x^4}{2} + \frac{x^2y^2}{2} + \frac{y^4}{2} + C_1 = C \text{ or } x^4 + x^2y^2 + y^4 = C.$$

According to Remark 2, the arbitrary constant C_1 that appears in the formula of $F(x, y)$ (due to the form of $g(y)$) can be discarded, because the form of the general solution already has a constant that can be joined with C_1.

It is interesting to note that the same equation can also be classified as homogeneous. In most cases, the exact differential method provides technically simpler solution. However, we recommend to the reader to solve this equation as homogeneous and compare the obtained solutions.

3. $\frac{x}{\sqrt{x^2+y^2}} + \frac{1}{x} + \frac{1}{y} = \left(\frac{x}{y^2} - \frac{1}{y} - \frac{y}{\sqrt{x^2+y^2}}\right) y'$.

This equation is not separable or homogeneous, nor one of those reducible to these two types. Therefore, let us try exact differential. Write the equation in the form $Adx + Bdy = 0$ with $A = \frac{x}{\sqrt{x^2+y^2}} + \frac{1}{x} + \frac{1}{y}$, $B = -\frac{x}{y^2} + \frac{1}{y} + \frac{y}{\sqrt{x^2+y^2}}$ and check that $A_y = -xy(x^2+y^2)^{-3/2} - \frac{1}{y^2} = B_x$, which means that the equation is exact. Then, there exists a function $F(x, y)$ such that $dF = Adx + Bdy$ and this function is found by solving the system $\begin{cases} F_x = A = \frac{x}{\sqrt{x^2+y^2}} + \frac{1}{x} + \frac{1}{y} \\ F_y = B = \frac{y}{\sqrt{x^2+y^2}} + \frac{1}{y} - \frac{x}{y^2} \end{cases}$.

Both relations have the same degree of difficulty of integration, so it does not matter which equation we start with. For a change, we can start with the second one. Integrating it with respect to y, we get

$$F = \int \frac{y}{\sqrt{x^2+y^2}} + \frac{1}{y} - \frac{x}{y^2} dy = \sqrt{x^2+y^2} + \ln|y| + \frac{x}{y} + g(x).$$

Differentiating the last expression of F with respect to x, we have: $F_x = \frac{x}{\sqrt{x^2+y^2}} + \frac{1}{y} + g_x(x)$. Comparing this formula for F_x with the one given in the first equation of the system, we conclude that $g_x(x) = \frac{1}{x}$, and consequently, $g(x) = \ln|x|$ (we chose the zero constant). Therefore, $F = \sqrt{x^2+y^2} + \ln|y| + \frac{x}{y} + \ln|x|$ and the general solution comes in the form

$$\sqrt{x^2+y^2} + \ln|y| + \frac{x}{y} + \ln|x| = C.$$

4. $(2x^3 - xy^2)dx + (2y^3 - x^2y)dy = 0$.

By verifying that $(2x^3 - xy^2)_y = -2xy = (2y^3 - x^2y)_x$, we ensure that the equation is exact. Next, we find the function F, whose differential is the left-hand side of the equation, by solving the system $\begin{cases} F_x = 2x^3 - xy^2 \\ F_y = 2y^3 - x^2y \end{cases}$.

The two equations are simple to integrate, so let us start with the first: $F = \frac{x^4}{2} - \frac{x^2y^2}{2} + g(y)$ and $F_y = -x^2y + g_y(y)$. Comparing the last expression for F_y with the second equation of the system, we conclude that $g_y(y) = 2y^3$, and consequently, $g(y) = \frac{y^4}{2}$ (we need only one function $g(y)$). Substituting this expression for $g(y)$ into the formula of F, we get $F = \frac{x^4}{2} - \frac{x^2y^2}{2} + \frac{y^4}{2}$. With the function F found, we have the general solution of the original equation in the form

$$\frac{x^4}{2} - \frac{x^2y^2}{2} + \frac{y^4}{2} = C \text{ or } x^4 - x^2y^2 + y^4 = C_1.$$

5. $(2x - y + 1)dx + (2y - x - 1)dy = 0$.

First, by calculating the partial derivatives and comparing the results ($2x -$

$y+1)_y = -1 = (2y-x-1)_x$, we verify that the equation is exact. Second, we find the function F, whose differential is the left-hand side of the equation, by solving the system $\begin{cases} F_x = 2x - y + 1 \\ F_y = 2y - x - 1 \end{cases}$. The two equations are simple to integrate, so let us start with the first: $F = x^2 - xy + x + g(y)$ and then $F_y = -x + g_y(y)$. Comparing the last expression for F_y with the second equation of the system, we conclude that $g_y(y) = 2y - 1$, and consequently, $g(y) = y^2 - y$ (we need only one function $g(y)$). Substituting this expression for $g(y)$ into the formula for F, we get $F = x^2 - xy + x + y^2 - y$. Finally, the general solution of the original equation is found in the form $x^2 - xy + x + y^2 - y = C$.

This equation is also reducible to homogeneous. We suggest to the reader to solve it using the corresponding algorithm and compare the solutions.

6. $(3x^2 - 3y^2 + 4x)dx - (6xy + 4y)dy = 0$.
First, we verify that the given equation is exact: $(3x^2 - 3y^2 + 4x)_y = -6y = -(6xy + 4y)_x$. Second, we find the function F, whose differential is the left-hand side of the equation, by solving the system $\begin{cases} F_x = 3x^2 - 3y^2 + 4x \\ F_y = -6xy - 4y \end{cases}$.
For a change, let us start with integration of the second equation: $F = -3xy^2 - 2y^2 + h(x)$ and then $F_x = -3y^2 + h_x(x)$. Comparing this expression for F_x with the first equation of the system, we conclude that $h_x(x) = 3x^2 + 4x$, and consequently, $h(x) = x^3 + 2x^2$ (we need only one function $h(x)$). Substituting this expression for $h(x)$ into the formula for F, we get $F = -3xy^2 - 2y^2 + x^3 + 2x^2$. Finally, the general solution of the original equation is found in the form $-3xy^2 - 2y^2 + x^3 + 2x^2 = C$.

7. $\frac{2x(1-e^y)}{(1+x^2)^2} dx + \frac{e^y}{1+x^2} dy = 0$.
First, we verify that the given equation is exact: $\left(\frac{2x(1-e^y)}{(1+x^2)^2} \right)_y = -\frac{2xe^y}{(1+x^2)^2} = \left(\frac{e^y}{1+x^2} \right)_x$. Second, we find the function F, whose differential is the left-hand side of the equation, by solving the system $\begin{cases} F_x = \frac{2x(1-e^y)}{(1+x^2)^2} \\ F_y = \frac{e^y}{1+x^2} \end{cases}$. Here, the expression in the second equation is noticeably simpler for integration, so we start with the second equation: $F = \frac{e^y}{1+x^2} + h(x)$ and then $F_x = -\frac{2xe^y}{(1+x^2)^2} + h_x(x)$. Comparing this expression for F_x with the first equation of the system, we conclude that $h_x(x) = \frac{2x}{(1+x^2)^2}$, and consequently, $h(x) = -\frac{1}{1+x^2}$ (we need only one function $h(x)$). Substituting this expression for $h(x)$ into the formula for F, we get $F = \frac{e^y}{1+x^2} - \frac{1}{1+x^2}$. Finally, the general

solution of the original equation is $\frac{e^y-1}{1+x^2} = C$.

8. $\left(\frac{\sin 2x}{y} + x\right) dx + \left(y - \frac{\sin^2 x}{y^2}\right) dy = 0$.

First, we verify that the given equation is exact: $\left(\frac{\sin 2x}{y} + x\right)_y = -\frac{\sin 2x}{y^2} = \left(y - \frac{\sin^2 x}{y^2}\right)_x$. Second, we find the function F, whose differential is the left-hand side of the equation, by solving the system $\begin{cases} F_x = \frac{\sin 2x}{y} + x \\ F_y = y - \frac{\sin^2 x}{y^2} \end{cases}$. The expressions in the two equations are of comparable difficulty for integration. We can start with the second equation: $F = \frac{y^2}{2} + \frac{\sin^2 x}{y} + h(x)$ and then $F_x = \frac{2\sin x \cos x}{y} + h_x(x)$. Comparing this expression for F_x with the first equation of the system, we conclude that $h_x(x) = x$, and consequently, $h(x) = \frac{x^2}{2}$. Substituting this expression for $h(x)$ into the formula for F, we get $F = \frac{y^2}{2} + \frac{\sin^2 x}{y} + \frac{x^2}{2}$. Finally, knowing the function F, we find the general solution of the original equation in the form $\frac{x^2+y^2}{2} + \frac{\sin^2 x}{y} = C$.

9. $\left(\frac{x}{\sqrt{x^2-y^2}} - 1\right) dx - \frac{y}{\sqrt{x^2-y^2}} dy = 0$.

First, we verify that the given equation is exact: $\left(\frac{x}{\sqrt{x^2-y^2}} - 1\right)_y = \frac{xy}{\sqrt{(x^2-y^2)^3}} = \left(-\frac{y}{\sqrt{x^2-y^2}}\right)_x$. Second, we find the function F, whose differential is the left-hand side of the equation, by solving the system $\begin{cases} F_x = \frac{x}{\sqrt{x^2-y^2}} - 1 \\ F_y = -\frac{y}{\sqrt{x^2-y^2}} \end{cases}$. The expressions in the two equations are of comparable difficulty to integrate. We can start integrating the first equation: $F = \sqrt{x^2 - y^2} - x + g(y)$ and then $F_y = -\frac{y}{\sqrt{x^2-y^2}} + g_y(y)$. Comparing this expression of F_y with the second equation of the system, we conclude that $g_y(y) = 0$, and consequently, $g(y) = 0$ (we can choose any constant). Therefore, $F = \sqrt{x^2 - y^2} - x$ and the general solution of the original equation has the form $\sqrt{x^2 - y^2} - x = C$.

10. $x(2x^2 + y^2) + y(x^2 + 2y^2)y' = 0$.

The given equation is exact, because $\left(x(2x^2 + y^2)\right)_y = 2xy = \left(y(x^2 + 2y^2)\right)_x$. So, we can find the function F, whose differential is the left-hand side of the equation, by solving the system $\begin{cases} F_x = 2x^3 + xy^2 \\ F_y = x^2y + 2y^3 \end{cases}$. The expressions in the two equations are of comparable difficulty for integration. We can start integrating the first equation: $F = \frac{x^4}{2} + \frac{x^2 y^2}{2} + g(y)$ and then $F_y = x^2 y + g_y(y)$. Comparing this expression of F_y with the second equation of the system, we conclude that $g_y(y) = 2y^3$, and consequently, $g(y) = \frac{y^4}{2}$ (we chose constant of integration equal to 0). Therefore, $F = \frac{x^4}{2} + \frac{x^2 y^2}{2} + \frac{y^4}{2}$ and the general solution of the original equation has the form $x^4 + x^2 y^2 + y^4 = C$.

11. $\left(3x^2 \tan y - 2\frac{y^3}{x^3}\right) dx + \left(\frac{x^3}{\cos^2 y} + 4y^3 + 3\frac{y^2}{x^2}\right) dy = 0$.

First, we verify that the given equation is exact: $\left(3x^2 \tan y - 2\frac{y^3}{x^3}\right)_y =$

$\frac{3x^2}{\cos^2 y} - 6\frac{y^2}{x^3} = \left(\frac{x^3}{\cos^2 y} + 4y^3 + 3\frac{y^2}{x^2}\right)_x$. Second, we find the function F, whose differential is the left-hand side of the equation, by solving the system

$\begin{cases} F_x = 3x^2 \tan y - 2\frac{y^3}{x^3} \\ F_y = \frac{x^3}{\cos^2 y} + 4y^3 + 3\frac{y^2}{x^2} \end{cases}$. We start with integration of the first equation:

$F = x^3 \tan y + \frac{y^3}{x^2} + g(y)$ and then $F_y = x^3 \frac{1}{\cos^2 y} + 3\frac{y^2}{x^2} + g_y(y)$. Comparing this expression of F_y with the second equation of the system, we conclude that $g_y(y) = 4y^3$, and consequently, $g(y) = y^4$ (we chose constant of integration equal to 0). Therefore, $F = x^3 \tan y + \frac{y^3}{x^2} + y^4$ and the general solution of the original equation has the form $x^3 \tan y + \frac{y^3}{x^2} + y^4 = C$.

12. $\left(2x + \frac{x^2 + y^2}{x^2 y}\right) dx = \frac{x^2 + y^2}{xy^2} dy$.

The given equation is exact, since $\left(2x + \frac{x^2 + y^2}{x^2 y}\right)_y = \frac{y^2 - x^2}{x^2 y^2} = \left(\frac{x^2 + y^2}{xy^2}\right)_x$. So,

we can find the function F that satisfies the system $\begin{cases} F_x = 2x + \frac{x^2 + y^2}{x^2 y} \\ F_y = -\frac{x^2 + y^2}{xy^2} \end{cases}$.

We start by integrating the second equation: $F = -\int \frac{x}{y^2} + \frac{1}{x} dy = \frac{x}{y} - \frac{y}{x} + h(x)$ and then $F_x = \frac{1}{y} + \frac{y}{x^2} + h_x(x)$. Comparing this expression of F_y with the first equation of the system, we conclude that $h_x(x) = 2x$, and consequently, $h(x) = x^2$ (we choose constant of integration 0). Therefore, $F = \frac{x}{y} - \frac{y}{x} + x^2$ and the general solution of the original equation has the form $\frac{x}{y} - \frac{y}{x} + x^2 = C$.

13. $\frac{y + \sin x \cdot \cos^2 (xy)}{\cos^2 (xy)} dx + \left(\frac{x}{\cos^2 (xy)} - \sin y\right) dy = 0$.

The given equation is exact, since

$\left(\frac{y + \sin x \cdot \cos^2 (xy)}{\cos^2 (xy)}\right)_y = \frac{1}{\cos^2 (xy)} + \frac{xy \sin(xy)}{\cos^3 (xy)} = \left(\frac{x}{\cos^2 (xy)} - \sin y\right)_x$.

So, we can find the function F, whose differential is the left-hand side of the equation, by solving the system $\begin{cases} F_x = \frac{y}{\cos^2 (xy)} + \sin x \\ F_y = \frac{x}{\cos^2 (xy)} - \sin y \end{cases}$. We start by integrating the first equation: $F = \tan(xy) - \cos x + g(y)$ and then $F_y = \frac{x}{\cos^2(xy)} + g_y(y)$. Comparing this expression of F_y with the second equation of the system, we conclude that $g_y(y) = -\sin y$, and consequently, $g(y) = \cos y$ (constant of integration is 0). Therefore, $F = \tan(xy) - \cos x + \cos y$ and the general solution of the original equation has the form $\tan(xy) - \cos x + \cos y = C$.

14. $\left(\frac{y}{2\sqrt{xy}} + 2xy\sin(x^2y) + 4 \right) dx + \left(\frac{x}{2\sqrt{xy}} + x^2\sin(x^2y) \right) dy = 0$.

The given equation is exact, because

$$\left(\frac{y}{2\sqrt{xy}} + 2xy\sin(x^2y) + 4 \right)_y = \frac{1}{4\sqrt{xy}} + 2x\sin(x^2y) + 2x^3y\cos(x^2y)$$

$$= \left(\frac{x}{2\sqrt{xy}} + x^2\sin(x^2y) \right)_x .$$

So, we can find the function F, whose differential is the left-hand side of the equation, by solving the system $\begin{cases} F_x = \frac{y}{2\sqrt{xy}} + 2xy\sin(x^2y) + 4 \\ F_y = \frac{x}{2\sqrt{xy}} + x^2\sin(x^2y) \end{cases}$. We start with integration of the second equation: $F = \sqrt{xy} - \cos(x^2y) + h(x)$ and then $F_x = \frac{x}{2\sqrt{xy}} + 2xy\sin(x^2y) + h_x(x)$. Comparing this expression of F_x with the first equation of the system, we conclude that $h_x(x) = 4$, and consequently, $h(x) = 4x$ (constant of integration is 0). Therefore, $F = \sqrt{xy} - \cos(x^2y) + 4x$ and the general solution of the original equation has the form $\sqrt{xy} - \cos(x^2y) + 4x = C$.

Exercises for section 3.1

Solve the following exact equations:

1. $(3x^2 - 2x - y) dx + (2y - x + 3y^2)dy = 0$.

2. $\left(\frac{y}{\sqrt{1-x^2y^2}} - 2x \right) dx + \frac{x}{\sqrt{1-x^2y^2}}dy = 0$.

3. $\left(\frac{x}{\sqrt{x^2-y^2}} - 1 \right) dx - \frac{y}{\sqrt{x^2-y^2}}dy = 0$.

4. $(1 + e^{x/y}) dx + e^{x/y} \left(1 - \frac{x}{y} \right) dy = 0$.

5. $2xy dx + (x^2 - y^2)dy = 0$.

6. $e^{-y}dx - (2y + xe^{-y})dy = 0$.

7. $\frac{3x^2+y^2}{y^2}dx - \frac{2x^3+5y}{y^3}dy = 0$.

8. $(3x^2 + 6xy^2)dx + (6x^2y + 4y^3)dy = 0$.

9. $\left(\frac{xy}{\sqrt{1+x^2}} + 2xy - \frac{y}{x} \right) dx + (\sqrt{1+x^2} + x^2 - \ln x)dy = 0$.

10. $\left(\sin y + y \sin x + \frac{1}{x}\right) dx + \left(x \cos y - \cos x + \frac{1}{y}\right) dy = 0.$

11. $(3x^2 + y - 1)dx + (x + 3y^2 - 1)dy = 0.$

12. $(y + \sin x)dx + (x + \cos y)dy = 0.$

13. $(y^2 + \ln x)dx + (2xy - \ln y)dy = 0.$

14. $(1 + 3x^2 \ln y)dx + (3y^2 + \frac{x^3}{y})dy = 0.$

15. $\left(2x - \frac{\sin^2 y}{x^2}\right) dx + \left(2y + \frac{\sin 2y}{x}\right) dy = 0.$

16. $\left(\frac{y}{x^2} + \frac{1}{y}\right) dx - \left(\frac{x}{y^2} + \frac{1}{x} + 2y\right) dy = 0.$

17. $\frac{y}{x}dx + (1 + \ln(xy)) dy = 0.$

18. $2x(1 + \sqrt{x^2 - y})dx - \sqrt{x^2 - y}dy = 0.$

3.2 Equations reducible to exact equation. Integrating factor

An equation $A(x, y)dx + B(x, y)dy = 0$ is the general form of a first order equation. With the condition $A_y = B_x$ it is an exact equation, without this condition it is not. We note that a form of functions A and B (simple or intricate) is not an indication of whether an equation is exact. In Example 1 of exact equations the functions A and B are relatively simple, while in Example 2 they are relatively complicated. The simple equation $ydx + xdy = 0$ is exact (with $F = xy$), but another, no less simple equation $xdx + xdy = 0$ is not, although it does not cost to make it an exact equation by dividing the equation by x. Similarly, multiplication of the equation in Example 2 by x (or by y) transforms the equation into a non-exact one.

Integrating factor

As noted, multiplication (or division) of a non-exact differential equation by a function can result in an exact differential equation, and vice versa, an exact equation multiplied (divided) by a function can become a non-exact equation. Then, a question arises whether for any non-exact equation there is a multiplier that leads to an exact equation. (Obviously, this function must not be a constant, because if it was, then the relationship between A_y and B_x would not change.) Theoretically, the answer to this question is positive (under the weak constraints on smoothness of A and B and non-vanishing of both), and the function used in this multiplication is called an integrating factor.

Definition. If a (non-exact) equation $A(x, y)dx + B(x, y)dy = 0$ is converted into exact one by multiplying by the function $\mu(x, y)$, then μ is called an *integrating factor*.

Let us show that a differential equation whose solution exists always has an integrating factor. In fact, write the equation $Adx + Bdy = 0$ in the form $y' = -\frac{A}{B}$ (under the assumption that $B \neq 0$) and consider a general solution $F(x, y) = C$ of this equation. Calculating the differential of F, we get $F_x dx + F_y dy = 0$ or $y' = -\frac{F_x}{F_y}$. Comparing the two expressions for y', we get the relation $\frac{F_x}{F_y} = \frac{A}{B}$ or $\frac{F_x}{A} = \frac{F_y}{B} = \mu$. Therefore, the equation $F_x dx + F_y dy = 0$ can be rewritten in the form

$$dF = \frac{F_x}{A} Adx + \frac{F_y}{B} Bdy = \mu Adx + \mu Bdy = 0,$$

where the left side of the equation $\mu Adx + \mu Bdy = 0$ is the differential of F, that is, multiplication by μ results in the exact equation.

As an additional point, we can also show that there is an infinite set of integrating factors. First, a multiple of an integrating factor is also an integrating factor. But the two factors can differ not only by a constant multiplier. Indeed, if $\mu Adx + \mu Bdy = F_x dx + F_y dy = dF$, then

$$e^F \mu Adx + e^F \mu Bdy = e^F F_x dx + e^F F_y dy = (e^F)_x dx + (e^F)_y dy = d(e^F),$$

that is, $e^F \mu$ is also an integrating factor for the equation $Adx + Bdy = 0$ (recall that $F(x, y) = C$ is the general solution of the equation $Adx + Bdy = 0$). Actually, if μ is an integrating factor, then any function in the form $h(F)\mu$, where h is a differentiable function of argument F, is an integrating factor of the equation $Adx + Bdy = 0$. In this case,

$$h(F)F_x dx + h(F)F_y dy = d(\int h(F)dF).$$

Method of solution.

In practice, finding an integrating factor can be more complicated than solving the original equation, because, in general, it requires to solve a partial differential equation. Even so, there are some particular cases when an integrating factor can be found effectively. Usually, this occurs when the integrating factor depends only on one variable.

Let us consider two main cases when the integrating factor can be found quite simply (except for possible technical problems of integration): first, when μ depends only on x, and second, when it depends only on y. The two cases are similar, so we discuss in detail only the first one. We deduce in parallel both the formula for $\mu(x)$ and the condition when there is an integrating factor as a function of x alone.

If we expect the equation $\mu(x)A(x,y)dx + \mu(x)B(x,y)dy = 0$ to be exact, then the condition $(\mu A)_y = (\mu B)_x$ must be satisfied or, taking into account that μ depends only on x: $\mu A_y = \mu_x B + \mu B_x$. Solving for μ, we get

$$\frac{\mu_x}{\mu} = \frac{A_y - B_x}{B} \quad \text{or} \quad (\ln \mu)_x = \frac{A_y - B_x}{B}.$$

On the right-hand side we have a function of x alone, which implies that the expression $\frac{A_y - B_x}{B}$ must also depend only on x. This is the condition for the validity of this assumption and execution of the algorithm. If this is true, the function $\ln \mu$ is found by direct integration of a known function of x (of course, in practice integration can be a tricky problem). Knowing μ, we transform the equation into exact and then follow the procedure of solution of exact equations.

If μ depends only on y, the reasoning is analogous and it leads to the equation $(\ln \mu)_y = \frac{B_x - A_y}{A}$. The condition for finding $\mu(y)$ is that the expression $\frac{B_x - A_y}{A}$ should depend only on y.

Sometimes we can solve examples where the integrating factor depends on both x and y, but the two variables can be combined into a single parameter that represents the only variable of the integrating factor. This combination of x and y must be indicated or derived from additional information about equation (such as physical or geometrical interpretation of the equation), which may represent a model of processes in applied sciences.

Remark. There is always a set of integrating factors, even when μ depends only on x or only on y (as has already been shown). However, we are interested in finding just one of them, in order to transform the given equation into an exact one. Therefore, we can always choose the function μ that has the simplest form.

Examples.

1. $(x^2 + y^2 + 1)dx - 2xydy = 0$.
First, we check the type of the equation: it is not separable or homogeneous or reducible to one these two types. It is also not exact, because $A_y = 2y \neq -2y = B_x$ (sign matters!). Let us try to find an integrating factor that depends only on x. We simply follow the algorithm of finding it, without memorizing the general formulas which have already been deduced. If μ is an integrating factor, then the relation $(\mu A)_y = (\mu B)_x$ must be satisfied, that is, $(\mu(x^2 + y^2 + 1))_y = (\mu(-2xy))_x$. If it depends only on x, then from this relation it follows that $\mu A_y = \mu_x B + \mu B_x$, that is, $\mu \cdot 2y = \mu_x \cdot (-2xy) + \mu \cdot (-2y)$. Sending all the terms with μ to the left-hand side, we get $(\ln \mu)_x = \frac{2y + 2y}{-2xy} = \frac{2}{-x}$. Since the right-hand side $\frac{A_y - B_x}{B} = -\frac{2}{x}$ depends

only on x, the assumption that μ can depend only on x is true and we find the following expression for μ: $\ln\mu = \int \frac{2}{-x} = -2\ln_x + C$. We are interested in only one integrating factor, so we can discard the constant (for simplicity) and choose $\mu = \frac{1}{x^2}$.

We verify now that the found function μ is really an integrating factor:

$$(\mu(x^2+y^2+1))_y = \left(\frac{1}{x^2}(x^2+y^2+1)\right)_y = \frac{2y}{x^2} = \left(\frac{1}{x^2}(-2xy)\right)_x = (\mu(-2xy))_x.$$

We can then solve the exact equation

$$\frac{1}{x^2}(x^2+y^2+1)dx - \frac{1}{x^2}2xydy = 0$$

using the corresponding algorithm. To begin with, we write the equation in the form $(1+\frac{y^2}{x^2}+\frac{1}{x^2})dx - \frac{2y}{x}dy = 0$. Now we solve the system for the function

$F:\begin{cases} F_x = 1 + \frac{y^2}{x^2} + \frac{1}{x^2} \\ F_y = -\frac{2y}{x} \end{cases}$. The integration of the second equation in y seems

to be a bit simpler, so we start with this operation: $F = \int -\frac{2y}{x}dy = -\frac{y^2}{x} + g(x)$. Deriving the last formula with respect to x, we have: $F_x = \frac{y^2}{x^2} + g_x(x)$. Comparing this expression of F_x with the first equation of the system, we conclude that $g_x(x) = 1 + \frac{1}{x^2}$, and consequently, $g(x) = x - \frac{1}{x}$ (we chose the zero constant). Therefore, $F = \frac{y^2}{x^2} + x - \frac{1}{x}$ and the general solution of the equation comes in the form $\frac{y^2}{x^2} + x - \frac{1}{x} = C$. Since multiplication by $\mu = \frac{1}{x^2}$ does not change the set of solutions, the original equation has the same solutions.

2. $(x - xy)dx + (y + x^2)dy = 0$.

First, we see that this is not exact equation because $A_y = -x \neq 2x = B_x$. Let us try to find the integrating factor that depends on x. If $\mu(x)$ is an integrating factor, then the relation $\mu A_y = \mu_x B + \mu B_x$ must be satisfied, that is, $\mu \cdot (-x) = \mu_x \cdot (y + x^2) + \mu \cdot 2x$. Joining all the terms with μ on the left side, we get $(\ln\mu)_x = \frac{-x-2x}{y+x^2} = \frac{-3x}{y+x^2}$. Since the right-hand side depends on both x and y, the assumption that μ can only depend on x is not valid. Then, let us try the second assumption that μ depends only on y. In this case, the relation $\mu_y A + \mu A_y = \mu B_x$ must be satisfied, that is, $\mu_y(x - xy) + \mu \cdot (-x) = \mu \cdot 2x$. We arrive at the following equation for μ: $(\ln\mu)_y = \frac{3x}{x-xy} = \frac{3}{1-y}$, whose right side depends only on y, and therefore, there exists an integrating factor that depends only on y. Solving the last relation, we find $\ln\mu = \int \frac{3}{1-y} = -3\ln(1-y) + C$. Choosing $C = 0$ (for simplicity), we get $\mu = \frac{1}{(1-y)^3}$.

Now we check that the found function μ is really an integrating factor:

$$\left(\mu(x - xy)\right)_y = \left(\frac{1}{(1 - y)^3}(x - xy)\right)_y = \left(\frac{x}{(1 - y)^2}\right)_y = \frac{2x}{(1 - y)^3}$$

$$= \left(\frac{1}{(1 - y)^3}(y + x^2)\right)_x = \left(\mu(y + x^2)\right)_x.$$

Then, we can solve the exact equation

$$\frac{1}{(1 - y)^3}(x - xy)dx + \frac{1}{(1 - y)^3}(y + x^2)dy = 0$$

using the studied method. To find the function F, we have to solve the
system: $\begin{cases} F_x = \frac{x}{(1-y)^2} \\ F_y = \frac{1}{(1-y)^3}(y + x^2) \end{cases}$. The integration of the first equation with
respect to x seems simpler, so let us start with this: $F = \int \frac{x}{(1-y)^2} dy = \frac{x^2}{2(1-y)^2} + g(y)$. Differentiating the last formula with respect to y, we have:
$F_y = \frac{x^2}{(1-y)^3} + g_y(y)$. Comparing this expression of F_y with the second
equation of the system, we conclude that $g_y(y) = \frac{y}{(1-y)^3}$, and consequently,

$$g(y) = \int \frac{y - 1 + 1}{(1 - y)^3} dy = \int \frac{-1}{(1 - y)^2} dy + \int \frac{1}{(1 - y)^3} dy = \frac{-1}{1 - y} + \frac{1}{2(1 - y)^2}$$

(we chose the zero constant). Therefore,

$$F = \frac{x^2}{2(1 - y)^2} - \frac{1}{1 - y} + \frac{1}{2(1 - y)^2} = \frac{x^2 + 1}{2(1 - y)^2} - \frac{1}{1 - y}$$

and the general solution of the equation (with μ and without μ) is $\frac{x^2+1}{2(1-y)^2} - \frac{1}{1-y} = C$.

3. $(2xy^2 - y)dx + (y^2 + x + y)dy = 0$.
First, we see that the equation is not exact, because

$$A_y = (2xy^2 - y)_y = 4xy - 1 \neq 1 = (y^2 + x + y)_x = B_x.$$

Let us try to find the integrating factor that depends on x. If $\mu(x)$ is an
integrating factor, then the relation $(\mu A)_y = \mu A_y = (\mu B)_x$ must be satisfied,
that is, $\mu(2xy^2 - y)_y = (\mu(y^2 + x + y))_x$. By differentiating we get $\mu(4xy - 1) = \mu_x(y^2 + x + y) + \mu$. Regrouping the terms, we have $\frac{\mu_x}{\mu} = \frac{4xy - 2}{y^2 + x + y}$. The
dependence of the expression on the right-hand side on y shows that there
is no integrating factor that depends only on x. Let us make an attempt to

find the integrating factor that depends on y. In this case, we have $(\mu A)_y = \mu B_x = (\mu B)_x$ and then, $\mu_y(2xy^2 - y) + \mu(4xy - 1) = \mu$. Again, bringing all the terms with μ to the left side, we get $\frac{\mu_y}{\mu} = -\frac{4xy-2}{2xy^2-y}$. The right side can be simplified to a form that does not contain x: $\frac{\mu_y}{\mu} = -\frac{2(xy-1)}{y(2xy-1)} = -\frac{2}{y}$, which means that there exists an integrating factor that depends only on y. To find it, we simply solve the equation of separable variables $\frac{\mu_y}{\mu} = -\frac{2}{y}$ whose integration gives $\ln|\mu| = -2\ln|y| + C$. Since we need only one integrating factor, we can choose $C = 0$. Eliminating the logarithms, we have $\mu = \frac{1}{y^2}$.

Now we verify that the found function μ is really an integrating factor:

$$(\mu(2xy^2 - y))_y = \left(\frac{1}{y^2}(2xy^2 - y)\right)_y = \left(2x - \frac{1}{y}\right)_y = \frac{1}{y^2}$$

$$= \left(\frac{1}{y^2}(y^2 + x + y)\right)_x = (\mu(y^2 + x + y))_x.$$

Thus, the corresponding partial derivatives are equal. Then, we can solve the exact equation

$$\frac{1}{y^2}(2xy^2 - y)dx + \frac{1}{y^2}(y^2 + x + y)dy = 0.$$

First, we write the equation in the form $(2x - \frac{1}{y})dx + (1 + \frac{x}{y^2} + \frac{1}{y})dy = 0$. Next, we solve the system for the function F: $\begin{cases} F_x = 2x - \frac{1}{y} \\ F_y = 1 + \frac{x}{y^2} + \frac{1}{y} \end{cases}$. Integrating the first equation with respect to x, we get $F = x^2 - \frac{x}{y} + g(y)$. Differentiating the last formula with respect to y, we obtain $F_y = \frac{x}{y^2} + g_y(y)$. Comparing this expression with the second equation of the system, we conclude that $g_y(y) = 1 + \frac{1}{y}$, and consequently, $g(y) = y + \ln|y|$ (we chose the zero constant). Therefore, $F = x^2 - \frac{x}{y} + y + \ln|y|$ and the general solution of the equation (both multiplied by μ and the original one) comes in the form $x^2 - \frac{x}{y} + y + \ln|y| = C$. Since multiplication by $\mu = \frac{1}{y^2}$ eliminates the function $y = 0$ from consideration, we have to check whether this function is a particular solution of the original equation. A simple calculation shows that, in fact, $y = 0$ is the particular solution not included in the general solution.

     4. $(1 - x^2y)dx + x^2(y - x)dy = 0$.

First, we see that the equation is not exact, because

$$A_y = (1 - x^2y)_y = -x^2 \neq 2xy - 3x^2 = (x^2y - x^3)_x = B_x.$$

Let us try to find an integrating factor that depends on x. If $\mu(x)$ is an integrating factor, then the relation $(\mu A)_y = \mu A_y = (\mu B)_x$ must be satisfied, that is, $\mu(1 - x^2y)_y = (\mu(x^2y - x^3))_x$. Differentiating, we get: $-\mu x^2 = \mu_x(x^2y - x^3) + \mu(2xy - 3x^2)$. Regrouping the terms, we obtain $\frac{\mu_x}{\mu} = \frac{2x^2 - 2xy}{x^2y - x^3} = -\frac{2}{x}$. The dependence of the expression on the right-hand side on x alone shows that the integrating factor can be found by solving the last equation. Integrating this separable equation, we obtain $\ln|\mu| = -2\ln|x|$ (we choose the integration constant $C = 0$), or eliminating logarithms, $\mu = \frac{1}{x^2}$.

We can now check that the found function μ is really an integrating factor: on the one hand

$$(\mu(1 - x^2y))_y = \left(\frac{1}{x^2}(1 - x^2y)\right)_y = \left(\frac{1}{x^2} - y\right)_y = -1$$

and on the other hand

$$(\mu(x^2y - x^3))_x = \left(\frac{1}{x^2}(x^2y - x^3)\right)_x = (y - x)_x = -1.$$

Thus, the corresponding partial derivatives are equal. Then, we can solve the exact equation $\frac{1}{x^2}(1 - x^2y)dx + \frac{1}{x^2}(x^2y - x^3)dy = 0$. We find the function F that satisfies the system: $\begin{cases} F_x = \frac{1}{x^2} - y \\ F_y = y - x \end{cases}$. Integrating the first equation with respect to x, we get $F = -\frac{1}{x} - yx + g(y)$. Differentiating the last formula with respect to y, we obtain $F_y = -x + g_y(y)$. Comparing this expression with the second equation of the system, we conclude that $g_y(y) = y$, and consequently, $g(y) = \frac{y^2}{2}$ (we chose the zero constant). Therefore, $F = -\frac{1}{x} - yx + \frac{y^2}{2}$ and the general solution of the equation (both multiplied by μ and the original) comes in the form $-\frac{1}{x} - yx + \frac{y^2}{2} = C$.

5. $(3x^2\cos y - \sin y)\cos y\, dx = x\, dy$.

First, we verify that the equation is not exact:

$$A_y = ((3x^2\cos y - \sin y)\cos y)_y = (-3x^2\sin y - \cos y)\cos y - (3x^2\cos y - \sin y)\sin y,$$

while $B_x = (-x)_x = -1$. Let us try to find an integrating factor that depends on x. If $\mu(x)$ is an integrating factor, then the relation $(\mu A)_y = \mu A_y = (\mu B)_x$ must be satisfied, that is,

$$\mu((-3x^2\sin y - \cos y)\cos y - (3x^2\cos y - \sin y)\sin y) = \mu_x \cdot (-x) + \mu \cdot (-1).$$

Regrouping the terms, we get

$$\frac{\mu_x}{\mu} = \frac{(3x^2\sin y + \cos y)\cos y + (3x^2\cos y - \sin y)\sin y - 1}{x}.$$

Since the variable y cannot be eliminated from the right-hand side, there is no integrating factor that depends only on x. Next, let us make the second attempt: to find the integrating factor that depends on y. In this case, we have $(\mu A)_y = \mu B_x = (\mu B)_x$, that is,

$$\mu_y(3x^2 \cos y - \sin y) \cos y$$

$$+\mu \left[(-3x^2 \sin y - \cos y) \cos y - (3x^2 \cos y - \sin y) \sin y\right] = \mu \cdot (-1).$$

Regrouping the terms and simplifying, we obtain

$$\frac{\mu_y}{\mu} = \frac{(3x^2 \sin y + \cos y) \cos y - 1}{(3x^2 \cos y - \sin y) \cos y} + \frac{(3x^2 \cos y - \sin y) \sin y}{(3x^2 \cos y - \sin y) \cos y}$$

$$= \frac{3x^2 \sin y \cos y - \sin^2 y}{(3x^2 \cos y - \sin y) \cos y} + \tan y = \frac{(3x^2 \cos y - \sin y) \sin y}{(3x^2 \cos y - \sin y) \cos y} + \tan y = 2 \tan y.$$

The fact that the right-hand side does not contain x means that the last equation can be solved and there exists an integrating factor that depends only on y. Solving the separable equation for μ, we find $\ln |\mu| = -2 \ln |\cos y|$, whence $\mu = \frac{1}{\cos^2 y}$.

Multiplying the original equation by $\mu = \frac{1}{\cos^2 y}$, we have $(3x^2 - \tan y)dx = \frac{x}{\cos^2 y}dy$. By calculating the partial derivatives

$$(3x^2 - \tan y)_y = -\frac{1}{\cos^2 y} \quad \text{and} \quad \left(-\frac{x}{\cos^2 y}\right)_x = -\frac{1}{\cos^2 y},$$

we can see that the new equation is exact. Then, we proceed to its solution by finding the function $F(x, y)$ that satisfies the system $\begin{cases} F_x = 3x^2 - \tan y \\ F_y = -\frac{x}{\cos^2 y} \end{cases}$.

Integrating the second equation with respect to y, we obtain $F = -x \tan y + g(x)$. Differentiating the last formula with respect to x, we get $F_x = -\tan y + g_x(x)$. Comparing this result with the first equation of the system, we conclude that $g_x(x) = 3x^2$, and consequently, $g(x) = x^3$. So, $F = -x \tan y + x^3$ and the general solution of the equation (both multiplied by μ and the original one) has the form $-x \tan y + x^3 = C$. In the transformation of the original equation into exact one, we multiply the former by $\frac{1}{\cos^2 y}$, which shows that the functions $y = \frac{\pi}{2} + k\pi, k \in \mathbb{Z}$, which vanish $\cos y$, were disregarded and are not included in the general solution. It is easy to check that each of these functions is a particular solution of the original equation. Thus, the set of all solutions of the original equation contains the general solution $-x \tan y + x^3 = C$ and also all the functions $y = \frac{\pi}{2} + k\pi, k \in \mathbb{Z}$.

6. $\left(2y + \frac{1}{(x+y)^2}\right)dx + \left(3y + x + \frac{1}{(x+y)^2}\right)dy = 0.$

We start by checking that the equation is not exact:

$$A_y = 2 - \frac{2}{(x+y)^3} \neq 1 - \frac{2}{(x+y)^3} = B_x.$$

Next, we see that there is no integrating factor μ that depends only on x or on y. In fact, the assumption that μ depends only on x implies that $(\mu A)_y = \mu A_y = (\mu B)_x$, which leads to the relation

$$\mu\left(2 - \frac{2}{(x+y)^3}\right) = \mu_x\left(3y + x + \frac{1}{(x+y)^2}\right) + \mu\left(1 - \frac{2}{(x+y)^3}\right),$$

that is, $\frac{\mu_x}{\mu} = \frac{1}{3y+x+(x+y)^{-2}}$. Since the function on the right side depends on y, the equation for $\mu(x)$ is not solvable. Similarly, assuming that μ depends only on y, we have the condition $(\mu A)_y = \mu B_x$, which leads to the relation

$$\mu_y\left(2y + \frac{1}{(x+y)^2}\right) + \mu\left(2 - \frac{2}{(x+y)^3}\right) = \mu\left(1 - \frac{2}{(x+y)^3}\right),$$

that is, $\frac{\mu_y}{\mu} = \frac{-1}{2y+(x+y)^{-2}}$. The last equation for $\mu(y)$ is also unsolvable, because the function on the right side depends on x.

Let us now look for the integrating factor that depends on the sum of the variables: $\mu(z) = \mu(x+y)$. In this case, the condition $(\mu A)_y = (\mu B)_x$ takes the form

$$\mu_z\left(2y + \frac{1}{(x+y)^2}\right) + \mu\left(2 - \frac{2}{(x+y)^3}\right) = \mu_z\left(3y + x + \frac{1}{(x+y)^2}\right) + \mu\left(1 - \frac{2}{(x+y)^3}\right).$$

Gathering the terms with μ_z and with μ and simplifying, we get $\mu_z \cdot (x+y) = \mu$, that is, $\frac{\mu_z}{\mu} = \frac{1}{x+y} = \frac{1}{z}$. Since the right-hand side depends only on z, the last equation can be solved and we find $\ln|\mu| = \ln|z|$ or $\mu = z = x + y$.

Multiplying the original equation by $\mu = x + y$, we get

$$(x+y)\left(2y + \frac{1}{(x+y)^2}\right)dx + (x+y)\left(3y + x + \frac{1}{(x+y)^2}\right)dy = 0$$

or

$$\left(2xy + 2y^2 + \frac{1}{x+y}\right)dx + \left(x^2 + 4xy + 3y^2 + \frac{1}{x+y}\right)dy = 0.$$

We can verify that this is an exact equation:

$$\left(2xy + 2y^2 + \frac{1}{x+y}\right)_y = 2x + 4y - \frac{1}{(x+y)^2} = \left(x^2 + 4xy + 3y^2 + \frac{1}{x+y}\right)_x.$$

Then, we solve the system $\begin{cases} F_x = 2xy + 2y^2 + \frac{1}{x+y} \\ F_y = x^2 + 4xy + 3y^2 + \frac{1}{x+y} \end{cases}$ for the function
F. Integrating the first equation with respect to x, we get $F = x^2y + 2y^2x + \ln|x+y| + g(y)$. Differentiating the last formula with respect to y, we get: $F_y = x^2 + 4xy + \frac{1}{x+y} + g_y(y)$. Comparing this expression with the second equation of the system, we conclude that $g_y(y) = 3y^2$, and consequently, $g(y) = y^3$. Therefore, $F = x^2y + 2y^2x + \ln|x+y| + y^3$ and the general solution of the equation (both multiplied by μ and original) comes in the form $x^2y + 2y^2x + \ln|x+y| + y^3 = C$.

7. $\left(y - \frac{1}{x}\right)dx + \frac{1}{y}dy = 0$.
First we verify that the equation is not exact: $A_y = 1 \neq 0 = B_x$. Next, we check that there is no integrating factor μ that depends only on x or on y. In fact, the assumption that μ depends only on x implies that $\mu A_y = (\mu B)_x$, which leads to the relation $\mu = \mu_x \frac{1}{y}$, that is, $\frac{\mu_x}{\mu} = y$. Since the function on the right side depends on y, the equation for μ is not solvable. Similarly, assuming that μ depends only on y, we arrive at the condition $(\mu A)_y = \mu B_x$, which leads to the relation $\mu_y(y - \frac{1}{x}) + \mu = 0$, that is, $\frac{\mu_y}{\mu} = \frac{-1}{y - \frac{1}{x}}$. The last equation for μ is also unsolvable, because the function on the right side depends on x.

Let us now look for the integrating factor that depends on the quotient of the variables: $\mu(z) = \mu(\frac{x}{y})$. In this case, the condition $(\mu A)_y = (\mu B)_x$ takes the form $\mu_z \cdot \frac{-x}{y^2} \cdot (y - \frac{1}{x}) + \mu = \mu_z \cdot \frac{1}{y} \cdot \frac{1}{y}$. Regrouping the terms and simplifying, we get $\mu_z \cdot \frac{-x}{y} = -\mu$, that is, $\frac{\mu_z}{\mu} = \frac{y}{x} = \frac{1}{z}$. Since the right-hand side depends only on z, the last equation can be solved and we find $\ln|\mu| = \ln|z|$ or $\mu = z = \frac{x}{y}$.

Multiplying the original equation by $\mu = \frac{x}{y}$, we have $(x - \frac{1}{y})dx + \frac{x}{y^2}dy = 0$. Comparing partial derivatives $\left(x - \frac{1}{y}\right)_y = \frac{1}{y^2} = \left(\frac{x}{y^2}\right)_x$, we conclude that the new equation is exact. Then, we proceed to finding the function F that satisfies the system: $\begin{cases} F_x = x - \frac{1}{y} \\ F_y = \frac{x}{y^2} \end{cases}$. Integrating the second equation with respect to y, we get $F = -\frac{x}{y} + g(x)$. Differentiating this relation with respect to x, we obtain $F_x = -\frac{1}{y} + g_x(x)$. Comparing this expression with the first equation of the system, we conclude that $g_x(x) = x$, and consequently, $g(x) = \frac{x^2}{2}$. Therefore, $F = -\frac{x}{y} + \frac{x^2}{2}$ and the general solution of the equation (both multiplied by μ and the original one) comes in the form $-\frac{x}{y} + \frac{x^2}{2} = C$. Since the integrating factor contains y in the denominator, the function $y = 0$ was eliminated from consideration in the modified equation. However, it is not admissible in the original equation either, since y is in the denominator of one of the coefficients. Therefore, no solution was lost in the solution procedure.

Exercises for section 3.2

Find an integrating factor, transform the given equation into exact one, and solve:

1. $\left(\frac{x}{y} + 1\right) dx + \left(\frac{x}{y} - 1\right) dy = 0.$
2. $\left(x^2 + y\right) dx - x dy = 0.$
3. $(xy^2 + y) dx - x dy = 0.$
4. $(x \cos y - y \sin y) dy + (x \sin y + y \cos y) dx = 0.$
5. $(x^2 + y^2 + x) dx + y dx = 0.$
6. $y dy = (x dy + y dx)\sqrt{1 + y^2}.$
7. $y^2 dx - (xy + x^3) dy = 0.$
8. $y(x + y) dx + (xy + 1) dy = 0.$
9. $(3x + 2y + y^2) dx + (x + 4xy + 5y^2) dy = 0.$
10. $2xy \ln y dx + (x^2 + y^2 \sqrt{y^2 + 1}) dy = 0.$
11. $(x + y^2) dx - 2xy dy = 0.$
12. $\left(1 - \frac{x}{y}\right) dx + \left(2xy + \frac{x}{y} + \frac{x^2}{y^2}\right) dy = 0.$
13. $\left(x^2 - \sin^2 y\right) dx + x \sin 2y dy = 0.$
14. $y dx - (x + x^2 + y^2) dy = 0.$
15. $(-xy \sin x + 2y \cos x) dx + 2x \cos x dy = 0.$
16. $(x^2 + 2xy - y^2) dx + (y^2 + 2xy - x^2) dy = 0.$
17. $y(x + y + 1) dx + (x + 2y) dy = 0.$
18. $x dx + (x^2 y + 4y) dy = 0, \ y(1) = 0.$

3.3   Formation of a differential

Sometimes it is possible to form a differential on the left-hand side of the equation, whether it is exact or not. When making a differential, the known formulas of differentiation (equivalent to calculating derivatives) are used, such as:

$$d(xy) = y dx + x dy, \ d(y^2) = 2y dy, \ d\left(\frac{y}{x}\right) = \frac{x dy - y dx}{x^2}, \ d \ln y = \frac{dy}{y},$$

etc. This approach does not have a structured algorithm, depending on simplicity of the terms on the left-hand side and on ability to manipulate them. Therefore, we will only present a few examples of this type. As we will see, in cases where the attempt to form a differential is successful, this can usually save technical work.

Examples.

1. $(x^3 + xy^2)dx + (x^2y + y^3)dy = 0$.
We regroup the terms as follows: $x^3dx + xy(ydx + xdy) + y^3dy = 0$ and note that

$$x^3dx = d\left(\frac{x^4}{4}\right), xy(ydx + xdy) = xyd(xy) = d\left(\frac{(xy)^2}{2}\right) \text{ and } y^3dy = d\left(\frac{y^4}{4}\right).$$

So, the original equation can be written as follows:

$$d\left(\frac{x^4}{4}\right) + d\left(\frac{(xy)^2}{2}\right) + d\left(\frac{y^4}{4}\right) = d\left(\frac{x^4}{4} + \frac{(xy)^2}{2} + \frac{y^4}{4}\right) = 0.$$

Therefore, the general solution is $\frac{x^4}{4} + \frac{(xy)^2}{2} + \frac{y^4}{4} = C$ or $x^4 + 2(xy)^2 + y^4 = C$.
We note that the original equation is exact, but the standard procedure for solving it requires more technical work.

2. $ydx - (4x^2y + x)dy = 0$.
We regroup the terms as follows: $ydx - xdy - 4x^2ydy = 0$ and note that $xdy - ydx$ represents the numerator of the differential of $\frac{y}{x}$. Then, we divide by $-x^2$ and obtain

$$\frac{xdy - ydx}{x^2} + 4ydy = d\left(\frac{y}{x}\right) + d\left(2y^2\right) = d\left(\frac{y}{x} + 2y^2\right) = 0.$$

Therefore, the general solution is $\frac{y}{x} + 2y^2 = C$.
We note that the original equation is not exact and that the employed procedure shows that one of its integrating factors is $\frac{1}{x^2}$. Obviously, the technical work required for finding this integrating factor and subsequent solution of the obtained exact equation using the standard procedure is much more involved.

3. $(2xy^2 - y)dx + (y^2 + x + y)dy = 0$.
First, we regroup the terms as follows:

$$2xy^2dx - (ydx - xdy) + (y^2 + y)dy = 0.$$

Note that $2xdx = d(x^2)$ and that the middle term represents the numerator of the differential of $\frac{x}{y}$. Then, we divide by y^2 and obtain

$$2xdx - \frac{ydx - xdy}{y^2} + \left(1 + \frac{1}{y}\right)dy$$

$$= d(x^2) - d\left(\frac{x}{y}\right) + d\left(y + \ln|y|\right) = d\left(x^2 - \frac{x}{y} + y + \ln|y|\right) = 0.$$

Therefore, the general solution is $x^2 - \frac{x}{y} + y + \ln|y| = C$. To achieve this result, we had to divide by y and the function $y = 0$ was not recovered in the form of the general solution. Therefore, we have to check whether $y = 0$ is the solution to the equation. Substitution of this function into the equation shows that it is a particular solution not included in the general one.

We note that the original equation is not exact and that the employed procedure reveals that one of its integrating factors is $\frac{1}{y^2}$.

4. $(y + xy^3)dx + (2x + x^2y^2)dy = 0$.
We divide by xy^2 and regroup the terms: $ydx + xdy + \frac{1}{xy}dx + \frac{2}{y^2}dy = 0$. In this way, we can form a differential of the first two terms:

$$d(xy) + \frac{1}{xy}dx + \frac{2}{y^2}dy = 0.$$

Now let us try to find a factor depending on xy (that allows us not to lose the possibility of forming the differential of the first term), which transforms the remaining terms into a differential. If we divide by xy, then the last two terms form a differential:

$$\frac{d(xy)}{xy} + \frac{1}{x^2y^2}dx + \frac{2}{xy^3}dy = d(\ln|xy|) - d\left(\frac{1}{xy^2}\right) = d\left(\ln|xy| - \frac{1}{xy^2}\right) = 0.$$

Therefore, the general solution is $\ln|xy| - \frac{1}{xy^2} = C$. To obtain this solution, we had to divide by y and the function $y = 0$ was not recovered in the form of the general solution. Substituting $y = 0$ into the original equation, we see that this is a particular solution not included in the general solution.

We note that the original equation is not exact and that the process of solution shows that one of its integrating factors is $\frac{1}{x^2y^3}$.

5. $\frac{xdy - ydx}{x^2 + y^2} = 0$.
We can see that this is an exact equation: $\left(\frac{-y}{x^2+y^2}\right)_y = \frac{y^2 - x^2}{(x^2+y^2)^2} = \left(\frac{x}{x^2+y^2}\right)_x$.
However, it is much simpler not to carry out this verification and subsequent standard procedure of solution, but to note that the equation can be simplified to the form $xdy - ydx = 0$. The latter is not exact, but it is separable. Still better, it enables a simple formation of a complete differential through the transformation $\frac{xdy - ydx}{x^2} = d\left(\frac{y}{x}\right) = 0$. Therefore, the general solution comes in the form $\frac{y}{x} = C$ or $y = Cx$. Since we only divided by x, no solution was disregarded in this procedure.

We suggest to the reader to solve $xdy - ydx = 0$ as a separable equation and compare the obtained solutions.

6. $xdx + ydy + \frac{ydx - xdy}{x^2 + y^2} = 0.$

It is easy to verify that this equation is exact:

$$\left(x + \frac{y}{x^2 + y^2}\right)_y = \frac{x^2 - y^2}{(x^2 + y^2)^2} = \left(y - \frac{x}{x^2 + y^2}\right)_x.$$

However, it is much simpler to solve this equation by forming exact differentials on the left-hand side. First, we note that $xdx = d\left(\frac{x^2}{2}\right)$ and $ydy = d\left(\frac{y^2}{2}\right)$. The third term can be used for the quotient differential:

$$\frac{ydx - xdy}{x^2 + y^2} = \frac{ydx - xdy}{y^2} \cdot \frac{1}{(x/y)^2 + 1} = d\left(\frac{x}{y}\right) \cdot \frac{1}{(x/y)^2 + 1} = d\left(\arctan\frac{x}{y}\right).$$

(In the last equality we have used the formula $\frac{dt}{t^2 + 1} = d(\arctan t)$.) Thus, the whole equation is transformed into the form

$$xdx + ydy + \frac{ydx - xdy}{x^2 + y^2} = d\left(\frac{x^2}{2}\right) + d\left(\frac{y^2}{2}\right) + d\left(\arctan\frac{x}{y}\right)$$

$$= d\left(\frac{x^2}{2} + \frac{y^2}{2} + \arctan\frac{x}{y}\right) = 0.$$

Consequently, the general solution is $\frac{x^2 + y^2}{2} + \arctan\frac{x}{y} = C.$

A task for the reader: apply the standard algorithm of exact equations and compare the obtained solutions.

7. $\frac{xdx + ydy}{\sqrt{x^2 + y^2}} + \frac{xdy - ydx}{x^2} = 0.$

It is easy to verify that this equation is exact:

$$\left(\frac{x}{\sqrt{x^2 + y^2}} - \frac{y}{x^2}\right)_y = -\frac{xy}{\sqrt{(x^2 + y^2)^3}} - \frac{1}{x^2} = \left(\frac{y}{\sqrt{x^2 + y^2}} + \frac{1}{x}\right)_x.$$

However, it is much simpler to solve this equation by forming exact differentials. First, we note that $xdx = d\left(\frac{x^2}{2}\right)$, $ydy = d\left(\frac{y^2}{2}\right)$, and therefore, the numerator of the first fraction can be written as $\frac{1}{2}d\left(x^2 + y^2\right)$. Furthermore, $\frac{dt}{\sqrt{t}} = 2d\left(\sqrt{t}\right)$. So, the whole first fraction takes the form

$$\frac{xdx + ydy}{\sqrt{x^2 + y^2}} = \frac{1}{2}\frac{d\left(x^2 + y^2\right)}{\sqrt{x^2 + y^2}} = d\left(\sqrt{x^2 + y^2}\right).$$

The second fraction is simply the differential of the quotient $\frac{y}{x}$: $\frac{xdy-ydx}{x^2} =$ $d\left(\frac{y}{x}\right)$. Thus, the original equation can be written as $d\left(\sqrt{x^2+y^2}+\frac{y}{x}\right)=0$, from which it follows that the general solution is $\sqrt{x^2+y^2}+\frac{y}{x}=C$.

A task for the reader: apply the standard algorithm of exact equations and compare the obtained solutions.

8. $(x^2+y^2+1)dx+2xydy=0$.
This equation is exact: $\left(x^2+y^2+1\right)_y=2y=(2xy)_x$. But it can also be solved by forming exact differentials. First, we regroup the terms:

$$(x^2+y^2+1)dx+2xydy=(x^2+1)dx+(y^2dx+2xydy).$$

Now we see that

$$(x^2+1)dx=d\left(\frac{x^3}{3}+x\right) \text{ and } y^2dx+2xydy=y^2dx+xd\left(y^2\right)=d\left(xy^2\right).$$

Therefore, we can write the equation in the form $d\left(\frac{x^3}{3}+x+xy^2\right)=0$, from which it follows that $\frac{x^3}{3}+x+xy^2=C$ is the general solution.

A task for the reader: apply the standard algorithm of exact equations and compare the obtained solutions.

9. $(x^2+y^2+1)dx-2xydy=0$.
This equation is not exact: $\left((x^2+y^2+1)\right)_y=2y\neq-2y=(-2xy)_x$. Nevertheless, it can be solved by forming exact differentials of its terms. First, we note that $(x^2+1)dx$ forms a differential together with any multiplier dependent on x. The remaining two terms, which pose the greatest problem, we will represent as follows:

$$y^2dx-2xydy=x^2\left[\frac{y^2}{x^2}dx-2\frac{y}{x}dy\right]=x^2\left[-y^2d\left(\frac{1}{x}\right)-\frac{1}{x}d(y^2)\right]=-x^2d\left(\frac{y^2}{x}\right).$$

So, the original equation can be written as $(x^2+1)dx-x^2d\left(\frac{y^2}{x}\right)=0$. Dividing by x^2 and forming another differential of the first two terms, we get

$$\left(1+\frac{1}{x^2}\right)dx-d\left(\frac{y^2}{x}\right)=d\left(x-\frac{1}{x}-\frac{y^2}{x}\right)=0.$$

Therefore, the general solution is $x-\frac{1}{x}-\frac{y^2}{x}=C$. We note that the original equation admits the integrating factor μ which depends only on x.

Task for the reader: find $\mu(x)$, solve the equation with this factor and compare the obtained solutions.

Exercises for section 3.3

Solve the following equations by forming exact differentials:
1. $(y - 4xy^3)dx = (2x^2y^2 + x)dy$.
2. $(x + y^2)dx - 2xydy = 0$.
3. $(x^2 + y)dx - xdy = 0$.
4. $(x + y^2)dx - 2xydy = 0$.
5. $(2x^2y + 2y + 5)dx + (2x^3 + 2x)dy = 0$.
6. $(2xy^2 - 3y^3)dx + (7 - 3xy^2)dy = 0$.
7. $2xydx = (x^2 - 2y^3)dy$.
8. $(y - 3x^2y^3)dx - (x + x^3y^2)dy = 0$.
9. $(2xy^2 + y)dx - (x^2y + 2x)dy = 0$.
10. $(2xy^3 + y)dx - 2xdy = 0$.
11. $x^3dy + 2(y - x^2)ydx = 0$.
12. $xdy = y(1 - ye^x)dx$.

4   Linear equations and reducible to them

4.1   Linear equations

Definition. The equation $y' + a(x)y = b(x)$, where $a(x)$ and $b(x)$ are functions only of the independent variable x, is called *linear*. This is the *normalized form* (or *canonical form*) of a linear equation with the leading coefficient (coefficient with the derivative) equal to 1, which is used to simplify considerations, without any loss of generality in the study of the equation. The *non-normalized form of linear equation* is $c(x)y' + a(x)y = b(x)$, $c(x) \neq 0$. If $b(x) \equiv 0$, the equation is called *homogeneous*, otherwise – *nonhomogeneous*.

   Remark 1. If $a(x) \equiv 0$ or $b(x) \equiv 0$, then the linear equation is at the same time separable equation, the simplest type that was solved before. Therefore, in what follows, we will assume that $a(x)b(x) \neq 0$.
   Remark 2. The terminology here is a bit ambiguous: a homogeneous linear equation is generally not homogeneous in the sense of the equation $y' = f\left(\frac{y}{x}\right)$. For example, $y' = e^x y$ is homogeneous linear equation, but it is not homogeneous like $y' = \frac{y}{x}$. On the other hand, the equation $y' = \cos\frac{y}{x}$ is homogeneous, but not linear homogeneous. Of course, there are also equations belonging to both types at the same time, such as $y' = \frac{y}{x}$. In the sections dedicated to linear equations, we will use the terms "homogeneous" and "nonhomogeneous" to refer to different cases of linear equations.
   Remark 3. It should be noted that sometimes an interchange of un-

known function and independent variable can modify the type of equation and simplify its resolution. This is not the case with equations of the three previous types and reducible to them: the equations $y' = f(x)g(y)$, $y' = f(ax+by+c)$, $y' = f\left(\frac{y}{x}\right)$, $y' = f\left(\frac{ax+by+c}{\alpha x+\beta y+\gamma}\right)$ are of the same type with respect to unknown function y (depending on variable x) and with respect to unknown function x (depending on variable y). Just rewrite these equations for the inverse function $x(y)$ to check this:

$$x' = \frac{1}{f(x)}\frac{1}{g(y)} = F(x)G(y), \quad x' = \frac{1}{f(ax+by+c)} = F(ax+by+c),$$

$$x' = \frac{1}{f(\frac{1}{x/y})} = F\left(\frac{x}{y}\right), \quad x' = \frac{1}{f\left(\frac{ax+by+c}{\alpha x+\beta y+\gamma}\right)} = F\left(\frac{ax+by+c}{\alpha x+\beta y+\gamma}\right)$$

(recall that, by the theorem on inverse function, $x' = \frac{1}{y'}$). The exact equation simply does not distinguish between unknown function and independent variable, neither for classification nor for solution. But the situation is different for linear equations. For example, the equation $y' = \frac{1}{xy+y^2}$ is not linear with respect to $y(x)$, but it becomes linear if we interchange the unknown function and independent variable: $x' = yx + y^2$ (note that the general form of the linear equation with respect to unknown function $x(y)$ is $x' + \alpha(y)x = \beta(y)$).

Methods of solution.

There are several methods of solution of linear equations. In what follows, we consider the three basic methods for linear equations of the first order: the first, called the method of variation of parameter or the Lagrange method, reduces the linear equation to the sequence of two separable equations; the second, called the method of integrating factor, transforms the linear equation into exact equation; and the third, a variation of the second method, transforms the left part of the equation (that is, the homogeneous part) into a differential.

Method 1 – method of variation of parameter (Lagrange method).

We start with the homogeneous equation $y' + a(x)y = 0$ which, at the same time, is separable one. Then we rewrite it in the form $\frac{dy}{y} = -a(x)dx$ and integrate each side with respect to its own variable to find the general solution: $\ln|y| = -\int a(x)dx + C$. Here we understand the integration symbol as one of the antiderivatives of the function $a(x)$. The last relation can be transformed into explicit form for y: $y = Ce^{-\int a(x)dx}$.

We now return to the complete (nonhomogeneous) linear equation and search for its general solution in the form: $y = C(x)e^{-\int a(x)dx}$, where $C(x)$ is a function to be determined. This means that in the structure of the solution of the homogeneous part, we replace the constant C with the function $C(x)$ (hence the name of the method – variation of parameter). To find $C(x)$, we substitute the proposed function into the equation and get

$$y' + a(x)y$$

$$= \left(C'(x)e^{-\int a(x)dx} + C(x)(-a(x))e^{-\int a(x)dx}\right) + a(x)C(x)e^{-\int a(x)dx} = b(x).$$

Simplifying, we have one more separable equation $C'(x)e^{-\int a(x)dx} = b(x)$, now for $C(x)$. Writing in the form $C'(x) = b(x)e^{\int a(x)dx}$ and integrating, we find $C(x) = \int b(x)e^{\int a(x)dx}dx + A$, where A is an arbitrary constant. Substituting this expression into the proposed formula of solution, we obtain the general solution of the original (nonhomogeneous) equation

$$y = \left(\int b(x)e^{\int a(x)dx}dx + A\right)e^{-\int a(x)dx}.$$

Remark. According to the method of solution, the obtained general solution contains all the solutions of the linear equation.

Method 2 – method of integrating factor.
We write the (complete) linear equation in the form $(a(x)y - b(x))dx + dy = 0$ and note that $A_y = (a(x)y - b(x))_y = a(x) \neq 0 = 1_x = B_x$, that is, the equation is not exact, unless $a(x) \equiv 0$, which is a trivial, uninteresting case. Assuming that $a(x) \neq 0$, let us try to find an integrating factor μ that depends only on x. In this case, we have the separable equation for $\mu(x)$:

$$(\mu A)_y = (\mu(a(x)y - b(x)))_y = \mu a(x) = \mu_x = (\mu \cdot 1)_x = (\mu B)_x.$$

Solving this equation, we get $\mu = e^{\int a(x)dx}$. Therefore, the equation

$$e^{\int a(x)dx}(a(x)y - b(x))dx + e^{\int a(x)dx}dy = 0$$

is exact. Next, we have to solve the system $\begin{cases} F_x = e^{\int a(x)dx}(a(x)y - b(x)) \\ F_y = e^{\int a(x)dx} \end{cases}$
for the function F. We start with the second equation, integrating it with respect to y: $F = e^{\int a(x)dx}y + g(x)$. Differentiating now with respect to x, we obtain $F_x = a(x)e^{\int a(x)dx}y + g_x(x)$. Comparing this expression with the first equation of the system, we conclude that $g_x(x) = -b(x)e^{\int a(x)dx}$, and consequently, $g(x) = -\int b(x)e^{\int a(x)dx}dx$. Substituting g into the formula for F, we get $F = e^{\int a(x)dx}y - \int b(x)e^{\int a(x)dx}dx$ and the general solution of

the equation is $e^{\int a(x)dx}y - \int b(x)e^{\int a(x)dx}dx = C$. Solving the last formula for y, we obtain the same general solution as in the first method:

$$y = \left(\int b(x)e^{\int a(x)dx}dx + C \right) e^{-\int a(x)dx}.$$

Method 3 – alternative method of integrating factor.

Let us find a function $\mu(x)$ that transforms the left-hand side of the original equation into the differential of a function, namely: $\mu(y' + ay) = (\mu y)'$. Writing this relation in the form $\mu y' + \mu ay = \mu y' + \mu' y$ and simplifying, we get $\mu a = \mu'$ which is a separable equation for unknown μ. Note that we get the equation of the integrating factor of Method 2, which has the solution $\mu = e^{\int adx}$. Although the found function $\mu(x)$ is an integrating factor for the whole equation, we do not follow the method of exact equation. Instead, we represent the left side of the original equation in the desired form $\mu y' + \mu ay = (\mu y)' = \mu b$ and simply integrate both sides with respect to x, arriving at the general solution of the original equation: $\mu y = \int \mu b dx$. Substituting the expression of μ into this formula, we obtain $e^{\int adx}y = \int e^{\int adx}b dx$. Finally, solving the last relation for y, we get the same general solution as in the first and second methods: $y = \left(\int be^{\int adx}dx + C \right) e^{-\int adx}$ (the constant C can always be separated from or added to the indefinite integral).

Remark. Any of the three methods can be used to solve linear equations. In this text, we will use the first method in most cases.

Examples.

1. $y' + 2y = x + 2$.
This equation is linear with $a(x) = 2$ and $b(x) = x + 2$. Let us solve it using the Lagrange method (the method of variation of parameter). First we solve the corresponding homogeneous equation $y' + 2y = 0$. This is a separable equation and its general solution is $y = Ce^{-2x}$. Returning to the original equation, we look for a general solution in the form $y = C(x)e^{-2x}$, where $C(x)$ should be found from the condition that the proposed expression satisfies the given equation. Substituting this function into the original equation, we have

$$y' + 2y = (C'(x)e^{-2x} - 2C(x)e^{-2x}) + 2C(x)e^{-2x} = x + 2.$$

Simplifying and isolating C', we obtain the equation $C'(x) = (x + 2)e^{2x}$ (which is even simpler than separable one). Integrating by parts, we find:

$$C(x) = \int (x+2)e^{2x}dx = (x+2)\frac{1}{2}e^{2x} - \int \frac{1}{2}e^{2x}dx = (x+2)\frac{1}{2}e^{2x} - \frac{1}{4}e^{2x} + A$$

or $C(x) = \frac{2x+3}{4}e^{2x} + A$. Substituting this expression into the formula for y, we find the general solution of the linear equation:

$$y = \left(\frac{2x+3}{4}e^{2x} + A\right)e^{-2x} = \frac{2x+3}{4} + Ae^{-2x}.$$

2. $y' = \frac{y}{2y\ln y + y - x}$.

This equation is not linear with respect to $y(x)$, but it is linear for $x(y)$:

$$x' = \frac{2y\ln y + y - x}{y} = -\frac{x}{y} + 2\ln y + 1.$$

Let us solve it using the Lagrange method. We start with the corresponding homogeneous equation $x' = -\frac{x}{y}$. This is a separable equation and its general solution is $x = \frac{C}{y}$. We now use the function $x = \frac{C(y)}{y}$ as a proposed general solution of the original equation for unknown $x(y)$. To specify $C(y)$, we substitute this function into the original equation:

$$x' + \frac{x}{y} = \frac{C'(y)}{y} - \frac{C(y)}{y^2} + \frac{C(y)}{y}\frac{1}{y} = 2\ln y + 1.$$

Simplifying and solving for C', we get $C'(y) = 2y\ln y + y$. Integrating by parts, we have

$$C(y) = \int 2y\ln y + y\,dy = y^2\ln y - \int y^2\frac{1}{y}dy + \frac{y^2}{2} = y^2\ln y + A.$$

Substituting this expression into the proposed form of the solution, we obtain

$$x = \frac{y^2\ln y + A}{y} = y\ln y + \frac{A}{y}.$$

This is the general solution in implicit form for the original equation.

3. $y' + y = 2e^x$.

This equation is linear with $a(x) = 1$ and $b(x) = 2e^x$. Let us solve it using the Lagrange method. We start with the corresponding homogeneous equation $y' + y = 0$. It has separable variables and its general solution is $y = Ce^{-x}$. Returning to the original equation, we look for the general solution in the form $y = C(x)e^{-x}$. Substituting this function into the original equation, we get

$$y' + y = (C'(x)e^{-x} - C(x)e^{-x}) + C(x)e^{-x} = 2e^x.$$

Simplifying and isolating C', we obtain the equation $C'(x) = 2e^{2x}$ whose integration gives $C(x) = e^{2x} + A$. Substituting this expression into the formula for y, we find the general solution of the linear equation:

$$y = \left(e^{2x} + C_1\right) e^{-x} = e^x + Ae^{-x}.$$

4. $xy' - y = xe^x$.
This equation is found in Table 1 of section 2 in Chapter 1 under number 5. This is a linear equation in non-normalized form with $c(x) = x$, $a(x) = -1$ and $b(x) = xe^x$ (or $\tilde{a}(x) = -\frac{1}{x}$ and $\tilde{b}(x) = e^x$ in the normalized form). Let us solve it using the Lagrange method (without a preliminary normalization of the equation). First we consider the corresponding homogeneous equation $xy' - y = 0$. This is a separable equation and its general solution is $y = Cx$. Returning to the original equation, we look for the general solution in the form $y = C(x)x$. Substituting this function into the original equation, we get $xy' - y = x(C'(x)x + C(x)) - C(x)x = xe^x$. Simplifying and solving for C', we obtain the equation $C'(x) = \frac{e^x}{x}$. The integral $\int \frac{e^x}{x} dx$ exists for any $x \neq 0$, but it cannot be expressed in elementary functions. Therefore, the general solution of the original equation can be written in the form $y = x \left(\int \frac{e^x}{x} dx + A\right)$. According to the algorithm of solution, this general solution contains all the particular ones.

5. $y' + y \tan x = \frac{1}{\cos x}$.
This equation is linear with $a(x) = \tan x$ and $b(x) = \frac{1}{\cos x}$. We solve it using the Lagrange method. First, let us consider the corresponding homogeneous equation $y' + y \tan x = 0$, which is a separable one. Separating variables and integrating, we get $\int \frac{dy}{y} = -\int \tan x dx$ or $\int \frac{dy}{y} = \int \frac{d(\cos x)}{\cos x}$. The result of the integration is $\ln|y| = \ln|\cos x| + C$ or $y = C \cos x$. Returning to the original equation, we look for the general solution in the form $y = C(x) \cos x$. Substituting this function into the original equation, we have

$$y' + y \tan x = (C'(x) \cos x - C(x) \sin x) + C(x) \cos x \tan x = \frac{1}{\cos x}.$$

Simplifying and solving for C', we get the equation $C'(x) = \frac{1}{\cos^2 x}$ whose integration gives $C(x) = \tan x + A$. Substituting this expression into the formula for y, we find the general solution of the linear equation:

$$y = (\tan x + A) \cos x = \sin x + A \cos x.$$

6. $xy' - 2y = 2x^4$, $y(1) = 0$.
This is the Cauchy problem for the linear equation with the coefficients

$a(x) = -2\frac{y}{x}$ and $b(x) = 2x^3$ in the normalized form. We solve it using the Lagrange method without preliminary normalization. First, let us consider the corresponding homogeneous equation $xy' - 2y = 0$, which is of separable variables. Separating variables and integrating, we get $\int \frac{dy}{y} = 2 \int \frac{dx}{x}$, whence $\ln|y| = 2\ln|x| + C$ or $y = Cx^2$. Returning to the original equation, we look for the general solution in the form $y = C(x)x^2$. Substituting this function into the original equation, we obtain

$$xy' - 2y = x(C'(x)x^2 + C(x) \cdot 2x) - 2C(x)x^2 = 2x^4.$$

Simplifying and solving for C', we get the equation $C'(x) = 2x$ whose integration gives $C(x) = x^2 + A$. Substituting this expression into the formula for y, we find the general solution of the linear equation:

$$y = (x^2 + C_1)x^2 = x^4 + Ax^2.$$

Applying the initial condition, we specify the constant A: $0 = 1 + A$, whence $A = -1$. Hence, the solution to the Cauchy problem is $y = x^4 - x^2$.

7. $y' + 2y = e^{-x}$.
This equation is linear with $a(x) = 2$ and $b(x) = e^{-x}$. Let us solved it using the method of integrating factor. We rewrite the equation in the form $(2y - e^{-x})dx + dy = 0$ and find the integrating factor $\mu(x)$ from the equation $(\mu(2y - e^{-x}))_y = (\mu \cdot 1)_x$ or, simplifying, $2\mu = \mu_x$. Solving this separable equation, we get $\mu = e^{2x}$ (we only need one particular non-zero solution). Therefore, the equation $e^{2x}(2y - e^{-x})dx + e^{2x}dy = 0$ is an exact one. We now find the function F by solving the system $\begin{cases} F_x = 2e^{2x}y - e^x) \\ F_y = e^{2x} \end{cases}$. Integrating the second equation in y, we get: $F = e^{2x}y + g(x)$. Deriving the last equation in x, we have $F_x = 2e^{2x}y + g_x(x)$. Comparing this expression with the first equation of the system, we conclude that $g_x(x) = -e^x$, whence $g(x) = -e^x$. Substituting g into the formula of F, we obtain $F = e^{2x}y - e^x$ and the general solution of the given equation takes the form $e^{2x}y - e^x = C$.

8. $y' + \frac{y}{x} = xe^{x/2}$.
This equation is linear with $a(x) = \frac{1}{x}$ and $b(x) = xe^{x/2}$. Let us solve it using the alternative method of integrating factor. First, we find the function $\mu(x)$ whose product with the left-hand side of the equation generates the derivative of a function, that is, we multiply the original equation by $\mu(x)$ and find such $\mu(x)$ that transforms the left-hand side to the form $\mu y' + \mu \frac{y}{x} = (\mu y)'$. This leads to following differential equation for μ: $\mu \frac{1}{x} = \mu'$, whose solution is $\mu = x$. So, the equation $xy' + y = x^2e^{x/2}$ can be written in the form $(xy)' = x^2 e^{x/2}$ and, integrating both sides with respect to x, we obtain

the general solution of the original equation:

$$xy = \int x^2 e^{x/2} dx = (2x^2 - 8x + 16)e^{x/2} + C.$$

9. $(x^2 + 1)y' + 4xy = 3$, $y(0) = 0$.
This is the Cauchy problem for the linear equation with the coefficients $a(x) = \frac{4x}{x^2+1}$ and $b(x) = \frac{3}{x^2+1}$ in the normalized form. Let us solve it using the alternative method of integrating factor. First, we find the function $\mu(x)$ whose product with the left-hand side of the equation, written in the normalized form $y' + \frac{4x}{x^2+1}y = \frac{3}{x^2+1}$, generates the derivative of a function, that is, $\mu y' + \mu\frac{4x}{x^2+1}y = (\mu y)'$. It follows from the last formula that μ satisfies the differential equation $\mu\frac{4x}{x^2+1} = \mu'$ whose solution is $\ln|\mu| = 2\ln(x^2 + 1)$ or $\mu = (x^2 + 1)^2$. So, the equation $y' + \frac{4x}{x^2+1}y = \frac{3}{x^2+1}$ after multiplication by $\mu = (x^2 + 1)^2$ becomes $((x^2 + 1)^2 y)' = 3(x^2 + 1)$. Integrating both sides, we obtain the general solution of the original equation: $(x^2+1)^2 y = x^3 + 3x + C$.
Substituting the initial condition into the last formula, we specify $C = 0$, and therefore, the solution to the Cauchy problem is $y = \frac{x^3+3x}{(x^2+1)^2}$.

10. $x^2 y' + xy + 1 = 0$, $y(1) = 0$.
This is the Cauchy problem for the linear equation whose normalized form is $y' + \frac{1}{x}y = -\frac{1}{x^2}$ with $a(x) = \frac{1}{x}$ and $b(x) = -\frac{1}{x^2}$. Let us solve it using the method of integrating factor. We rewrite the equation in the form $(xy + 1)dx + x^2 dy = 0$ and find the integrating factor $\mu(x)$ by solving the separable equation $(\mu(xy + 1))_y = (\mu \cdot x^2)_x$ or $x\mu = x^2\mu_x + 2x\mu$ or still $x\mu_x = -\mu$. Separating the variables and integrating, we get $\int \frac{d(\mu}{\mu} = -\int \frac{dx}{x}$. Then we find the integrating factor in the form $\ln|\mu| = -\ln|x|$ or $\mu = \frac{1}{x}$ (we only need one factor). Multiplying the original equation by μ, we get the exact differential equation $\left(y + \frac{1}{x}\right)dx + xdy = 0$. We now find the complete differential of the left-hand side by solving the system $\begin{cases} F_x = y + \frac{1}{x} \\ F_y = x \end{cases}$.
Integrating the second equation in y, we get $F = xy + g(x)$. Deriving the last equation in x, we have $F_x = y + g_x(x)$. Comparing this expression with the first equation of the system, we conclude that $g_x(x) = \frac{1}{x}$, whence $g(x) = \ln|x|$. Substituting g into the formula of F, we obtain $F = xy + \ln|x|$ and the general solution of the equation takes the form $xy + \ln|x| = C$.
To find the solution of the Cauchy problem, we substitute the initial condition into the general solution and get $0 + \ln 1 = C$. So, the solution of the problem is $xy + \ln|x| = 0$.

11. $(2x + y)dy = ydx + 4\ln y dy$.
This equation is not separable or homogeneous or exact or linear for $y(x)$.

However, if we interchange the independent variable and unknown function, then for the function $x(y)$ we have the linear equation $x' - \frac{2}{y}x = \frac{y-4\ln y}{y}$. Let us solve this equation using the alternative method of integrating factor. First, we find the function $\mu(y)$, whose product with the left-hand side of the equation represents the derivative of a function, that is, $\mu x' - \mu\frac{2}{y}x = (\mu x)'$ (here the differentiation is made with respect to y). It follows from this relation that μ satisfies the differential equation $-\mu\frac{2}{y} = \mu'$ whose solution is $\ln|\mu| = -2\ln|y|$ or $\mu = \frac{1}{y^2}$. So, the equation $x' - \frac{2}{y}x = \frac{y-4\ln y}{y}$ after multiplication by $\mu = \frac{1}{y^2}$ becomes $\left(\frac{x}{y^2}\right)' = \frac{y-4\ln y}{y^3}$. Integrating, we obtain the general solution of the original equation:

$$\frac{x}{y^2} = \int \frac{y-4\ln y}{y^3}dy = -\frac{1}{y} - 4\left(\ln y \cdot \frac{-1}{2y^2} - \int \frac{1}{y} \cdot \frac{-1}{2y^2}dy\right) = -\frac{1}{y} + \frac{2\ln y}{y^2} + \frac{1}{y^2} + C.$$

12. $y' = \frac{y}{3x-y^2}$.

This equation is not separable or homogeneous or exact or linear for $y(x)$, but if we interchange the independent variable and unknown function, then for the function $x(y)$ we have the linear equation $x' = \frac{3}{y}x - y$. Let us solve it using the alternative method of integrating factor. First, we find the function $\mu(y)$ such that $\mu x' - \mu\frac{3}{y}x = (\mu x)'$ (here the differentiation is made with respect to y). Therefore, μ satisfies the differential equation: $-\mu\frac{3}{y} = \mu'$ whose solution is $\ln|\mu| = -3\ln|y|$ or $\mu = \frac{1}{y^3}$. So, the equation $x' - \frac{3}{y}x = -y$ after multiplication by $\mu = \frac{1}{y^3}$ becomes $\left(\frac{x}{y^3}\right)' = -\frac{1}{y^2}$. Integrating, we get the general solution of the original equation: $\frac{x}{y^3} = \frac{1}{y} + C$.

13. $(1 - 2xy)y' = y(y - 1)$, $y(0) = 1$.

This is the Cauchy problem for the equation which is not separable or homogeneous or exact or linear for $y(x)$. However, if we interchange the independent variable and unknown function, then for the function $x(y)$ we have the linear equation

$$x' = \frac{1-2xy}{y(y-1)} = -\frac{2}{y-1}x + \frac{1}{y(y-1)}.$$

Let us solve it using the alternative method of integrating factor. First, we find the function $\mu(y)$ such that $\mu x' + \mu\frac{2y}{y(y-1)}x = (\mu x)'$ (here the differentiation is made with respect to y). It follows from this relation that μ satisfies the differential equation $\mu\frac{2}{y-1} = \mu'$ whose solution is $\ln|\mu| = 2\ln|y-1|$ or $\mu = (y-1)^2$. So, the equation $x' = -\frac{2}{y-1}x + \frac{1}{y(y-1)}$ multiplied by

$\mu = (y-1)^2$ takes the form $\left((y-1)^2 x\right)' = \frac{y-1}{y}$. Integrating, we obtain the general solution of the original equation: $(y-1)^2 x = y - \ln|y| + C$.

Substituting the initial condition into this solution, we have $0 = 1-0+C$, whence $C = -1$. Thus, the solution to the Cauchy problem is $(y-1)^2 x = y - \ln|y| - 1$.

14. $(xy' - 1)\ln x = 2y$, $y(e) = 0$.

This is the Cauchy problem for the linear equation whose normalized form is $y' - \frac{2}{x\ln x}y = \frac{1}{x}$, where $a(x) = -\frac{2}{x\ln x}$ and $b(x) = \frac{1}{x}$. Let us use the alternative method of integrating factor. First, we find the function $\mu(x)$ such that $\mu y' - \mu \frac{2}{x\ln x}y = (\mu y)'$. This leads to the following equation for μ: $-\mu \frac{2}{x\ln x} = \mu'$ whose solution is $\ln|\mu| = -2\ln|\ln x|$ or $\mu = \frac{1}{\ln^2 x}$. Therefore, multiplication of the equation $y' - \frac{2}{x\ln x}y = \frac{1}{x}$ by $\mu = \frac{1}{\ln^2 x}$ transforms it into the form $\left(\frac{1}{\ln^2 x}y\right)' = \frac{1}{x\ln^2 x}$. Integrating, we obtain the general solution of the original equation: $\frac{1}{\ln^2 x}y = -\frac{1}{\ln x} + C$ or $y = -\ln x + C\ln^2 x$.

Applying the initial condition, we get $0 = -\ln e + C\ln^2 e$, whence $C = 1$. Thus, the solution to the Cauchy problem is $y = -\ln x + \ln^2 x$.

15. $(\sin^2 y + x\cot y)y' = 1$, $y(0) = \frac{\pi}{2}$.

This is the Cauchy problem for the equation which is not separable or homogeneous or exact or linear for $y(x)$. However, interchanging the independent variable and unknown function, we obtain the linear equation $x' = x\cot y + \sin^2 y$ for the function $x(y)$. Let us solve it using the alternative method of integrating factor. First, we find the function $\mu(y)$ such that $\mu x' - \mu x\cot y = (\mu x)'$ (here the differentiation is performed with respect to y). It follows from this relation that μ satisfies the differential equation $-\mu\cot y = \mu'$ whose solution is $\ln|\mu| = -\ln|\sin y|$ or $\mu = \frac{1}{\sin y}$. So the equation $x' = x\cot y + \sin^2 y$ multiplied by $\mu = \frac{1}{\sin y}$ takes the form $\left(\frac{1}{\sin y}x\right)' = \sin y$. Integrating, we obtain the general solution of the original equation: $\frac{1}{\sin y}x = -\cos y + C$.

Applying the initial condition, we have $\frac{1}{\sin(\pi/2)} \cdot 0 = -\cos\frac{\pi}{2} + C$, whence $C = 0$. Thus, the solution of the Cauchy problem is $\frac{1}{\sin y}x = -\cos y$ or $2x = -\sin 2y$.

Exercises for section 4.1

Solve the following linear equations (with respect to y or x):

1. $x^2 y' = 2xy + 3$.
2. $y\,dx = (3x - y^2)\,dy$.
3. $xy' + (x+1)y = 3x^2 e^{-x}$.

4. $xy' - 2y + x^2 = 0$.

5. $y' - y = e^x$.

6. $y' + \frac{1}{x}y = 3x$.

7. $xdy + (x^2 - y)dx = 0$.

8. $2ydx + (y^2 - 2x)dy = 0$.

9. $y' - y\sin x = \sin x \cos x$.

10. $(1 + x^2)y' - 2xy = (1 + x^2)^2$.

11. $(x - 2xy - y^2)y' + y^2 = 0$.

12. $dx = (2x + e^y)dy$.

13. $y' + y\tan x = e^x \cos x$.

14. $x(y - \sqrt{1 + x^2})dx + (1 + x^2)dy = 0$.

15. $(1 + y^2)dx + (xy - y^3)dy = 0$.

16. $(\sin x - 1)y' + y\cos x = \sin x$.

4.2 Bernoulli equation

Definition. The equation $y' + a(x)y = b(x)y^n$, where n is a real number, is called Bernoulli equation.

Remark 1. In the cases $n = 0$ and $n = 1$ we have the simplest type of equation, the linear one. In the case $n = 1$ we even have a homogeneous linear equation. Therefore, Bernoulli equations are usually considered to be those with $n \neq 0$ and $n \neq 1$.

Remark 2. Just like the linear equation, the Bernoulli equation can change its type when the meaning of x and y is interchanged.

Methods of solution.

Method 1 - reduction to linear equation.

The Bernoulli equation is reduced to linear by the following change of the unknown function: $z = \frac{1}{y^{n-1}}$. For a simpler implementation of this procedure, the original equation is divided by y^n and written in the form

$$\frac{1}{y^n}y' + a(x)\frac{1}{y^{n-1}} = b(x).$$

Then, the middle term indicates the change of function that must be made. Since $z' = -\frac{n-1}{y^n}y'$, the equation for z comes in the form

$$-\frac{1}{n-1}z' + a(x)z = b(x) \text{ or } z' + (1-n)a(x)z = (1-n)b(x),$$

which is a linear equation.

Method 2 - alternative method of integrating factor.

This method works in the same way as for a linear equation. The original equation is written in the form $\frac{1}{y^n}y' + a(x)\frac{1}{y^{n-1}} = b(x)$ and we are looking for a function $\mu(x)$ that transforms the left side to the form

$$\mu\left(\frac{1}{y^n}y' + a(x)\frac{1}{y^{n-1}}\right) = \left(\mu\frac{y^{1-n}}{1-n}\right)'.$$

Simplifying this relationship, we get the following equation for $\mu(x)$: $\mu' = (1-n)a\mu$, whose solution is $\mu = e^{\int(1-n)a\,dx}$. Then, the left-hand side of the original equation is represented in the desired form: $\left(\mu\frac{y^{1-n}}{1-n}\right)' = \mu b$ and integration of both sides in x leads to the general solution of the original equation: $\mu\frac{y^{1-n}}{1-n} = \int \mu b\,dx$.

Examples.

1. $3xy' - 2y = \frac{x^3}{y^2}$.

This is the Bernoulli equation with $n = -2$. Multiplying by y^2, we get $3xy^2y' - 2y^3 = x^3$. The middle term indicates the change of function that must be made: $z = y^3$. Then the equation takes the form $xz' - 2z = x^3$. For the homogeneous part we have $xz' = 2z$, and consequently, $\int \frac{dz}{z} = \int 2\frac{dx}{x}$. Therefore, $z = Cx^2$ and we look for a general solution of the nonhomogeneous equation in the form $z = C(x)x^2$. Substituting into the equation, we get

$$xz' - 2z = x(C'(x)x^2 + C(x)\cdot 2x) - 2C(x)x^2 = C'(x)x^3 = x^3 \text{ or } C'(x) = 1.$$

So, $C(x) = x + A$ and $z = (x+A)x^2$. Returning to the function y, we have the general solution of the original equation $y^3 = (x+A)x^2$.

2. $y' = \frac{2x}{x^2+y+1}$.

This is not a Bernoulli equation (neither linear nor one of the previous types), but interchanging the meaning of the independent variable and the unknown function, we have a Bernoulli equation for $x(y)$:

$$x' = \frac{x^2+y+1}{2x} = \frac{x}{2} + \frac{y+1}{2}x^{-1}$$

(here $n = -1$). Multiplying by x, we get $xx' = \frac{x^2}{2} + \frac{y+1}{2}$. The middle term indicates the change of the function that should be made: $z = x^2$. So, we get the linear equation $\frac{z'}{2} = \frac{z}{2} + \frac{y+1}{2}$ or $z' = z + y + 1$. Solving the corresponding homogeneous equation $z' = z$, we find $z = Ce^y$. Therefore, the general solution for the nonhomogeneous linear equation can be found in the form $z = C(y)e^y$. Substituting into the equation, we obtain

$$C'(y)e^y + C(y)e^y = C(y)e^y + y + 1$$

and, after simplification, $C'(y) = (y+1)e^{-y}$. Integrating by parts, we find $C(y)$:

$$C(y) = \int (y+1)e^{-y}dy = -(y+1)e^{-y} + \int e^{-y}dy = -(y+2)e^{-y} + A.$$

So, the solution $z(y)$ takes the form

$$z = (A - (y+2)e^{-y})e^y = Ae^y - y - 2.$$

Returning to $x(y)$ we obtain $x^2 = Ae^y - y - 2$ which is the general solution of the primitive equation.

3. $y' + y = x\sqrt{y}$.

This is the Bernoulli equation with $n = \frac{1}{2}$. Dividing by $\sqrt{y}$ we get $\frac{y'}{\sqrt{y}} + \sqrt{y} = x$, where the middle term indicates the change of function $z = \sqrt{y}$. So for z we have the linear equation $2z' + z = x$. For the homogeneous part we have $2z' + z = 0$ and then $\int \frac{dz}{z} = -\frac{1}{2}\int dx$. Therefore, $z = Ce^{-x/2}$ and for nonhomogeneous equation we look for a general solution in the form $z = C(x)e^{-x/2}$. Substituting into the equation, we obtain

$$2z' + z = 2C'(x)e^{-x/2} - C(x)e^{-x/2} + C(x)e^{-x/2} = x \text{ or } C'(x) = \frac{x}{2}e^{x/2}.$$

Then, $C(x) = (x-2)e^{x/2} + A$ and

$$z = ((x-2)e^{x/2} + A)e^{-x/2} = x - 2 + Ae^{-x/2}.$$

Returning to the function y, we have the general solution of the original equation: $\sqrt{y} = x - 2 + Ae^{-x/2}$.

4. $\cos^2 y(ydx + 2xdy) = 2y\sqrt{x}dy$.

This equation is not separable or homogeneous or exact or linear, nor of a type directly reducible to one of these four known types of equations for $y(x)$. However, if we interchange the independent variable and unknown function, then for the function $x(y)$ we have the Bernoulli equation

$$x' = \frac{-2x\cos^2 y + 2y\sqrt{x}}{y\cos^2 y} = -\frac{2}{y}x + \frac{2}{\cos^2 y}\sqrt{x} \text{ with } n = \frac{1}{2}.$$

Dividing by $\sqrt{x}$ we get $\frac{x'}{\sqrt{x}} = -\frac{2}{y}\sqrt{x} + \frac{2}{\cos^2 y}$, where the middle term indicates the change of the function $z = \sqrt{x}$. So for z we have the linear equation $2z' = -\frac{2}{y}z + \frac{2}{\cos^2 y}$ or $z' = -\frac{1}{y}z + \frac{1}{\cos^2 y}$. For the homogeneous part we get $z' = -\frac{1}{y}z$, whence $z = \frac{C}{y}$, and for the complete linear equation we look for a general solution in the form $z = \frac{C(y)}{y}$. Substituting into the equation, we obtain $\frac{C'}{y} - \frac{C}{y^2} = -\frac{1}{y}\frac{C}{y} + \frac{1}{\cos^2 y}$ or simplifying $C' = \frac{y}{\cos^2 y}$. Integrating by parts, we get

$$C = \int \frac{y}{\cos^2 y}\,dy = y\tan y - \int \tan y\,dy = y\tan y + \int \frac{d(\cos y)}{\cos y} = y\tan y + \ln|\cos y| + A.$$

Thus, the general solution of the original equation is found in the form

$$\sqrt{x} = z = \frac{C(y)}{y} = \frac{y\tan y + \ln|\cos y| + A}{y}.$$

5. $2y' - \frac{x}{y} = \frac{xy}{x^2-1}$.

By rearranging the terms in the form $2y' - \frac{x}{x^2-1}y = \frac{x}{y}$, we can see that this is the Bernoulli equation with $n = -1$. Multiplying by y we get $2yy' - \frac{x}{x^2-1}y^2 = x$, where the middle term shows the change of the function to be made: $z = y^2$. So, for z we obtain the linear equation $z' - \frac{x}{x^2-1}z = x$. For the homogeneous part we have $z' = \frac{x}{x^2-1}z$. Separating variables and integrating, we get $\int \frac{dz}{z} = \int \frac{x}{x^2-1}dx$, and after integration, $\ln|z| = \frac{1}{2}\ln|x^2 - 1| + C$ or $z = C\sqrt{x^2-1}$. The solution of the complete linear equation is found in the form $z = C(x)\sqrt{x^2-1}$. Substituting into the equation, we obtain

$$C'\sqrt{x^2-1} + C\frac{x}{\sqrt{x^2-1}} - \frac{x}{x^2-1}C\sqrt{x^2-1} = x \quad \text{or} \quad C' = \frac{x}{\sqrt{x^2-1}}.$$

Integrating, we have

$$C = \int \frac{x}{\sqrt{x^2-1}}\,dx = \frac{1}{2}\sqrt{x^2-1} + A.$$

Thus, the general solution of the original equation is found in the form $y^2 = z = \frac{1}{2}\sqrt{x^2-1} + A$.

6. $xy^2 y' = x^2 + y^3$.

Rewriting the equation in the form $xy' = y + \frac{x^2}{y^2}$ we see that this is the Bernoulli equation with $n = -2$. The original form already shows the change of function that should be made: $z = y^3$. Then, we obtain the linear equation $\frac{1}{3}xz' = z + x^2$ or $xz' = 3z + 3x^2$. To solve the homogeneous part $xz' = 3z$, we separate variables and integrate: $\int \frac{dz}{z} = \int \frac{3dx}{x}$, that results in $\ln|z| = 3\ln|x| + C$ or $z = Cx^3$. The solution of the complete linear equation we seek in the form $z = C(x)x^3$. Substituting into the equation, we get $x(C'x^3 + 3Cx^2) = 3Cx^3 + 3x^2$ or simplifying $C' = \frac{3}{x^2}$. Integrating, we obtain $C = \frac{-3}{x} + A$. Thus, the general solution of the original equation has the form $y^3 = z = \left(-\frac{3}{x} + A\right)x^3$ or $y^3 = -3x^2 + Ax^3$.

7. $y'x + y = -xy^2$.

This is the Bernoulli equation with $n = 2$. We divide it by y^2 and change the function $z = \frac{1}{y}$, arriving at the linear equation $-xz' + z = -x$. To solve the homogeneous part $-xz' + z = 0$, we separate variables and integrate: $\int \frac{dz}{z} = \int \frac{dx}{x}$, that results in $\ln|z| = \ln|x| + C$ or $z = Cx$. The solution of the complete linear equation we seek in the form $z = C(x)x$. Substituting into the equation, we get $-x(C'x + C) + Cx = -x$ or simplifying $C' = \frac{1}{x}$. Integrating, we obtain $C = \ln|x| + A$. Thus, the general solution of the original equation has the form $\frac{1}{y} = z = (\ln|x| + A)x$.

8. $y' + xy = x^3y^3$.

This is the Bernoulli equation with $n = 3$. We divide it by y^3 and change the function $z = \frac{1}{y^2}$, arriving at the linear equation $-\frac{1}{2}z' + xz = x^3$. To solve the homogeneous part $-\frac{1}{2}z' + xz = 0$, we separate variables and integrate: $\int \frac{dz}{z} = 2\int xdx$, that results in $\ln|z| = x^2 + C$ or $z = Ce^{x^2}$. The solution of the complete linear equation we seek in the form $z = C(x)e^{x^2}$. Substituting into the equation, we get

$$-\frac{1}{2}(C'e^{x^2} + C \cdot 2xe^{x^2}) + xCe^{x^2} = x^3 \text{ or } C' = -2e^{-x^2}x^3.$$

Integrating, we obtain

$$C = -\int 2e^{-x^2}x^3dx = -\int e^{-x^2}x^2d(x^2) = -\int e^{-t}tdt = e^{-t}t - \int e^{-t} \cdot 1dt$$

$$= e^{-t}t + e^{-t} + A = e^{-x^2}x^2 + e^{-x^2} + A.$$

Thus, the general solution of the original equation is found in the form

$$\frac{1}{y^2} = z = \left(e^{-x^2}x^2 + e^{-x^2} + A\right)e^{x^2} \text{ or } \frac{1}{y^2} = x^2 + 1 + Ae^{x^2}.$$

9. $2yy' + \frac{3}{x}y^2 = -\frac{1}{x^3}$.

This is the Bernoulli equation with $n = -1$, which becomes clear if we write it in the normalized form $y' + \frac{3}{2x}y = -\frac{1}{2x^3y}$. Starting with the original form, let us apply the alternative method of integrating factor. First, we look for a function $\mu(x)$ such that $\mu\left(2yy' + \frac{3}{x}y^2\right) = (\mu \cdot y^2)'$. It follows from this relation that μ satisfies the differential equation $\mu\frac{3}{x} = \mu'$, whose solution is $\ln|\mu| = 3\ln|x|$ or $\mu = x^3$. Therefore, the multiplication of the equation $2yy' + \frac{3}{x}y^2 = -\frac{1}{x^3}$ by $\mu = x^3$ transforms it into the form $\left(x^3 y^2\right)' = -1$. Integrating, we obtain the general solution of the original equation: $x^3 y^2 = -x + C$.

10. $y' - 2xy = 2x^3 y^2$, $y(0) = 1$.
This is the Cauchy problem for the Bernoulli equation with $n = 2$. We divide the equation by y^2 and change the function $z = \frac{1}{y}$, which gives us the linear equation $z' + 2xz = -2x^3$. Solving the homogeneous part, we get $\ln|z| = -x^2 + C$ or $z = Ce^{-x^2}$. The solution of the complete linear equation can be found in the form $z = C(x)e^{-x^2}$. Substituting into the equation, we obtain $C'e^{-x^2} - C \cdot 2xe^{-x^2} + 2xCe^{-x^2} = -2x^3$ or simplifying $C' = -2x^3 e^{x^2}$. Integrating, we find $C = (1 - x^2)e^{x^2} + A$ and

$$z = ((1 - x^2)e^{x^2} + A)e^{-x^2} = 1 - x^2 + Ae^{-x^2}.$$

Thus, the general solution of the original equation has the form $\frac{1}{y} = 1 - x^2 + Ae^{-x^2}$. Applying the initial condition, we have $\frac{1}{1} = 1 - 0 + Ae^0$, whence $A = 0$. Therefore, the solution of the Cauchy problem is $\frac{1}{y} = 1 - x^2$.

We can also solve this equation using the alternative method of integrating factor. To do this, we rewrite the original equation in the modified form $\frac{1}{y^2}y' - 2x\frac{1}{y} = 2x^3$ and find

$$\mu = e^{\int (1-n)a\,dx} = e^{\int -1 \cdot (-2x)\,dx} = e^{x^2}.$$

(Recall that $\mu(x)$ is such a function that $\mu\left(\frac{1}{y^2}y' - 2x\frac{1}{y}\right) = \left(\mu \cdot \frac{-1}{y}\right)'$ from which comes the equation $\mu' = 2x\mu$ whose solution is $\ln|\mu| = x^2$ or $\mu = e^{x^2}$.)
Multiplying the modified equation by μ, we find $\left(e^{x^2} \cdot \frac{-1}{y}\right)' = 2x^3 e^{x^2}$ and then integrate both sides in x:

$$-\frac{e^{x^2}}{y} = \int 2x^3 e^{x^2}\,dx = \int x^2 e^{x^2}\,d(x^2) = \int te^t\,dt = (t-1)e^t + C = (x^2-1)e^{x^2} + C.$$

Finally, we get the same general solution as in the first algorithm $\frac{1}{y} = 1 - x^2 + Ce^{-x^2}$. Consequently, we find the same solution of the Cauchy problem $\frac{1}{y} = 1 - x^2$.

11. $y' + \frac{x}{1-x^2}y = x\sqrt{y}$.
This is the Bernoulli equation with $n = \frac{1}{2}$. Dividing by $\sqrt{y}$ we obtain

$\frac{y'}{\sqrt{y}} + \frac{x}{1-x^2}\sqrt{y} = x$, where the middle term indicates the change of the function $z = \sqrt{y}$. Then, for z we have the linear equation $2z' + \frac{x}{1-x^2}z = x$. Solving the homogeneous part we find $\ln|z| = \frac{1}{4}\ln|1-x^2| + C$ or $z = C\sqrt[4]{1-x^2}$. For the complete linear equation we seek general solution in the form $z = C(x)\sqrt[4]{1-x^2}$. Substituting into the equation, we get

$$2z' + \frac{x}{1-x^2}z = 2C'\sqrt[4]{1-x^2} - Cx(1-x^2)^{-3/4} + \frac{x}{1-x^2}C\sqrt[4]{1-x^2} = x$$

or, after simplification, $C' = \frac{x}{2}\frac{1}{\sqrt[4]{1-x^2}}$. Then, $C(x) = -\frac{1}{3}(1-x^2)^{3/4} + A$ and

$$z = \left(A - \frac{1}{3}(1-x^2)^{3/4}\right)\sqrt[4]{1-x^2} = A\sqrt[4]{1-x^2} - \frac{1}{3}(1-x^2).$$

Returning to the function y, we have the general solution of the original equation: $\sqrt{y} = A\sqrt[4]{1-x^2} - \frac{1}{3}(1-x^2)$.

In parallel, we can solve this equation using the alternative method of integrating factor. To do this, we rewrite the original equation in the modified form $\frac{y'}{\sqrt{y}} + \frac{x}{1-x^2}\sqrt{y} = x$ and find

$$\mu = e^{\int (1-n)a\,dx} = e^{\int \frac{1}{2}\cdot\frac{x}{1-x^2}\,dx} = e^{-\frac{1}{4}\ln|1-x^2|} = \frac{1}{\sqrt[4]{1-x^2}}.$$

(Recall that $\mu(x)$ satisfies the equation

$$(\mu \cdot 2\sqrt{y})' = \mu\left(\frac{y'}{\sqrt{y}} + \frac{x}{1-x^2}\sqrt{y}\right),$$

whence $\mu' = \frac{1}{2}\frac{x}{1-x^2}\mu$, whose solution is $\ln|\mu| = \int \frac{1}{2}\frac{x}{1-x^2}\,dx$ or $\mu = e^{\int \frac{1}{2}\cdot\frac{x}{1-x^2}\,dx}$.)

Multiplying the modified equation by μ, we find $\left(\frac{1}{\sqrt[4]{1-x^2}}\cdot 2\sqrt{y}\right)' = x\frac{1}{\sqrt[4]{1-x^2}}$ and then integrate both sides in x:

$$\frac{1}{\sqrt[4]{1-x^2}}\cdot 2\sqrt{y} = -\frac{2}{3}(1-x^2)^{3/4} + C.$$

Finally, we arrive at the same general solution as in the first algorithm: $\sqrt{y} = -\frac{1}{3}(1-x^2) + C\sqrt[4]{1-x^2}$.

12. $y' - 9x^2y = (x^5 + x^2)y^{2/3}$.

This is a Bernoulli equation with $n = \frac{2}{3}$. Dividing by $y^{2/3}$ we get $\frac{y'}{y^{2/3}} - 9x^2y^{1/3} = x^5 + x^2$, where the middle term shows the change of the function to be performed: $z = y^{1/3}$. Then, for z we have the linear equation $3z' - 9x^2z =$

$x^5 + x^2$ or $z' - 3x^2z = \frac{1}{3}(x^5 + x^2)$. Solving the homogeneous part we find $\ln|z| = x^3 + C$ or $z = Ce^{x^3}$. For the complete linear equation we seek the general solution in the form $z = C(x)e^{x^3}$. Substituting into the equation, we obtain

$$z' - 3x^2z = C'e^{x^3} + C \cdot 3x^2e^{x^3} - 3x^2 \cdot Ce^{x^3} = \frac{1}{3}(x^5 + x^2)$$

or $C' = \frac{1}{3}(x^5 + x^2)e^{-x^3}$. The integral can be calculated as follows:

$$\int(x^5 + x^2)e^{-x^3}dx = \int x^5 e^{-x^3}dx + \frac{1}{3}\int x^2 e^{-x^3}dx = \frac{1}{3}\int(x^3 + 1)e^{-x^3}d(x^3)$$

$$= \frac{1}{3}\int(t+1)e^{-t}dt = \frac{1}{3}\left(-(t+1)e^{-t} + \int e^{-t}dt\right) = \frac{1}{3}\left(-(t+2)e^{-t} + A\right)$$

$$= \frac{1}{3}\left(-(x^3+2)e^{-x^3} + A\right).$$

So, $C(x) = \frac{1}{9}\left(-(x^3+2)e^{-x^3} + A\right)$ and

$$z = \frac{1}{9}\left(-(x^3+2)e^{-x^3} + A\right)e^{x^3} = Be^{x^3} - \frac{1}{9}(x^3+2).$$

Returning to the function y, we have the general solution of the original equation: $y^{1/3} = Be^{x^3} - \frac{x^3}{9} - \frac{2}{9}$.

Exercises for section 4.2

Solve the following Bernoulli equations (with respect to y or x):
1. $y' - \frac{y}{x} = \frac{1}{2y}$.
2. $(xy + x^2y^3)y' = 1$.
3. $y' + 2xy = 2x^3y^3$.
4. $3y^2y' + y^3 + x = 0$.
5. $(1 + x^2)y' - 2xy = 4\sqrt{y(1 + x^2)}\arctan x$.
6. $x^3\sin yy' + 2y = xy'$.
7. $y' = y^4\cos x + y\tan x$.
8. $(x+1)(y' + y^2) = -y$.
9. $xy' - 2x^2\sqrt{y} = 4y$.
10. $2y' - \frac{x}{y} = \frac{xy}{x^2-1}$.
11. $y'x^3\sin y = xy' - 2y$.
12. $(2x^2y\ln y - x)y' = y$.
13. $8y' + 3x^2y(y^2 - 4) = 0$.
14. $(y^2 - 1)dx - y(x + (y^2 - 1)\sqrt{x})dy = 0$.

Exercises for Chapter 3

Solve the following explicit equations of the first order and the corresponding initial value problems; if possible, use different methods for the same equation and compare the obtained solutions:

1. $3e^x \sin y dx + (1 - e^x) \cos y dy = 0$.
2. $y' = \frac{y}{x} + e^{-y/x}$.
3. $\frac{2x}{y^3} dx + \frac{y^2 - 3x^2}{y^4} dy = 0$, $y(1) = 1$.
4. $(x + 2y)y' = 1$, $y(0) = -1$.
5. $xy' + y = \sin x$, $y(\frac{\pi}{2}) = \frac{2}{\pi}$.
6. $(x^2 - \sin^2 y)dx + x \sin 2y dy = 0$.
7. $(1 - x)(y' + y) = e^{-x}$, $y(0) = 0$.
8. $y' = \frac{y+2}{x+1} + \tan \frac{y-2x}{x+1}$.
9. $xy' + y = y^2 \ln x$, $y(1) = 1$.
10. $(x + y - 1)^2 y' = 2(y + 2)^2$.
11. $y' = 2x(x^2 + y)$, $y(0) = 0$.
12. $xy^2 dx + (x^2 y - x)dy = 0$, $y(-1) = 1$.
13. $y' + 2y = e^x y^2$, $y(1) = 0$.
14. $(4x^2 - xy + y^2)dx + (x^2 - xy + 4y^2)dy = 0$.
15. $y(1 + xy)dx + (5y - x + y^2 \sin y)dy = 0$.
16. $\cos y dx = (x + 2 \cos y) \sin y dy$, $y(0) = \frac{\pi}{4}$.
17. $(x + 2)(1 + y^2)dx + (x + 1)y^2 dy = 0$.
18. $\left(\frac{1}{x} - \frac{y^2}{(x-y)^2}\right) dx + \left(\frac{x^2}{(x-y)^2} - \frac{1}{y}\right) dy = 0$.
19. $(x^2 + y^2 - x)dx - ydy = 0$.
20. $ydx + (3x - xy + 2)dy = 0$.

Chapter 4

Implicit equations of the first order: methods of solution

1 Polynomial equations with respect to the derivative

Definition. A *polynomial equation with respect to the derivative* has the form

$$a_n(x,y)(y')^n + a_{n-1}(x,y)(y')^{n-1} + \ldots + a_1(x,y)y' + a_0(x,y) = 0.$$

Remark. All coefficients $a_k(x,y)$ cannot depend on y' and the leading coefficient $a_n(x,y)$ cannot be identically zero.

Method of solution.

First, we solve the polynomial equation with respect to unknown $p = y'$, considering that the point (x,y) is fixed (in generic form) and the coefficients are real numbers (as in any polynomial equation):

$$a_n p^n + a_{n-1} p^{n-1} + \ldots + a_1 p + a_0 = 0.$$

Let us assume that for each pair (x,y) we are able to find k real roots $f_1(x,y), \ldots, f_k(x,y)$, $k \leq n$ of the last equation. So, the polynomial equation can be written in the form

$$(p - f_1) \cdot \ldots \cdot (p - f_k) \cdot Q_{n-k}(p) = 0,$$

where $Q_{n-k}(p) = (b_{n-k}p^{n-k} + \ldots + b_1 p + b_0)$ is the polynomial of order $n - k$. Correspondingly, the original equation takes the form

$$(y' - f_1(x, y)) \cdot \ldots \cdot (y' - f_k(x, y)) \cdot Q_{n-k}(y') = 0.$$

The first k factors produce k explicit equations of the first order $y' = f_1(x, y), \ldots, y' = f_k(x, y)$ to which we can try to apply the methods already studied in Chapter 3. The solution of any of these equations is also the solution of the original equation. The remaining part of the equation $b_{n-k}(x, y)(y')^{n-k} + \ldots + b_1(x, y)y' + b_0(x, y) = 0$ can no longer be simplified to linear factors and all that remains is to try some special method of solution (if one exists).

Examples.

1. $yy'^2 + (x - y)y' - x = 0$.
This is a second-degree polynomial equation. Introducing the notation $p = y'$ and considering (x, y) fixed, we solve the quadratic equation $yp^2 + (x - y)p - x = 0$. The roots are

$$p = \frac{y - x \pm \sqrt{(y - x)^2 + 4xy}}{2y} = \frac{y - x \pm (y + x)}{2y},$$

that is, $p_1 = 1$, $p_2 = -\frac{x}{y}$. Returning to the function y, we have, respectively, the two explicit equations $y' = 1$, $y' = -\frac{x}{y}$. The general solution of the first is $y = x + A$ and of the second $y^2 + x^2 = B^2$. Both families are solutions of the original equation.

2. $y'^3 - yy'^2 - x^2 y' + x^2 y = 0$.
This is a third-degree polynomial equation. Introducing the notation $p = y'$ and fixing (x, y), we solve the cubic equation $p^3 - yp^2 - x^2 p + x^2 y = 0$. To do this, we group the terms

$$p^2(p - y) - x^2(p - y) = (p - y)(p^2 - x^2) = 0.$$

So, we have the three real roots $p_1 = y$, $p_2 = x$, $p_3 = -x$. Returning to the function y, we have, respectively, the three explicit equations $y' = y$, $y' = x$, $y' = -x$, all of separable variables. The general solution of the first is $y = Ae^x$, of the second $y = \frac{x^2}{2} + B$, of the third $y = -\frac{x^2}{2} + C$. They are all solutions of the original equation.

3. $y'^2 - 4x^2 = 0$.
This is a second-degree polynomial equation. Introducing the notation $p = y'$ and fixing (x, y), we solve the quadratic equation $p^2 - 4x^2 = 0$. The roots are $p = \pm 2x$. Returning to the function y, we have the two explicit equations of

separable variables $y' = 2x$ and $y' = -2x$. The general solution of the first is $y = x^2 + A$ and of the second $y = -x^2 + +B$. Both families are solutions of the original equation.

4. $yy'^2 - (xy + 1)y' + x = 0$, $y(1) = 1$.

This is the Cauchy problem for a second-degree polynomial equation. Solving the quadratic equation with respect to the derivative, we have the two explicit equations of separable variables

$$y' = \frac{xy + 1 \pm \sqrt{(xy + 1)^2 - 4xy}}{2y} = \frac{xy + 1 \pm \sqrt{(xy - 1)^2}}{2y} = \frac{xy + 1 \pm (xy - 1)}{2y}.$$

Simplifying, we get $y' = \frac{1}{y}$ and $y' = x$. The solutions are $\frac{y^2}{2} = x + A$ and $y = \frac{x^2}{2} + B$, respectively. Applying the initial condition to both families, we get $\frac{1}{2} = 1 + A$ (in the first) and $1 = \frac{1}{2} + B$ (in the second), whence $A = -\frac{1}{2}$ and $B = \frac{1}{2}$. So, the solutions to the Cauchy problem are $\frac{y^2}{2} = x - \frac{1}{2}$ (from the first family) and $y = \frac{x^2}{2} + \frac{1}{2}$ (from the second family).

5. $y'^2 - 4y = 0$, $y(1) = 1$.

This is the Cauchy problem for a second-degree polynomial equation. Solving the quadratic equation with respect to the derivative, we get two explicit equations of separable variables: $y' = \pm 2\sqrt{y}$. The general solution of the first is $\sqrt{y} = x + A$ and of the second $\sqrt{y} = -x + B$. In addition, we have the special solution $y = 0$ which is not included in these two families. The special solution does not satisfy the initial condition. Applying the initial condition to the two families, we get $\sqrt{1} = 1 + A$ (in the first) and $\sqrt{1} = -1 + B$ (in the second), whence $A = 0$ and $B = 2$. So, the solutions of the Cauchy problem are $\sqrt{y} = x$, $x \geq 0$ (from the first family) and $\sqrt{y} = 2 - x$, $x \leq 2$ (from the second family).

6. $y^2(1 + y'^2) = a^2$, $a \in \mathbb{R}$.

This is a second-degree polynomial equation. First, we solve for the square of the derivative: $y'^2 = \frac{a^2 - y^2}{y^2}$. Now solving the quadratic equation for the derivative, we find the two explicit equations of separable variables: $y' = \pm \frac{1}{y}\sqrt{a^2 - y^2}$. The general solution of the two equations has the form $\sqrt{a^2 - y^2} = \pm x + C$. Eliminating the root, we get $y^2 + (x + C)^2 = a^2$. In addition, it has two particular solutions $y = \pm a$ not included in the general solutions.

7. $y'^3 - \frac{1}{4x}y' = 0$.

This a third-degree polynomial equation. Rewriting it in the form $y'(y'^2 - \frac{1}{4x}) = 0$, we get three explicit equations of separable variables $y' = 0$, $y' =$

$\pm \frac{1}{2\sqrt{x}}$. Their general solutions are $y = A$, $y = \sqrt{x} + B$ and $y = -\sqrt{x} + C$, respectively.

8. $y'^3 - xy'^2 - 4yy' + 4xy = 0$, $y(0) = 1$.

This is the Cauchy problem for a third-degree polynomial equation. Rewriting it in the form $(y' - x)(y'^2 - 4y) = 0$, we get three explicit equations of separable variables $y' = x$, $y' = \pm 2\sqrt{y}$. Their general solutions are $y = \frac{x^2}{2} + A$, $\sqrt{y} = x + B$ and $\sqrt{y} = -x + C$. In addition, we have the special solution $y = 0$ not included in these three families. By substituting the initial condition into these families, we get three solutions of the Cauchy problem: $y = \frac{x^2}{2} + 1$, $\sqrt{y} = x + 1, x \geq -1$ and $\sqrt{y} = -x + 1, x \leq 1$.

9. $yy' + y'^2 = x^2 + xy$.

This is a second-degree polynomial equation. Solving the quadratic equation for the derivative, we find the two explicit equations

$$y' = \frac{-y \pm \sqrt{y^2 + 4(x^2 + xy)}}{2} = \frac{-y \pm \sqrt{(y + 2x)^2}}{2} = \frac{-y \pm (y + 2x)}{2}.$$

Simplifying, we arrive at the separable equation $y' = x$ and another reducible to the separable $y' = -x - y$. The first is trivial and gives the solution $y = \frac{x^2}{2} + A$. To solve the second, we introduce the function $z = x + y$ and arrive at the separable equation $z' - 1 = -z$, whose solution is $z = 1 + Ce^{-x}$. Returning to y, we find the second family of solutions of the original equation $y = -x + 1 + Ce^{-x}$.

10. $x^2y'^2 + 3xyy' + 2y^2 = 0$.

This is a second-degree polynomial equation. Solving the quadratic equation for the derivative, we find the two explicit separable equations

$$y' = \frac{-3xy \pm \sqrt{9x^2y^2 - 8x^2y^2}}{2x^2} = \frac{-3xy \pm xy}{2x^2}.$$

Simplifying, we get $y' = -2\frac{y}{x}$ and $y' = -\frac{y}{x}$. The general solutions are $y = \frac{A}{x^2}$ and $y = \frac{B}{x}$, respectively.

11. $y'^2 + y(y - x)y' - xy^3 = 0$.

This is a second-degree polynomial equation. By solving the quadratic equation for the derivative or, equivalently, regrouping the terms in the form

$$y'(y' + y^2) - xy(y' + y^2) = (y' + y^2)(y' - xy) = 0,$$

we reduce the original equation to the two separable equations $y' = -y^2$ and $y' = xy$. Their general solutions are $\frac{1}{y} = x + A$ and $y = Be^{x^2/2}$, respectively.

The particular solution $y = 0$ of the first equation, disregarded in its general solution, is included in the general solution of the second. Thus, all the solutions of the original equation are found.

12. $y'^2 + y(\sin x - 2xy)y' - 2xy \sin x = 0$.

This is a second-degree polynomial equation. By regrouping its terms in the form

$$y'(y' + \sin x) - 2xy(y' + \sin x) = (y' + \sin x)(y' - 2xy) = 0,$$

we reduce the original equation to the two separable equations $y' = -\sin x$ and $y' = 2xy$. Their general solutions are $y = \cos x + A$ and $y = Be^{x^2}$, respectively. These are all solutions of the original equation.

13. $y'^3 - 2xy'^2 + y' = 2x$.

This is a third-degree polynomial equation. By rewriting the equation in the factored form

$$y'^3 - 2xy'^2 + y' - 2x = y'(y'^2 + 1) - 2x(y'^2 + 1) = (y'^2 + 1)(y' - 2x) = 0$$

we conclude that the original equation is equivalent to the single explicit equation $y' = 2x$. Solving the latter, we find the general solution of the original equation in the form $y = x^2 + C$. These are all solutions of the given equation.

14. $y'^2 + y^2(\ln^2 y - 1) = 0$.

This is a second-degree polynomial equation. Solving for the derivative, we get the two equations with separable variables $y' = \pm y\sqrt{1 - \ln^2 y}$. Separating the variables and integrating, we get

$$\int \frac{dy}{y\sqrt{1 - \ln^2 y}} = \int \frac{d(\ln y)}{\sqrt{1 - \ln^2 y}} = \arcsin(\ln y) = \pm x + C.$$

The same general solution can also be written in the form $\ln y = \pm \sin(x+C)$. To perform the separation of variables it was necessary to divide by y and $\sqrt{1 - \ln^2 y}$, and the functions $y = 0$ and $\ln y = \pm 1$ are not included in the general solution. Since the original equation has the term $\ln y$, the function $y = 0$ cannot be considered, but the functions $y = e^{\pm 1}$ are admissible and represent special solutions of the original equation.

Exercises for section 1.1

Solve the following polynomial equations:

1. $y'^2 = y^3 - y^2$.

2. $y'^2 - y^2 = 0$.
3. $(y' + 1)^3 = 27(x + y)^2$.
4. $y'^2 = 4y^3(1 - y)$.
5. $y'^3 + y^2 = yy'(y' + 1)$.
6. $y'^2 + xy = y^2 + xy'$.
7. $y'^3 + (x + 2)e^y = 0$.
8. $y'(2y - y') = y^2 \sin^2 x$.
9. $y(xy' - y)^2 = y - 2xy'$.
10. $yy'(yy' - 2x) = x^2 - 2y^2$.
11. $y'^2 + 4xy' - y^2 - 2x^2y = x^4 - 4x^2$.
12. $xy'(xy' + y) = 2y^2$.

2   Explicit equations in y

2.1   Explicit equations in y in general form

Definition. An *explicit equation with respect to unknown function y* has the form

$$y = f(x, y').$$

Method of solution.

We introduce the parameter $p = y'$ and rewrite the original equation in the form of the equivalent system $\begin{cases} y' = p \\ y = f(x, p) \end{cases}$. Differentiating the second equation of the system with respect to x and using the first to substitute p for y', we get $y' = p = f_x + f_p p_x$, which is the explicit first order equation for the function $p(x)$. If it is possible to find the solution of the last equation (using one of the known methods), then we have the relation $G(p, x, C) = 0$ which we combine with the second equation of the system to obtain the parametric form of the general solution of the original equation $\begin{cases} G(p, x, C) = 0 \\ y = f(x, p) \end{cases}$. If possible, the parameter p is eliminated from the system to obtain a simpler form of the solution. For example, if the relation $G(p, x, C) = 0$ can be transformed into the explicit form $p = g(x, C)$, then the parametric form can be reduced to the explicit $y = f(x, g(x, C))$.

Remark. The described algorithm of deriving an explicit equation for the new function $p(x)$ requires the elimination of y, since this is the unknown function of x. Therefore, after introducing the parameter p, the original equation must be written in the form where y is set separately from other terms: $y = f(x, p)$. Only in this case the function y will disappear after differentiation (by replacing its derivative with p). If the original equation is in the form $g(x)y = f(x, y')$, it must be normalized to the form $y = \frac{f(x,y')}{g(x)}$ before applying the proposed method of solution (unless $g(x)$ is a constant).

Examples.

1. $x^2 y'^4 + 2xy' - y = 0$.

Although this equation is polynomial with respect to y', solving it as such will be technically complicated because of finding the roots of the fourth-degree equation. It is much simpler to treat this equation as explicit in y. By introducing the parameter $p = y'$, we have the equivalent system $\begin{cases} y' = p \\ y = 2xp + x^2 p^4 \end{cases}$. Differentiating the second equation of the system with respect to x, we obtain the first order explicit equation for the function p: $p = 2p + 2xp' + 2xp^4 + x^2 4p^3 p'$. We simplify and regroup the terms of this equation: $p' \cdot 2x(1 + 2xp^3) = -p(1 + 2xp^3)$. Obviously, there are two options to satisfy this equation: $p' \cdot 2x = -p$ and $1 + 2xp^3 = 0$. Solving the first relation (the differential equation for p), we get $p^2 = \frac{C}{x}$. So the general solution of the original equation has the following parametric form: $\begin{cases} xp^2 = C \\ y = 2xp + x^2 p^4 \end{cases}$.

We can still eliminate the parameter and obtain the following implicit form $(y - C^2)^2 = 4Cx$ (for instance, noting that $x^2 p^4 = C^2$ from the first equation and substituting it in the second, we get $y = 2xp + C^2$; then we transfer C^2 to the left-hand side and square it, which gives $(y - C^2)^2 = 4x^2 p^2$; and finally we substitute p^2 from the first relation to arrive at the shown implicit solution). From the second relation for p we have another function $\begin{cases} 1 + 2xp^3 = 0 \\ y = 2xp + x^2 p^4 \end{cases}$, which can be transformed into the implicit form $16y^3 = -3x^2$ (from the first equation the parameter p is expressed via x and then all that remains is to substitute this expression into the second equation and simplify). Substituting this function into the original equation, we see that this is another (special) solution of the original equation.

2. $y = x + y' - \ln y'$.

This equation is explicit with respect to y. By introducing the parameter $p = y'$, we have the equivalent system $\begin{cases} y' = p \\ y = x + p - \ln p \end{cases}$. Differentiating the second equation of the system with respect to x, we get the first order explicit equation for the function p: $p = 1 + p' - \frac{1}{p}p'$. Regrouping the terms, we obtain $p - 1 = \frac{p-1}{p}p'$. There are two options to satisfy this equation: $\frac{p'}{p} = 1$ and $p - 1 = 0$. Solving the first relation (the differential equation for p), we get $p = Ce^x$. Then, the general solution of the original equation has the following parametric form: $\begin{cases} p = Ce^x \\ y = x + p - \ln p \end{cases}$. It is easy to eliminate p and obtain the explicit form

$$y = x + Ce^x - x \ln C = Ce^x - \ln C.$$

From the second relation for p we have another function $\begin{cases} p = 1 \\ y = x + p - \ln p \end{cases}$ whose explicit form is $y = x + 1$. Substituting this in the original equation, we see that this function is a (special) solution of the original equation.

3. $y'^2 - xy' + \frac{x^2}{2} - y = 0$.

This equation is polynomial with respect to y', but solving it as such is technically complicated due to the form of the explicit equations: $y' = \frac{x \pm \sqrt{4y - x^2}}{2}$. However, by treating the equation as explicit with respect to y, we are able to find the solution without any problem. By introducing the parameter $p = y'$, we have the equivalent system $\begin{cases} y' = p \\ y = p^2 - xp + \frac{x^2}{2} \end{cases}$.
Differentiating the second equation with respect to x, we get the first order explicit equation for p: $p = 2pp' - p - xp' + x$, or, simplifying and regrouping the terms, $(p' - 1)(2p - x) = 0$. So, we have two options: $p' = 1$ and $2p = x$. Solving the first equation, we get $p = x + C$. This leads to the general solution of the original equation in parametric form: $\begin{cases} p = x + C \\ y = p^2 - xp + \frac{x^2}{2} \end{cases}$.
Eliminating p, we find the explicit form of the general solution:

$$y = (x + C)^2 - x(x + C) + \frac{x^2}{2} = Cx + C^2 + \frac{x^2}{2}.$$

From the second relation we have another function $\begin{cases} p = x/2 \\ y = p^2 - xp + \frac{x^2}{2} \end{cases}$ whose explicit form is $y = \frac{x^2}{4} - x\frac{x}{2} + \frac{x^2}{2} = \frac{x^2}{4}$. Substituting into the original equation, we see that this is another (special) solution of the original equation.

4. $y = \frac{xy'}{2} + \frac{y'^2}{x^2}$.

Using the parameter $p = y'$, we get the equivalent system $\begin{cases} y' = p \\ y = \frac{xp}{2} + \frac{p^2}{x^2} \end{cases}$.
Differentiating the second equation with respect to x, we obtain the first order explicit equation for p: $p = \frac{p}{2} + \frac{xp'}{2} + \frac{2pp'}{x^2} - \frac{2p^2}{x^3}$, or, simplifying and regrouping the terms, $p\left(\frac{1}{2} + \frac{2p}{x^3}\right) = p'x\left(\frac{1}{2} + \frac{2p}{x^3}\right)$. Then, we have two options: $p'x = p$ and $\frac{2p}{x^3} = -\frac{1}{2}$. Solving the first equation, we get $p = Cx$

and, joining this expression with the second equation of the system, we find the general solution of the original equation: $y = \frac{Cx^2}{2} + C^2$. From the second relation we get another function $p = -\frac{x^3}{4}$ which, together with the second equation of the system, results in the function $y = -\frac{x^4}{16}$. Substitution into the original equation shows that this is another (special) solution of the original equation.

The given equation can also be treated as polynomial, but in this case the procedure of solution becomes more complicated.

5. $6x^2y - 6y'^2 + (12x^2 - 3x^3)y' - 6x^4 + x^5 = 0$.

First, we write the equivalent system $\begin{cases} y' = p \\ y = \frac{p^2}{x^2} - (2 - \frac{x}{2})p + x^2 - \frac{x^3}{6} \end{cases}$ by introducing the parameter $p = y'$. (Note that it is mandatory to divide by x^2, because y has to disappear after the differentiation on the next step.) Next, differentiating the second equation with respect to x, we obtain the first order explicit equation for p:

$$p = \frac{2pp'}{x^2} - \frac{2p^2}{x^3} + \frac{p}{2} - \left(2 - \frac{x}{2}\right)p' + 2x - \frac{x^2}{2}.$$

Simplifying and regrouping the terms, we have

$$p'\left(\frac{2p}{x^2} - 2 + \frac{x}{2}\right) = \frac{p}{2} + \frac{2p^2}{x^3} - 2x + \frac{x^2}{2} = \left(\frac{2p}{x} - 2x + \frac{x^2}{2}\right) + \left(-\frac{2p}{x} + \frac{2p^2}{x^3} + \frac{p}{2}\right)$$

$$= x\left(\frac{2p}{x^2} - 2 + \frac{x}{2}\right) + \frac{p}{x}\left(\frac{x}{2} + \frac{2p}{x^2} - 2\right) = \left(\frac{2p}{x^2} - 2 + \frac{x}{2}\right)\left(x + \frac{p}{x}\right).$$

There are two options to satisfy the last relation: $p' = x + \frac{p}{x}$ and $\frac{2p}{x^2} - 2 + \frac{x}{2} = 0$. The first option is a linear equation for p that can be solved using the Lagrange method. We start with the homogeneous linear equation $p' = \frac{p}{x}$ whose solution is $p = Cx$, and then seek the solution of the complete linear equation in the form $p = C(x)x$. Substituting the latter in the linear equation, we obtain, after simplification, $C' = 1$, whence $C = x + A$. Therefore, $p = (x + A)x$. Combining this expression with the second equation of the system, we find the general solution of the original equation:

$$y = (x + A)^2 - \left(2 - \frac{x}{2}\right)(x + A)x + x^2 - \frac{x^3}{6} = A^2 + A\frac{x^2}{2} + \frac{x^3}{3}.$$

From the second relation $\frac{2p}{x^2} - 2 + \frac{x}{2} = 0$ we get $p = x^2 - \frac{x^3}{4}$ and, bringing this expression into the second equation of the system, we find the function

$$y = \frac{\left(x^2 - \frac{x^3}{4}\right)^2}{x^2} - \left(2 - \frac{x}{2}\right)\left(x^2 - \frac{x^3}{4}\right) + x^2 - \frac{x^3}{6} = \frac{x^3}{3} - \frac{x^4}{16}.$$

Substitution in the original equation shows that this is another (special) solution of the original equation.

The given equation can also be considered as polynomial, but in this case a solution becomes more complicated or even technically impossible.

6. $y = y'^2 e^{y'}$, $y(1) = 0$.

This is the Cauchy problem for the equation explicit with respect to y. Using the parameter $p = y'$, we get the equivalent system $\begin{cases} y' = p \\ y = p^2 e^p \end{cases}$. Differentiating the second equation with respect to x, we obtain the first order explicit equation for p: $p = 2pp'e^p + p^2 e^p p'$, which results in the separable equation $p'e^p(2 + p) = 1$ and the additional relation $p = 0$. Solving the equation, we get $e^p(p + 1) = x + C$ and combining this expression with the second equation of the system, we find the general solution of the original equation in parametric form: $\begin{cases} x = e^p(p + 1) + C \\ y = p^2 e^p \end{cases}$. The relation $p = 0$, substituted into the second equation of the system, generates another (special) solution $y = 0$.

Substituting the initial condition into the general solution, we obtain $\begin{cases} 1 = e^p(p + 1) + C \\ 0 = p^2 e^p \end{cases}$. From the second relation it follows that $p = 0$ and then the first (with $p = 0$) specifies the value of the constant: $1 = 1 + C$, that is, $C = 0$. So in the general solution there is one, with constant $C = 0$, which satisfies the initial condition. But this is not the only solution to the problem, because the special solution also satisfies the same initial condition. So, we have the two solutions to the Cauchy problem: $\begin{cases} x = e^p(p + 1) \\ y = p^2 e^p \end{cases}$ and $y = 0$.

7. $y' \sin y' + \cos y' - y = 0$.

Introducing $p = y'$, we get the equivalent system $\begin{cases} y' = p \\ y = p \sin p + \cos p \end{cases}$. Differentiating the second equation with respect to x, we obtain the separable equation for the function p: $p = p \cos p\, p'$. Simplifying, we get the equation $p' \cos p = 1$ and the additional relation $p = 0$. Solving the equation, we have $x = \sin p + C$ and, combining this expression with the second equation of the system, we find the general solution of the original equation in parametric form: $\begin{cases} x = \sin p + C \\ y = p \sin p + \cos p \end{cases}$. The relation $p = 0$, substituted into the second equation of the system, generates another (special) solution $y = 1$.

8. $y'^2 + (x + a)y' - y = 0$, $a \in \mathbb{R}$.

For the parameter $p = y'$, we get the equivalent system $\begin{cases} y' = p \\ y = p^2 + (x+a)p \end{cases}$.
Differentiating the second equation with respect to x, we obtain the explicit equation for the function p: $p = 2pp' + (x+a)p' + p$ or simplifying $(2p + x + a)p' = 0$. Then we have two options: $p = C$ and $2p = -x - a$. Joining the first relation with the second equation of the system, we find the general solution of the original equation: $y = C^2 + (x+a)C$. Using the second relation, we arrive at one more function $\begin{cases} 2p = -x - a \\ y = p^2 + (x+a)p \end{cases}$ which can be rewritten in explicit form

$$ y = \frac{(x+a)^2}{4} - (x+a)\frac{x+a}{2} = -\frac{(x+a)^2}{4}. $$

Substituting this function into the original equation, we see that this is another solution, not included in the general solution.

9. $x^4 y'^2 - xy' - y = 0$.

Using the parameter $p = y'$, we get the equivalent system $\begin{cases} y' = p \\ y = x^4 p^2 - xp \end{cases}$.
Differentiating the second equation with respect to x, we obtain an explicit equation for the function p: $p = 4x^3 p^2 + 2x^4 pp' - p - xp'$ or, simplifying and regrouping the terms, $xp'(2x^3 p - 1) = 2p(1 - 2x^3 p)$. So, we have two options: $xp' = -2p$ and $2x^3 p = 1$. The solution to the first equation is $p = \frac{C}{x^2}$. Combining it with the second equation of the system, we find the general solution of the original equation: $y = x^4 \frac{C^2}{x^4} - x\frac{C}{x^2} = C^2 - \frac{C}{x}$. Using the second relation, we arrive at one more function $y = x^4 \frac{1}{4x^6} - x\frac{1}{2x^3} = -\frac{1}{4x^2}$. Substituting this function into the original equation, we see that this is another solution, not included in the general solution.

10. $xy'^2 + xy' - y = 0$, $y(1) = 2$.
This is the Cauchy problem for the equation explicit in y. Using the parameter $p = y'$, we get the equivalent system $\begin{cases} y' = p \\ y = xp^2 + xp \end{cases}$. Differentiating the second equation with respect to x, we obtain a separable equation for the function p: $p = p^2 + 2xpp' + p + xp'$ or, simplifying and regrouping the terms, $xp'(2p + 1) = -p^2$. Separating variables and integrating, we obtain

$$ \int \frac{2p + 1}{p^2} dp = 2\ln|p| - \frac{1}{p} = -\ln|x| + C $$

or $x = \frac{C}{p^2} e^{1/p}$. Combining this function with the second equation of the system, we find the general solution of the original equation: $\begin{cases} x = \frac{C}{p^2} e^{1/p} \\ y = xp^2 + xp \end{cases}$.

In the separation of variables, the function $p = 0$ has been eliminated from consideration. Substituting $p = 0$ into the second equation of the system, we have $y = 0$, which is another solution, not included in the general solution.

Substituting the initial condition into the general solution, we obtain
$$\begin{cases} 1 = \frac{C}{p^2}e^{1/p} \\ 2 = p^2 + p \end{cases}.$$ Solving the second relation for unknown p, we find the two
roots $p = -2$ and $p = 1$. If $p = -2$, the first relation determines the constant $C = 4\sqrt{e}$. For the second root $p = 1$, the first relation gives the value $C = e^{-1}$. We also note that the special solution $y = 0$ does not satisfy the initial condition. Thus, we have two solutions to the Cauchy problem, both
contained in the general solution: $\begin{cases} x = \frac{4\sqrt{e}}{p^2}e^{1/p} \\ y = xp^2 + xp \end{cases}$ and $\begin{cases} x = \frac{e^{-1}}{p^2}e^{1/p} \\ y = xp^2 + xp \end{cases}$.

Exercises for section 2.1

Solve the following explicit with respect to y equations:
1. $y = y'^2 + 2y'^3$.
2. $y = \ln(1 + y'^2)$.
3. $y = (y' - 1)e^{y'}$.
4. $y'^4 - y'^2 = y^2$.
5. $5y + y'^2 = x(x + y')$.
6. $y'^3 + y^2 = xyy'$.
7. $y'\sin y' + \cos y' - y = 0$.
8. $x^4 y'^2 - xy' - y = 0$.
9. $y = xy' - x^2 y'^3$.
10. $y = 2xy' + y^2 y'^3$.

2.2   Lagrange and Clairaut equations

Definition. The explicit equation with respect to y that has the form

$$y = xg(y') + h(y'), g(p) \neq p$$

is called *Lagrange equation.*

Method of solution.

The main distinction of the Lagrange equation among other equations explicit with respect to y is the guaranteed possibility to solve this equation using the technique presented below. As with any explicit in y equation, we start with introduction of the parameter $p = y'$ and representation of the original equation in the form of the equivalent system $\begin{cases} y' = p \\ y = xg(p) + h(p) \end{cases}$.

Differentiating the second equation of the system with respect to x, we get $p = g(p) + xg_p(p)p' + h_p(p)p'$. At this point, to ensure that the last equation can be solved using the known techniques, we interchange the independent variable x and the unknown function p, which leads to the equation

$$x'(p - g(p)) = xg_p(p) + h_p(p) \quad \text{or} \quad x' = x\frac{g_p(p)}{p - g(p)} + \frac{h_p(p)}{p - g(p)}$$

(recall that $g(p) \neq p$). In this way, we have found the linear equation with respect to $x(p)$ whose method of solution is already known. After finding the general solution of this equation $x = F(p, C)$, we combine it with the second equation of the system and obtain the solution of the original equation in parametric form $\begin{cases} x = F(p, C) \\ y = xg(p) + h(p) \end{cases}$. If possible, the parameter p is eliminated from the system to obtain a simpler form of the solution.

Definition. The explicit equation with respect to y that has the form

$$y = xy' + h(y')$$

is called *Clairaut equation*. Notice that this equation deals with the case $g(p) = p$ which was omitted in the definition of the Lagrange equation.

Method of solution.
Algorithm of solution of Clairaut equation is even simpler. Again, we start by introducing the parameter $p = y'$ and rewriting the original equation in the form of the equivalent system $\begin{cases} y' = p \\ y = xp + h(p) \end{cases}$. Differentiating the second equation of the system with respect to x, we get $p = p + xp' + h_p(p)p'$ or $p'(x + h_p(p)) = 0$. The first option is $p' = 0$ that gives $p = C$ and the general solution of the primitive equation in the explicit form $y = Cx + h(C)$. The second option is $x + h_p(p) = 0$ (which is not a differential equation, since x is an independent variable and $h(p)$ is a given function). Together with the second equation of the system, this relationship can give another solution to Clairaut equation.

Examples.

1. $y = 2xy' + \ln y'$.
This is the Lagrange equation with $g(p) = 2p$ and $h(p) = \ln p$. Using the parameter $p = y'$, we write the equivalent system $\begin{cases} y' = p \\ y = 2xp + \ln p \end{cases}$. Differentiating the second equation of the system with respect to x, we get

$$p = 2p + 2xp' + \frac{1}{p}p' \quad \text{or} \quad -p = 2xp' + \frac{1}{p}p'.$$

Interchanging x and p, we have the linear equation for $x(p)$: $px' = -2x - \frac{1}{p}$. Solving the homogeneous part $px' = -2x$, we obtain $x = \frac{C}{p^2}$. Substituting the function $x = \frac{C(p)}{p^2}$ into the complete linear equation, we get the equation $C'(p) = -1$ whose solution is $C(p) = -p + A$ and, consequently, $x = -\frac{1}{p} + \frac{A}{p^2}$. Therefore, the general solution of the primitive equation is

$$\begin{cases} x = -\frac{1}{p} + \frac{A}{p^2} \\ y = 2xp + \ln p \end{cases}.$$

2. $y = xy' - y'^2$.

This is the Clairaut equation. By introducing the parameter $p = y'$, we have the equivalent system $\begin{cases} y' = p \\ y = xp - p^2 \end{cases}$. Differentiating the second equation of the system, we get

$$p = p + xp' - 2pp' \quad \text{or} \quad p'(x - 2p) = 0.$$

The solution of $p' = 0$ gives $p = C$ and the corresponding general solution of the original equation: $y = Cx - C^2$. The relation $x = 2p$ generates another function $y = x\frac{x}{2} - \frac{x^2}{4} = \frac{x^2}{4}$. Substituting it into the original equation, we see that it is another solution of this equation.

3. $y = 2xy' - y'^2$.

This is the Lagrange equation with $g(p) = 2p$ and $h(p) = -p^2$. Using the parameter $p = y'$, we have the equivalent system $\begin{cases} y' = p \\ y = 2xp - p^2 \end{cases}$. Differentiating the second equation of the system with respect to x, we get

$$p = 2p + 2xp' - 2pp' \quad \text{or} \quad p = 2p'(p - x).$$

Interchanging x and p, we get the linear equation for $x(p)$: $x' = -\frac{2}{p}x + 2$. Solving the homogeneous part $x' = -\frac{2}{p}x$, we obtain $x = \frac{C}{p^2}$. Substituting the function $x = \frac{C(p)}{p^2}$ into the complete linear equation, we get the equation $\frac{C'}{p^2} = 2$ or $C' = 2p^2$ whose solution is $C(p) = \frac{2p^3}{3} + A$, and consequently, $x = \frac{1}{p^2}\left(\frac{2p^3}{3} + A\right) = \frac{2p}{3} + \frac{A}{p^2}$. So, the general solution of the primitive equation is $\begin{cases} x = \frac{2p}{3} + \frac{A}{p^2} \\ y = 2xp - p^2 \end{cases}$. In the course of the solution, we divided by p, so we have to test whether $p = 0$ generates another solution. Substitution of $p = 0$ into the second equation of the system gives $y = 0$, which is the solution of the original equation (not included in the general solution).

4. $y' + y = xy'^2$.

This is the Lagrange equation with $g(p) = p^2$ and $h(p) = -p$. Using the parameter $p = y'$, we have the equivalent system $\begin{cases} y' = p \\ y = xp^2 - p \end{cases}$. Differentiating the second equation of the system with respect to x, we get

$$p = p^2 + 2xpp' - p' \quad \text{or} \quad p(1-p) = p'(2xp - 1).$$

Interchanging x and p, we arrive at the linear equation for $x(p)$: $x' = \frac{2}{1-p}x - \frac{1}{p(1-p)}$. Solving the homogeneous part $x' = \frac{2}{1-p}x$, we obtain $x = \frac{C}{(p-1)^2}$. Substituting the function $x = \frac{C(p)}{(p-1)^2}$ into the complete linear equation, we get the equation $\frac{C'}{(p-1)^2} = \frac{1}{p(p-1)}$ or $C' = \frac{p-1}{p}$ whose solution is $C(p) = p - \ln|p| + A$, and consequently, $x = \frac{p-\ln|p|+A}{(p-1)^2}$. So, the general solution of the primitive equation is $\begin{cases} x = \frac{p-\ln|p|+A}{(p-1)^2} \\ y = xp^2 - p \end{cases}$. In the process of solution, it was necessary to divide by p and $p - 1$, so we have to test whether $p = 0$ and $p = 1$ generate other solutions. Substitution of $p = 0$ and $p = 1$ into the second equation of the system gives $y = 0$ and $y = x - 1$, respectively. Both of these functions are solutions of the original equation (not included in the general solution).

5. $y = x(1 + y') + y'^3$.

To solve this Lagrange equation, we introduce the parameter $p = y'$ and arrive at the equivalent system $\begin{cases} y' = p \\ y = x(1+p) + p^3 \end{cases}$. Differentiating the second equation of the system with respect to x, we get

$$p = 1 + p + xp' + 3p^2p' \quad \text{or} \quad p'(x + 3p^2) = -1.$$

Interchanging x and p, we have the linear equation for $x(p)$: $x' = -x - 3p^2$. Solving the homogeneous part $x' = -x$, we obtain $x = Ce^{-p}$. Substituting the function $x = C(p)e^{-p}$ into the complete linear equation, we get the equation $C'e^{-p} = -3p^2$ or $C' = 3p^2e^p$ whose solution is $C(p) = (3p^2 - 6p + 6)e^p + A$, and consequently,

$$x = [(3p^2 - 6p + 6)e^p + A]e^{-p} = 3p^2 - 6p + 6 + Ae^{-p}.$$

So, the general solution of the given equation is

$$\begin{cases} x = 3p^2 - 6p + 6 + Ae^{-p} \\ y = x(1+p) + p^3 \end{cases}.$$

6. $2y(y' + 2) = xy'^2$.

To solve this Lagrange equation, we introduce the parameter $p = y'$ and set up the equivalent system $\begin{cases} y' = p \\ 2y = \frac{xp^2}{p+2} \end{cases}$. Differentiating the second equation

of the system with respect to x, we get

$$2p = \frac{(p^2+2xpp')(p+2)-xp^2p'}{(p+2)^2} \quad \text{or} \quad 2p(p+2)^2 = p^2(p+2)+p'(2xp(p+2)-xp^2).$$

Simplifying and regrouping, we arrive at $(p+1)(p^2+4p) = p'x(p^2+4p)$. Therefore, we have the separable equation $p'x = p+2$ and the relation $p^2+4p = 0$. In this case, to solve the equation there is no need to interchange x and p. Separating variables and integrating, we get $\ln|p+2| = \ln|x| + C$ or $p+2 = Cx$. Combining this function with the second equation of the system, we obtain the general solution of the primitive equation: $2y = \frac{(Cx-2)^2}{C}$. The additional relation $p^2+4p = 0$ results in $p = 0$ and $p = -4$. Substituting these values into the second equation of the system, we get two more functions $y = 0$ and $y = -4x$. Both are solutions of the original equation, not included in the general solution.

7. $y = x + y'^2 - y'$.

To solve this Lagrange equation, we introduce the parameter $p = y'$ and set up the equivalent system $\begin{cases} y' = p \\ y = x + p^2 - p \end{cases}$. Differentiating the second equation of the system with respect to x, we get

$$p = 1 + 2pp' - p' \quad \text{or} \quad p - 1 = p'(2p - 1).$$

This is the separable equation, and consequently, there is no need to interchange x and p. Separating variables $\frac{2p-1}{p-1}dp = dx$ and integrating, we get $x = 2p+\ln|p-1|+C$. Together with the second equation of the system, this gives the general solution of the primitive equation: $\begin{cases} x = 2p + \ln|p-1| + C \\ y = x + p^2 - p \end{cases}$.

In the separation of variables, it was necessary to divide by $p-1$ and, consequently, the value $p = 1$ was eliminated from the general solution. This relation, substituted into the second equation of the system, generates another function $y = x$, which is a special solution.

8. $8y'^3 - 12y'^2 = 27(y - x)$.

We put $27y$ separately $27y = 8y'^3 - 12y'^2 + 27x$ and solve this Lagrange equation using the standard method. We introduce the parameter $p = y'$ and set up the equivalent system $\begin{cases} y' = p \\ 27y = 8p^3 - 12p^2 + 27x \end{cases}$. Differentiating the second equation with respect to x, we obtain

$$27p = 24p^2p' - 24pp' + 27 \quad \text{or} \quad 27(p - 1) = 24p'p(p - 1).$$

To solve the separable equation $8p'p = 9$ there is no need to interchange x and p. Integrating, we get $4p^2 = 9x + C$. Together with the second equation of the system, this gives the general solution of the primitive equation:
$\begin{cases} 9x = 4p^2 + C \\ 27y = 8p^3 - 12p^2 + 27x \end{cases}$. The additional relation $p - 1 = 0$ leads to one more function $27y = -4 + 27x$, which is a special solution of the given equation.

9. $\sqrt{y'^2 + 1} + xy' - y = 0$.
This is the Clairaut equation. By introducing the parameter $p = y'$, we get the equivalent system $\begin{cases} y' = p \\ y = xp + \sqrt{p^2 + 1} \end{cases}$. Differentiating the second equation of the system, we obtain

$$p = p + xp' + \frac{pp'}{\sqrt{p^2 + 1}} \quad \text{or} \quad p'\left(x + \frac{p}{\sqrt{p^2 + 1}}\right) = 0.$$

The solution of $p' = 0$ is $p = C$ and the corresponding general solution of the original equation is $y = Cx + \sqrt{C^2 + 1}$. The relation $x = -\dfrac{p}{\sqrt{p^2+1}}$ generates another function, whose explicit form is $y = \sqrt{1 - x^2}$. Substituting it into the original equation, we see that this is another solution of the given equation (not included in the general solution).

10. $y = xy' + y'^2$.
This is the Clairaut equation. By introducing the parameter $p = y'$, we get the equivalent system $\begin{cases} y' = p \\ y = xp + p^2 \end{cases}$. Differentiating the second equation of the system, we obtain

$$p = p + xp' + 2pp' \quad \text{or} \quad p'(x + 2p) = 0.$$

The solution of $p' = 0$ is $p = C$ and the general solution of the original equation is $y = Cx + C^2$. The relation $x = -2p$ generates one more function $y = x\frac{-x}{2} + \frac{x^2}{4} = -\frac{x^2}{4}$, which is a special solution of the original equation.

11. $y = xy' + \sqrt{1 - y'^2}$.
This is the Clairaut equation. By introducing the parameter $p = y'$, we get the equivalent system $\begin{cases} y' = p \\ y = xp + \sqrt{1 - p^2} \end{cases}$. Differentiating the second equation of the system, we obtain

$$p = p + xp' - \frac{pp'}{\sqrt{1 - p^2}} \quad \text{or} \quad p'\left(x - \frac{p}{\sqrt{1 - p^2}}\right) = 0.$$

The solution of $p' = 0$ is $p = C$ and the corresponding general solution of the original equation is $y = Cx + \sqrt{1 - C^2}$. The relation $x = \frac{p}{\sqrt{1-p^2}}$ generates another function, which can be expressed in the implicit form $y^2 - x^2 = 1$. Substituting it into the original equation, we see that this is a special solution of the given equation.

12. $x = \frac{y}{y'} + \frac{1}{y'^2}$.

This is the Clairaut equation for the function $x(y)$. In fact, by interchanging x and y, we get the explicit Clairaut equation with respect to x: $x = yx' + x'^2$. Introducing the parameter $p = x'$, we obtain the equivalent system $\begin{cases} x' = p \\ x = yp + p^2 \end{cases}$. Differentiating the second equation of the system with respect to independent variable y, we get

$$p = p + yp' + 2pp' \quad \text{or} \quad p'(y + 2p) = 0.$$

The solution of $p' = 0$ is $p = C$ and the general solution of the original equation is $x = Cy + C^2$. The relation $y = -2p$ generates another function $x = y\frac{-y}{2} + \frac{y^2}{4} = -\frac{y^2}{4}$, which is a special solution of the original equation.

Exercises for section 2.2

Solve the following Lagrange and Clairaut equations:
1. $2yy' = x(y'^2 + 4)$.
2. $y = -xy' + y'^2$.
3. $y = -xy' - a\sqrt{1 + y'^2}$, $a \in \mathbb{R}$.
4. $y'^3 + xy'^2 - y = 0$.
5. $y'^2 - 2xy' + y = 0$.
6. $y + xy' = 4\sqrt{y'}$.
7. $xy' - y = \ln y'$.
8. $xy'(y' + 2) = y$.
9. $2y'^2(y - xy') = 1$.
10. $y = xy' - (2 + y')$.

3 Explicit equations in x

Definition. An *explicit equation with respect to independent variable x* has the form

$$x = f(y, y').$$

Method of solution.

We introduce the parameter $p = y'$ and rewrite the original equation in the form of the equivalent system $\begin{cases} y' = p \\ x = f(y,p) \end{cases}$. From this point on, the algorithm of solution follows a slightly different path than that applied to explicit equations with respect to y. Now we interchange the meaning of x and y, and differentiate the second equation of the system with respect to y. Then, we obtain $x' = \frac{1}{p} = f_y + f_p p_y$ which is the explicit first-order equation for the function $p(y)$. If it is possible to find the solution (using one of the known methods), then we have the relation $G(p, y, C) = 0$, which we combine with the second equation of the system to obtain the parametric form of the general solution of the original equation $\begin{cases} G(p, y, C) = 0 \\ x = f(y,p) \end{cases}$. If possible, the parameter p is eliminated from the system to obtain a simpler form of the solution.

Remark. The algorithm of derivation of an explicit equation for the new function $p(y)$ requires the elimination of x, since this is the unknown function of y. Therefore, after introduction of the parameter p, the original equation must be written in the form where x is set separately: $x = f(y, p)$. Only in this case the function $x(y)$ will disappear after differentiation (its derivative will be replaced by $\frac{1}{p}$). If the original equation has the form $g(y)x = f(y, y')$, it must be normalized to the form $x = \frac{f(y,y')}{g(y)}$ before applying the described method of solution (unless $g(y)$ is a constant).

Examples.

1. $y'^3 - 2xyy' + 4y^2 = 0$.
This equation can be considered polynomial for y', but treating it this way is technically complicated. It is much simpler to solve the equation as explicit with respect to x: $2x = \frac{y'^3 + 4y^2}{yy'}$. Using the parameter $p = y'$, we set up the equivalent system $\begin{cases} y' = p \\ 2x = \frac{p^3 + 4y^2}{yp} \end{cases}$. Interchanging x and y, we differentiate the second equation with respect to y to get

$$2\frac{1}{p} = \frac{2p'p^3 y + 4y^2 p - p^4 - 4y^3 p'}{y^2 p^2}.$$

Simplifying and regrouping the terms, we obtain $(2p'y - p)(p^3 - 2y^2) = 0$. Solving the equation $2p'y - p = 0$, we have $p^2 = Cy$. Using this relationship together with the second expression of the system, we get the general solution in parametric form $\begin{cases} p^2 = Cy \\ 2x = \frac{p^3 + 4y^2}{yp} \end{cases}$. The parameter can still be eliminated, leading to the explicit form $y = \frac{C(2x-C)^2}{16}$ (rewrite the second expression in the form $2xy = p^2 + \frac{4y^2}{p} = Cy + \frac{4y^2}{p}$; then isolate p by regrouping $2xy - Cy = \frac{4y^2}{p}$ and cut y to obtain $2x - C = \frac{4y}{p}$; square both sides of the last relation and substitute p^2 for y once again: $(2x - C)^2 = \frac{16y^2}{Cy}$ and then $16y = C(2x - C)^2$). The second relation $p^3 - 2y^2 = 0$ together with $2x = \frac{p^3 + 4y^2}{yp}$ generates another function $2x = \frac{2y^2 + 4y^2}{y(2y^2)^{1/3}} = \frac{6}{2^{1/3}}y^{1/3}$ or $27y = 2x^3$. Substitution of this function into the primitive equation shows that it is another (special) solution of the given equation.

2. $x = \ln y' + \sin y'$.

This is the explicit equation with respect to x. Using the parameter $p = y'$, we set up the equivalent system $\begin{cases} y' = p \\ x = \ln p + \sin p \end{cases}$. Interchanging x and y, we differentiate the second equation with respect to y to get $\frac{1}{p} = \frac{1}{p}p' + \cos p\, p'$. Considering this equation for the function $y(p)$ (making p an independent variable and y an unknown function), we obtain the equation $y_p = 1 + p \cos p$, whose general solution is $y = p + p \sin + \cos p + C$. This relation together with the second equation of the system represent the general solution of the primitive equation: $\begin{cases} y = p + p \sin + \cos p + C \\ x = \ln p + \sin p \end{cases}$.

3. $y'^3 - 4xyy' + 8y^2 = 0$.

This equation is polynomial with respect to y', but solving it as such is technically complicated due to the expressions for the roots of the cubic equation. Therefore, it is simpler to solve this equation as explicit with respect to x. Rewriting the equation in the form $x = \frac{y'^3 + 8y^2}{4yy'}$ and using the parameter $p = y'$, we set up the equivalent system $\begin{cases} y' = p \\ x = \frac{p^3 + 8y^2}{4yp} \end{cases}$. Interchanging x and y, we differentiate the second equation with respect to y to get

$$\frac{1}{p} = \frac{3p^2 p' + 16y)yp - (p^3 + 8y^2)(p + yp')}{4y^2 p^2},$$

or simplifying, $4y^2 p = 2yp^3 p' - 8y^3 p' + 8y^2 p - p^4$. Simplifying once again and factoring, we arrive at $(2yp' - p)(p^3 - 4y^2) = 0$. First, we solve the equation $2yp' - p = 0$ and find $p^2 = Cy$. Using this relationship together with the second expression of the system, we obtain the general solution in parametric form $\begin{cases} p^2 = Cy \\ x = \frac{p^3 + 8y^2}{4yp} \end{cases}$. The parameter can still be eliminated, arriving at the implicit form $x = \frac{p^2}{4y} + \frac{2y}{p} = \frac{C}{4} + \frac{2\sqrt{y}}{\sqrt{C}}$. By isolating y, we can also obtain the explicit form of the general solution $y = A(x - A)^2$ (here $A = \frac{C}{4}$). Second, we consider the relation $p^3 = 4y^2$ together with $x = \frac{p^3 + 8y^2}{4yp}$, which represent another parametric function. Eliminating p, we get $x = \frac{3}{4^{1/3}} y^{1/3}$ or, isolating y: $y = \frac{4}{27} x^3$. Substituting this function into the primitive equation, we see that this is one more (special) solution of the given equation.

4. $x = \frac{y}{y'} \ln y - \frac{y'^2}{y^2}$, $y(0) = 1$.

This is the Cauchy problem for explicit in x equation. Using the parameter $p = y'$, we set up the equivalent system $\begin{cases} y' = p \\ x = \frac{y \ln y}{p} - \frac{p^2}{y^2} \end{cases}$. Interchanging x and y, we differentiate the second equation in y and obtain the equation for the function $p(y)$:

$$\frac{1}{p} = \frac{\ln y + y\frac{1}{y}}{p} - \frac{y \ln y}{p^2} p' - \frac{2pp'}{y^2} + \frac{2p^2}{y^3}.$$

Simplification and rearrangement of the terms leads to $p' \left(\frac{y \ln y}{p^2} + \frac{2p}{y^2} \right) = \frac{p}{y} \left(\frac{y \ln y}{p^2} + \frac{2p}{y^2} \right)$. So, we have two options. First, we solve the equation $p' = \frac{p}{y}$ and find $p = Cy$. Substituting this expression into the second equation of the system, we get the general solution $x = \frac{y \ln y}{Cy} - \frac{C^2 y^2}{y^2} = \frac{\ln y}{C} - C^2$. Second, we consider the relation $\frac{y \ln y}{p^2} + \frac{2p}{y^2} = 0$ which can be written in the form $p^3 = -\frac{y^3 \ln y}{2}$. Together with the second equation of the system, this produces another function

$$x = y \ln y \cdot \left(\frac{-2}{y^3 \ln y} \right)^{1/3} - \frac{1}{y^2} \cdot \left(\frac{y^3 \ln y}{2} \right)^{2/3}$$

$$= (-2)^{1/3} (\ln y)^{2/3} - \frac{1}{2^{2/3}} (\ln y)^{2/3} = -\frac{3}{2^{2/3}} (\ln y)^{2/3}.$$

Substitution of this function into the primitive equation shows that this is one more (special) solution of the given equation.

Substituting the initial condition into the general solution, we get $0 = \frac{\ln 1}{C} - C^2$. Formally, the first term vanishes, and from the equation $0 = -C^2$ we get $C = 0$, but the constant C is in the denominator of the first term of the general solution, and consequently, cannot be zero. Therefore, there is no solution to the problem among the functions of the general solution. However, we note that the additional solution satisfies the initial condition: $0 = -\frac{3}{2^{2/3}} (\ln 1)^{2/3}$. Thus, the only solution to the Cauchy problem is $x = -\frac{3}{2^{2/3}} (\ln y)^{2/3}$.

5. $y' + \sin y' - x = 0$, $y(0) = 0$.

This is the Cauchy problem for explicit in x equation. Using the parameter $p = y'$, we get the equivalent system $\begin{cases} y' = p \\ x = p + \sin p \end{cases}$. Interchanging x and y, we differentiate the second equation with respect to y and obtain the separable equation for the function $p(y)$: $\frac{1}{p} = p' + \cos p \cdot p'$. Simplifying and integrating, we get

$$y = \int (p + p\cos p)dp = \frac{p^2}{2} + p\sin p + \cos p + C.$$

So, the general solution of the original equation in parametric form is

$$\begin{cases} y = \frac{p^2}{2} + p\sin p + \cos p + C \\ x = p + \sin p \end{cases}.$$

Substituting the initial condition into the general solution, we obtain $\begin{cases} 0 = \frac{p^2}{2} + p\sin p + \cos p + C \\ 0 = p + \sin p \end{cases}$. The second relation determines the specific values of p that must be used in the first to specify the constant C. Obviously, $p = 0$ satisfies the second relation. We also note that the function $x(p) = p + \sin p$ is strictly increasing, because its derivative $x_p = 1 + \cos p$ is positive at all points except the isolated points $p = \frac{3\pi}{2} + 2k\pi$, $k \in \mathbb{Z}$, where the derivative vanishes. Therefore, $p = 0$ is the only root of the second relation. Using this value in the first equation of the system, we have $C = -1$. Therefore, the only solution to the Cauchy problem is the function

$$\begin{cases} y = \frac{p^2}{2} + p\sin p + \cos p - 1 \\ x = p + \sin p \end{cases}.$$

6. $x^2 y'^2 - 2xyy' - x^2 = 0$.

First, we divide the equation by x and set x separately in the obtained equation: $x = \frac{2yy'}{y'^2 - 1}$. We solve this equation as explicit for x. Using the parameter $p = y'$, we get the equivalent system $\begin{cases} y' = p \\ x = \frac{2yp}{p^2 - 1} \end{cases}$. Interchanging x and y, we differentiate the second equation with respect to y and obtain the explicit equation for $p(y)$:

$$\frac{1}{p} = \frac{(2p + 2yp')(p^2 - 1) - 2yp \cdot 2pp'}{(p^2 - 1)^2}.$$

Simplifying and regrouping the terms, we get

$$\frac{(p^2 - 1)^2}{p} - 2p(p^2 - 1) = 2yp'(p^2 - 1) - 4yp^2 p'.$$

Again simplifying, we have

$$\frac{p^4 - 1}{p} = 2yp'(p^2 + 1) \quad \text{or} \quad \frac{(p^2 - 1)(p^2 + 1)}{p} = 2yp'(p^2 + 1).$$

Cutting $p^2 + 1$, we get the separable equation $2yp' = \frac{p^2 - 1}{p}$. Separating variables and integrating, we find $y = C(p^2 - 1)$. Thus, the general solution of the original equation has the form $\begin{cases} y = C(p^2 - 1) \\ x = \frac{2yp}{p^2 - 1} \end{cases}$. By performing a few algebraic manipulations, we can transform parametric form of the solution into explicit $y = C(\frac{x^2}{4C^2} - 1)$ (substitute $p^2 - 1 = \frac{y}{C}$ into the formula for x and get $x = \frac{2yp}{y/C} = 2Cp$; then, insert $p = \frac{x}{2C}$ in the formula for y and find $y = C(\frac{x^2}{4C^2} - 1)$).

7. $y' \ln \frac{y'}{4} = 4x$.

This equation is explicit with respect to x and it can be found in Table 1 in section 2 of Chapter 1 under number 8. Using the parameter $p = y'$, we set up the equivalent system $\begin{cases} y' = p \\ 4x = p \ln \frac{p}{4} \end{cases}$. Considering x as an unknown function of y, we differentiate the second equation with respect to y and obtain the explicit equation for $p(y)$: $\frac{4}{p} = p' \ln \frac{p}{4} + p'$. Interchanging y and p, we find a separable equation for the function $y(p)$: $4y' = p \ln \frac{p}{4} + p$. It remains to integrate the left-hand side with respect to p to find $y(p)$:

$$4y = \int p \ln \frac{p}{4} + p \, dp = \frac{p^2}{2} + \frac{p^2}{2} \ln \frac{p}{4} - \int \frac{p^2}{2} \frac{1}{p} dp = \frac{p^2}{4} + \frac{p^2}{2} \ln \frac{p}{4} + C.$$

Together with the second equation of the system, it gives the general solution: $\begin{cases} y = \frac{p^2}{16} + \frac{p^2}{8} \ln \frac{p}{4} + C \\ 4x = p \ln \frac{p}{4} \end{cases}$. By introducing the new parameter $t = \frac{p}{4}$, we can arrive at a slightly simpler representation: $\begin{cases} y = t^2 + 2t^2 \ln t + C \\ x = t \ln t \end{cases}$.

8. $y'^2 + e^{y'} = x$.

This equation is explicit with respect to x and it can be found in Table 1 in section 2 of Chapter 1 under number 9. Using the parameter $p = y'$, we set up the equivalent system $\begin{cases} y' = p \\ x = p^2 + e^p \end{cases}$. Considering x as an unknown function of y, we differentiate the second equation with respect to y and obtain the explicit equation for $p(y)$: $\frac{1}{p} = 2pp' + e^p p'$. Interchanging y and p, we find the separable equation for the function $y(p)$: $y' = 2p^2 + pe^p$. It remains to integrate the left-hand side with respect to p to find $y(p)$:

$$y = \int 2p^2 + pe^p dp = \frac{2}{3}p^3 + (p-1)e^p + C.$$

Together with the second equation of the system, it gives the general solution:

$$\begin{cases} y = \int 2p^2 + pe^p dp = \frac{2}{3}p^3 + (p-1)e^p + C \\ x = p^2 + e^p \end{cases}.$$

Exercises for section 3

Solve the following explicit with respect to x equations:

1. $3y'^3 - xy' + 1 = 0$.
2. $x = \frac{y}{2y'} + e^{yy'}$.
3. $y'^2 - 4xyy' + 8y^2 = 0$.
4. $x = \frac{y}{y'} \ln y - \frac{y'^2}{y^2}$.
5. $e^{y'} + y' = x$.
6. $xy'^3 = 1 + y'$.
7. $y'^3 - y' = x + 1$.
8. $x = y'\sqrt{y'^2 + 1}$.

Exercises for Chapter 4

Solve the following implicit equations of the first order and the corresponding initial values problems; if possible, use different methods for the same equation and compare the obtained solutions:

1. $y = xy'^2 - 2y'^3$.
2. $y'^2 - 2yy' = y^2(e^x - 1)$, $y(0) = 1$.
3. $y'^3 - xy^4y' - y^5 = 0$.
4. $y'^2 - 2xy' = x^2 - 4y$.
5. $y'^3 = 3(xy' - y)$.
6. $x = y' \sin y'$.
7. $y = (2 + y')\sqrt{1 - y'}$.
8. $y'^2 - (y + x^2)y' + x^2y = 0$, $y(0) = 0$.
9. $y = xy'^2 - 2y'^3$.
10. $y'^4 = 2yy' + y^2$.
11. $y'^2 + 2yy' \cot x - y^2 = 0$.
12. $(3x + 5)y'^2 - (3y + x)y' + y = 0$.

Chapter 5

First order equations: applications

In this Chapter we consider different application problems, geometric and scientific (mostly physical). The focus is on the construction of differential equations based on the descriptive conditions of the problem. A solution of generated equations will be treated in abbreviated form, using techniques already studied in the previous Chapters.

1 Geometric problems

Apart from a few special cases, the construction of the differential equation from a geometric description of curves consists of the following steps. First, we fix a point and consider a given geometric property at that specific point (but in general form). Second, we translate the property, usually defined in a descriptive form, into a relation involving unknown function and its derivatives. Next, we unfix the specific point and obtain the differential equation we are looking for. The last step is to solve the derived equation, whose solution will represent the curves that satisfy the given property. Frequently, in order to better understand the conditions of the problem and be able to transform the geometric condition into a differential equation, we need to make a sketch that reflects the indicated property of the curves.

Problem 1. Find the curves whose slope at any point is three times greater than the slope of the line passing through the same point and the origin.

Solution.

First, recall that the slope of a curve (at any chosen point) is defined as the slope of the tangent line passing through that point. (The slope of a line, as usual, is a shorthand term for the tangent of the angle that this line forms with the positive direction of the x-axis.) Recall that the slope of the tangent line is given by the value of the derivative at the considered point. So, using notation $y(x)$ for the sought curves, their slope at any point x_0 is given by $y'(x_0)$. On the other hand, the line passing through the points (x_0, y_0) and $(0,0)$ has the equation $\tilde{y} = \frac{y_0}{x_0} x$ and its slope is $\frac{y_0}{x_0}$. According to the condition of the problem, $y'(x_0) = 3\frac{y_0}{x_0}$. Since this is valid for any point of the curve we are looking for, we unfix x_0 (and correspondingly y_0) and obtain the equation $y' = 3\frac{y}{x}$. Solving this separable equation, we find the family of curves $y = Cx^3$ (see Fig.5.1). Note that at the origin the found curves are defined (as well their slope which is 0), but the relationship between their slope and the indicated line is not true, because the quotient $\frac{y}{x}$ is not defined at $x = 0$.

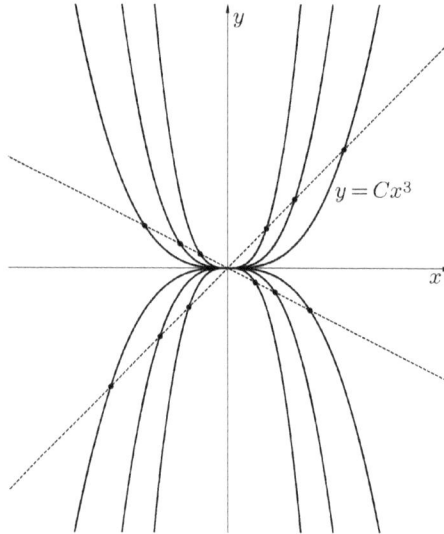

Figure 5.1: Problem 1: the found curves.

Problem 2. Find the curves with the following property: the segment of the tangent line located between the coordinate axes is bisected by the point of contact.

Solution.

Let us denote the point of contact by $P_0 = (x_0, y_0)$. According to the problem, the point P_0 divides the segment $P_1 P_2$ in half, that is, the lengths of the two segments are equal: $|P_1 P_0| = |P_0 P_2|$ (see Fig.5.2). Therefore, the segments $P_1 A$ and AO also have equal lengths: $|P_1 A| = |AO|$ (see Fig.5.2). The coordinates of the last three points are: $P_1 = (0, y_1)$, $y_1 = y_0 - y_0' x_0$ (since P_1 is the point where the y-axis intersects the tangent line $y - y_0 = y_0'(x - x_0)$), $A = (0, y_0)$ and $O = (0,0)$. Then, $|P_1 A| = |y_1 - y_0| = |y_0' x_0|$ and $|AO| = |y_0 - 0| = |y_0|$. Therefore, the condition $|P_1 A| = |AO|$ can be written as $|y_0' x_0| = |y_0|$ or $|\frac{y_0' x_0}{y_0}| = 1$.

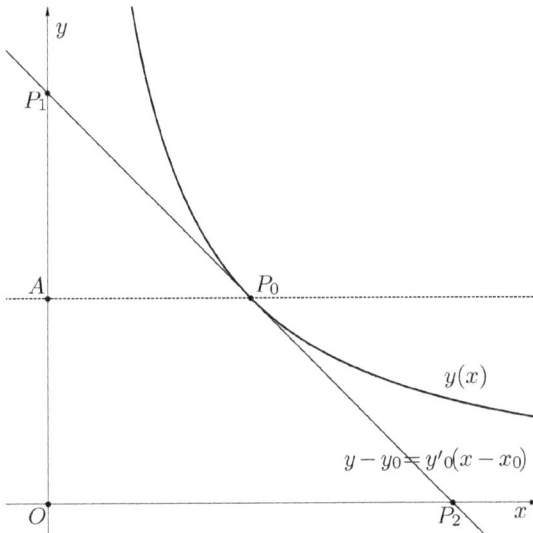

Figure 5.2: Problem 2: the geometric condition

Taking into account the location of P_0, we can also drop the sign of the absolute value. Indeed, if P_0 is in the first quadrant, then for it to fall within the segment $P_1 P_2$, the slope y_0' must be negative. This means that if $x_0 > 0, y_0 > 0$ then $y_0' < 0$, and consequently, $\frac{y_0' x_0}{y_0} < 0$. If P_0 is in the second quadrant, for it to fall within the segment $P_1 P_2$, the slope y_0' must be positive. This means that if $x_0 < 0, y_0 > 0$ then $y_0' > 0$, and consequently, $\frac{y_0' x_0}{y_0} < 0$. Similarly, we can show that $\frac{y_0' x_0}{y_0} < 0$ for any location of P_0. Therefore, the derived formula takes the form $|\frac{y_0' x_0}{y_0}| = -\frac{y_0' x_0}{y_0} = 1$. Since the property is satisfied for any point of contact, we can unfix P_0 and obtain the differential equation $\frac{y'x}{y} = -1$ or $y' = -\frac{y}{x}$. Solving this separable equation, we get the family of hyperbolas $y = \frac{C}{x}$ (see Fig.5.3).

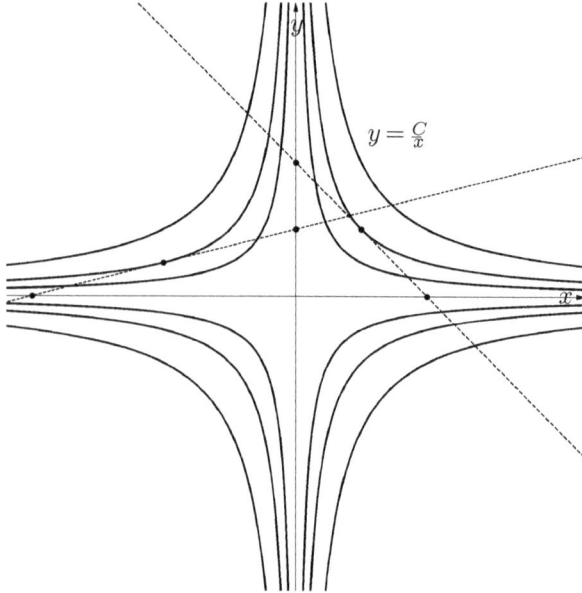

Figure 5.3: Problem 2: the found curves.

Problem 3. Find the curves such that the length of the segment between the origin and the point where the tangent line intersects the y-axis is equal to the length of the radius vector of the point of contact.

Solution.

We denote the point of contact by $P_0 = (x_0, y_0)$ and the point of intersection of the tangent line with the y-axis by $P_1 = (x_1, y_1)$ (see Fig.5.4). The abscissa x_1 equals 0, while the ordinate y_1 is found from the equation of the tangent line $y - y_0 = y_0'(x - x_0)$ by substituting $x_1 = 0$: $y_1 = y_0 - y_0'x_0$. The condition of the problem says that $|OP_1| = |OP_0|$, whence $|y_0 - y_0'x_0| = \sqrt{x_0^2 + y_0^2}$.

Unfixing P_0 and dropping the absolute value, we have the two homogeneous equations $y' = \frac{y}{x} \pm \sqrt{\frac{y^2}{x^2} + 1}$. We solve the equation with the sign $+$ following the standard procedure. By changing $z = \frac{y}{x}$, we get the separable equation $xz' = \sqrt{z^2 + 1}$, whose general solution has the form

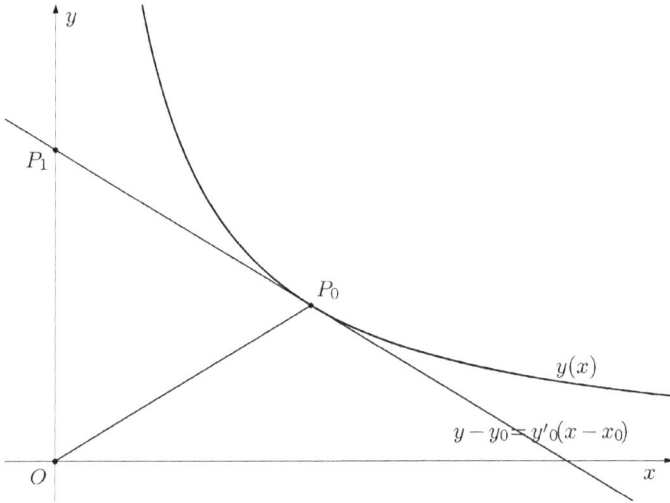

Figure 5.4: Problem 3: the geometric condition.

$\ln(z + \sqrt{z^2 + 1}) = \ln|x| + C$ (the argument of the first logarithm is positive and, therefore, does not need the absolute value). In the equivalent form, we have $\ln(z + \sqrt{z^2 + 1}) = \ln|Cx|$ or, eliminating the logarithm, $z + \sqrt{z^2 + 1} = |Cx|$. Although formally we have the two relations $z + \sqrt{z^2 + 1} = \pm Cx$, but due to arbitrariness of the constant C, it can always absorb the negative sign. So, actually, the only relationship to be considered is $z + \sqrt{z^2 + 1} = Cx$. Returning to y, we have $\frac{y}{x} + \sqrt{(\frac{y}{x})^2 + 1} = Cx$ or $\sqrt{(\frac{y}{x})^2 + 1} = Cx - \frac{y}{x}$. Squaring both sides and simplifying, we arrive at the equation of a family of parabolas $1 = C^2 x^2 - 2Cy$ or, isolating y: $y = \frac{C^2 x^2 - 1}{2C}$ (see Fig.5.5). We leave it to the reader to show that another family of curves, originated by the equation with the negative sign, also represents a set of parabolas, in the form $y = \frac{C^2 - x^2}{2C}$.

Problem 4. Find the curves whose normal lines intersect at the same point.

Solution.

We denote the point of contact by $P_0 = (x_0, y_0)$ (this is the point where the corresponding normal line passes) and the point where all the normal lines intersect by $P_1 = (x_1, y_1)$. Recall that the equation of the normal line to the graph of $y(x)$ at a point P_0 has the form $y - y_0 = -\frac{1}{y_0'}(x - x_0)$. According to the condition of the problem, the point P_1 belongs to each

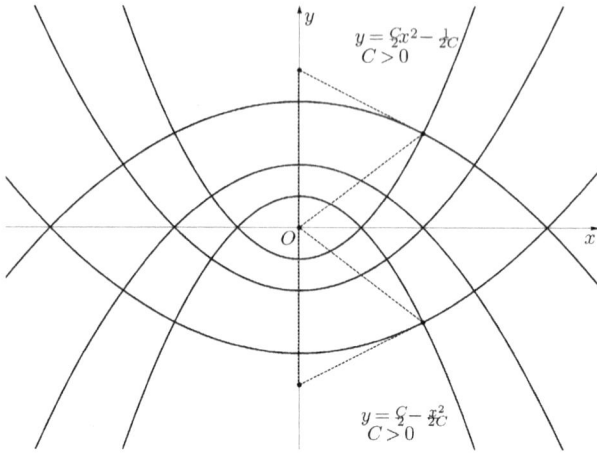

Figure 5.5: Problem 3: the found curves.

normal line, that is, the coordinates of P_1 satisfy the equation of the normal line: $y_1 - y_0 = -\frac{1}{y_0'}(x_1 - x_0)$. Since this property holds for any point P_0, we have the differential equation $y_1 - y = -\frac{1}{y'}(x_1 - x)$ or $y'(y - y_1) = -(x - x_1)$. Solving this separable equation, we find a family of circles with the centerpoint P_1: $(x - x_1)^2 + (y - y_1)^2 = C^2$ (see Fig.5.6).

Problem 5. Find the shape of a plane mirror that reflects all the rays emitted from a point source in parallel rays.

Solution.

We introduce the coordinate system in such a way that the source-emitter is set at the origin and the y-axis is oriented in the direction of the reflected rays. We denote a point on the sought curve by $P_0 = (x_0, y_0)$ and consider the following three lines: $y - y_0 = y_0'(x - x_0)$ – the line R_0 tangent to the mirror at the point P_0; $y = \frac{y_0}{x_0}x$ – the line R_1 of the ray emitted from the source; $x = x_0$ – the line R_2 of the ray reflected from the mirror (see Fig. 5.7). Recall that the tangent of the angle between two lines $A_1x + B_1y = C_1$ and $A_2x + B_2y = C_2$ is determined by the formula $\tan \varphi = \frac{A_1 B_2 - A_2 B_1}{A_1 A_2 + B_1 B_2}$. Then, the angle between the lines R_0 and R_1 is calculated by the formula $\tan \varphi_1 = \frac{-y_0' \cdot 1 + \frac{y_0}{x_0} \cdot 1}{-y_0' \cdot (-\frac{y_0}{x_0}) + 1 \cdot 1} = \frac{-y_0' x_0 + y_0}{y_0' y_0 + x_0}$, and for the angle between the lines R_0 and R_2 we have $\tan \varphi_2 = \frac{1 \cdot 1}{1 \cdot (-y_0')}$. According to the optical properties of a mirror, the angle of incidence is equal to the angle of reflection, which means

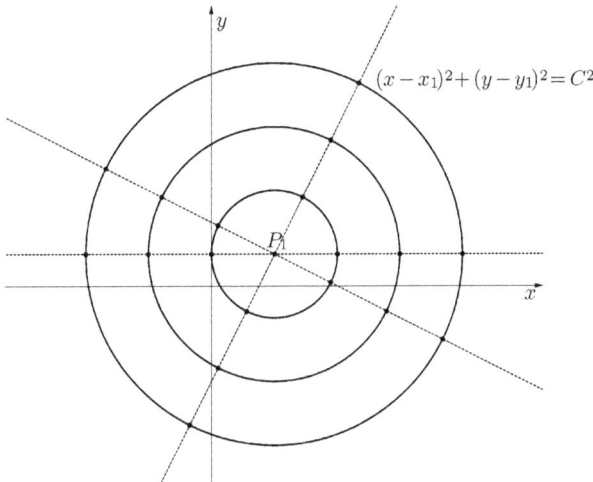

Figure 5.6: Problem 4: the found curves.

that $\tan\varphi_1 = \tan\varphi_2$, that is, $\frac{-y_0'x_0+y_0}{y_0'y_0+x_0} = \frac{-1}{y_0'}$. Since this property holds for any point P_0 of the mirror, we can unfix P_0 and arrive at the differential equation $\frac{-y'x+y}{y'y+x} = \frac{-1}{y'}$, which can be rewritten in the form $y'(y'x - y) = y'y + x$ or $y'^2 - 2\frac{y}{x}y' - 1 = 0$. This is an implicit (polynomial) equation. Solving the quadratic equation with respect to the derivative, we get the two explicit equations $y' = \frac{y}{x} \pm \sqrt{\frac{y^2}{x^2} + 1}$. The same pair of equations was already found in the course of the solution of Problem 3. So, from that moment on, we can follow the solution of that problem and find the two families of parabolas: $y = \frac{C^2x^2-1}{2C}$ and $y = \frac{C^2-x^2}{2C}$ (see Fig.5.5).

Problem 6. Find the curves that pass through the point $(1,2)$ and have the following property: the area of the triangle formed by the radius vector of a point on the curve, the tangent line at that point and the x-axis is equal to 2.

Solution.

We denote the point of contact by $P_0 = (x_0, y_0)$ and the point of intersection of the tangent line with the x-axis by $A = (x_1, y_1)$ (see Fig.5.8). The ordinate y_1 is 0 and the abscissa x_1 is found from the equation of the tangent line $y - y_0 = y_0'(x - x_0)$ by substituting $y_1 = 0$: $x_1 = x_0 - \frac{y_0}{y_0'}$. So, the base of the triangle OP_0A has the length $|OA| = |x_0 - \frac{y_0}{y_0'}|$ and its height is $|BP_0| = |y_0|$. Therefore, according to the problem, the area of the triangle

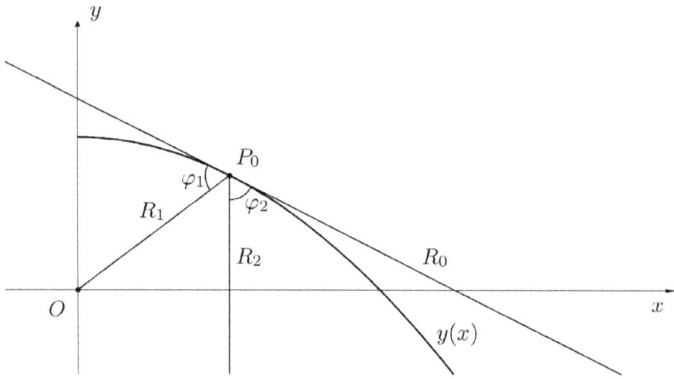

Figure 5.7: Problem 5: geometric condition.

is equal to $\frac{1}{2}|OA| \cdot |BP_0| = \frac{1}{2}|x_0 - \frac{y_0}{y_0'}| \cdot |y_0| = 2$.

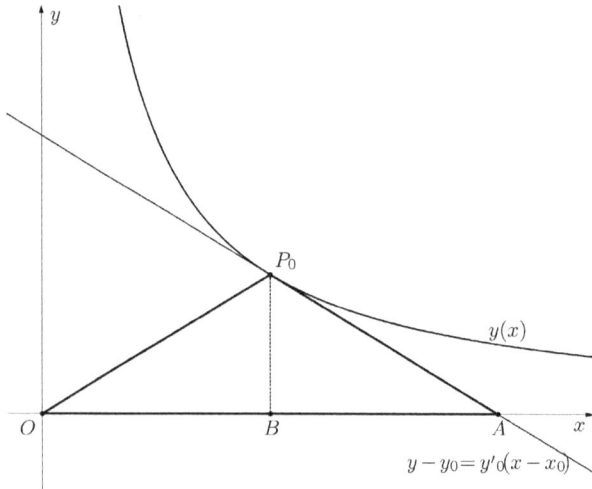

Figure 5.8: Problem 6: geometric condition.

Unfixing P_0 and dropping the absolute value, we have the two equations $(x - \frac{y}{y'}) \cdot y = \pm 4$. Interchanging the unknown function y and the independent variable x, we get the two linear equations with respect to the function $x(y)$: $x' - \frac{x}{y} = \pm \frac{4}{y^2}$. Let us solve the equation with the positive sign following the standard procedure. We start with the corresponding homogeneous equation

$x' - \frac{x}{y} = 0$ whose general solution is $x = Cy$. Next, we look for the solution of the nonhomogeneous equation in the form $x = C(y)y$, which leads to the following equation for $C(y)$: $C'y = \frac{4}{y^2}$, whence $C(y) = -\frac{2}{y^2} + C_1$, and consequently, $x(y) = C_1 y - \frac{2}{y}$. Similarly, the solution of the equation with the negative right-hand side is found in the form $x(y) = C_2 y + \frac{2}{y}$. Since the curve must pass through the point $(1, 2)$, we specify the two functions: from the first set we get $x = y - \frac{2}{y}$ and from the second $x = \frac{2}{y}$ (see Fig.5.9).

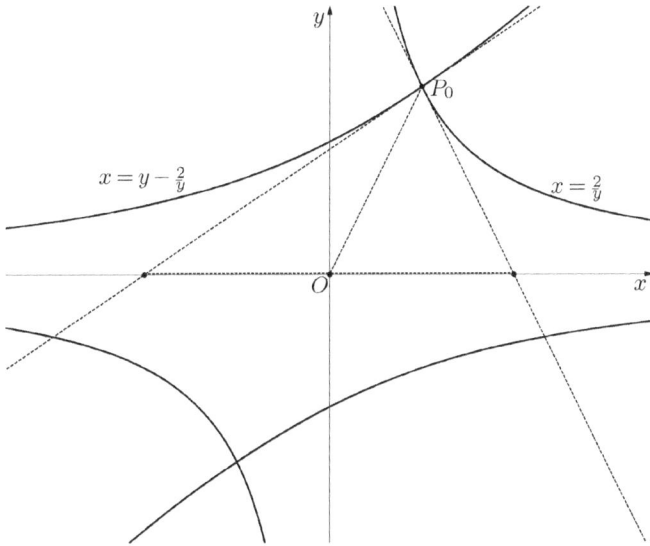

Figure 5.9: Problem 6: the found curves.

Problems for section 1

1. Find the curves that have the following property: the area of the triangle formed by the tangent line to the curve, the line perpendicular to the x-axis, which passes through the point of contact, and the x-axis is a constant quantity equal to a^2.

2. Find the curves such that the point of intersection of the tangent line and the x-axis has the same distance from the point of contact and from the origin of the coordinates.

3. Find the curves with the following property: the area of the trapezoid bounded by the coordinate axes, the tangent line and the line perpendicular to the x-axis, which passes through the point of contact, is a constant quan-

tity equal to $3a^2$.

4. Find the curves that have the following property: the area of the triangle bounded by the tangent line, the x-axis and the segment connecting the origin of the coordinates with the point of contact, is a constant quantity equal to a^2.

5. Find the curves such that the distance from the tangent line to the origin of the coordinates is equal to the abscissa of the point of contact.

6. Find the curves such that the point of intersection of the tangent line with the x-axis has the abscissa equal to $\frac{2}{3}$ of the abscissa of the point of contact.

7. Find the curves with the following property: the length of the segment of the x-axis located between the tangent line and the normal line is equal to $2a$.

8. Find the curves that pass through the point $(1, 5)$ and have the following property: the length of the segment of the y-axis located between the tangent line and the origin is three times greater than the abscissa of the point of contact.

9. Find the curves that pass through the point $(2, -1)$ and have the following property: the slope of the curve at any point is six times greater than the square of the ordinate of the point of contact.

10. Find the curves that pass through the point $(1, 2)$ and have the following property: the product of the slope and the sum of the coordinates of the contact point is twice as great as the ordinate of the point of contact.

11. Find the curves with the following property: the length of the perpendicular to the tangent line, drawn from the origin, is equal to the abscissa of the point of contact.

12. Find the curves with the following property: the segment of the tangent line located between the coordinate axes is divided in the ratio $1 : 2$ at the point of contact.

2   Science problems

In this section we consider various science problems which can be solved by constructing the corresponding differential equation. These problems can be divided into the six groups: cooling/heating body, growth of a population/species, radioactive decay, problems of motion (finding the distance/velocity/acceleration), flow of mixtures and personal finance.

2.1 Cooling/heating a body

From experiments it can be observed that the rate at which a body cools/heats in an environment, which preserves a constant temperature, is proportional to the difference between the temperature of the body and that of the environment. This relationship is called Newton's empirical law of cooling and its mathematical expression has the form $T_t = k(T - T_0)$, where $T(t)$ is the temperature of the body and T_0 is that of the environment.

Problem 1. When a cake is taken out of the oven, its temperature is 300^0C. Three minutes later the temperature drops to 200^0C. How long will it take for the cake to cool down to 50^0C, if the air temperature is 25^0C?

Solution.

According to the conditions of the problem, let us measure the temperature in degrees Celsius and the time in minutes. By Newton's law, the equation of cooling is $T_t = k(T - T_0)$, where $T(t)$ is the temperature of the cake and T_0 is the temperature of the air (considered constant). Solving the differential equation (of separable variables), we get $T = T_0 + Ce^{kt}$, where $T_0 = 25$ is a given constant, but two other parameters – k and C – must still be defined according to the additional conditions of the problem. Let us assume that $t = 0$ at the moment when the cake was taken out of the oven. So, the first additional condition can be written in the form $T(0) = 300$, and substituting it in the general solution we find $300 = 25 + Ce^0$, whence $C = 275$. The second condition corresponds to the instant $t = 3$ when $T(3) = 200$. Taking this into the expression for $T(t)$, we get $200 = 25 + 275e^{3k}$, whence $e^{3k} = \frac{175}{275} = \frac{7}{11}$ or $k = \frac{1}{3}\ln\frac{7}{11} \approx -0.151$. Now, with all the parameters of the general law specified, we can answer the question of the problem: by substituting the temperature $T = 50$ into the formula $T = T_0 + Ce^{kt}$, we find the expression for the required time: $t = \frac{1}{k}\ln\frac{T-T_0}{C} = \frac{3}{\ln\frac{7}{11}}\ln\frac{50-25}{275} = \frac{3}{\ln\frac{7}{11}}\ln\frac{1}{11} \approx 15.88$ minutes.

Problem 2. The temperature of a body was 35^0C when it was found at 5 o'clock in the morning. After an hour, the coroner took another temperature measurement, which came out at 32^0C. Assuming that the temperature at the time of death was 37^0C and the air temperature is 22^0C, find the time of death.

Solution.

According to the conditions of the problem, let us measure the temperature in degrees Celsius and the time in hours. By Newton's law, the temperature of the body $T(t)$ satisfies the equation $T_t = k(T - T_0)$, where T_0 is the temperature of the air (considered constant). Solving the differential equation (of separable variables) we get $T = T_0 + Ce^{kt}$, where $T_0 = 22$

is a given quantity, but the other two parameters – k and C – still need to be specified according to the additional conditions of the problem. Let us measure the time backward, considering that $t = 0$ at 6 o'clock in the morning when $T = 32^0C$. So $T(0) = 32 = 22 + Ce^0$, whence $C = 10$. To find k, we apply the condition that $T(1) = 35$ and, consequently, $35 = 22 + 10e^k$, whence $e^k = \frac{13}{10}$ or $k = \ln \frac{13}{10} \approx 0.262$. Now, with the parameters specified, we can answer the question of the problem, using the condition that at the required time $T(t) = 37$: $37 = 22 + 10e^{kt}$, whence $t = \frac{1}{k} \ln \frac{15}{10} \approx 1.55$. This means that the death occurred at 1.55 hours, that is, 93 minutes before 6 o'clock in the morning, that is, at 4 hours and 27 minutes.

2.2   Growth of a population/species

One of the simplest models of population dynamics is the exponential growth model (also called the Malthus model). In this model it is assumed that there are no limitations for the reproduction of a population, presuming that individuals of a given species do not face challenges to survival, which are normally found in nature in the form of restricted living space, limited food reserves, the existence of other competing species and predators, etc. Under this simplifying assumption, the rate at which the population growth is proportional to the current number of individuals of the population, that is, $y_t = ky$, where y is the total population at time t and k is the coefficient of proportionality (specific for each species).

A more real-world model takes into account that the population of a species cannot grow without limits and, normally, there is a specific number for the size of the population when it is in equilibrium with the environment (the last term here includes all the limitations of growth that are found in real life). This model is called logistic (also called the Verhulst model) and it is defined by the equation $y_t = (k - ay)y$, where k and a are two positive constants, different for each specific model. The first is responsible for growth, like in the exponential model, and the second for growth limitations: when population exceeds certain number, the first factor becomes negative and population starts to decline. In this equation, $y_e = \frac{k}{a}$ is the equilibrium value (assuming this value, the size of the population does not change anymore, since $y_t = 0$ at this point).

Problem 1. Within 1 hour, the number of bacteria doubles. Knowing that the growth rate of the number of bacteria is proportional to their current number, find the time when the initial number is tripled.

Solution.

Here we use the measure of time in hours. The corresponding equation is

$N_t = kN$, where $N(t)$ is the current number of bacteria. The solution of this separable equation is $N = N_0 e^{kt}$, where N_0 is the initial number of bacteria. Since their number has doubled after 1 hour, $2N_0 = N_0 e^k$, whence $k = \ln 2$. Therefore, the law of N can be written in the form $N = N_0 e^{\ln 2 t} = N_0 2^t$. For $N = 3N_0$ we have the relation $3N_0 = N_0 2^t$ or $t = \log_2 3 \approx 1.58$ hours.

Problem 2. According to Bertalanoff's empirical law, the growth rate of the length of a fish is proportional to the difference between the maximum length of the species and the current length of the fish. Find the length of the fish after 2 years, if its initial length was 5 cm, after 1 year it was 25 cm and the maximum length of the species is half a meter.

Solution.

In this problem, we measure length in centimeters and time in years. Translating the given law into the mathematical statement, we have $P_t = k(P_m - P)$, where $P(t)$ is the current length of the fish, P_m is its maximum length and k is the proportionality coefficient, which is different for each species. To answer the question, we first solve the equation (with separable variables) and find the general solution $P = P_m - Ce^{-kt}$. The next step is to determine the coefficients C and k. Considering the initial length at time $t = 0$, we get $5 = 50 - Ce^0$, that is, $C = 45$. The next condition, at one year, determines k: $25 = 50 - 45e^{-k}$, whence $-k = \ln \frac{5}{9}$ or $k = \ln \frac{9}{5}$. Finally, we find the asked length: $P = 50 - 45e^{-2k} = 50 - 45(\frac{5}{9})^2 \approx 36.1$ cm.

Problem 3. According to demographic data, the population of Brazil, measured in thousands of people, in the year 1960 was 70119 (that is, 70 million 119 thousand), in the year 1970 – 93139, in the year 1980 – 119071 and in the year 2000 – 169544. Using the first two measurements, make the population forecast for the years 1980 and 2000 according to the exponential model. Using the first three measurements and the logistic model, predict the population for the year 2000. Compare the results of the forecasts with the actual data. (We took the data from these years to avoid the strong influence of immigration waves that happened before that period).

Solution.

Let us denote the number of people at time t by $P(t)$ and let us measure t in years and P in thousands of people. Initial instant $t = 0$ is set to be the year 1960. According to the exponential model, we have $P_t = \alpha P$ (we use the proportionality coefficient α here to distinguish it from the coefficient k used in the logistic model). Solving this separable equation, we find $P = P_0 e^{\alpha t}$, where $P_0 = P(0) = 70119$ is the number of inhabitants at the initial moment. Using the data for the year 1970, $P_1 = P(10) = 93139$, we find that $P_1 = P_0 e^{10\alpha}$ or $e^{10\alpha} = \frac{P_1}{P_0} \approx 1.32830$, whence $\alpha = \frac{1}{10} \ln \frac{P_1}{P_0} \approx 0.0283900$.

Actually, the value of α itself is not necessary to make a forecast for the years 1980 and 2000, it is sufficient to know the value of $e^{10\alpha}$. In fact, according to the exponential model, $P_2 = P(20) = P_0 e^{20\alpha} = P_0(e^{10\alpha})^2 \approx 123660$ and $P_3 = P(40) = P_0 e^{40\alpha} = P_0(e^{10\alpha})^4 \approx 218086$. It is seen that this forecast overestimates the population, because it doesn't take into account restrictions on the population growth .

Now let us use the logistic model, whose equation is $P_t = (k - aP)P$, where k and a are two positive constants specific to the considered data. Bearing in mind that the equilibrium value of this model is $P_e = \frac{k}{a}$, the same equation can be rewritten in the form $P_t = kP(1 - \frac{P}{P_e})$. Since this model has the two parameters k and P_e (in addition to the initial value P_0), to specify the model (that is, to find the specific values of k and P_e) we will need data from three years (instead of two years as in the exponential model). First, we find the general solution of the logistic equation $P_t = kP(1 - \frac{P}{P_e})$, keeping the parameters k and P_e in the general form. The equation is separable and we can rewrite it in the form $\frac{dP}{P(1-\frac{P}{P_e})} = kdt$. Integrating the two sides and adjusting the terms, we get $\int \frac{P_e dP}{P(P-P_e)} = -\int kdt$. The integral on the left-hand side can be calculated using simple fractions:

$$\int \frac{P_e dP}{P(P - P_e)} = \int \frac{1}{P - P_e} - \frac{1}{P} dP = \ln|P-P_e| - \ln|P| + C = \ln\left|\frac{P - P_e}{P}\right| + C.$$

Therefore, $\ln\left|\frac{P-P_e}{P}\right| = -kt + C$ or $\frac{P-P_e}{P} = Ae^{-kt}$. Solving for P, we get $P = \frac{P_e}{1-Ae^{-kt}}$. Applying the initial condition $P(0) = P_0$, we express the integration constant A as a function of P_e: $P_0 = \frac{P_e}{1-A}$, that is, $A = 1 - \frac{P_e}{P_0}$. The two parameters of the model, k and P_e, we find using the conditions of the years 1970 and 1980: $P_1 = P(10) = \frac{P_e}{1-Ae^{-10k}}$ and $P_2 = P(20) = \frac{P_e}{1-Ae^{-20k}}$. To eliminate the parameter k, we rewrite the two relations in the form $e^{-10k} = \frac{1}{A}(1 - \frac{P_e}{P_1})$ and $e^{-20k} = \frac{1}{A}(1 - \frac{P_e}{P_2})$ and note that the left-hand side of the latter is the square of the left-hand side of the former. Then, $\frac{1}{A^2}(1 - \frac{P_e}{P_1})^2 = \frac{1}{A}(1 - \frac{P_e}{P_2})$. In this equation, only one unknown P_e remains and the other values are given. To find P_e, we simplify the equation to the form $(1 - \frac{P_e}{P_1})^2 = A(1 - \frac{P_e}{P_2})$ and substitute the expression for A: $(1 - \frac{P_e}{P_1})^2 = (1 - \frac{P_e}{P_0})(1 - \frac{P_e}{P_2})$. Cutting 1 on both sides, we get $-2\frac{P_e}{P_1} + \frac{P_e^2}{P_1^2} = -\frac{P_e}{P_0} - \frac{P_e}{P_2} + \frac{P_e^2}{P_0 P_2}$. Cutting P_e once and grouping the remaining terms with P_e and without, we get $\frac{P_e}{P_1^2} - \frac{P_e}{P_0 P_2} = \frac{2}{P_1} - \frac{1}{P_0} - \frac{1}{P_2}$. Therefore,

$$P_e = \frac{\frac{2}{P_1} - \frac{1}{P_0} - \frac{1}{P_2}}{\frac{1}{P_1^2} - \frac{1}{P_0 P_2}} = P_1 \frac{2P_0 P_2 - P_1 P_0 - P_1 P_2}{P_0 P_2 - P_1^2}.$$

Knowing P_e, we find the expression for e^{-10k} (we don't need to know k itself):

$$e^{-10k} = \frac{1}{A}\left(1 - \frac{P_e}{P_1}\right) = \left(1 - \frac{P_e}{P_0}\right)\left(1 - \frac{P_e}{P_1}\right).$$

Applying the values $P_0 = 70119$, $P_1 = 93139$ and $P_2 = 119071$, we calculate $P_e \approx 263830$, $A \approx -2.76260$, $e^{-10k} \approx 0.663377$ and the corresponding value $k \approx 0.0410412$. With all the parameters specified, we find the population in the year 2000 by the formula $P(40) = \frac{P_e}{1 - Ae^{-40k}} = \frac{P_e}{1 - A(e^{-10k})^4} \approx 171875$. As expected, the logistic model offers the forecast much closer to reality than the exponential one.

2.3 Radioactive decay

From experiments it is known that the rate of disintegration of a radioactive substance is proportional to its current amount (the number of nuclei or the mass). Denoting by $m(t)$ the mass at time t, we have the corresponding equation for the radioactive decay $m_t = km$. The half-life of a radioactive substance is the time required to halve its mass. This information is determined experimentally for each radioactive substance and used to specify the parameter k.

Problem 1. The half-life of plutonium is 1620 years. How long will it take before $1/4$ of the initial mass of plutonium remains?

Solution.

In this problem we will measure time in years. The separable equation $m_t = km$ has the general solution $m(t) = m_0 e^{kt}$, where m_0 is the initial mass of plutonium. To determine the parameter k, we use the half-life condition: $\frac{m_0}{2} = m(1620) = m_0 e^{1620k}$, whence $e^{1620k} = \frac{1}{2}$ or $k = -\frac{\ln 2}{1620} \approx -0.000428$. Using this value of k, we find the answer to the problem by solving the equation $\frac{m_0}{4} = m_0 e^{kt}$, that is, $e^{kt} = \frac{1}{4}$, or, isolating t: $t = -\frac{\ln 4}{k} \approx \frac{\ln 4}{0.000428} \approx$ 3239 years.

Problem 2. The theory of carbon dating is based on the fact that the isotope carbon-14 is produced in the atmosphere by the action of cosmic radiation on nitrogen. The ratio between the amount of carbon-14 and ordinary carbon in the atmosphere is approximately constant and, as a result, the proportion of the isotope in all living organisms is the same as in the atmosphere. When an organism dies, the absorption of carbon-14 ceases and it only decreases in quantity due to radioactive disintegration. Given that a fossilized bone contains one-thousandth of the original amount of carbon-14 and that the half-life of carbon-14 is 5600 years, find the age of the fossil.

Solution.

In this problem we measure time in years. To solve the problem, first we find the general solution of the separable equation $m_t = km$: $m(t) = m_0 e^{kt}$, where m_0 is the initial mass of carbon-14. Next, we determine the parameter k knowing the half-life: $\frac{m_0}{2} = m(5600) = m_0 e^{5600k}$, whence $e^{5600k} = \frac{1}{2}$ or $k = -\frac{\ln 2}{5600} \approx -0.000124$. Using this k, we find the answer to the problem by solving the equation $\frac{m_0}{1000} = m_0 e^{kt}$, that is, $e^{kt} = \frac{1}{1000}$, or isolating t: $t = -\frac{\ln 1000}{k} \approx \frac{\ln 1000}{0.000124} \approx 55700$ years.

2.4   Velocity/Acceleration

This type of problem uses the definition of instantaneous velocity $v(t) = x_t(t)$, where $x(t)$ is the position of a body in rectilinear motion as a function of time t, and also the definition of instantaneous acceleration $a(t) = v_t(t)$. In addition, Newton's second law $F = ma$ is applied, where F is the force exerted on a body of mass m that causes it to move with acceleration a. In particular, if F is the gravitational force, then $a = g \approx 10 m/s^2$ is the gravitational acceleration.

Problem 1. Two racing cars start from the same point with accelerations of 4 and 6 m/s^2. When will the distance between the two reach 400 meters?

Solution.

In this problem we will measure the distance in meters and the time in seconds. Instead of calculating the characteristics of the movement of each car, it is simpler to consider their relative motion: the distance $d = d_2 - d_1$, the speed $v = v_2 - v_1$ and the acceleration $a = a_2 - a_1$. For acceleration we have $a = 6 - 4 = 2$. Therefore, $v = 2t + C$ and the constant $C = 0$ since at the initial instant $v_1 = v_2$, that is, $v = 0$. Then, for the distance we have $d = \int 2t dt = t^2 + B$, where $B = 0$, because initially the cars are at the same point. Finally, we find the answer to the problem by solving the equation $d(t) = 400 = t^2$ which gives $t = 20$ seconds.

Problem 2. A stone fell from the roof and, when it hit the ground, it had velocity of $40 m/s$. Considering only the gravitational force with a constant acceleration of $10 m/s^2$, find the height of the building.

Solution.

Let us measure the distance in meters from the ground and the time in seconds, starting from the instant when the fall began. Since the height $h(t)$ of the stone decreases during the fall, the relationship between h and the velocity of the fall comes in the form $h_t = -v$. In turn, the velocity is increasing due to the gravitational force, so $v_t = g = 10m/s^2$. Since g is constant, from the second equation we get $v = gt + C$, where C is determined from the condition that at the initial instant the velocity was 0: $v(0) = 0 = C$. Moving on to the first equation, for $h(t)$, we get $h = -\int v dt = -\int gt dt = -g\frac{t^2}{2} + B$, where $B = h(0) = h_0$ is the initial height of the stone (that is, the height of the roof). To find B we use the condition that the final velocity is $40m/s$, which allows us to find the time t_1 spent falling: $v(t_1) = 40 = 10t_1$, whence $t_1 = 4$. Substituting the final instant into the formula for $h(t)$ and noting that $h(t_1) = 0$ (at the final instant t_1 the stone was already on the ground), we get $h(t_1) = 0 = h_0 - g\frac{t_1^2}{2}$, whence $h_0 = 10\frac{4^2}{2} = 80$ meters.

Problem 3 (escape velocity). A rocket is shot vertically upward from the Earth's surface. Considering only the gravitational force, but taking into account its variation with altitude, find the speed of the rocket. Determine the initial velocity required to take the rocket to the given altitude and find the minimum initial velocity required for the rocket not to return to the Earth's surface.

Solution.

Let us direct the x-axis vertically upward with the origin point on the Earth's surface. According to Newton's law of gravitation, the gravitational force is inversely proportional to the square of the distance to the center of the Earth and is given by the formula $F(x) = -\frac{mgR^2}{(R+x)^2}$, where m is the mass of the rocket, $g \approx 10m/s^2$ is the gravitational acceleration at the Earth's surface (at sea level) and R is the radius of the Earth (considered to be a ball). The negative sign is used for convenience to indicate that the force is contrary to the movement of the rising rocket. According to Newton's second law, $ma = F$, where a is acceleration, and therefore $mv_t = -\frac{mgR^2}{(R+x)^2}$ or $v_t = -\frac{gR^2}{(R+x)^2}$, where v is the velocity. In this equation we have two variables, in addition to the unknown function v: altitude x and time t. To eliminate one of them, let us express the velocity v as a function of x, using the chain rule: $v_t = v_x \cdot x_t = v_x \cdot v$. Substituting this expression into the obtained equation, we get the ordinary differential equation $vv_x = -\frac{gR^2}{(R+x)^2}$. This is a separable equation whose general solution is $\frac{v^2}{2} = \frac{gR^2}{R+x} + C$. By adding the initial condition $v(0) = v_0$ (the launch velocity v_0 at the Earth's surface), we specify the constant $C = \frac{v_0^2}{2} - gR$, and therefore, we find the velocity of motion $\frac{v^2}{2} = \frac{gR^2}{R+x} + \frac{v_0^2}{2} - gR$ or $v = \pm\sqrt{\frac{gR^2}{R+x} + \frac{v_0^2}{2} - gR}$ (the positive sign corresponds to the ascent phase and the negative sign to the descent phase).

To determine the initial velocity needed to bring the rocket to altitude x_0, we specify $v(x_0) = 0$ and substitute this condition into the velocity formula: $0 = \frac{gR^2}{R+x_0} + \frac{v_0^2}{2} - gR$, whence $v_0 = \sqrt{2gR - 2\frac{gR^2}{R+x_0}} = \sqrt{2gR\frac{x_0}{R+x_0}}$. The minimum initial velocity v_e required for the rocket not to return to the Earth's surface is found by considering the limit of the last expression when $x_0 \to +\infty$: $v_e = \lim\limits_{x_0 \to +\infty} \sqrt{2gR\frac{x_0}{R+x_0}} = \sqrt{2gR}$. This velocity is called the escape velocity. Using the approximate values of $g \approx 10m/s^2$ and $R \approx 6370 \cdot 10^3 m$, we find $v_e \approx 11300 m/s = 11.3 km/s$.

Problem 4 (airplane route). An airplane is flying from city A to city B, located at the same latitude, at a constant speed. Knowing that throughout the flight the wind is blowing from the south toward the north and that the pilot is always steering the plane straight toward the destination city, find the plane path.

Solution.

For convenience, let us place the x-axis along the latitude of the cities with the destination city B located at the origin of the coordinates and the city A at the point $(a, 0)$, $a > 0$. The y-axis is directed upward, in the direction of the wind (see Fig.5.10). At instant t, the position of the plane is $\mathbf{p}(t) = (x(t), y(t))$ and its velocity is $\mathbf{w}(t) = \mathbf{p}_t(t) = (x_t(t), y_t(t)) \equiv (u, v)$. The displacement vector of the plane (without the effect of the wind) is directed toward the city B (the origin of the coordinates) and forms the angle θ with the x-axis (see Fig.5.10). For the tangent of this angle we have the relation $\tan \theta = \frac{y}{x}$. The wind does not affect the movement along the x-axis, so $x_t = u = -W \cos \theta = -W\frac{x}{\sqrt{x^2+y^2}}$, where $W = const$ is the scalar velocity (speed) of the plane (without the presence of wind) and the negative sign reflects the movement in the opposite direction to the direction of the x-axis. Since the wind is blowing in the positive direction of the y-axis, $y_t = v = V - W \sin \theta = V - W\frac{y}{\sqrt{x^2+y^2}}$, where $V = const$ is the wind speed. Recalling from differential calculus the formula for the derivative of a parametric function $y_x = \frac{y_t}{x_t}$, we find the following equation for the trajectory $y(x)$: $y_x = \frac{y_t}{x_t} = \frac{V - W\frac{y}{\sqrt{x^2+y^2}}}{-W\frac{x}{\sqrt{x^2+y^2}}}$ or $y_x = \frac{Wy - V\sqrt{x^2+y^2}}{Wx}$ or still $y_x = \frac{y}{x} - k\sqrt{1 + (\frac{y}{x})^2}$, where $k = \frac{V}{W}$ is a constant coefficient. The last equation is homogeneous and can be transformed into a separable equation by changing the function $z = \frac{y}{x}$: $xz_x + z = z - k\sqrt{1 + z^2}$ or, after simplification, $xz_x = -k\sqrt{1 + z^2}$. Separating the variables and integrating, we get $\int \frac{dz}{\sqrt{1+z^2}} =$

$-k \int \frac{dx}{x}$. Calculating integrals, we find $\ln(z + \sqrt{1 + z^2}) = -k \ln|x| + C$ or $z + \sqrt{1 + z^2} = \frac{C}{x^k}$.

At the initial moment, the plane is positioned in the city $A = (a, 0)$ and directed toward $B = (0, 0)$, that is, we have the initial condition of the trajectory $y(a) = 0$ or, in terms of z, $z(a) = 0$. Substituting this condition into the solution for z, we determine the constant C: $C = a^k$. The solution z then takes the form $z + \sqrt{1 + z^2} = \left(\frac{a}{x}\right)^k$. For convenience, we denote $\left(\frac{a}{x}\right)^k \equiv q$ and solve the equation $z + \sqrt{1 + z^2} = q$ with respect to z: $\sqrt{1 + z^2} = q - z$ or $1 + z^2 = (q - z)^2$ or finally $z = \frac{q^2 - 1}{2q}$. Recalling that $z = \frac{y}{x}$, we get $\frac{y}{x} = \frac{1}{2}(q - \frac{1}{q})$ or $y = \frac{x}{2}\left(\left(\frac{x}{a}\right)^k - \left(\frac{a}{x}\right)^{-k}\right)$. In the case $k = \frac{1}{2}$ the trajectory of the airplane is shown in Fig.5.10.

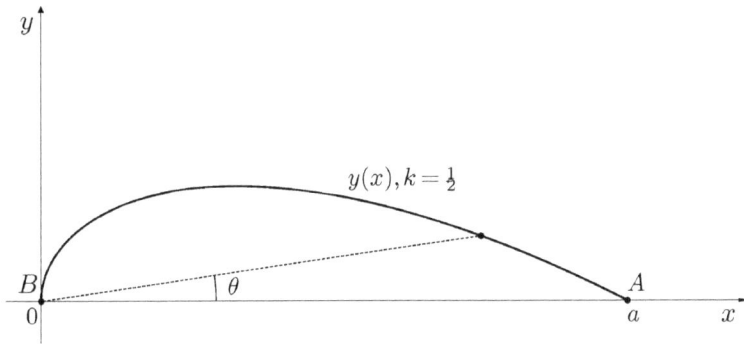

Figure 5.10: Problem 4: the trajectory of the plane $y(x)$ in the case $k = \frac{1}{2}$.

2.5 Flow of mixtures

In problems of this type, we consider deposits (tanks) of liquid/gas of constant volume. The content of deposit is altered due to injection into the deposit of another liquid/gas and pumping out the current misture of the tank at the same speed. For simplicity, it is assumed that the two different liquids/gases are stirred in the tank uniformly and immediately.

Problem 1. A saline solution is pumped continuously at a rate of 2 liters per minute in a tank which initially contains 10 liters of pure water. At the same time, the mixture obtained in a tank is drained at the same rate. Knowing that the concentration of salt in the incoming solution is 3 kilograms per liter, and assuming that this solution mixes immediately and evenly with the water in the tank, find the amount of salt in the tank after 5 minutes.

Solution.

In the solution of this problem, we will measure the quantity (mass) of salt in kilograms and the time in minutes. We denote the mass of salt in the tank at time t by $m(t)$ and evaluate the variation of $m(t)$ over the time interval $[t, t + \Delta t]$, where t is a fixed instant in time. This variation is due to the difference between the amount A of salt entering the tank and the amount B leaving it, that is, $\Delta m \equiv m(t + \Delta t) - m(t) = A - B$. Calculation of the first quantity is simple, since the salt concentration in the incoming solution is constant: during Δt minutes, $2\Delta t$ liters of salt solution enter, containing $0.3 \cdot 2\Delta t$ kilograms of salt, that is, $A = 0.6\Delta t$. The amount of salt pumped out varies in time due to the change in the amount of salt in the tank. If the mass of salt in the tank during the interval $[t, t + \Delta t]$ was constant and equal to $m(t)$ (the mass at the beginning of this interval), then the concentration of salt (mass of salt in 1 liter) in the tank would be constant $\frac{m(t)}{10}$ and, therefore, during Δt minutes the same $2\Delta t$ liters containing $\frac{m(t)}{10} \cdot 2\Delta t$ kilograms of salt would come out. However, the mass of salt in the tank is continuously increasing, so a slightly larger amount will come out, which we can express as $B = \frac{m(t) + \beta}{10} \cdot 2\Delta t$, where β is an unknown quantity. However, it is clear that the quantity $m(t) + \beta$ lies between $m(t)$ and $m(t + \Delta t)$, that is, $0 < \beta < \Delta m$, and, therefore, the function β satisfies the property $\lim\limits_{\Delta t \to 0} \beta = 0$.

Thus, the variation of $m(t)$ in the time interval $[t, t + \Delta t]$ can be represented as $\Delta m = A - B = 0.6\Delta t - (m(t) + \beta) \cdot 0.2\Delta t$. Dividing by Δt and taking limit when $\Delta t \to 0$, we get $\lim\limits_{\Delta t \to 0} \frac{\Delta m}{\Delta t} = \lim\limits_{\Delta t \to 0} 0.6 - (m(t) + \beta) \cdot 0.2$, that is, we obtain the following differential equation: $m_t = 0.6 - 0.2m$. This is a separable equation, whose general solution is $m(t) = 3 - Ce^{-0.2t}$. Using the initial condition $m(0) = 0$, we specify the constant $C = 3$ and get the solution $m(t) = 3(1 - e^{-0.2t})$. Therefore, the mass of salt in the tank after 5 minutes of this process will be $m(5) = 3(1 - e^{-0.2 \cdot 5}) = 3(1 - e^{-1}) \approx 1.9 kg$.

Problem 2. A tank contains 100 liters of a solution with 10 kilograms of uniformly dissolved salt. A pure water is continuously pumped in this tank at the rate of 5 liters per minute. This water immediately mixes with the solution in the tank, and this mixture is drained at the same rate. Find the amount of salt in the tank after 1 hour. What time is necessary to halve the amount of salt in the tank?

Solution.

In the solution of this problem, we will measure the amount (mass) of salt in kilograms and the time in minutes. We denote the mass of salt in the tank at time t by $m(t)$ and evaluate the variation of $m(t)$ over the time interval $[t, t + \Delta t]$, where t is a fixed instant of time. This variation is caused by the difference between the amount A of salt entering the tank and the amount B leaving it, that is, $\Delta m \equiv m(t + \Delta t) - m(t) = A - B$. Since a pure water enters the tank, $A = 0$. The amount of salt coming out varies in time due to the change in the amount of salt in the tank. If the mass of salt in the tank during the interval $[t, t + \Delta t]$ was constant and equal to $m(t)$ (mass at the beginning of this interval), then the concentration of salt in the tank would be constant $\frac{m(t)}{100}$ and, therefore, during Δt minutes, $5\Delta t$ liters containing $\frac{m(t)}{100} \cdot 5\Delta t$ kilograms of salt would come out. However, the mass of salt in the tank is continuously decreasing, so a slightly smaller amount will be pumped out, which we can express as $B = \frac{m(t) + \beta}{100} \cdot 5\Delta t$, where β is an unknown negative quantity. However, it is clear that the quantity $m(t) + \beta$ lies between $m(t)$ and $m(t + \Delta t)$, that is, $\Delta m < \beta < 0$, and, therefore, the function β satisfies the property $\lim_{\Delta t \to 0} \beta = 0$.

Thus, the variation of $m(t)$ in the time interval $[t, t + \Delta t]$ can be represented as $\Delta m = A - B = 0 - (m(t) + \beta) \cdot 0.05\Delta t$. Dividing by Δt and taking the limit when $\Delta t \to 0$, we get $\lim_{\Delta t \to 0} \frac{\Delta m}{\Delta t} = \lim_{\Delta t \to 0} (-(m(t) + \beta) \cdot 0.05)$, that is, we obtain the following differential equation: $m_t = -0.05m$. This is a separable equation, whose general solution is $m(t) = Ce^{-0.05t}$. Using the initial condition $m(0) = 10$, we specify the constant $C = 10$ and have the solution $m(t) = 10e^{-0.05t}$. Therefore, the mass of salt in the tank after 1 hour of this process will be $m(60) = e^{-0.05 \cdot 60} = 10e^{-3} \approx 0.5kg$. To answer the second question, we have to solve the equation $5 = 10e^{-0.05t}$ for t, which gives, $e^{-0.05t} = \frac{1}{2}$ or $-0.05t = \ln\frac{1}{2}$, whence $t = 20\ln 2 \approx 13.9$ minutes.

2.6   Personal finance

Suppose that a money deposit is made in a bank that pays interest at an annual rate r. The value $S(t)$ of the investment at any time t depends on the initial capital (the principal) S_0, on the interest rate r and on the frequency with which interest is compounded. Depending on type of deposit and bank rules, interest can be compounded quarterly, monthly or even daily. If interest is compounded n times per year at an annual rate r, then after one year we get $S(1) = S_0(1 + \frac{r}{n})^n$ and after t years – $S(t) = S_0(1 + \frac{r}{n})^{nt}$. This is a discrete equation for an investment (savings) account.

Sometimes it is useful to consider an approximated model, assuming, somewhat artificially, that compounding takes place continuously. In this case, interest $\Delta S = S(t + \Delta t) - S(t)$ accrued in the time period $[t, t + \Delta t]$

can be evaluated as follows. If deposit was constant S_c during the period $[t, t + \Delta t]$, the interest would be found by the formula $\Delta S = S_c \cdot r \cdot \Delta t$. However, in the continuous model, the deposit is an increasing function, that is, $S(t) < S_c < S(t+\Delta t)$, which implies that we have the following evaluation for interest: $S(t) \cdot r \cdot \Delta t < \Delta S < S(t+\Delta t) \cdot r \cdot \Delta t$. Dividing this inequality by Δt and taking limit when $\Delta t \to 0$, we get (under the natural supposition that $S(t)$ is a smooth function): $\lim_{\Delta t \to 0} S(t) \cdot r \leq \lim_{\Delta t \to 0} \frac{\Delta S}{\Delta t} \leq \lim_{\Delta t \to 0} S(t+\Delta t) \cdot r$. Since both left-hand and right-hand side limits are equal to $rS(t)$ and the mid-term limit is the derivative S_t, we obtain the following differential equation for $S(t)$: $S_t(t) = rS(t)$. This is a continuous model for an investment (savings) account. We already met the same kind of equation in previous problems. Its solution is easily found (the equation is separable): $S(t) = S_0 e^{rt}$, where S_0 is the principal (initial capital).

Of course, the supposition about continuously calculated compound does not lead to the same result as the calculations of real interest, occurred at finite time intervals. However, comparing the two solution, we can see that $\lim_{n \to \infty} S_0 (1 + \frac{r}{n})^{nt} = S_0 e^{rt}$. Therefore, the constructed continuous model approximates well the real growth of deposit if interest is compounded quite frequently. For example, if interest is calculated annually at the rate $r = 6\%$, then the relative interest $\frac{S(t)}{S_0}$ equals 1.0600 after one year, 1.3382 after five years and 1.7909 after ten years. If interest is calculated monthly at the same annual rate $r = 6\%$, then the relative interest equals 1.0617 after one year, 1.3489 after five years and 1.8194 after ten years. The continuous model gives the following results: 1.0618 after one year, 1.3499 after five years and 1.8221 after ten years. Thus, if one needs an approximate value of the investment for different frequencies of compounding, then the solution of differential equations can be used.

Suppose now that an amount W is withdrawn every year from an investment account that pays interest at the rate r per year (assume also that the account never will be negative). If interest is calculated and withdrawal happens only once at the end of a year, then after one year the amount on the account is $S(1) = S_0(1 + r) - W$, after two years $S(2) = (S_0(1+r) - W)(1+r) - W = S_0(1+r)^2 - W[1 + (1+r)]$ and after t years $S(t) = S_0(1+r)^t - W[1 + (1+r) + \ldots + (1+r)^{t-1}] = S_0(1+r)^t - W\frac{(1+r)^t - 1}{r}$. If interest is calculated and equal withdrawals are made n times per year (keeping the same annual values of the rate r and withdrawal W), then $S(\frac{1}{n}) = S_0(1 + \frac{r}{n}) - W\frac{1}{n}$, $S(\frac{2}{n}) = S_0(1 + \frac{r}{n})^2 - W\frac{1}{n}[1 + (1 + \frac{r}{n})]$, after one year the amount is $S(1) = S_0(1 + \frac{r}{n})^n - \frac{W}{n}[1 + (1 + \frac{r}{n}) + \ldots + (1 + \frac{r}{n})^{n-1}] = S_0(1 + \frac{r}{n})^n - W\frac{(1 + \frac{r}{n})^n - 1}{r}$ and after t years $S(t) = S_0(1 + \frac{r}{n})^{nt} - W\frac{(1 + \frac{r}{n})^{nt} - 1}{r}$. This is a discrete equation for an investment account with withdrawing.

For an approximated model with continuous compounding and withdrawing the balance equations during the period $[t, t+\Delta t]$ is $\Delta S = S_c \cdot r \cdot \Delta t - W \cdot \Delta t$ or $\frac{\Delta S}{\Delta t} = S_c \cdot r - W$. The first term on the right side was already evaluated in the problem without withdrawal and the second term is constant. Then, taking limit when $\Delta t \to 0$, we get: $S_t = \lim_{\Delta t \to 0} \frac{\Delta S}{\Delta t} = \lim_{\Delta t \to 0} (S_c \cdot r - W) = rS(t) - W$. This is a differential equation (continuous model) of an investment account with withdrawing.

Using the same reasoning, we can derive the discrete and continuous models for an investment account with depositing. Assuming that interest is compounded n times per year at annual rate r and annual deposit D is equally distributed in n installments, we obtain the following discrete equation: $S(t) = S_0(1 + \frac{r}{n})^{nt} + D\frac{(1+\frac{r}{n})^{nt}-1}{r}$. The corresponding differential equation takes the form $S_t = rS(t) + D$.

Let us solve the linear equation $S_t = rS + C$, where constant C can be depositing D or withdrawing $-W$. The solution of the corresponding homogeneous equation $S_t = rS$ is $S = Ae^{rt}$, and substituting the function $S = A(t)e^{rt}$ in the original equation we have $A_t = Ce^{-rt}$, whence $A = B - \frac{C}{r}e^{-rt}$, where B is an arbitrary constant. Therefore, the general solution is $S = Be^{rt} - \frac{C}{r}$. Applying the initial condition $S(0) = S_0$, we get $S = S_0e^{rt} + \frac{C}{r}(e^{rt} - 1)$. Obviously, for $C = 0$ we return to the solution of the continuous model of investment account without withdrawing/depositing.

Problem 1. Suppose a person deposit \$2000 in savings account with the annual rate of 5%. What will be the balance after 20 years?

Solution.

In this problem the time is measured in years and the amount of money in thousands of dollars. The corresponding equation of the differential model is $S_t = rS(t)$ with $r = 0.05$. The general solution was found in the form $S = S_0e^{rt}$, where S_0 is the principal. Specifying the parameters of the problem, we have $S = 2e^{0.05t}$, which means that $S(20) = 2e^{0.05 \cdot 20} \approx 5.437$. So, after 20 years it will be \$5437.

Problem 2. Suppose someone decided to save \$3000 each year during 30 years until retirement. Assuming that he/she has no previous savings and the proposed annual rate is 6%, how much will this person accumulate after 30 years?

Solution.

The corresponding equation of the differential model is $S_t = rS(t) + D$ with $r = 0.06$ and $D = 3$ (the time is measured in years and the amount of money in thousands of dollars). The general solution was found in the form

$S = Be^{rt} - \frac{D}{r}$, where B is an arbitrary constant. Specifying the parameters of the problem, we have $S = Be^{0.06t} - \frac{3}{0.06} = Be^{0.06t} - 50$. Applying the initial condition $S(0) = 0$, we get $S(0) = 0 = B - 50$, whence $B = 50$ and $S = 50e^{0.06t} - 50$. To answer the question, we use $t = 30$ in the last formula: $S(30) = 50e^{0.06 \cdot 30} - 50 \approx 252.482$. So, after 30 years the retirement balance will be \$252482.

Problem 3. Suppose someone decided that will need \$40000 each year to live after retirement during 20 years. Assuming that his/her retirement account will earn 5% interest, how much money must be in the account at the time of retirement?

Solution.

The corresponding equation of the differential model is $S_t = rS(t) - W$ with $r = 0.05$ and $W = 40$ (the time is measured in years and the amount of money in thousands of dollars). The general solution was found in the form $S = Be^{rt} + \frac{W}{r}$, where B is an arbitrary constant. Specifying the parameters of the problem, we have $S = Be^{0.05t} + \frac{40}{0.05} = Be^{0.05t} + 800$. The additional condition is $S(20) = 0$, that determines the constant B: $S(20) = 0 = Be^{0.05 \cdot 20} + 800 = Be + 800$, whence $B = -\frac{800}{e} \approx -294.3$. To find the required amount of money it remains to calculate $S(0) = Be^{0.05 \cdot 0} + 800 \approx -294.3 + 800 = 505.7$. Thus, it is required approximately \$505700.

Problems for section 2

1. A 20-liter tank contains a mixture of nitrogen and oxygen in the ratio 8/2. 0.1 liter of nitrogen is pumped into this tank every second and the same volume of the mixture is pumped out. Find the time when there will be 99% of nitrogen in the tank. (Assume that the nitrogen is immediately mixed evenly with the content of the tank).

2. In a room of $200m^3$, the air contains 0.15% of carbon dioxide. The air conditioning unit is injecting $20m^3$ of air containing 0.04% of carbon dioxide into the room and the same volume of air is coming out through the windows. When will the amount of carbon dioxide in the room decrease threefold? (Assume that the air injected by the appliance is immediately mixed evenly with the air in the room).

3. A lake of constant volume V contains pollutants, uniformly distributed with concentration c. A creek carries water, containing pollutant of concentration k, into the lake at a rate r, and the same amount of water leaves the lake. Assuming that the incoming pollutant is immediately mixed uniformly in the lake, find the general formula for the pollutant concentration $c(t)$ at an instant t, expressed in terms of k, r, V and the initial concentration c_0. If

the stream entering the lake contains no more pollutants, determine the time needed for the initial pollutant concentration in the lake to halve. Specify this time if $V = 4900km^3$ and $r = 160km^3/year$.

4. A body has cooled from 80^0C to 60^0C in 10 minutes. If the air temperature is 20^0C, find when the temperature of the body will drop to 30^0C?

5. It is well known that a mixture of two liquids of different temperatures almost immediately assumes a temperature equal to the average of the two temperatures with the weights proportional to the volume of each liquid: if the first liquid has volume V_1 with temperature T_1 and the second V_2 with T_2, then the temperature of the mixture will be $T = T_1 \cdot \frac{V_1}{V} + T_2 \cdot \frac{V_2}{V}$, $V = V_1 + V_2$. Assuming that $200ml$ of coffee of temperature 100^oC should be mixed with $20ml$ of milk of temperature 10^oC and the mixture should be cooled to 30^oC in an environment of temperature 20^oC, determine which process will take less time: adding the milk first and waiting for it to cool, or waiting for the coffee without milk to cool to a certain temperature and then adding the milk.

6. A boat slows down its motion under the action of water resistance which is proportional to the velocity of the boat. If the initial velocity of the boat is $10m/s$ and after $5s$ its velocity has been reduced to $8m/s$, when will the speed be $1m/s$? How far will the boat travel until it stops?

7. Determine the distance traveled by a body in 20 seconds, if its velocity is proportional to the distance traveled. It is known that in 10 seconds the body passed 100 meters and in 15 seconds 200 meters.

8. A bullet hits a plate of $10cm$ thickness with the velocity of $200m/s$ and leaves the plate with the velocity of $80m/s$. Assuming that the resistance of the plate to the bullet's passage is proportional to the square of its velocity, find the time it takes for the bullet to pass through the plate.

9. From the experiments it was determined that in one year 0.44 milligrams are disintegrated from each gram of radium. How many years does it take to halve the initial amount of the radium?

10. From the experiments it was determined that a half of the initial amount of uranium is disintegrated within 1600 years. What part of the uranium will be disintegrated in 100 years?

11. A body is falling from a height of 5 meters. Calculate its maximum velocity. (Disconsider the air resistance.)

12. The air resistance during the descent of a skydiver is proportional to the square of its velocity. Find the law of fall and the maximum speed the skydiver can reach, considering that the coefficient of proportionality is 0.004.

13. A ball was thrown upward from a balcony at the height of $20m$ with the velocity $15m/s$. Find the height of the ball with respect to the ground as a

function of time. When does the ball reach its maximum height? When will it hit the ground? (Disconsider the air resistance.)

14. A missile has been launched from the ground with the velocity $300m/s$ and the angle of elevation 30°. Find the point at which the missile reaches the ground. Find the angle of elevation that guarantees the longest range of the missile and the hitting point. What will change if the missile is launched from an elevation of $10m$? (Disregard the air resistance.)

15. A bullet is fired from the ground with the angle of elevation 60°. What initial velocity must the bullet have in order to reach a point at the height $150m$ on a tower located $250m$ away. (Disregard the air resistance.)

16. A car is moving with the velocity $90km/h$ when the driver start braking that produces a constant deceleration of $10m/s^2$. What distance will be covered until the car stops?

17. Find the trajectory of the airplane from Problem 4 of section 2.4 in the case when $k = \frac{1}{10}$. Explain what happens in the cases $k = 0$ and $k = 1$.

18. The population of mosquitoes in an isolated area increases at a rate proportional to their current number and doubles within a week. Determine the number of mosquitoes after 4 weeks, if their initial number is 20.

19. The number of bacteria increases according to the exponential growth model and doubles every 8 hours. If initially there was 20000 bacteria, when their number will reach 100000? If, in addition, a biologist is collecting 2000 bacteria per hour, the population will continue to grow or will decrease? In the latter case, how long will it take before all the bacteria disappear? Under the same conditions, what would happen if the initial amount of bacteria was 25000?

20. The number of bacteria in a population follows the logistic equation. It was observed that within 10 hours the number of bacteria doubled and that the population reached equilibrium when their number is 20000. Knowing that the initial number of bacteria was 1000, find their number after 25 hours. How long will it take to reach half of the equilibrium population?

21. Suppose that $1600 are invested at the annual rate of 6% compounded continuously. Assuming that there is no withdrawals or deposits, how much will be in the account after 10 years? How long will it take to reach $10000?

22. Suppose that $4000 are invested at the annual rate of 5% compounded continuously. Additionally, deposits of $1000 are made each year. Assuming that there is no withdrawals, how much will be in the account after 6 years? How long will it take to reach $20000?

23. On the tenth birthday of the son, parents decided to make an investment paying 8% of annual interest, compounded continuously, with the goal to reach $40000 on his eighteenth birthday for college education. Assuming that there is no withdrawals or deposits, what initial deposit should be made

to achieve this goal?

24. Twenty years before retirement a man decided to make annually deposits in savings account that pays 5% of annual interest, compounded continuously, with the goal to reach $600000 to his retirement. Assuming that there is no initial deposit and no withdrawals, what amount should be deposited each year to achieve this goal?

Chapter 6

Higher order equations: introduction

Let us briefly recall some concepts of higher order equations introduced in the first part and add the new notions of theory used in this part.

1   Basic definitions: equation and its solution

Definition of an ordinary differential equation of order n. The equation in the form

$$F(x, y, y', \ldots y^{(n)}) = 0,$$

where F is a given function of $n + 2$ arguments and $y(x)$ is an unknown function of variable x, is called an *ordinary differential equation (ODE) of order* n.

Definition of the explicit/normal form of ODE of order n. The equation in the form

$$y^{(n)} = f(x, y, y', \ldots y^{(n-1)}),$$

where f is a given function of $n + 1$ arguments and $y(x)$ is an unknown function of variable x, is called *ordinary differential equation (ODE) of order* n *in explicit/normal form*.

Definition of a particular solution. A function $y(x)$ is a *particular solution* of an ODE if, substituted into this ODE, it transforms the equation into an identity.

Definition of a general solution. A function $y(x, C_1, \ldots, C_n)$ of independent variable x and n independent parameters $C_1, \ldots, C_n$ is a *general* *solution* of ODE of order n if for any particular choice of $C_1, \ldots, C_n$ it represents a particular solution of the same ODE.

2   Cauchy problem (initial value problem)

While an equation of the first order usually requires one initial condition to determine a single solution, the equation of order n is linked with n complementary conditions of a specific form, which can be interpreted physically as specific values of the object location, its velocity, acceleration, etc. all at the same instant (in which case the unknown function is interpreted as the object position in linear motion). These additional conditions are called initial conditions.

Definition of initial conditions. Given a differential equation of order n, the following conditions are called *initial conditions*:

$$y(x_0) = y_0, y'(x_0) = y_1, \ldots, y^{(n-1)}(x_0) = y_{n-1}.$$

Definition of the Cauchy problem (initial value problem). The differential equation of order n together with n initial conditions form the *Cauchy problem*:

$$\begin{cases} F(x, y, y', \ldots y^{(n)}) = 0 \\ y(x_0) = y_0, y'(x_0) = y_1, \ldots, y^{n-1}(x_0) = y_{n-1} \end{cases}.$$

Here, $y_0, y_1, \ldots, y_{n-1}$ are the given values. This problem is also called the *initial condition problem* or the *initial value problem*.

If the Cauchy problem is formulated for an explicit equation, then there exists an important result, called the *Cauchy theorem* (also the *Picard theorem* or *Picard-Lindelöf* theorem) which guarantees the existence and uniqueness of a solution of the problem

$$\begin{cases} y^{(n)} = f(x, y, y', \ldots, y^{(n-1)}) \\ y(x_0) = y_0, y'(x_0) = y_1, \ldots, y^{n-1}(x_0) = y_{n-1} \end{cases}.$$

Cauchy theorem (theorem of existence and uniqueness of solutions). If there exists a neighborhood of the point $(x_0, y_0, y_1, \ldots, y_{n-1}) \in \mathbb{R}^{n+1}$ where the function f and its partial derivatives $f_y, f_{y'}, \ldots, f_{y^{n-1}}$ are continuous functions, then there exists a neighborhood of x_0 in $\mathbb{R}$ where the solution of the Cauchy problem exists and is unique.

3 Other problems with higher order equations

Unlike first order equations, for which there is only one natural way to add a complementary condition, which leads to the Cauchy problem (although other, much rarer problems can also arise), for higher order equations there are several natural ways to add complementary conditions, in addition to setting the initial conditions. This leads to important problems that are different from the Cauchy problem and do not have such a simple result of existence and uniqueness of solutions as guaranteed by the Cauchy theorem.

Due to a greater complexity of the theoretical results on solvability of these problems and also the greater difficulties of solving them in practice, we will only consider the case of second order equations. There are two types of classical problems (besides the Cauchy problem) related to second order equations: the boundary value problem and the Sturm-Liouville problem (the problem of eigenvalues and eigenfunctions). Their formulations are as follows.

Definition of the boundary value problem. The problem of solving the equation

$$F(x, y, y', y'') = 0$$

on an interval $[x_1, x_2]$ together with the two *boundary conditions* placed at the endpoints of the interval

$$\alpha_1 y(x_1) + \beta_1 y'(x_1) = \gamma_1, \ \alpha_2 y(x_2) + \beta_2 y'(x_2) = \gamma_2,$$

where $\alpha_1, \beta_1, \gamma_1$ and $\alpha_2, \beta_2, \gamma_2$ are given constants such that $\alpha_1^2 + \beta_1^2 \neq 0$, $\alpha_2^2 + \beta_2^2 \neq 0$, is called the *boundary value problem*.

Definition of the Sturm-Liouville problem. The problem of solving the equation

$$y'' + a(x)y' = \lambda y$$

on an interval $[x_1, x_2]$, where $y \neq 0$ and λ is a pair of unknowns (a function and a parameter independent of x), together with the two conditions imposed at the endpoints of the interval

$$\alpha_1 y(x_1) + \beta_1 y'(x_1) = 0, \alpha_2 y(x_2) + \beta_2 y'(x_2) = 0,$$

where α_1, β_1 and α_2, β_2 are given constants such that $\alpha_1^2 + \beta_1^2 \neq 0$, $\alpha_2^2 + \beta_2^2 \neq 0$, is called the *Sturm-Liouville problem*. The unknowns λ and y are called the *eigenvalue and eigenfunction*, respectively, and so the problem is also called the *eigenvalue and eigenfunction problem*.

Chapter 7

Higher order equations: methods of reduction of order

In this section, we will consider three different types of equations whose order can be reduced. None of them has its own name, so each one is referred to by its form.

1 Equation $y^{(n)} = f(x)$

Method of solution.

To solve the *equation*

$$y^{(n)} = f(x)$$

we apply consecutive integration n times and obtain the general solution.

Examples.

1. $y''' = x + \cos x$.

Using successive integration three times, we get

$$y'' = \int x + \cos x dx = \frac{x^2}{2} + \sin x + A;$$

$$y' = \int \frac{x^2}{2} + \sin x + A dx = \frac{x^3}{6} - \cos x + Ax + B;$$

$$y = \int \frac{x^3}{6} - \cos x + Ax + Bdx = \frac{x^4}{24} - \sin x + A\frac{x^2}{2} + Bx + C.$$

2. $y'' = xe^{-x}$, $y(0) = y'(0) = 0$.
Integrating for the first time we get

$$y' = \int xe^{-x}dx = -xe^{-x} + \int e^{-x}dx = -(x+1)e^{-x} + A.$$

The initial conditions can be applied both at each step of reduction of order or after finding the general solution. In this problem we use the first way and apply the second initial condition to the already calculated derivative: $y'(0) = -1 + A = 0$, that is, $A = 1$. Integrating the second time (with the constant A already specified) we have

$$y = \int -(x+1)e^{-x} + 1dx = (x+1)e^{-x} - \int e^{-x}dx + x = (x+2)e^{-x} + x + B.$$

Applying the first initial condition, we get $y(0) = 2 + B = 0$, that is, $B = -2$. Thus, the solution to the Cauchy problem is $y = (x+2)e^{-x} + x - 2$.

3. $y''' = \sin x$, $y(0) = 1, y'(0) = 0, y''(0) = 0$.
First, we integrate successively three times:

$$y'' = \int \sin xdx = -\cos x + A; \quad y' = \int -\cos x + Adx = -\sin x + Ax + B;$$

$$y = \int -\sin x + Ax + Bdx = \cos x + A\frac{x^2}{2} + Bx + C.$$

Now, we apply the initial conditions: $y(0) = 1 + C = 1$, whence $C = 0$; $y'(0) = B = 0$; $y''(0) = -1 + A = 0$, whence $A = 1$. So the solution to the problem is $y = \cos x + \frac{x^2}{2}$.

Exercises for section 1

Solve the following initial value problems:
1. $y'' = x + \sin x$, $y(0) = -3, y'(0) = 0$.
2. $y'' = \frac{x}{e^{2x}}$, $y(0) = \frac{1}{4}, y'(0) = -\frac{1}{4}$.
3. $y''' = \frac{\ln x}{x^2}$, $y(1) = 0, y'(1) = 1, y''(1) = 2$.

2   Equation $F(x, y^{(k)}, \dots, y^{(n)}) = 0$, $k > 0$ (equation without y and lower derivatives)

The equation

$$F(x, y^{(k)}, \dots, y^{(n)}) = 0, k > 0$$

does not possess the unknown function and its lower-order derivatives.

Method of solution.

Introducing the new function $z = y^{(k)}$, we lower the order of the original equation k units, obtaining the equation of order $n-k$: $F(x, z, \ldots, z^{(n-k)}) = 0$. We solve, if possible, the last equation for the unknown z and then apply repeated integration k times to restore y from the equation $y^{(k)} = z$.

Examples.

1. $xy''' = y'' - xy''$.

We start with solving the first order equation for the function $z = y''$: $xz' = z - xz$. Separating variables, we get $\int \frac{1}{z} dp = \int \frac{1-x}{x} dx$, and after integration, $\ln|z| = \ln|x| - x + C$ or $z = Cxe^{-x}$. Returning to y, we integrate twice more to obtain the general solution of the original equation:

$$y' = \int Cxe^{-x} dx = -C(x+1)e^{-x} + A,$$

$$y = \int -C(x+1)e^{-x} + A dx = C(x+2)e^{-x} + Ax + B.$$

2. $xy'' = y' \ln \frac{y'}{x}$.

We reduce the order by one using $z = y'$: $xz' = z \ln \frac{z}{x}$. In this homogeneous equation we change the function $u = \frac{z}{x}$ and arrive at the separable equation $xu' + u = u \ln u$. This leads to the following integration: $\int \frac{1}{u(\ln u - 1)} du = \int \frac{1}{x} dx$. The left integral is calculated as follows:

$$\int \frac{1}{u(\ln u - 1)} dz = \int \frac{d(\ln u)}{\ln u - 1} = \ln|\ln u - 1| + C.$$

Together with the integral on the right side, this gives the following solution for u: $\ln|\ln u - 1| = \ln|x| + C$, which we can still represent in the form $\ln u - 1 = Cx$. Returning to the function z, we get $\ln \frac{z}{x} = Cx + 1$ or $z = xe^{Cx+1}$. Finally, for y we have the first-order equation $y' = xe^{Cx+1}$, whose general solution is obtained via simple integration by parts on the right-hand side in the case $C \neq 0$:

$$y = e \int xe^{Cx} dx = e \left(x\frac{1}{C}e^{Cx} - \frac{1}{C} \int e^{Cx} dx \right) = e \left(\frac{x}{C}e^{Cx} - \frac{1}{C^2}e^{Cx} \right) + B.$$

This integration is not possible when $C = 0$, but in this case we have the simpler equation $y' = xe$, whose integration gives another set of solutions $y = \frac{e}{2}x^2 + A$.

3. $(1 - x^2)y'' - xy' = 2$.

By changing the function $z = y'$ we lower the order of the equation to the first: $(1 - x^2)z' - xz = 2$. The last equation is linear, so we know how to solve it. We start with the homogeneous linear equation: separating variables and integrating, we get $\int \frac{1}{z} dz = \int \frac{x}{1-x^2} dx$ and, after integration, $\ln|z| = -\frac{1}{2} \ln|1-x^2| + C$ or $z = \frac{C}{\sqrt{1-x^2}}$. Then we look for the solution of the nonhomogeneous equation in the form $z = \frac{C(x)}{\sqrt{1-x^2}}$, where C is specified by substituting into the equation: $(1 - x^2) \left(\frac{C'}{\sqrt{1-x^2}} + \frac{Cx}{\sqrt{(1-x^2)^3}} \right) - x \frac{C}{\sqrt{1-x^2}} = 2$ or, simplifying, $C' = \frac{2}{\sqrt{1-x^2}}$. Therefore, $C = 2 \arcsin x + A$ and so $z = \frac{2 \arcsin x + A}{\sqrt{1-x^2}}$. Returning to y, we perform one more integration to obtain the general solution:

$$y = \int z dx = \int \frac{2 \arcsin x + A}{\sqrt{1 - x^2}} dx = 2 \int \arcsin x d(\arcsin x) + A \int \frac{1}{\sqrt{1 - x^2}} dx$$

$$= \arcsin^2 x + A \arcsin x + B.$$

4. $2xy'y'' = (y')^2 - 1$.

By changing the function $z = y'$ we reduce the order of the equation to the first: $2xzz' = z^2 - 1$. The last equation is separable and we have $\int \frac{2z}{z^2-1} dz = \int \frac{dx}{x}$ and, after integration, $\ln|z^2 - 1| = \ln|x| + C$ or $z^2 - 1 = Cx$. Returning to y, we get $(y')^2 - 1 = Cx$, which gives two equations $y' = \pm\sqrt{Cx + 1}$. We can solve them in parallel: $y = \pm \int \sqrt{Cx + 1} dx = \pm \frac{2}{3C}(Cx+1)^{\frac{3}{2}} + B$. These are the two general solutions. In addition, the functions corresponding to the constant $C = 0$ were eliminated in the last integration. In this case, we have $(y')^2 - 1 = 0$, which results in the two more families $y = \pm x + A$. It is easy to check that they are also solutions of the original equation.

5. $\cos 2x \cdot y''' + 2 \sin 2x \cdot y'' = 1$.

By changing the function $z = y''$ we lower the order of the equation to the first: $\cos 2x \cdot z' + 2 \sin 2x \cdot z = 1$. The last equation is linear. We start with its homogeneous part: $\cos 2x \cdot z' + 2 \sin 2x \cdot z = 0$. Separating variables and integrating, we get $\int \frac{dz}{z} = -\int \frac{2 \sin 2x}{\cos 2x} dx$. The second integral can be calculated as follows:

$$\int \frac{-2 \sin 2x}{\cos 2x} dx = \int \frac{d(\cos 2x)}{\cos 2x} = \ln|\cos 2x| + C,$$

and then, we find z in the form $\ln|z| = \ln|\cos 2x| + C$ or $z = C \cos 2x$. The solution to the nonhomogeneous equation is sought in the form $z = C(x) \cos 2x$, where $C(x)$ is found by substituting this form into the equation:

$$\cos 2x(C' \cos 2x - 2C \sin 2x) + 2 \sin 2x \cdot C(x) \cos 2x = 1.$$

Simplifying, we have $C' = \frac{1}{\cos^2 2x}$ and integrating $C = \frac{1}{2}\tan 2x + A$. Then we find the function z:

$$z = \left(\frac{1}{2}\tan 2x + A\right)\cos 2x = \frac{1}{2}\sin 2x + A\cos 2x.$$

Returning to y, we have to solve the second order equation $y'' = \frac{1}{2}\sin 2x + A\cos 2x$. Integrating twice, we find the general solution $y = -\frac{1}{8}\sin 2x - A\frac{1}{4}\cos 2x + Bx + C$.

6. $y'' = y' + x$.

By changing the function $z = y'$ we reduce the order of the equation to the first: $z' = z + x$. The last equation is linear and we solve it using the alternative method of integrating factor. To do this, we find the factor μ from the equation $\mu' = -\mu$ whose solution is $\mu = e^{-x}$. Then, multiplying the linear equation by e^{-x}, we get $e^{-x}z' - e^{-x}z = (e^{-x}z)' = e^{-x}x$. Integrating this relation, we find $e^{-x}z = (-x-1)e^{-x}+C$ or $z = -x-1+Ce^x$. Returning to y, we perform one more integration:

$$y = \int z\,dx = \int -x - 1 + Ce^x dx = -\frac{x^2}{2} - x + Ce^x + B.$$

7. $xy'' = y'\ln\frac{y'}{x}$.

By changing the function $z = y'$ we lower the order of the equation to the first: $xz' = z\ln\frac{z}{x}$. The last equation is homogeneous, and by using the function $u = \frac{z}{x}$ we reduce it to a separable equation: $xu' + u = u\ln u$. Separating variables and integrating, we get $\int\frac{du}{u(\ln u - 1)} = \int\frac{dx}{x}$. Calculating integrals, we obtain

$$\int\frac{du}{u(\ln u - 1)} = \int\frac{d(\ln u - 1)}{\ln u - 1} = \ln|\ln u - 1| = \ln|x| + C$$

or $\ln u - 1 = Cx$. Then, for z we have $\ln\frac{z}{x} = Cx + 1$ or $z = xe^{Cx+1}$. Returning to y, we perform one more integration:

$$y = \int xe^{Cx+1}dx = x\frac{1}{C}e^{Cx+1} - \frac{1}{C}\int e^{Cx+1}dx = \frac{x}{C}e^{Cx+1} - \frac{1}{C^2}e^{Cx+1} + B.$$

8. $y'' - 2y'\cot x = \sin^2 x$.

By changing the function $z = y'$ we lower the order of the equation to the first: $z' - 2z\cot x = \sin^2 x$. The last equation is a linear one that we can solve using the variation of parameter. For the homogeneous part, we find

$$\ln|z| = 2\int\cot x\,dx = 2\int\frac{d(\sin x)}{\sin x} = 2\ln|\sin x| + C$$

or $z = C \sin^2 x$. Now we seek the solution of the nonhomogeneous equation in the form $z = C(x) \sin^2 x$, where C is found by substituting into the equation:

$$(C' \sin^2 x + C \cdot 2 \sin x \cos x) - 2C \sin^2 x \cot x = \sin^2 x$$

or, simplifying, $C' = 1$. So $C(x) = x + A$ and $z = (x + A) \sin^2 x$. Going back to y, we perform one more integration:

$$y = \int (x+A) \sin^2 x \, dx = \int (x+A) \frac{1 - \cos 2x}{2} dx = \frac{1}{2} \int x + A - A \cos 2x - x \cos 2x \, dx$$

$$= \frac{1}{2} \left(\frac{x^2}{2} + Ax - \frac{A}{2} \sin 2x - \left(\frac{x}{2} \sin 2x - \int \frac{1}{2} \sin 2x \, dx \right) \right)$$

$$= \frac{x^2}{4} + \frac{Ax}{2} - \frac{A}{4} \sin 2x - \frac{x}{4} \sin 2x + \frac{1}{8} \cos 2x + B.$$

Exercises for section 2

Solve the following equations:
1. $y'' + 2xy'^2 = 0$
2. $x^2 y'' + xy' = 1$
3. $x^2 y'' = y'^2$
4. $y''(e^x + 1) + y' = 0$
5. $y''' = 2(y'' - 1) \cot x$
6. $y''' = \sqrt{1 + y''^2}$
7. $yy'' = y'^2 - y'^3$
8. $xy''' = y'' - xy''$

3   Equation $F(y, y', \ldots, y^{(n)}) = 0$ (equation without x)

In the equation

$$F(y, y', \ldots, y^{(n)}) = 0$$

the function y is present, which makes it impossible to reduce the order using the previous method. But the independent variable is not in the equation, which allows the following procedure to be applied.

Method of solution.

We introduce a new function $z(y) = y'$ which is considered to be a function of the variable y. Consequently, the derivatives of y with respect to x and of z with respect to y are related by the following formulas:

$$y'' = z_y \cdot y_x = z_y z, \quad y''' = (z_y z)_y \cdot y_x = (z_y z)_y z, \quad \ldots ,$$

$$y^{(n)} = (\ldots((z_y z)_y)_y \ldots z)_y \cdot y_x = (\ldots((z_y z)_y)_y \ldots z)_y z,$$

where the differentiation in y in the last expression is repeated $n-1$ times. Then, the original equation of order n is reduced to $G(z, z', \ldots, z^{(n-1)}) = 0$ which is the equation of order $n-1$ for the function z of independent variable y. If possible, we solve the last equation and return to $y' = z(y)$ to find the original function y.

Examples.

1. $y''y^3 = -1$.
Using the function $z(y) = y'$, we transform the primitive equation into the first order equation: $z_y z y^3 = -1$. The general solution of this separable equation is $\frac{z^2}{2} = \frac{1}{2y^2} + A$ or $z = \pm\sqrt{\frac{1}{y^2} + A}$. Now we have to solve the separable equation for y (actually, the two equations joined into one formula): $y' = \pm\sqrt{\frac{1}{y^2} + A}$. Separation of variables leads to the following integrals: $\int \frac{y}{\sqrt{Ay^2+1}}dy = \pm\int dx$. Integration is simple on both sides and we get the general solution in the form $\sqrt{Ay^2 + 1} = A(B \pm x)$.

2. $3y'y'' = 2y$, $y(0) = y'(0) = 1$.
Using the function $z(y) = y'$, we transform the primitive equation into the first order equation: $3zz_y z = 2y$ or $z_y = \frac{2}{3}\frac{y}{z^2}$. The general solution of this separable equation has the form $z^3 = y^2 + A$ or $z = \sqrt[3]{y^2 + A}$. Returning to y, we have the separable equation $y' = \sqrt[3]{y^2 + A}$, which is simple to solve theoretically, arriving at the integral formula $\int \frac{1}{\sqrt[3]{y^2+A}}dy = \int dx$, but the integral on the left-hand side is not calculated in terms of elementary functions for an arbitrary value of A. However, we can try to simplify this integral by specifying constant A from the initial conditions. Since y and its derivative are connected via the formula $y' = \sqrt[3]{y^2 + A}$ for any x, then for $x = 0$ we have (using the initial conditions) $y'(0) = 1 = \sqrt[3]{1^2 + A}$, whence $A = 0$. Then, the integral on the right-hand side is radically simplified and its calculation is elementary: $\int \frac{1}{\sqrt[3]{y^2}}dy = 3\sqrt[3]{y}+C$. Therefore, together with the right-hand side, we have the solution $3\sqrt[3]{y} = x + C$. Now applying the initial condition $y(0) = 1$, we find $3\sqrt[3]{1} = 0 + C$, that is, $C = 3$. Finally, the solution to the Cauchy problem comes in the form $3\sqrt[3]{y} = x + 3$ or $y = (\frac{x}{3} + 1)^3$.

3. $y'' = y'e^y$, $y(0) = 0, y'(0) = 1$.
Using the function $z(y) = y'$, we transform the original equation into the first order equation: $z_y z = ze^y$ or $z_y = e^y$. The general solution of this separable

equation is $z = e^y + C$. Now we return to y: $y' = e^y + C$. To make integration easier, we first apply the initial conditions: $y'(0) = 1 = 1 + C = e^{y(0)} + C$, whence $C = 0$. Now, integrating $y' = e^y$, we get $\int e^{-y} dy = \int dx$, where $-e^{-y} = x + B$. Applying the first initial condition, we specify B: $-e^0 = 0 + B$, that is, $B = -1$. Thus, the solution to the problem is $e^{-y} = 1 - x$.

4. $(y')^2 + 2yy'' = 0$, $y(0) = 1, y'(0) = 1$.
Using the function $z(y) = y'$, we transform the original equation into the first order equation: $z^2 + 2yzz_y = 0$. This is a separable equation and solving it we get $\int \frac{dz}{z} = -\frac{1}{2} \int \frac{dy}{y}$, or after integration, $\ln|z| = -\frac{1}{2}\ln|y| + C$. Eliminating the logarithm and returning to y' we get another separable equation $y' = \frac{C}{\sqrt{y}}$. Integrating the last one, $\int \sqrt{y}\, dy = \int C\, dx$ and calculating integrals, we get the general solution $\frac{2}{3}\sqrt{y^3} = Cx + B$. Now we apply the initial conditions: $\frac{2}{3}\sqrt{1} = 0 + B$ and $\sqrt{1} = C$, whence $B = \frac{2}{3}$ and $C = 1$. Thus, the solution to the Cauchy problem is $\frac{2}{3}\sqrt{y^3} = x + 1$.

5. $2yy'' - y'^2 + 1 = 0$, $y(0) = 2, y'(0) = 1$.
Using the function $z(y) = y'$, we transform the original equation into the first order equation: $2yzz_y - z^2 + 1 = 0$. This is a separable equation and solving it we get $\int \frac{2zdz}{z^2-1} = \int \frac{d(z^2-1)}{z^2-1} = \int \frac{dy}{y}$, or after integration, $\ln|z^2-1| = \ln|y| + C$. Eliminating the logarithm and returning to y' we have $y'^2 - 1 = Cy$. To simplify further calculations, we apply the initial conditions and specify C: $1^2 - 1 = C \cdot 2$, whence $C = 0$. Next, we have to solve $y'^2 = 1$, that is, $y' = \pm 1$. Integrating, we find the solutions $y = \pm x + B$. Applying the first initial condition, we get $2 = 0 + B$, that is, $B = 2$. So we have two solutions to the Cauchy problem $y = \pm x + 2$.

6. $y'' = \frac{1}{y^3}$, $y(0) = 1, y'(0) = 0$.
Using the function $z(y) = y'$, we transform the original equation into the first order equation: $zz_y = \frac{1}{y^3}$. Separating variables and integrating, we get $\frac{z^2}{2} = -\frac{1}{2y^2} + C$ or $z^2 = -\frac{1}{y^2} + C$. To simplify calculations, we apply the initial conditions:

$$z(0)^2 = y'^2(0) = 0 = -\frac{1}{y(0)^2} + C = -\frac{1}{1} + C,$$

whence $C = 1$. Going back to y, we have to solve two more explicit equations $y' = \pm\sqrt{1 - \frac{1}{y^2}}$, that we can do in parallel. First, we calculate the integral

$$\int \frac{dy}{\sqrt{1 - \frac{1}{y^2}}} = \int \frac{ydy}{\sqrt{y^2 - 1}} = \frac{1}{2}\int \frac{d(y^2-1)}{\sqrt{y^2-1}} = \sqrt{y^2 - 1} + B.$$

Using this result to solve the two separable equations $y' = \pm\sqrt{1 - \frac{1}{y^2}}$, we obtain $\sqrt{y^2 - 1} = \pm x + B$. It remains to substitute these two solutions into the first initial condition: $\sqrt{1^2 - 1} = \pm 0 + B$, whence $B = 0$. So we have two solutions to the original problem: $\sqrt{y^2 - 1} = \pm x$.

7. $y'' + \frac{2}{1-y}y'^2 = 0$, $y(0) = 0$, $y'(0) = 1$.

Using the function $z(y) = y'$, we transform the original equation into the first order equation: $zz_y + \frac{2}{1-y}z^2 = 0$ or simplifying $z_y = -\frac{2}{1-y}z$. Separating variables and integrating, we get $\ln|z| = 2\ln|1 - y| + C$ or $z = C(1-y)^2$. To simplify calculations, we apply the initial conditions:

$$z(0) = y'(0) = 1 = C(1 - y(0))^2 = C(1 - 0)^2,$$

whence $C = 1$ and, therefore, $z = (1 - y)^2$. Returning to y, we have to solve one more equation $y' = (1 - y)^2$. Integrating, we obtain the solution $\int \frac{dy}{(1-y)^2} = \frac{1}{1-y} = x + B$. All that remains is to substitute this solution into the first initial condition: $\frac{1}{1-0} = 0 + B$, whence $B = 1$. This gives us the solution to the Cauchy problem: $\frac{1}{1-y} = x + 1$.

Exercises for section 3

Solve the following equations and initial value problems:
1. $2y'^2 = (y-1)y''$, $y(0) = 2$, $y'(0) = 2$.
2. $y'' + y'^2 = 2e^{-y}$.
3. $y'' = 2y^3$, $y(0) = 1, y'(0) = 1$
4. $2yy'' = y^2 + y'^2$
5. $y^4 - y^3y'' = 1$
6. $yy'' = 1 + y'^2$
7. $y'' = e^{2y}$, $y(0) = 0, y'(0) = 1$
8. $2yy'' - 3y'^2 = 4y^2$

Chapter 8

Linear equations: theoretical properties

In this section we will introduce the theory of linear equations.

1 Definitions and basic results

Definition of a linear equation of order n. The equation in the form

$$a_n(x)y^{(n)} + \ldots + a_1(x)y' + a_0(x)y = f(x),$$

where $a_n, \ldots, a_1, a_0$, $a_n \neq 0$ and f are given functions of x is called *linear equation of order* n. If $f(x) \equiv 0$, the equation is called *homogeneous linear*, otherwise – *nonhomogeneous linear*. If $a_n \equiv 1$, the form of the linear equation is called *canonical or normalized*.

  Remark 1. In the context of linear equations, homogeneous linear equations are simply called homogeneous and nonhomogeneous linear equations – nonhomogeneous.

  Remark 2. The functions $a_n, \ldots, a_1, a_0$ are called *coefficients of the equation*, and the function $f(x)$ is called *right-hand side*.

  In what follows, we will use the operator notation

$$Ly \equiv a_n(x)y^{(n)} + \ldots + a_1(x)y' + a_0(x)y$$

for the entire left-hand side of the linear equation. We will show the two main properties of the *operator* L.

Theorem 1. Linearity of the operator L. L is a linear operator, that is,

$$L(\alpha_1 y_1 + \alpha_2 y_2) = \alpha_1 L y_1 + \alpha_2 L y_2,$$

for any functions y_1, y_2 (n times differentiable) and any real or complex constants α_1, α_2.

Proof. For better understanding, we demonstrate this result using real constants α_1, α_2, but the complex case is treated in the same way. Using the definition of the operator L and the properties of derivatives, we have:

$$L(\alpha_1 y_1 + \alpha_2 y_2)$$

$$= a_n(x)(\alpha_1 y_1 + \alpha_2 y_2)^{(n)} + \ldots + a_1(x)(\alpha_1 y_1 + \alpha_2 y_2)' + a_0(x)(\alpha_1 y_1 + \alpha_2 y_2)$$

$$= a_n(x)\left(\alpha_1 y_1^{(n)} + \alpha_2 y_2^{(n)}\right) + \ldots + a_1(x)\left(\alpha_1 y_1' + \alpha_2 y_2'\right) + a_0(x)\left(\alpha_1 y_1 + \alpha_2 y_2\right)$$

$$= \alpha_1\left(a_n(x)y_1^{(n)} + \ldots + a_1(x)y_1' + a_0(x)y_1\right)$$

$$+ \alpha_2\left(a_n(x)y_2^{(n)} + \ldots + a_1(x)y_2' + a_0(x)y_2\right) = \alpha_1 L y_1 + \alpha_2 L y_2.$$

Theorem 2. Property of linear combination of solutions. If y_1 is a solution of the linear equation with the right-hand side f_1 and y_2 is a solution of the same equation with the right-hand side f_2, then $\alpha_1 y_1 + \alpha_2 y_2$ is the solution of the same equation with the right-hand side $\alpha_1 f_1 + \alpha_2 f_2$, where α_1, α_2 are arbitrary constants. Expressing this result in formulas we have:

$$L y_1 = f_1, L y_2 = f_2 \implies L(\alpha_1 y_1 + \alpha_2 y_2) = \alpha_1 f_1 + \alpha_2 f_2, \forall f_1, f_2, \forall \alpha_1, \alpha_2.$$

Proof. This result follows directly from Theorem 1: $L(\alpha_1 y_1 + \alpha_2 y_2) = \alpha_1 L y_1 + \alpha_2 L y_2 = \alpha_1 f_1 + \alpha_2 f_2$.

Corollary 1. If y_1 and y_2 are solutions of the homogeneous equation, then $\alpha_1 y_1 + \alpha_2 y_2$ is the solution of the same homogeneous equation, whatever the constants α_1, α_2 are chosen, that is,

$$L y_1 = 0, L y_2 = 0 \implies L(\alpha_1 y_1 + \alpha_2 y_2) = 0, \forall \alpha_1, \alpha_2.$$

This result follows immediately from Theorem 2 with $f_1 = f_2 = 0$.

Corollary 2. If y_h is a solution of the homogeneous equation and y_n is a solution of the nonhomogeneous equation with the right-hand side f, then $y_h + y_n$ is the solution of the same nonhomogeneous equation with the right-hand side f, that is,

$$L y_h = 0, L y_n = f \implies L(y_h + y_n) = f.$$

The result follows immediately from Theorem 2 with $f_1 = 0, f_2 = f$ and $\alpha_1 = \alpha_2 = 1$.

Corollary 3. If y_1 and y_2 are solutions of the same nonhomogeneous equation, then $y_1 - y_2$ is the solution of the corresponding homogeneous equation, that is,

$$Ly_1 = f, Ly_2 = f \;\Rightarrow\; L(y_1 - y_2) = 0.$$

The result follows immediately from Theorem 2 with $f_1 = f_2 = f$ and $\alpha_1 = 1, \alpha_2 = -1$.

Theorem 3. The function $y = u + iv$ is a solution of the nonhomogeneous equation with the right-hand side $h = f + ig$ if and only if $u = Re(y)$ is the solution of the same equation with the right-hand side $f = Re(h)$ and $v = Im(y)$ is the solution of the same equation with the right-hand side $g = Im(h)$. In formulas, we have

$$L(u + iv) = f + ig \;\Leftrightarrow\; L(u) = f, Lv = g.$$

Proof. The converse (right-to-left) implication follows immediately from Theorem 2 with $f_1 = f, f_2 = g$ and $\alpha_1 = 1, \alpha_2 = i$. The direct (left-to-right) implication follows from the properties of the operator L and the complex numbers. Indeed, for real functions u and v and real functions f and g we have, according to the linearity of the L operator, $L(u + iv) = Lu + iLv = f + ig$. Since the expressions Lu and Lv, as well as the functions f and g, are real, we use the property that two complex expressions are equal if and only if their real and imaginary parts are equal to each other. Therefore, we have $L(u) = f$ and $Lv = g$.

Corollary. The function $y = u + iv$ is a solution of the homogeneous equation if and only if its real part $u = Re(y)$ and imaginary part $v = Im(y)$ are solutions of the same equation. In formulas, we have

$$L(u + iv) = 0 \;\Leftrightarrow\; L(u) = 0, Lv = 0.$$

This result follows directly from Theorem 3 with the choice $f = g = 0$.

Theorem 4. Cauchy theorem for linear equations. The Cauchy problem for a linear equation

$$\begin{cases} a_n(x)y^{(n)} + \ldots + a_1(x)y' + a_0(x)y = f(x), a_n(x) \neq 0 \\ y(x_0) = y_0, y'(x_0) = y_1, \ldots, y^{n-1}(x_0) = y_{n-1} \end{cases}$$

has a unique solution in a neighborhood of x_0 provided that the coefficients $a_n, \ldots, a_1, a_0$ and the right-hand side $f(x)$ are continuous functions in a neighborhood of x_0.

Additionally, if $a_n, \ldots, a_1, a_0$ and $f(x)$ are continuous functions on an interval I, then the Cauchy problem has a unique solution on I for any $x_0 \in I$.

Remark. In part, this result follows directly from the Cauchy general theorem. Indeed, noting that the linear equation can be rewritten in the form $y^{(n)} = -\frac{a_{n-1}}{a_n} y^{(n-1)} - \ldots - \frac{a_1}{a_n} y' - \frac{a_1}{a_n} y + f \equiv g(x, y, \ldots y^{(n-1)})$ (recall that $a_n \neq 0$), we can see that the function g and its partial derivatives $g_y = -\frac{a_1}{a_n}, g_{y'} = -\frac{a_1}{a_n}, \ldots, g_{y^{n-1}} = -\frac{a_{n-1}}{a_n}$ are continuous functions (according to the conditions of Theorem 4), and therefore, the Cauchy general theorem guarantees the existence and uniqueness of the solution.

In what follows, we assume that the conditions of the Cauchy Theorem are satisfied for each Cauchy equation/problem we encounter in this part of the text.

2 Structure of solutions of linear equations

Definition. Linear independence of functions. Functions $y_1, \ldots, y_n$ are *linearly independent* on a set I if their linear combination $C_1 y_1 + \ldots C_n y_n$ is zero on I only when all the constants $C_1, \ldots, C_n$ are zero. In other words, the equation

$$C_1 y_1 + \ldots C_n y_n = 0$$

with respect to the unknowns $C_1, \ldots, C_n$ has the only solution $C_1 = \ldots = C_n = 0$. Otherwise, functions $y_1, \ldots, y_n$ are *linearly dependent*.

Remark 1. This concept is usually introduced in Linear Algebra for general vector spaces and it is used here for the space of smooth functions over the field of real numbers $\mathbb{R}$.

Remark 2. If the set I is not explicitly specified, we consider I to be the domain of the set of functions $y_1, \ldots, y_n$.

Remark 3. The property of linear dependence in this definition is equivalent to the information that one of the functions in the set $y_1, \ldots, y_n$ can be expressed as a linear combination of the others. In fact, let us assume that y_1 is a linear combination of the other functions: $y_1 = a_2 y_2 + \ldots a_n y_n$. Then $-y_1 + a_2 y_2 + \ldots a_n y_n = 0$, i.e. the linear combination of all the functions is zero when $C_1 = -1, C_2 = a_2, \ldots, C_n = a_n$, that is, not all the constants C_i are zero. On the other hand, if $y_1, \ldots, y_n$ are linearly dependent, then there are constants $C_1, \ldots, C_n$, not all zero, such that the equation $C_1 y_1 + \ldots C_n y_n = 0$ is satisfied. Without loss of generality we can assume that one of the non-zero constants is C_1. Then y_1 can be expressed as a linear combination of the other functions: $y_1 = -\frac{C_2}{C_1}y_2 - \ldots - \frac{C_n}{C_1}y_n$. Of course, linear independence means that none of the functions in the set can be expressed as a linear combination of the other functions.

Remark 4. Obviously, the zero function together with any others always form a set of linearly dependent functions.

Theorem 5. General solution of homogeneous linear equation.
The general solution of a homogeneous linear equation

$$Ly = a_n(x)y^{(n)} + \ldots + a_1(x)y' + a_0(x)y = 0$$

can be found in the form

$$y_{gh} = C_1 y_1 + \ldots C_n y_n,$$

where $y_1, \ldots, y_n$ are linearly independent particular solutions of this equation and $C_1, \ldots, C_n$ are arbitrary constants. The general solution y_{gh} contains all the particular solutions.

Proof. The fact that $y_{gh} = C_1 y_1 + \ldots C_n y_n$ is the general solution follows from the properties of solutions of linear equations and the linear independence of functions $y_1, \ldots, y_n$. Indeed, if $y_1, \ldots, y_n$ are particular solutions of the linear equation, then, according to Corollary 1 to Theorem 2, their linear combination $C_1 y_1 + \ldots C_n y_n$ is also a solution of this equation. In addition, if $y_1, \ldots, y_n$ are linearly independent, then none of these solutions can be expressed as a linear combination of the others and therefore the number of arbitrary constants $C_1, \ldots, C_n$ cannot be reduced. In this case, since the number of arbitrary parameters $C_1, \ldots, C_n$ is equal to the order of the equation, then the definition of the general solution is satisfied.

The proof that the general solution y_{gh} contains all the particular solutions is more complicated. The following proof is therefore optional. Let us consider a particular solution y_p of the equation $Ly = 0$, that is, $Ly_p = 0$ is an identity. Choose a point, for example $x = 0$, and calculate the values of y_p and its derivatives up to order $n - 1$ at that point: $y_p(0) = y_0, \ldots y_p^{(n-1)} = y_{n-1}$. Then, y_p satisfies the following Cauchy problem
$$\begin{cases} Ly = 0 \\ y(0) = y_0, \ldots, y^{n-1}(0) = y_{n-1} \end{cases}$$
(we call it here the C problem). Consider now the same Cauchy problem for the function $y_{gh} = C_1 y_1 + \ldots C_n y_n$. First, according to the properties of solutions of homogeneous equations, y_{gh} satisfies the equation of the C problem whatever coefficients $C_1, \ldots, C_n$ are chosen. Second, by substituting the expression for y_{gh} into the initial

conditions, we obtain the following algebraic linear system

$$\begin{cases} C_1 y_1(0) + \ldots + C_n y_n(0) = y_0 \\ \quad \ldots \\ C_1 y_1^{(n-1)}(0) + \ldots + C_n y_n^{(n-1)}(0) = y_{n-1} \end{cases}$$

with respect to the unknowns $C_1, \ldots, C_n$. If the determinant of this system

$$W = \begin{vmatrix} y_1(0) & \cdots & y_n(0) \\ & \cdots & \\ y_1^{(n-1)}(0) & \cdots & y_n^{(n-1)}(0) \end{vmatrix}$$

is different from 0, then the system has the only solution.

So now the problem is to show that $W \neq 0$. To do this, we use the contradiction method: let us assume, for contradiction, that $W = 0$. Consider now the Cauchy problem with the same equation, but with the homogeneous initial conditions: $\begin{cases} Ly = 0 \\ y(0) = 0, \ldots, y^{n-1}(0) = 0 \end{cases}$. As before, y_{gh} satisfies the equation of the problem whatever coefficients $C_1, \ldots, C_n$ are chosen. Application of the homogeneous initial conditions leads to the following algebraic linear system for unknowns $C_1, \ldots, C_n$:

$$\begin{cases} C_1 y_1(0) + \ldots + C_n y_n(0) = 0 \\ \quad \ldots \\ C_1 y_1^{(n-1)}(0) + \ldots + C_n y_n^{(n-1)}(0) = 0 \end{cases}.$$

By the assumption, the determinant of this system is zero, which means that this system has an infinite number of solutions $C_1, \ldots, C_n$. In other words, the Cauchy problem with homogeneous conditions has many solutions, which contradicts the result of the Cauchy theorem about the existence and uniqueness of the solution. Therefore, the assumption that $W = 0$ is false, and therefore, $W \neq 0$.

With this result, we return to the C system. We can conclude that this system has a single solution $C_1, \ldots, C_n$. So, the Cauchy problem C has the solution $y_{gh} = C_1 y_1 + \ldots C_n y_n$ under a certain choice of coefficients $C_1, \ldots, C_n$. But the C problem also satisfies the conditions of the Cauchy theorem and therefore has a unique solution. Since y_p and y_{gh} (under prescribed values of $C_1, \ldots, C_n$) are solutions of the same C problem, they coincide, which means that the particular solution y_p lies in the family of functions $y_{gh} = C_1 y_1 + \ldots C_n y_n$. Due to the arbitrariness of y_p, the statement is proven.

Definition. The set $y_1, \ldots, y_n$ of n linearly independent particular solutions of an homogeneous linear equation of order n is called the *fundamental set of solutions*.

Remark. With the concept of the fundamental set of solutions, Theorem 5 can be reformulated as follows: the general solution of homogeneous linear equation is the linear combination of solutions of the fundamental set. This linear combination contains all the particular solutions.

Definition. The determinant W that was defined in the proof of the second part of Theorem 5 is called the *Wronski determinant* or the *Wronskian*.

Theorem 6. General solution of nonhomogeneous linear equation. The general solution of a nonhomogeneous linear equation

$$Ly = a_n(x)y^{(n)} + \ldots + a_1(x)y' + a_0(x)y = f(x)$$

can be found in the form

$$y_{gn} = y_{gh} + y_{pn},$$

where y_{gh} is the general solution of the corresponding homogeneous equation and y_{pn} is a particular solution of the nonhomogeneous equation. The general solution y_{gn} contains all the particular solutions.

Proof. The fact that $y_{gn} = y_{gh} + y_{pn}$ is the general solution follows from the properties of solutions of linear equations, more precisely, from Corollary 2 to Theorem 2.

The proof that the general solution y_{gn} contains all the particular solutions is the immediate consequence of the corresponding property of the solution of the homogeneous equation y_{gh}. In fact, let us consider any particular solution y_{pq} of the equation $Ly = f$. Then, from Corollary 3 to Theorem 2, it follows that $y_h = y_{pq} - y_{pn}$ is a particular solution of the homogeneous equation: $Ly_h = 0$. Therefore, it is contained in the general solution of the homogeneous y_{gh} under a certain choice of parameters C_i: $y_h = y_{gh}(\tilde{C}_1, \ldots, \tilde{C}_n)$. Therefore, $y_{pq} = y_h + y_{pn} = y_{gh}(\tilde{C}_1, \ldots, \tilde{C}_n) + y_{pn}$, that is, y_{pq} is included in the general solution y_{gn} when the parameters are equal to $\tilde{C}_1, \ldots, \tilde{C}_n$.

Examples.

1. Show that the functions $1, x$ are linearly independent.
Consider the equation $C_1 \cdot 1 + C_2 x = 0$ in $\mathbb{R}$. We have to show that this equation is satisfied for all $x \in \mathbb{R}$ only when $C_1 = C_2 = 0$. First, choosing $x = 0$ we conclude that $C_1 = 0$. With the first constant specified, we choose $x = 1$ and get $C_2 = 0$.

2. Show that the functions $1, x, x^2$ are linearly independent.

Consider the equation $C_1 \cdot 1 + C_2 x + C_3 x^2 = 0$ in $\mathbb{R}$. We have to show that this equation is satisfied for all $x \in \mathbb{R}$ only when $C_1 = C_2 = C_3 = 0$. Like in the previous example, we start by choosing $x = 0$ and specifying $C_1 = 0$. Next, we differentiate the remaining equation $C_2 x + C_3 x^2 = 0$ with respect to x and get $C_2 + 2 C_3 x = 0$. Again using $x = 0$, we get $C_2 = 0$. With the first two constants zero, we choose $x = 1$ in the remaining equation $C_3 x^2 = 0$ and find $C_3 = 0$. Therefore, $C_1 = C_2 = C_3 = 0$ is the only solution.

3. Show that the functions $e^{\alpha x}, e^{\beta x}, \alpha \neq \beta$ are linearly independent. Consider the equation $C_1 e^{\alpha x} + C_2 e^{\beta x} = 0$ in $\mathbb{R}$. We have to show that this equation is satisfied for all $x \in \mathbb{R}$ only if $C_1 = C_2 = 0$. We divide the equation by $e^{\alpha x}$ and get $C_1 + C_2 e^{(\beta - \alpha)x} = 0$. Next, we differentiate the last equation with respect to x and obtain $(\beta - \alpha) C_2 e^{(\beta - \alpha)x} = 0$, whence $C_2 = 0$ (because $\alpha \neq \beta$). Then, we return to the equation $C_1 + C_2 e^{(\beta - \alpha)x} = 0$ and applying $C_2 = 0$ we conclude that $C_1 = 0$.

4. Show that the functions $\cos \alpha x, \sin \alpha x, \alpha \neq 0$ are linearly independent. Consider the equation $C_1 \cos \alpha x + C_2 \sin \alpha x = 0$ in $\mathbb{R}$. We have to show that this equation is satisfied for all $x \in \mathbb{R}$ only if $C_1 = C_2 = 0$. Using $x = 0$, we get that $C_1 = 0$. Applying this result and using $x = \frac{\pi}{\alpha}$ in the original equation, we find that $C_2 = 0$.

5. Show that the functions $e^{\alpha x}, x e^{\alpha x}, \ldots, x^k e^{\alpha x}, k \in \mathbb{N}$ are linearly independent. Consider the equation

$$C_1 e^{\alpha x} + C_2 x e^{\alpha x} + \ldots + C_{k-1} x^{k-1} e^{\alpha x} + C_k x^k e^{\alpha x} = 0$$

in $\mathbb{R}$. First, cut the common factor $e^{\alpha x}$ to simplify the equation:

$$C_1 + C_2 x + \ldots + C_{k-1} x^{k-1} + C_k x^k = 0.$$

Next, differentiate the last equation successively k times to obtain the following consequences:

$$C_2 + \ldots + C_{k-1}(k-1)x^{k-2} + C_k k x^{k-1} = 0, \ldots, C_{k-1}(k-1)! + C_k k! x = 0, \ C_k k! = 0.$$

Now, move backward, specifying one by one all the coefficients: the last equation implies that $C_k = 0$, the penultimate (together with $C_k = 0$) implies that $C_{k-1} = 0$, and so on, the first in the last row of the formulas (together with $C_k = C_{k-1} = \ldots = C_3 = 0$) shows that $C_2 = 0$, and finally, the original equation shows that $C_1 = 0$. Therefore, the functions are linearly independent.

Exercises for section 2

1. Show that the following functions are linearly independent:
1) $e^{\alpha x}, xe^{\alpha x}$;
2) $\cos \beta x e^{\alpha x}, \sin \beta x e^{\alpha x}, \beta \neq 0$;
3) $\cos \beta x, x \cos \beta x$;
4) $x \cos \alpha x, x \sin \alpha x, \alpha \neq 0$;
5) $x^2 e^{\alpha x}, x^2 e^{\beta x}, \alpha \neq \beta$.

2. Show that the given functions form the fundamental set of solutions of the homogeneous equation and write the general solution of this equation:
1) $1, x, x^2, y''' = 0$;
2) $e^x, e^{-x}, y'' - y = 0$;
3) $\cos 2x, \sin 2x, y'' + 4y = 0$;
4) $e^x, e^{2x}, e^{3x}, y''' - 6y'' +'' y' - 6y = 0$;
5) $\cos xe^x, \sin xe^x, y'' - 2y' + 2y = 0$.

3. Show that pairwise linear independence does not guarantee linear independence of the total set of functions. (Hint: consider the set of functions $1, x, 1 + x$.)

Chapter 9

Linear equations with constant coefficients: methods of solution

1 Homogeneous equations

Definition. A *homogeneous linear equation with constant coefficients* has the form

$$Ly = a_n y^{(n)} + \ldots + a_1 y' + a_0 y = 0,$$

where $a_n, \ldots, a_1, a_0$, $a_n \neq 0$ are constant coefficients and n is the order of the equation.

Method of solution.

Let us start with an elementary observation that for exponential function $e^{\lambda x}$ the differentiation is reduced to multiplication by λ and the derivative of order k results in multiplication by λ^k: $(e^{\lambda x})' = \lambda e^{\lambda x}, \ldots, (e^{\lambda x})^{(n)} = \lambda^n e^{\lambda x}$. Then, for $y = e^{\lambda x}$ the linear differential equation with constant coefficients becomes an algebraic equation with respect to the unknown λ: $a_n y^{(n)} + \ldots + a_1 y' + a_0 y = a_n \lambda^n e^{\lambda x} + \ldots + a_1 \lambda e^{\lambda x} + a_0 e^{\lambda x} = e^{\lambda x}(a_n \lambda^n + \ldots + a_1 \lambda + a_0) = 0$ or $a_n \lambda^n + \ldots + a_1 \lambda + a_0 = 0$. By finding the roots of the last equation, we will have the corresponding solutions of the differential equation, and if the number of independent solutions is n, then we will have the fundamental system of solutions and can form the general solution of the differential equation. This is the simple and general idea behind the method of solution of homogeneous equations with constant coefficients. Because of its importance, the algebraic equation for λ has its own name.

Definition. Given the equation with constant coefficients

$$a_n y^{(n)} + \ldots + a_1 y' + a_0 y = 0,$$

the corresponding algebraic equation

$$a_n \lambda^n + \ldots + a_1 \lambda + a_0 = 0$$

is called a *characteristic equation* and the corresponding polynomial

$$P_n(\lambda) = a_n \lambda^n + \ldots + a_1 \lambda + a_0$$

is called a *characteristic polynomial.*

Let us see what can happen in this general solution algorithm under different conditions regarding the roots of a characteristic equation.

Case 1: All the roots of the characteristic equation are real and simple.

In this case, we have n different roots $\lambda_1, \ldots, \lambda_n$, each of which gives rise to the corresponding solution $y_1 = e^{\lambda_1 x}, \ldots, y_n = e^{\lambda_n x}$. In section 2 of Chapter 8 (see Example 3), it was already shown that the exponential functions $e^{\alpha x}$ and $e^{\beta x}$ are linearly independent, and, in general way, $e^{\lambda_1 x}, \ldots, e^{\lambda_n x}$ form a system of n linearly independent functions, that is, a fundamental system of solutions of the differential equation. Therefore, the general solution (which contains all the particular ones) is found in the form

$$y_{gh} = C_1 y_1 + \ldots + C_n y_n = C_1 e^{\lambda_1 x} + \ldots + C_n e^{\lambda_n x}.$$

Examples.

1. $y'' = y$.
This is equation 4 from Table 1 in section 2 of Chapter 1. Its characteristic equation is $\lambda^2 - 1 = 0$ with the roots $\lambda_1 = -1$ and $\lambda_2 = 1$. Therefore, the two linearly independent solutions are $y_1 = e^{-x}$ and $y_2 = e^x$, and the general solution has the form

$$y_{gh} = C_1 e^{-x} + C_2 e^x.$$

2. $y'' - 2y' - 3y = 0$.
The characteristic equation is $\lambda^2 - 2\lambda - 3 = 0$ with the roots $\lambda_1 = -1$ and

$\lambda_2 = 3$. Therefore, the two linearly independent solutions are $y_1 = e^{-x}$ and $y_2 = e^{3x}$, and the general solution has the form

$$y_{gh} = C_1 e^{-x} + C_2 e^{3x}.$$

3. $y''' - 6y'' + 11y' - 6y = 0$.
The characteristic equation $\lambda^3 - 6\lambda^2 + 11\lambda - 6 = 0$ has the roots $\lambda_1 = 1$, $\lambda_2 = 2$ and $\lambda_3 = 3$. Therefore, the three linearly independent solutions are $y_1 = e^x$, $y_2 = e^{2x}$ and $y_3 = e^{3x}$. Consequently, the general solution is

$$y_{gh} = C_1 e^x + C_2 e^{2x} + C_3 e^{3x}.$$

4. $y''' - 7y'' + 6y' = 0$, $y(0) = 0, y'(0) = 0, y''(0) = 30$.
As usual, we find first the general solution of the equation and then apply the initial conditions. The characteristic equation is $\lambda^3 - 7\lambda^2 + 6\lambda = 0$ with the roots $\lambda_1 = 0$, $\lambda_2 = 1$ and $\lambda_3 = 6$. Since all the roots are real and simple, the three linearly independent solutions are $y_1 = 1$, $y_2 = e^x$ and $y_3 = e^{6x}$. So the general solution has the form $y_{gh} = C_1 + C_2 e^x + C_3 e^{6x}$. Substituting the initial conditions into y_{gh}, we get the system $C_1 + C_2 + C_3 = 0, C_2 + 6C_3 = 0, C_2 + 36C_3 = 30$, whose solution is $C_1 = 5, C_2 = -6, C_3 = 1$. So, the solution of the Cauchy problem is

$$y = 5 - 6e^x + e^{6x}.$$

5. $y'' + 3y' + 2y = 0$.
The characteristic equation is $\lambda^2 + 3\lambda + 2 = 0$, whose roots $\lambda_1 = -2, \lambda_2 = -1$ are real and simple. So, the two linearly independent solutions are $y_1 = e^{-2x}$ and $y_2 = e^{-x}$, and the general solution has the form

$$y_{gh} = C_1 e^{-2x} + C_2 e^{-x}.$$

6. $y^{(4)} - 5y'' + 4y = 0$, $y(0) = -2, y'(0) = 1, y''(0) = 2, y'''(0) = 0$.
First, we find the general solution of the equation. The characteristic equation $\lambda^4 - 5\lambda^2 + 4 = 0$ can be represented in the form $(\lambda^2 - 1)(\lambda^2 - 4) = 0$ which simplifies the finding of four distinct real roots: $\lambda_{1,2} = \pm 1, \lambda_{3,4} = \pm 2$. Consequently, the general solution is $y_{gh} = C_1 e^{-x} + C_2 e^x + C_3 e^{-2x} + C_4 e^{2x}$. Next, we calculate the derivatives of this solution

$$y'_{gh} = -C_1 e^{-x} + C_2 e^x - 2C_3 e^{-2x} + 2C_4 e^{2x},$$

$$y''_{gh} = C_1 e^{-x} + C_2 e^x + 4C_3 e^{-2x} + 4C_4 e^{2x},$$

$$y_{gh}''' = -C_1 e^{-x} + C_2 e^x - 8C_3 e^{-2x} + 8C_4 e^{2x}$$

and substitute the initial conditions into the found expressions:

$$y(0) = C_1 + C_2 + C_3 + C_4 = -2, \ y'(0) = -C_1 + C_2 - 2C_3 + 2C_4 = 1,$$

$$y''(0) = C_1 + C_2 + 4C_3 + 4C_4 = 2, \ y'''(0) = -C_1 + C_2 - 8C_3 + 8C_4 = 0.$$

Solving this system, we find $C_1 = -\frac{7}{3}, C_2 = -1, C_3 = \frac{3}{4}, C_4 = \frac{7}{12}$. Therefore, the solution of the Cauchy problem is

$$y = -\frac{7}{3} e^{-x} - e^x + \frac{3}{4} e^{-2x} + \frac{7}{12} e^{2x}.$$

Case 2: All the roots of the characteristic equation are real, but there are multiple roots.

The simple roots are treated in the same way as in Case 1 and, in addition, each multiple root (as well as a simple one) is treated independently from other roots. So let us take one of the multiple roots and see how to construct the corresponding independent solutions. Let λ_1 be a root of multiplicity $k \le n$, that is, $\lambda_1 = \lambda_2 = \ldots = \lambda_k$. Formally, we have k associated solutions $y_1 = e^{\lambda_1 x}, y_2 = e^{\lambda_2 x}, \ldots, y_k = e^{\lambda_k x}$, but all of them are equal (or one is a multiple of the other, since $Ce^{\lambda_1 x}$ is also a solution for any constant C). Obviously, these functions are linearly dependent; even worse, any pair of these functions is linearly dependent. Thus, there is only one function $y_1 = e^{\lambda_1 x}$ that can be taken from this entire family. The remaining $k-1$ functions corresponding to λ_1 must be obtained in another form.

It turns out that this form exists and is not much more complicated than the form of the first function: the other linearly independent functions can be found in the form $y_2 = x e^{\lambda_1 x}, \ldots, y_k = x^{k-1} e^{\lambda_1 x}$. In section 3 of Chapter 8 (see Example 5) it was already shown that the functions of this type are linearly independent. It remains to verify that each of them is a solution to the given differential equation. We will do this for the second function $y_2 = x e^{\lambda_1 x}$, since the procedure for the other functions is similar, albeit more tedious. Calculating the derivatives of y_2 successively:

$$y_2' = e^{\lambda_1 x} + \lambda_1 x e^{\lambda_1 x}, \ y_2'' = 2\lambda_1 e^{\lambda_1 x} + \lambda_1^2 x e^{\lambda_1 x}, \ \ldots,$$

$$y_2^{(n)} = n\lambda_1^{n-1} e^{\lambda_1 x} + \lambda_1^n x e^{\lambda_1 x},$$

and substituting them into the original equation, we get:

$$Ly_2 = a_n y_2^{(n)} + \ldots + a_1 y_2' + a_0 y_2$$

$$= a_n(n\lambda_1^{n-1}e^{\lambda_1 x} + \lambda_1^n xe^{\lambda_1 x}) + \ldots + a_1(e^{\lambda_1 x} + \lambda_1 xe^{\lambda_1 x}) + a_0 xe^{\lambda_1 x}$$

$$= e^{\lambda_1 x}\left[x(a_n\lambda_1^n + \ldots + a_1\lambda_1 + a_0) + (na_n\lambda_1^n + \ldots + a_1)\right]$$

$$= e^{\lambda_1 x}\left[xP(\lambda_1) + P_\lambda(\lambda_1)\right] = 0.$$

Therefore, y_2 is the solution of the original equation. Note that in the last formula we have used the two properties of the characteristic polynomial $P(\lambda) = a_n\lambda^n + \ldots + a_1\lambda + a_0$: first, $P(\lambda_1) = 0$ because λ_1 is a root of $P(\lambda)$, and second, $P_\lambda(\lambda_1) = na_n\lambda_1^{n-1} + \ldots + a_1 = 0$ because λ_1 is a multiple root of $P(\lambda)$. The same result can be shown for the other functions $y_3 = x^2 e^{\lambda_1 x}, \ldots, y_k = x^{k-1}e^{\lambda_1 x}$. Thus, the set of functions $y_1 = e^{\lambda_1 x}, y_2 = xe^{\lambda_1 x}, \ldots, y_k = x^{k-1}e^{\lambda_1 x}$ is linearly independent and each of these functions is a solution of the original equation. Thus, a root of multiplicity k generates a set of k linearly independent solutions. The same happens with other multiple roots. Gathering all the solutions, we find a fundamental system of solutions of the differential equation and the corresponding general solution.

Examples.

1. $y''' - y'' - y' + y = 0$.
The characteristic equation $\lambda^3 - \lambda^2 - \lambda + 1 = 0$ has the double root $\lambda_1 = \lambda_2 = 1$ and the simple root $\lambda_3 = -1$. According to the theory, the double root produces two independent solutions in the form $y_1 = e^x$ and $y_2 = xe^x$. The single root generates the solution $y_3 = e^{-x}$ which is independent of the first two. So the general solution has the form

$$y_{gh} = C_1 e^x + C_2 xe^x + C_3 e^{-x}.$$

2. $4y^{(4)} + 4y'' + y'' = 0$.
The characteristic equation $4\lambda^4 + 4\lambda^3 + \lambda^2 = 0$ has the two double roots: $\lambda_1 = \lambda_2 = \frac{1}{2}$ and $\lambda_3 = \lambda_4 = 0$. The first two independent solutions take the form $y_1 = e^{x/2}, y_2 = xe^{x/2}$ and the second pair is $y_3 = 1, y_4 = x$. Thus, the general solution has the form

$$y_{gh} = C_1 e^{x/2} + C_2 xe^{x/2} + C_3 + C_4 x.$$

3. $y''' - y'' - y' + y = 0$.
The characteristic equation $\lambda^3 - \lambda^2 - \lambda + 1 = 0$ has the double root $\lambda_1 = \lambda_2 = 1$ and a simple root $\lambda_3 = -1$. According to the theory, the double root produces two independent solutions in the form $y_1 = e^x$ and $y_2 = xe^x$. The

single root generates the solution $y_3 = e^{-x}$ which is independent of the first two. So the general solution has the form

$$y_{gh} = C_1 e^x + C_2 x e^x + C_3 e^{-x}.$$

Case 3. There are complex roots, but all of them are simple.

As we have seen before, in the case of a real root, each root is treated separately from the others, whether it is simple or multiple. The case of complex roots is a bit different: each complex root is considered together with its complex conjugate, but separately from all other roots. (Recall that in a polynomial equation with real coefficients, a complex root always appears together with its complex conjugate with equal multiplicity.) So, let us take a pair of complex conjugate roots $\lambda_1 = \alpha + i\beta$ and $\lambda_2 = \overline{\lambda}_1 = \alpha - i\beta$, $\alpha, \beta \in \mathbb{R}$ and find the form of the corresponding solution.

First, we show that a function $y_1 = e^{\lambda_1 x}$ of a real variable with a complex image is the solution of the differential equation (likewise the function with a real exponent), and so is $y_2 = e^{\overline{\lambda}_1 x}$. Indeed, if $y(x) = u(x) + iv(x)$, where $u(x)$ and $v(x)$ are real functions, then, by the definition, $y' = u' + iv'$ and the same simple rule is valid for derivatives of any order, that is, $y^{(k)} = u^{(k)} + iv^{(k)}$. To represent the function $y_1 = e^{\lambda_1 x}$ through its real and imaginary parts, we use Euler's formula $e^{i\gamma} = \cos\gamma + i\sin\gamma$ valid for any $\gamma \in \mathbb{R}$. Then, we have

$$y_1 = e^{\lambda_1 x} = e^{(\alpha + i\beta)x} = e^{\alpha x} e^{i\beta x} = e^{\alpha x} \cos\beta x + i e^{\alpha x} \sin\beta x.$$

Therefore,

$$y_1' = (e^{\lambda_1 x})' = (e^{\alpha x} \cos\beta x + i e^{\alpha x} \sin\beta x)' = (e^{\alpha x} \cos\beta x)' + i(e^{\alpha x} \sin\beta x)'$$

$$= (\alpha e^{\alpha x} \cos\beta x - \beta e^{\alpha x} \sin\beta x) + i(\alpha e^{\alpha x} \sin\beta x + \beta e^{\alpha x} \cos\beta x)$$

$$= \alpha e^{\alpha x}(\cos\beta x + i\sin\beta x) + \beta e^{\alpha x}(-\sin\beta x + i\cos\beta x)$$

$$= \alpha e^{\alpha x} e^{i\beta x} + i\beta e^{\alpha x} e^{i\beta x} = (\alpha + i\beta)e^{(\alpha + i\beta)x} = \lambda_1 e^{\lambda_1 x}.$$

Thus, we have the same simple formula for differentiation of the function $e^{\lambda_1 x}$ both for the real and complex λ_1: $(e^{\lambda_1 x})' = \lambda_1 e^{\lambda_1 x}$. The same is true for the function $y_2 = e^{\overline{\lambda}_1 x}$. Therefore, both functions are solutions of the original equation. In this way, a pair of simple complex conjugate roots generates a pair of linearly independent solutions. If the functions with complex image are admissible under the conditions of the problem, then these functions enter into the formation of the fundamental set of solutions.

Sometimes it is necessary or preferable to represent the solution in real form (for instance, if the equation is originated from the physical problem, then the real and imaginary parts of the solution have often a clear physical interpretation). In this case, using the Corollary to Theorem 3 in Chapter 8, we can replace the two solutions with complex image $y_1 = e^{\lambda_1 x}$ and $y_2 = e^{\overline{\lambda}_1 x}$ by the two real solutions $u = Re(y_1) = e^{\alpha x} \cos \beta x$ and $v = Im(y_1) = e^{\alpha x} \sin \beta x$ (recall that $Re(y_1)$ and $Im(y_1)$ stand for the real and imaginary part of the complex function y_1), which are also linearly independent and can be included in the fundamental system of solutions.

We note that if the requirement of the real form of solutions appears in the problem formulation, the two linearly independent solutions corresponding to the pair of (simple) complex conjugate roots $\lambda_{1,2} = \alpha \pm i\beta$ can be found directly in the form $u = e^{\alpha x} \cos \beta x$ and $v = e^{\alpha x} \sin \beta x$, without using the complex form.

Examples.

1. $y'' + 9y = 0$.
The characteristic equation $\lambda^2 + 9 = 0$ has two conjugate complex roots $\lambda_1 = 3i$ and $\lambda_2 = \overline{\lambda}_1 = -3i$. The corresponding solutions in the complex form are $y_1 = e^{3ix}$ and $y_2 = e^{-3ix}$, which generate the general solution $y_{gh} = C_1 e^{3ix} + C_2 e^{-3ix}$. If we need the solution in real form, then we replace y_1 and y_2 by $u = Re(y_1) = \cos 3x$ and $v = Im(y_1) = \sin 3x$. The general solution will be

$$y_{gh} = C_1 \cos 3x + C_2 \sin 3x.$$

2. $y'' + \omega^2 y = 0$, $\omega \in \mathbb{R}$.
The situation with the roots of the characteristic equation $\lambda^2 + \omega^2 = 0$ differs if $\omega = 0$ or $\omega \neq 0$. In the first case, we have a double real root $\lambda_{1,2} = 0$ and the linearly independent particular solutions are $y_1 = 1$ and $y_2 = x$, resulting in the general solution

$$y_{gh} = C_1 + C_2 x.$$

If $\omega \neq 0$, the equation has the two complex conjugate roots $\lambda_1 = i\omega$ and $\lambda_2 = \overline{\lambda}_1 = -i\omega$. The corresponding solutions in the complex form are $y_1 = e^{i\omega x}$ and $y_2 = e^{-i\omega x}$, which lead to the general solution $y_{gh} = C_1 e^{i\omega x} + C_2 e^{-i\omega x}$. To obtain the solutions in real form, we replace y_1 and y_2 by $u = Re(y_1) = \cos \omega x$ and $v = Im(y_1) = \sin \omega x$. So the general solution will be

$$y_{gh} = C_1 \cos \omega x + C_2 \sin \omega x.$$

3. $y'' - 4y' + 13y = 0$.

The characteristic equation $\lambda^2 - 4\lambda + 13 = 0$ has the two complex conjugate roots $\lambda_1 = 2 + 3i$ and $\lambda_2 = \overline{\lambda}_1 = 2 - 3i$. The corresponding solutions in the complex form are $y_1 = e^{(2+3i)x}$ and $y_2 = e^{(2-3i)x}$, which generate the general solution $y_{gh} = C_1 e^{(2+3i)x} + C_2 e^{(2-3i)x}$. If we need the solution in real form, then we replace y_1 and y_2 with $u = Re(y_1) = e^{2x}\cos 3x$ and $v = Im(y_1) = e^{2x}\sin 3x$. The general solution will be

$$y_{gh} = (C_1 \cos 3x + C_2 \sin 3x)e^{2x}.$$

4. $y^{(4)} + 3y'' - 4y = 0$.

The characteristic equation $\lambda^4 + 3\lambda^2 - 4 = 0$ has the two simple real roots $\lambda_1 = 1$, $\lambda_2 = -1$ and two complex conjugate roots $\lambda_3 = 2i$ and $\lambda_4 = \overline{\lambda}_3 = -2i$. The solutions corresponding to the real roots are $y_1 = e^x$ and $y_2 = e^{-x}$. The solutions associated with the complex roots are $y_3 = e^{2ix}$ and $y_4 = e^{-2ix}$. If we keep the complex form of solutions, then the general solution will be $y_{gh} = C_1 e^x + C_2 e^{-x} + C_3 e^{2ix} + C_4 e^{-2ix}$. If we prefer the real form, then we replace y_3 and y_4 by $u = Re(y_3) = \cos 2x$ and $v = Im(y_3) = \sin 2x$, and the general solution is

$$y_{gh} = C_1 e^x + C_2 e^{-x} + C_3 \cos 2x + C_4 \sin 2x.$$

Case 4: There are multiple complex roots.

Since complex roots appear in conjugate pairs and these pairs are treated independently from the other roots, whether simple or multiple, then we consider a pair of complex conjugate roots of multiplicity k, where $2k \le n$: $\lambda_1 = \alpha + i\beta$ and $\lambda_2 = \overline{\lambda}_1 = \alpha - i\beta$, $\alpha, \beta \in \mathbb{R}$.

It has already been shown (see Case 3) that the rule of differentiation of the exponential function $e^{\lambda_1 x}$ with a complex exponent λ_1 is the same as for real exponents. Therefore, the linearly independent solutions related to the complex root λ_1 can be found in the form similar to the case of multiple real roots, that is, $y_1 = e^{\lambda_1 x}$, $y_2 = xe^{\lambda_1 x}, \ldots, y_k = x^{k-1}e^{\lambda_1 x}$. The same is true for the root $\lambda_2 = \overline{\lambda}_1$: $z_1 = e^{\lambda_2 x}$, $z_2 = xe^{\lambda_2 x}, \ldots, z_k = x^{k-1}e^{\lambda_2 x}$. In this way, $2k$ complex roots generate $2k$ linearly independent solutions that are part of the fundamental system of solutions.

If there is a need to find the real form of the fundamental system and the general solution, then we transform solutions $y_1 = e^{\lambda_1 x}$, $y_2 = xe^{\lambda_1 x}, \ldots, y_k = x^{k-1}e^{\lambda_1 x}$ into the real solutions using the Corollary to Theorem 3 of Chapter 8 (as in Case 3): the functions y_1 and z_1 are replaced by the pair $u_1 = Re(y_1) = e^{\alpha x}\cos \beta x$ and $v_1 = Im(y_1) = e^{\alpha x}\sin \beta x$, the functions y_2 and z_2

are replaced by $u_2 = Re(y_2) = xe^{\alpha x} \cos \beta x$ and $v_2 = Im(y_2) = xe^{\alpha x} \sin \beta x$, and so on; finally, in place of y_k and z_k we use $u_k = Re(y_k) = x^{k-1}e^{\alpha x} \cos \beta x$ and $v_k = Im(y_k) = x^{k-1}e^{\alpha x} \sin \beta x$. Thus, $2k$ complex roots generate $2k$ linearly independent real solutions. We note that these solutions in the real form can be used directly, without going through the complex form.

Examples.

1. $y^{(4)} + 2y'' + y = 0$.
The characteristic equation $\lambda^4 + 2\lambda^2 + 1 = 0$ has the two complex conjugate roots of multiplicity 2: $\lambda_1 = i$ and $\lambda_2 = \overline{\lambda}_1 = -i$. The corresponding solutions in the complex form are $y_1 = e^{ix}$, $y_2 = xe^{ix}$ and $z_1 = e^{-ix}$, $z_2 = xe^{-ix}$. They form the general solution $y_{gh} = C_1e^{ix} + C_2e^{-ix} + C_3xe^{ix} + C_4xe^{-ix}$. To get the solution in the real form, we replace y_1 and z_1 by $u_1 = Re(y_1) = \cos x$ and $v_1 = Im(y_1) = \sin x$, and also y_2 and z_2 by $u_2 = Re(y_2) = x\cos x$ and $v_2 = Im(y_2) = x\sin x$. Then, the general solution will be

$$y_{gh} = C_1 \cos x + C_2 \sin x + C_3 x \cos x + C_4 x \sin x.$$

2. $y^{(5)} - y^{(4)} + 8y''' - 8y'' + 16y' - 16y = 0$.
The characteristic polynomial $P(\lambda) = \lambda^5 - \lambda^4 + 8\lambda^3 - 8\lambda^2 + 16\lambda - 16$ can be factored into the form $P(\lambda) = (\lambda - 1)(\lambda^2 + 4)^2$ from which it follows that the characteristic equation has the real root $\lambda_1 = 1$ and the two complex conjugate roots of multiplicity 2: $\lambda_2 = 2i$ and $\lambda_3 = \overline{\lambda}_2 = -2i$. So, the first solution is $y_1 = e^{-x}$ and the remaining four have the complex form $y_2 = e^{2ix}$, $y_3 = xe^{2ix}$ and $z_2 = e^{-2ix}$, $z_3 = xe^{-2ix}$. They form the general solution $y_{gh} = C_1e^x + C_2e^{2ix} + C_3e^{-2ix} + C_4xe^{2ix} + C_5xe^{-2ix}$. To obtain the solution in real form, we replace y_2 and z_2 by $u_2 = Re(y_2) = \cos 2x$ and $v_2 = Im(y_2) = \sin 2x$, and also y_3 and z_3 by $u_3 = Re(y_3) = x\cos 2x$ and $v_3 = Im(y_3) = x\sin 2x$. Then, the general solution will be

$$y_{gh} = C_1e^x + C_2 \cos 2x + C_3 \sin 2x + C_4 x \cos 2x + C_5 x \sin 2x.$$

Exercises for section 1

Solve the following homogeneous equations:
1. $y'' + y' - 2y = 0$.
2. $y'' + 4y' + 3y = 0$.
3. $y'' - 2y' = 0$.
4. $2y'' - 5y' + 2y = 0$.
5. $y'' - 3y' + 10y = 0$.

6. $y'' - 2y' + y = 0$.
7. $4y'' + 4y' + y = 0$.
8. $y^{(5)} - 6y^{(4)} + 9y''' = 0$.
9. $y''' - 3y'' + 3y' - y = 0$.
10. $y''' - 3y' + 2y = 0$.
11. $y'' - 4y' + 5y = 0$.
12. $y'' + 2y' + 10y = 0$.
13. $y'' + 4y = 0$.
14. $y''' - 8y = 0$.
15. $y^{(4)} - y = 0$.
16. $y^{(4)} + 4y = 0$.
17. $y'' + 4y' + 20y = 0$.
18. $y''' - y'' + 4y' - 4y = 0$.
19. $y^{(4)} + 2y'' + y = 0$.
20. $y^{(5)} + 8y''' + 16y' = 0$.

2 Nonhomogeneous equations: method of undetermined coefficients

Definition. A *nonhomogeneous linear equation with constant coefficients* has the form

$$Ly = a_n y^{(n)} + \ldots + a_1 y' + a_0 y = f,$$

where $a_n, \ldots, a_1, a_0$, $a_n \neq 0$, are constant coefficients, $f \neq 0$ is a function of x and n is the order of the equation.

There are different methods of solution of nonhomogeneous equations. Let us start with the method of undetermined coefficients.

This method is based on the fact that the general solution of a nonhomogeneous equation can be represented as the sum of the general solution of the corresponding homogeneous equation and a particular (any) solution of the nonhomogeneous equation (see Theorem 6 in Chapter 8): $y_{gn} = y_{gh} + y_{pn}$. The problem of finding the general solution of the homogeneous equation was solved in the previous section. So, if we can find any particular solution of the nonhomogeneous one, the current problem will also be solved. The method of undetermined coefficients is only applicable to some forms of the right-hand side, in which case a particular solution can be sought in a form similar to $f(x)$. This similarity between the forms of the right-hand side f and the particular solution y_{pn} makes it possible to suggest that the search for a particular solution in a given form involves some constants that must be found when substituting this form into the original equation. In what follows, we consider the two forms of the right-hand side, involving important and fairly generic functions, which admit the use of the method of undetermined coefficients.

2.1 Case 1: Right side $f = P_m(x)e^{\alpha x}$

Let us consider the equation $Ly = P_m(x)e^{\alpha x}$, where $P_m(x)$ is a polynomial of degree m and $\alpha \in \mathbb{C}$. Although we are currently interested in the right-hand sides with the real exponent α, in order to be able to use this method in other situations, we indicate that it also works when α is a complex number. For this (quite general) right-hand side, the particular solution can be found in the form $y_{pn} = x^s Q_m(x)e^{\alpha x}$, where s is the multiplicity of α as a root of the characteristic equation (that is, $s = 0$ if α is not a root, $s = 1$ if α is a simple root, and so on) and $Q_m(x)$ is a polynomial of degree m whose coefficients must be determined by substituting y_{pn} into the equation.

Let us establish a natural order in which the general solution of homogeneous equation and a particular solution of nonhomomeneous one are found in the method of undetermined coefficients. As can be seen, the construction of the general solution of the nonhomogeneous equation requires a knowledge of the general solution of the homogeneous part and, in addition, the form of the particular solution of the nonhomogeneous part itself depends on the values of the roots of the characteristic equation (used in the formation of the solutions of the homogeneous part). On the other hand, the finding of the general solution of the corresponding homogeneous equation does not depend on the form of a particular solution of the nonhomogeneous equation. Therefore, the algorithm of solution of a nonhomogeneous equation usually starts with its homogeneous part and then goes on finding a particular solution of the nonhomogeneous equation.

Examples.

1. $y'' - 8y' + 16y = (1 - x)e^x$.
We start by finding the general solution y_{gh} of the corresponding homogeneous equation $y'' - 8y' + 16y = 0$. The characteristic equation $\lambda^2 - 8\lambda + 16 = 0$ has the double real root $\lambda_1 = \lambda_2 = 4$. The corresponding solutions $y_1 = e^{4x}$ and $y_2 = xe^{4x}$ generate the general solution $y_{gh} = (C_1 + C_2 x)e^{4x}$. The particular solution of the nonhomogeneous equation can be sought in the form $y_{pn} = x^0(ax + b)e^x$ ($s = 0$ because $\alpha = 1 \neq 4$, and the polynomial $ax + b$ has degree 1 because $1 - x$ is the polynomial of degree 1), where the coefficients a and b must be found by substituting y_{pn} into the nonhomogeneous equation.

Taking y_{pn} into the original equation, we get

$$(ax + b + 2a)e^x - 8(ax + b + a)e^x + 16(ax + b)e^x = (1 - x)e^x$$

Simplifying and dividing by the exponential function e^x, we arrive at the polynomial equation $9ax + 9b - 6a = 1 - x$. The last equation is satisfied if and only if the coefficients with the corresponding powers are equal, which leads to the following system for a and b: $\begin{cases} 9a = -1 \\ -6a + 9b = 1 \end{cases}$. Then, we have $a = -\frac{1}{9}$, $b = \frac{1}{27}$. Therefore, the particular solution is found in the form $y_{pn} = \left(-\frac{1}{9}x + \frac{1}{27}\right)e^x$ and the general solution of the original equation is

$$y_{gn} = y_{gh} + y_{pn} = (C_1 + C_2 x)e^{4x} + \left(-\frac{1}{9}x + \frac{1}{27}\right)e^x.$$

2. $y'' - 8y' + 16y = (1 - x)e^{4x}$.

The homogeneous part of this equation is the same as in the previous example, so $y_{gh} = (C_1 + C_2 x)e^{4x}$. The particular solution of the nonhomogeneous equation should be sought in the different form $y_{pn} = x^2(ax + b)e^{4x}$ ($s = 2$ because $\alpha = 4$ is the double root of the characteristic equation), where the coefficients a and b are determined by substituting y_{pn} into the nonhomogeneous equation. Taking y_{pn} into the original equation, we get

$$(16ax^3 + 16bx^2 + 24ax^2 + 16bx + 6ax + 2b)e^{4x}$$

$$-8(4ax^3 + 4bx^2 + 3ax^2 + 2bx)e^{4x} + 16(ax^3 + bx^2)e^{4x} = (1 - x)e^{4x}.$$

Dividing by e^{4x} and simplifying, we obtain the polynomial equation $6ax + 2b = 1 - x$, which is equivalent to the following system of decoupled equations: $\begin{cases} 6a = -1 \\ 2b = 1 \end{cases}$, whence $a = -\frac{1}{6}$ and $b = \frac{1}{2}$. Therefore, the particular solution comes in the form $y_{pn} = \left(-\frac{1}{6}x^3 + \frac{1}{2}x^2\right)e^{4x}$ and the general solution of the original equation takes the form

$$y_{gn} = y_{gh} + y_{pn} = (C_1 + C_2 x)e^{4x} + \left(-\frac{1}{6}x^3 + \frac{1}{2}x^2\right)e^{4x}.$$

3. $y'' - 8y' + 16y = (3x - 3)e^x + (2 - 2x)e^{4x}$.

The homogeneous part of this equation is the same as in Examples 1 and 2, so $y_{gh} = (C_1 + C_2 x)e^{4x}$. But the right-hand side $f = (3x - 3)e^x + (2 - 2x)e^{4x}$ does not fit directly the type of the right-hand side $f = P_m(x)e^{\alpha x}$ that we know how to handle. So, to find the particular solution of the nonhomogeneous equation, we have to separate f into the two parts $f = f_1 + f_2$, where $f_1 = (3x-3)e^x$ and $f_2 = (2-2x)e^{4x}$ are the functions in the form admissible in the employed method. Having found the solution y_1 of f_1 and y_2 of f_2, we add the two solutions and, by the properties of linear equations (Theorem 2 in Chapter 8), the function $y_{pn} = y_1 + y_2$ will be a particular solution of the original equation. To find y_1 and y_2, we simply observe that, in this specific case, the function f_1 is the right-hand side of Example 1 multiplied by -3, and the function f_2 is the right-hand side of Example 2 multiplied by 2. So, again using the properties of linear equations (Theorem 2 in Chapter 8), we conclude that $y_1 = -3\left(-\frac{1}{9}x + \frac{1}{27}\right)e^x$ and $y_2 = 2\left(-\frac{1}{6}x^3 + \frac{1}{2}x^2\right)e^{4x}$. Then, $y_{pn} = y_1 + y_2 = \left(\frac{1}{3}x - \frac{1}{9}\right)e^x + \left(-\frac{1}{3}x^3 + x^2\right)e^{4x}$. Consequently,

$$y_{gn} = y_{gh} + y_{pn} = (C_1 + C_2 x)e^{4x} + \left(\frac{1}{3}x - \frac{1}{9}\right)e^x + \left(-\frac{1}{3}x^3 + x^2\right)e^{4x}.$$

4. $y'' + y' = 2x + 1$.

We start by finding the general solution y_{gh} of the corresponding homogeneous equation $y'' + y' = 0$. The characteristic equation $\lambda^2 + \lambda = 0$ has the simple real roots $\lambda_1 = 0, \lambda_2 = -1$. The corresponding solutions $y_1 = 1$ and $y_2 = e^{-x}$ form the general solution $y_{gh} = C_1 + C_2 e^{-x}$. Since the right-hand side of the nonhomogeneous equation consists only of a polynomial of the first degree, the exponential e^0 has the coefficient 0 which is the simple root of the characteristic equation. Under these conditions, we look for the particular solution of the nonhomogeneous equation in the form $y_{pn} = x(ax+b)$, where coefficients a and b must be found by substituting y_{pn} into the original equation. Calculating the derivatives and substituting into the equation, we get $2a + (2ax + b) = 2x + 1$, whence $a = 1$, $b = -1$. So, the particular solution comes in the form $y_{pn} = x^2 - x$, and the general solution of the original equation takes the form

$$y_{gn} = y_{gh} + y_{pn} = C_1 + C_2 e^{-x} + x^2 - x.$$

5. $y'' - 8y' + 17y = 10e^{2x}$.

We start with the homogeneous part $y'' - 8y' + 17y = 0$. Solving the characteristic equation $\lambda^2 - 8\lambda + 17 = 0$ we find the complex conjugate roots $\lambda_{1,2} = 4 \pm i$ and the corresponding solutions $y_{1,2} = e^{(4\pm i)x}$, which form the general solution $y_{gh} = C_1 e^{(4-i)x} + C_2 e^{(4+i)x}$ or in real form $y_{gh} = (A_1 \cos x + A_2 \sin x)e^{4x}$. Since the coefficient 2 of the exponential function on the right side does not coincide with the roots of the characteristic polynomial and the polynomial 10 has degree 0, we look for the particular solution of the nonhomogeneous equation in the form $y_{pn} = ae^{2x}$. Substituting into the original equation, we find $4ae^{2x} - 8 \cdot 2ae^{2x} + 17ae^{2x} = 10e^{2x}$, whence $a = 2$. Therefore, $y_{pn} = 2e^{2x}$ and

$$y_{gn} = y_{gh} + y_{pn} = (A_1 \cos x + A_2 \sin x)e^{4x} + 2e^{2x}.$$

6. $y'' - 6y' + 9y = 9x^2 - 39x + 65$, $y(0) = -1$, $y'(0) = 1$.
We start by finding the general solution y_{gh} of the homogeneous equation $y'' - 6y' + 9y = 0$. The characteristic equation $\lambda^2 - 6\lambda + 9 = 0$ has the double real root $\lambda_{1,2} = 3$. In this case, the general solution has the form $y_{gh} = (C_1 + C_2 x)e^{3x}$. Since the right-hand side of the nonhomogeneous equation consists only of a polynomial of the second degree, the exponential e^0 has the coefficient 0 which is not a root of the characteristic equation, and so the particular solution of the nonhomogeneous equation can be found in the form $y_{pn} = ax^2 + bx + c$. The coefficients a, b, c are determined by substituting y_{pn} into the original equation: $2a - 6(2ax + b) + 9(ax^2 + bx + c) = 9x^2 - 39x + 65$, whence $a = 1$, $b = -3$, $c = 5$. So the particular solution comes in the form $y_{pn} = x^2 - 3x + 5$ and the general solution of the original equation takes the form

$$y_{gn} = y_{gh} + y_{pn} = (C_1 + C_2 x)e^{3x} + x^2 - 3x + 5.$$

It remains to find the specific solution that satisfies the initial conditions. Substituting these conditions into y_{gn}, we get $C_1 + 5 = -1$ and $3C_1 + C_2 - 3 = 1$, whence $C_1 = -6$ and $C_2 = 22$. Therefore, the solution to the Cauchy problem is

$$y = (22x - 6)e^{3x} + x^2 - 3x + 5.$$

7. $y'' + y' - 6y = (6x + 1)e^{3x}$.
We start with the homogeneous part $y'' + y' - 6y = 0$. Solving the characteristic equation $\lambda^2 + \lambda - 6 = 0$, we find the simple real roots $\lambda_1 = -3$, $\lambda_2 = 2$. The two linearly independent solutions are $y_1 = e^{-3x}$ and $y_2 = e^{2x}$ and the general solution is $y_{gh} = C_1 e^{-3x} + C_2 e^{2x}$. Since the coefficient 3 of the exponential function on the right side does not coincide with the roots of the characteristic polynomial and the polynomial $6x + 1$ has degree 1, we look for the particular solution of the nonhomogeneous equation in the form $y_{pn} = (ax + b)e^{3x}$. Substituting into the original equation, we find the equation for a and b:

$$(9ax + 9b + 6a)e^{3x} + (3ax + 3b + a)e^{3x} - 6(ax + b)e^{3x} = (6x + 1)e^{3x}.$$

Eliminating the exponential and equating the coefficients of linear terms and constants, we have $6a = 6$ and $7a + 6b = 1$, whence $a = 1$ and $b = -1$. Therefore, $y_{pn} = (x - 1)e^{3x}$ and

$$y_{gn} = y_{gh} + y_{pn} = C_1 e^{-3x} + C_2 e^{2x} + (x - 1)e^{3x}.$$

8. $y'' - 12y' + 36y = 14e^{6x}$.

We start with the homogeneous equation $y'' - 12y' + 36y = 0$. The characteristic equation $\lambda^2 - 12\lambda + 36 = 0$ has the double real root $\lambda_{1,2} = 6$. So the two linearly independent solutions are $y_1 = e^{6x}$ and $y_2 = xe^{6x}$ and the general solution is $y_{gh} = (C_1 + C_2x)e^{6x}$. Since the coefficient 6 of the exponential function on the right side coincides with the double root of the characteristic polynomial and the polynomial 14 has degree 0, we look for the particular solution of the nonhomogeneous equation in the form $y_{pn} = ax^2e^{6x}$. Substituting into the original equation, we get:

$$a(36x^2 + 24x + 2)e^{6x} - 12a(6x^2 + 2x)e^{6x} + 36ax^2e^{6x} = 14e^{6x}.$$

Eliminating the exponential and simplifying, we find $2a = 14$ or $a = 7$. Therefore, $y_{pn} = 7x^2e^{6x}$ and

$$y_{gn} = y_{gh} + y_{pn} = (C_1 + C_2x)e^{6x} + 7x^2e^{6x}.$$

9. $y'' - 2y' = 6 + 12x - 24x^2$.

We start with the homogeneous equation $y'' - 2y' = 0$. The characteristic equation $\lambda^2 - 2\lambda = 0$ has the two simple real roots $\lambda_1 = 0$, $\lambda_2 = 2$ and, therefore, the two linearly independent solutions are $y_1 = 1$ and $y_2 = e^{2x}$ and the general solution is $y_{gh} = C_1 + C_2e^{2x}$. Since the exponential function is not present on the right-hand side, this means that it is e^0. The coefficient 0 of the exponential coincides with the simple root of the characteristic equation and the polynomial on the right-hand side has degree 2, which implies that the particular solution can be sought in the form $y_{pn} = x(ax^2 + bx + c)$. Substituting into the original equation, we obtain:

$$(6ax + 2b) - 2(3ax^2 + 2bx + c) = 6 + 12x - 24x^2.$$

Equating the coefficients of the same powers, we get $-6a = -24$, $6a - 4b = 12$, $2b - 2c = 6$, whence $a = 4, b = 3, c = 0$. Therefore, $y_{pn} = 4x^3 + 3x^2$ and

$$y_{gn} = C_1 + C_2e^{2x} + 4x^3 + 3x^2.$$

10. $y'' - 2y' + y = 4e^x$.

We start with the homogeneous part $y'' - 2y' + y = 0$. The characteristic equation $\lambda^2 - 2\lambda + 1 = 0$ has the double real root $\lambda_{1,2} = 1$. The corresponding solutions are $y_1 = e^x$ and $y_2 = xe^x$ and the general solution is $y_{gh} = (C_1 + C_2x)e^x$. Since the coefficient 1 of the exponential on the right-hand side coincides with the double root of the characteristic and the polynomial has degree 0, the particular solution of the nonhomogeneous equation can be found in the form $y_{pn} = ax^2e^x$. Substituting this form into the original equation, we obtain

$$a(x^2 + 4x + 2)e^x - 2a(x^2 + 2x)e^x + ax^2e^x = 4e^x,$$

whence $2a = 4$ or $a = 2$. So, $y_{pn} = 2x^2e^x$ and

$$y_{gn} = (C_1 + C_2x)e^x + 2x^2e^x.$$

2.2   Case 2: Right side $f = P_m(x)e^{\alpha x}\{\cos\beta x, \sin\beta x\}$

In this section we solve the equation $Ly = P_m(x)e^{\alpha x}\begin{cases}\cos\beta x \\ \sin\beta x\end{cases}$, where $P_m(x)$ is the polynomial of degree m, $\alpha \in \mathbb{R}$ and $\beta \in \mathbb{R}$. The key brackets mean that we consider in parallel either of the two cases $f_c = P_m(x)e^{\alpha x}\cos\beta x$ and $f_s = P_m(x)e^{\alpha x}\sin\beta x$. The structure of the general solution remains the same $y_{gn} = y_{gh} + y_{pn}$ (see Theorem 6 in Chapter 8), and the method of finding the general solution of the homogeneous part y_{gh} was already studied. So, we focus on finding a particular solution y_{pn}.

Compared to Case 1 (with $\alpha \in \mathbb{R}$), now we have a more general situation. However, the most efficient method of solution reduces Case 2 to Case 1 with $\alpha \in \mathbb{C}$. This is performed as follows. First, we introduce the auxiliary right-hand side $\tilde{f} = P_m(x)e^{(\alpha+i\beta)x}$ such that $f_c = Re(\tilde{f})$ and $f_s = Im(\tilde{f})$, and look for a particular solution of the auxiliary equation $Ly = \tilde{f}$. This is made in the same way as in Case 1, looking for the solution in the form $\tilde{y}_{pn} = x^sQ_m(x)e^{(\alpha+i\beta)x}$, where the (complex) coefficients of the polynomial $Q_m(x)$ are found by substituting $\tilde{y}_{pn}$ into the auxiliary equation. Then, after finding $\tilde{y}_{pn}$, we determine the solutions y_c of the equation $Ly = f_c$ and y_s of the equation $Ly = f_s$ by applying Theorem 3 of Chapter 8: $y_c = Re(\tilde{y})$ and $y_s = Im(\tilde{y})$. This completes the search for the particular solution of the original equation.

Alternatively, we can work only with the real form of particular solutions. In this case, regardless of whether the right-hand side is $f_c = P_m(x)e^{\alpha x}\cos\beta x$ or $f_s = P_m(x)e^{\alpha x}\sin\beta x$, we have to look for the particular solution in the form $y_{pn} = x^se^{\alpha x}(Q_m(x)\cos\beta x + S_m(x)\sin\beta x)$, where s is the multiplicity of the number $\alpha + i\beta$ as the root of the characteristic equation, and the functions $Q_m(x)$ and $S_m(x)$ are polynomials of degree m (usually different), whose (real) coefficients must be found by substituting y_{pn} into the original equation. This form of the particular solution usually requires more technical work, especially when the degree of the polynomial is high, but, on the other hand, it eliminates the need to work with the functions of complex image. Of course, the final form of the real solutions obtained in this way and those obtained using the complex form is the same.

In the six examples below, we solve the equations using both methods of finding particular solutions, but in subsequent problems we will use the procedure with the complex exponent to facilitate technical work.

Examples.

1. $y'' - 6y' + 9y = 25e^x \sin x$.

We start with the homogeneous part $y'' - 6y' + 9y = 0$. Solving the characteristic equation $\lambda^2 - 6\lambda + 9 = 0$, we find the double real root $\lambda_1 = \lambda_2 = 3$ and the corresponding solutions $y_1 = e^{3x}$ and $y_2 = xe^{3x}$, which form the general solution $y_{gh} = (C_1 + C_2 x)e^{3x}$. To find the particular solution of the nonhomogeneous equation, we transform the primitive equation to the auxiliary one $y'' - 6y' + 9y = 25e^{(1+i)x} = \tilde{f}$, the right-hand side of which is related to $f_s = 25e^x \sin x$ by the formula $f_s = Im(\tilde{f})$. The particular solution of the auxiliary equation is found in the form $\tilde{y} = ae^{(1+i)x}$ ($s = 0$ because $\gamma = 1+i$ is not a root of the characteristic equation and the degree of the polynomial is 0 because the polynomial in $\tilde{f}$ is the constant 25). The coefficient a must be found by substituting $\tilde{y}$ into the auxiliary equation. Recalling the simple rule of differentiation of exponential functions with complex exponent, we get the following equation for a:

$$a(1+i)^2 e^{(1+i)x} - 6a(1+i)e^{(1+i)x} + 9ae^{(1+i)x} = 25e^{(1+i)x}.$$

Dividing by the exponential and simplifying we get the equation $a(3 - 4i) = 25$, whence $a = 3 + 4i$. So, the auxiliary solution is found in the form $\tilde{y} = (3+4i)e^{(1+i)x}$. Therefore, the particular solution of the original equation is given by the formula

$$y_{pn} = Im(\tilde{y}) = Im\left((3+4i)e^{(1+i)x}\right) = Im\left((3+4i)e^x(\cos x + i\sin x)\right)$$

$$= Im\left(e^x[(3\cos x - 4\sin x) + i(3\sin x + 4\cos x)]\right) = e^x(3\sin x + 4\cos x).$$

Finally, the general solution of the original equation is found in the form

$$y_{gn} = y_{gh} + y_{pn} = (C_1 + C_2 x)e^{3x} + (3\sin x + 4\cos x)e^x.$$

In parallel, we find the particular solution via the alternative method using only the real form: $y_{pn} = x^0 e^x(a\cos x + b\sin x)$. Substituting into the original equation, we get

$$y''_{pn} - 6y'_{pn} + 9y_{pn} = e^x(-2a\sin x + 2b\cos x)$$

$$-6e^x(a\cos x + b\sin x - a\sin x + b\cos x) + 9e^x(a\cos x + b\sin x) = 25e^x\sin x.$$

Canceling the factor e^x and regrouping the terms on the left-hand side, we obtain

$$(2b - 6a - 6b + 9a)\cos x + (-2a - 6b + 6a + 9b)\sin x = 25\sin x$$

or simplifying $(3a - 4b)\cos x + (4a + 3b)\sin x = 25\sin x$. Solving the system $3a - 4b = 0, 4a + 3b = 25$, we find $a = 4, b = 3$, that is, the particular solution has the form $y_{pn} = e^x(4\cos x + 3\sin x)$, the same as found via the complex exponent. Therefore, we arrive at the same general solution as in the previous method.

2. $y'' + y = x\cos x$.

We start with the homogeneous part $y'' + y = 0$. Solving the characteristic equation $\lambda^2 + 1 = 0$ we find a pair of the complex conjugate roots $\lambda_{1,2} = \pm i$ and the corresponding solutions $y_{1,2} = e^{\pm ix}$, which form the general solution in the complex form $y_{gh} = C_1 e^{ix} + C_2 e^{-ix}$. Turning to the real form, we have $y_{gh} = C_1\cos x + C_2\sin x$. To find the particular solution of the original, we transform the given equation into the auxiliary equation $y'' + y = xe^{ix} = \tilde{f}$, the right-hand side of which is related to $f_c = x\cos x$ by the formula $f_c = Re(\tilde{f})$. The particular solution of the auxiliary equation we find in the form $\tilde{y} = x(ax + b)e^{ix}$ ($s = 1$ because $\gamma = i$ is the simple root of the characteristic equation and the degree of the polynomial is 1 because the polynomial x in $\tilde{f}$ is of the 1th degree). The coefficients a and b are found by substituting $\tilde{y}$ into the auxiliary equation:

$$(-ax^2 - bx + 2a + i(4ax + 2b))e^{ix} + (ax^2 + bx)e^{ix} = xe^{ix}.$$

Canceling the exponential and simplifying we arrive at the equation $2a + i(4ax + 2b) = x$, which is equivalent to the following system for the coefficients: $4ia = 1, a + ib = 0$. From the first equation it follows that $a = -\frac{i}{4}$ and using this value in the second equation we get $b = ia = \frac{1}{4}$. Therefore, the auxiliary solution is found in the form $\tilde{y} = \frac{1}{4}(-ix^2 + x)e^{ix}$. Then, the particular solution of the original equation is defined by the formula

$$y_{pn} = Re(\tilde{y}) = Re\left(\frac{1}{4}(-ix^2 + x)e^{ix}\right) = Re\left(\frac{1}{4}(-ix^2 + x)(\cos x + i\sin x)\right)$$

$$= \frac{1}{4}Re\left((x\cos x + x^2\sin x) + i(x\sin x - x^2\cos x)\right) = \frac{1}{4}(x\cos x + x^2\sin x).$$

Finally, the general solution of the original equation is found in the form

$$y_{gn} = y_{gh} + y_{pn} = C_1\cos x + C_2\sin x + \frac{1}{4}(x\cos x + x^2\sin x).$$

We can also find the particular solution using only the real functions. In this case, $\alpha = 0$ and $\beta = 1$ in the representation $P_m(x)e^{\alpha x} \cos \beta x$ of the right-hand side. So, the number $\alpha + i\beta = i$ is a simple root of the characteristic equation, and then, in the formula $y_{pn} = x^s e^{\alpha x} (Q_m(x)\cos \beta x + S_m(x)\sin \beta x)$ of the particular solution, the parameter s is equal to 1. Furthermore, $P_m(x) = x$, which indicates that $Q_m(x)$ and $S_m(x)$ are polynomials of degree 1. So we look for the particular solution in the form

$$y_{pn} = x\left((ax + b) \cos x + (cx + d) \sin x\right).$$

Substituting into the original equation, we get

$$y_{pn}'' + y_{pn} = [(-ax^2 - bx + 2cx + d + 2cx + 2a + d) \cos x$$

$$-(cx^2 + dx + 2ax + b + 2ax + b - 2c)\sin x] + [(ax^2 + bx)\cos x + (cx^2 + dx)\sin x] = x \cos x.$$

Regrouping the terms and simplifying, we get

$$(4cx + 2d + 2a) \cos x - (4ax + 2b - 2c) \sin x = x \cos x.$$

Then, we arrive at the system of two polynomial equations $4cx + 2d + 2a = x$, $4ax + 2b - 2c = 0$. Equating the coefficients with the same powers of x, we find from the second equation that $a = 0$ and $b = c$, and from the first (substituting $a = 0$) that $d = 0$ and $c = \frac{1}{4}$. Therefore, the particular solution takes the form $y_{pn} = \frac{1}{4}x \cos x + \frac{1}{4}x^2 \sin x$. The same was obtained via the complex exponent. Therefore, we arrive at the same general solution as in the previous method.

3. $y'' + 2y' + 5y = 8e^{-x} \cos 2x$.
We start with the homogeneous part $y'' + 2y' + 5y = 0$, whose characteristic equation $\lambda^2 + 2\lambda + 5 = 0$ has the complex conjugate roots $\lambda_{1,2} = -1 \pm 2i$. The corresponding solutions are $y_{1,2} = e^{(-1 \pm 2i)x}$, and the general solution has the complex form $y_{gh} = C_1 e^{(-1-2i)x} + C_2 e^{(-1+2i)x}$. In the real form, we obtain $y_{gh} = (C_1 \cos 2x + C_2 \sin 2x)e^{-x}$. To find the particular solution of the original equation, we consider the auxiliary equation $y'' + 2y' + 5y = 8e^{(-1+2i)x} = \tilde{f}$, the right-hand side of which is related to $f_c = 8e^{-x} \cos 2x$ by the formula $f_c = Re(\tilde{f})$. Since the coefficient of the complex exponent $\gamma = -1 + 2i$ is the simple root of the characteristic equation, we look for the auxiliary solution in the form $\tilde{y} = axe^{(-1+2i)x}$. The coefficient a is found by substituting $\tilde{y}$ into the auxiliary equation:

$$a[2(-1 + 2i) + (-1 + 2i)^2 x]e^{(-1+2i)x} + 2a[1 + (-1 + 2i)x]e^{(-1+2i)x}$$

$$+5axe^{(-1+2i)x} = 8e^{(-1+2i)x}.$$

By eliminating the exponential and simplifying, we get the equation $4ia = 8$, whence $a = -2i$. Therefore, $\tilde{y} = -2ixe^{(-1+2i)x}$ is the auxiliary solution and the particular solution of the original equation is defined by the formula

$$y_{pn} = Re(\tilde{y}) = Re\left(-2ix(\cos 2x + i\sin 2x)e^{-x}\right) = 2x\sin 2xe^{-x}.$$

Finally, the general solution of the original equation is found in the form

$$y_{gn} = y_{gh} + y_{pn} = (C_1\cos 2x + C_2\sin 2x)e^{-x} + 2x\sin 2xe^{-x}.$$

Alternatively, we can find the particular solution using only the real form. In this case, $\alpha = 1$ and $\beta = 2$ in the right-hand side $8e^{\alpha x}\cos\beta x$, and the number $\alpha + i\beta = 1 + 2i$ is the simple root of the characteristic equation. So, we look for the particular solution in the form $y_{pn} = x\left(a\cos 2x + b\sin 2x\right)e^{-x}$. Substituting into the original equation, we obtain

$$y''_{pn} + 2y'_{pn} + 5y_{pn} = [(-2a - 3ax + 4b - 4bx)\cos 2x$$

$$+(-2b - 3bx - 4a + 4ax)\sin 2x]e^{-x} + 2[(a - ax + 2bx)\cos 2x$$

$$+(b - bx - 2ax)\sin 2x]e^{-x} + 5[ax\cos 2x + bx\sin 2x]e^{-x} = 8\cos 2xe^{-x}.$$

Dividing by e^{-x} and simplifying, we get $4b\cos 2x - 4a\sin 2x = 8\cos 2x$, whence $b = 2$ and $a = 0$. Thus, the particular solution is found in the form $y_{pn} = 2x\sin 2xe^{-x}$, the same obtained using the complex exponent. Therefore, we arrive at the same general solution as in the previous method.

4. $y'' - 2y' + y = -12\cos 2x - 9\sin 2x$, $y(0) = -2, y'(0) = 0$.
We start with the homogeneous part $y'' - 2y' + y = 0$. The characteristic equation $\lambda^2 - 2\lambda + 1 = (\lambda - 1)^2 = 0$ has the double root $\lambda_{1,2} = 1$ and the corresponding solutions are $y_1 = e^x$, $y_2 = xe^x$ which form the general solution $y_{gh} = (C_1 + C_2x)e^x$. Since the polynomials together with the trigonometric functions are constant and the number $0 + 2i$ is not a root of the characteristic equation, the particular solution of the nonhomogeneous equation can be found in the form $y_{pn} = a\cos 2x + b\sin 2x$. Substituting into the original equation we get:

$$(-4a\cos 2x - 4b\sin 2x) - 2(-2a\sin 2x + 2b\cos 2x) + (a\cos 2x + b\sin 2x)$$

$$= -12\cos 2x - 9\sin 2x,$$

which results in the system $-3a - 4b = -12$, $4a - 3b = -9$. Its solution is $a = 0, b = 3$, and then $y_{pn} = 3\sin 2x$. Consequently,

$$y_{gn} = y_{gh} + y_{pn} = (C_1 + C_2x)e^x + 3\sin 2x.$$

It remains to satisfy the initial conditions. Applying these conditions to y_{gn}, we have $C_1 = -2$ and $C_1 + C_2 + 6 = 0$. So, $C_2 = -4$ and the solution of the Cauchy problem is

$$y_{gn} = (-2 - 4x)e^x + 3\sin 2x.$$

5. $y'' - 2y' + 5y = 10e^{-x}\cos 2x.$

We start with the homogeneous part $y'' - 2y' + 5y = 0$. Solving the characteristic equation $\lambda^2 - 2\lambda + 5 = 0$, we find the complex conjugate roots $\lambda_{1,2} = 1 \pm 2i$ and the corresponding solutions $y_{1,2} = e^{(1\pm 2i)x}$, which form the general solution $y_{gh} = C_1 e^{(1-2i)x} + C_2 e^{(1+2i)x}$. The real form of the general solution is $y_{gh} = (C_1\cos 2x + C_2\sin 2x)e^x$. To find the particular solution of the nonhomogeneous equation, we consider the auxiliary equation $y'' - 2y' + 5y = 10e^{(-1+2i)x} = \tilde{f}$, the right-hand side of which is related to $f_c = 10e^{-x}\cos 2x$ by the formula $f_c = Re(\tilde{f})$. The particular solution of the auxiliary equation is found in the form $\tilde{y} = ae^{(-1+2i)x}$ ($s = 0$ because $\gamma = -1 + 2i$ is not a root of the characteristic equation and the degree of the polynomial is 0 because the polynomial in $\tilde{f}$ is the constant 10). The coefficient a is found by substituting $\tilde{y}$ into the auxiliary equation:

$$a(-1 + 2i)^2 e^{(-1+2i)x} - 2a(-1 + 2i)e^{(-1+2i)x} + 5ae^{(-1+2i)x} = 10e^{(-1+2i)x}.$$

Dividing by the exponential and simplifying, we get the equation $a(4 - 8i) = 10$, whence $a = \frac{1}{2}(1 + 2i)$. So, the auxiliary solution is found in the form $\tilde{y} = \frac{1}{2}(1 + 2i)e^{(-1+2i)x}$. Therefore, the particular solution of the original equation is defined by the formula

$$y_{pn} = Re(\tilde{y}) = \frac{1}{2}e^{-x}Re\left((1 + 2i)(\cos 2x + i\sin 2x)\right) = \left(\frac{1}{2}\cos 2x - \sin 2x\right)e^{-x}.$$

Finally, the general solution of the original equation has the form

$$y_{gn} = y_{gh} + y_{pn} = (C_1\cos 2x + C_2\sin 2x)e^x + \left(\frac{1}{2}\cos 2x - \sin 2x\right)e^{-x}.$$

The particular solution of the nonhomogeneous equation can also be found using only the real form $y_{pn} = (A\cos 2x + B\sin 2x)e^{-x}$. Substituting this expression into the original equation, we get

$$(-3A\cos 2x - 3B\sin 2x - 4B\cos 2x + 4A\sin 2x)e^{-x} - 2(-A\cos 2x - B\sin 2x$$

$$-2A\sin 2x + 2B\cos 2x)e^{-x} + 5(A\cos 2x + B\sin 2x)e^{-x} = 10e^{-x}\cos 2x.$$

Equating the coefficients together with $\cos 2x$ and $\sin 2x$, we obtain the system of the two equations $4A - 8B = 10$, $8A + 4B = 0$. The solution of this system is $A = \frac{1}{2}, B = -1$ and then $y_{pn} = \left(\frac{1}{2}\cos 2x - \sin 2x\right) e^{-x}$. Thus, we arrive at the same particular solution as in the previous algorithm and, consequently, at the same general solution $y_{gn} = (C_1 \cos 2x + C_2 \sin 2x)e^x + \left(\frac{1}{2}\cos 2x - \sin 2x\right) e^{-x}$.

6. $y'' - 6y' + 25y = 9\sin 4x - 24\cos 4x$, $y(0) = 2, y'(0) = -2$.
We start with the homogeneous part $y'' - 6y' + 25y = 0$. The characteristic equation $\lambda^2 - 6\lambda + 25 = 0$ has the complex conjugate roots $\lambda_{1,2} = 3 \pm 4i$ and the corresponding solutions $y_{1,2} = e^{(3\pm 4i)x}$, which form the general solution $y_{gh} = C_1 e^{(3-4i)x} + C_2 e^{(3+4i)x}$. The real form of the general solution is $y_{gh} = (C_1 \cos 4x + C_2 \sin 4x)e^{3x}$. The particular solution of the nonhomogeneous equation can be found in the form $y_{pn} = a\cos 4x + b\sin 4x$. Substituting this form into the original equation, we get

$$(-16a\cos 4x - 16b\sin 4x) - 6(-4a\sin 4x + 4b\cos 4x)$$

$$+25(a\cos 4x + b\sin 4x) = 9\sin 4x - 24\cos 4x.$$

Equating the coefficients together with $\cos 4x$ and $\sin 4x$, we get the system of the two equations $9a - 24b = -24$, $24a + 9b = 9$, whence $a = 0$ and $b = 1$. So, $y_{pn} = \sin 4x$ and

$$y_{gn} = (C_1 \cos 4x + C_2 \sin 4x)e^{3x} + \sin 4x.$$

To find the solution of the Cauchy problem, we substitute the initial conditions into the general solution y_{gn} : $y(0) = C_1 = 2$ and $y'(0) = 3C_1 + 4C_2 + 4 = -2$. So $C_2 = -3$ and the solution sought is

$$y = (2\cos 4x - 3\sin 4x)e^{3x} + \sin 4x.$$

7. $y'' - 3y' + 2y = 3\cos x + 19\sin x$.
We start with the homogeneous part $y'' - 3y' + 2y = 0$. The characteristic equation $\lambda^2 - 3\lambda + 2 = 0$ has the simple real roots $\lambda_1 = 1, \lambda_2 = 2$. The corresponding solutions $y_1 = e^x$ and $y_2 = e^{2x}$ form the general solution $y_{gh} = C_1 e^x + C_2 e^{2x}$. The particular solution of the nonhomogeneous equation can be found in the form $y_{pn} = a\cos x + b\sin x$. Substituting this expression into the original equation, we get

$$(-a\cos x - b\sin x) - 3(-a\sin x + b\cos x) + 2(a\cos x + b\sin x) = 3\cos x + 19\sin x.$$

Equating the coefficients together with $\cos x$ and $\sin x$, we obtain the system of the two equations $a - 3b = 3$, $3a + b = 19$, whence $a = 6$ and $b = 1$. So $y_{pn} = 6\cos x + \sin x$ and

$$y_{gn} = C_1 e^x + C_2 e^{2x} + 6\cos x + \sin x.$$

Alternatively, we can find the particular solution using the complex form. To do this, we introduce the auxiliary equation $y'' - 3y' + 2y = e^{ix}$ and look for its particular solution in the form $\tilde{y} = ae^{ix}$. Substituting this function and its derivatives into the equation, we get $-ae^{ix} - 3iae^{ix} + 2ae^{ix} = e^{ix}$ or, canceling e^{ix} and simplifying, $a(1 - 3i) = 1$. Therefore, $a = \frac{1+3i}{10}$ and $\tilde{y} = \frac{1+3i}{10}e^{ix}$. So, the particular solution of the original equation has the form

$$y_{pn} = 3Re(\tilde{y}) + 19Im(\tilde{y}) = 3\left(\frac{1}{10}\cos x - \frac{3}{10}\sin x\right) + 19\left(\frac{3}{10}\cos x + \frac{1}{10}\sin x\right)$$

$$= 6\cos x + \sin x.$$

8. $y'' - 6y' + 13y = 39e^{-3x}\sin 2x$.

We start with the homogeneous part $y'' - 6y' + 13y = 0$. Solving the characteristic equation $\lambda^2 - 6\lambda + 13 = 0$, we find the complex conjugate roots $\lambda_{1,2} = 3 \pm 2i$ and the corresponding solutions $y_{1,2} = e^{(3\pm 2i)x}$, which form the general solution $y_{gh} = C_1 e^{(3-2i)x} + C_2 e^{(3+2i)x}$. The real form of the general solution is $y_{gh} = (C_1\cos 2x + C_2\sin 2x)e^{3x}$. To find a particular solution of the nonhomogeneous equation, we solve the auxiliary equation $y'' - 6y' + 13y = 39e^{(-3+2i)x} = \tilde{f}$, the right-hand side of which is related to $f_s = 39e^{-3x}\sin 2x$ by the formula $f_s = Im(\tilde{f})$. The particular solution of the auxiliary equation can be found in the form $\tilde{y} = ae^{(-3+2i)x}$ ($s = 0$ because $\gamma = 3 + 2i$ is not a root of the characteristic equation and the degree of the polynomial is 0 because the polynomial in $\tilde{f}$ is the constant 39). The coefficient a is found by substituting $\tilde{y}$ into the auxiliary equation:

$$a(-3+2i)^2 e^{(-3+2i)x} - 6a(-3+2i)e^{(-3+2i)x} + 13ae^{(-3+2i)x} = 39e^{(-3+2i)x}.$$

Dividing by the exponential and simplifying we get the equation $a(36-24i) = 39$, whence $a = \frac{3+2i}{4}$. Then, the auxiliary solution is found in the form $\tilde{y} = \frac{3+2i}{4}e^{(-3+2i)x}$. Therefore, the particular solution of the original equation is defined by the formula

$$y_{pn} = Im(\tilde{y}) = Im\left(\frac{3+2i}{4}e^{-3x}(\cos 2x + i\sin 2x)\right) = \frac{1}{4}e^{-3x}(2\cos 2x + 3\sin 2x).$$

Finally, the general solution of the original equation is found in the form

$$y_{gn} = y_{gh} + y_{pn} = (C_1\cos 2x + C_2\sin 2x)e^{3x} + \frac{1}{4}e^{-3x}(2\cos 2x + 3\sin 2x).$$

The particular solution of the nonhomogeneous equation can also be found using the real form $y_{pn} = (A\cos 2x + B\sin 2x)e^{-3x}$. Substituting this into the original equation, we get

$$(5A\cos 2x - 12B\cos 2x + 12A\sin 2x + 5B\sin 2x)e^{-3x} - 6(-3A\cos 2x + 2B\cos 2x$$

$$-2A \sin 2x - 3B \sin 2x)e^{-3x} + 13(A \cos 2x + B \sin 2x)e^{-3x} = 39e^{-3x} \cos 2x.$$

Equating the coefficients together with $\cos 2x$ and $\sin 2x$, we get the two-equation system $36A - 24B = 0$, $24A + 36B = 39$. The solution of this system is $A = \frac{1}{2}, B = \frac{3}{4}$ and then $y_{pn} = \left(\frac{1}{2}\cos 2x + \frac{3}{4}\sin 2x\right)e^{-3x}$. Hence, we arrive at the same particular solution as in the previous algorithm and, consequently, at the same general solution. However, we notice that this approach is more technically involved.

9. $y'' - 6y' + 13y = 39e^{3x} \sin 2x$.

We start with the homogeneous part $y'' - 6y' + 13y = 0$. Solving the characteristic equation $\lambda^2 - 6\lambda + 13 = 0$, we find the complex conjugate roots $\lambda_{1,2} = 3 \pm 2i$ and the corresponding solutions $y_{1,2} = e^{(3\pm 2i)x}$, which form the general solution $y_{gh} = C_1 e^{(3-2i)x} + C_2 e^{(3+2i)x}$. The real form of the general solution is $y_{gh} = (C_1 \cos 2x + C_2 \sin 2x)e^{3x}$. To find the particular solution of the nonhomogeneous equation, we solve the auxiliary equation $y'' - 6y' + 13y = 39e^{(3+2i)x} = \tilde{f}$, the right-hand side of which is related to $f_s = 39e^{3x} \sin 2x$ by the formula $f_s = Im(\tilde{f})$. The particular solution of the auxiliary equation can be found in the form $\tilde{y} = axe^{(3+2i)x}$ ($s = 1$ because $\gamma = 3 + 2i$ is a simple root of the characteristic equation and the degree of the polynomial is 0 because the polynomial in $\tilde{f}$ is the constant 39). The coefficient a is found by substituting $\tilde{y}$ into the auxiliary equation:

$$a[(3+2i)^2 x + 2(3+2i)]e^{(3+2i)x} - 6a[(3+2i)x + 1]e^{(-3+2i)x}$$
$$+13axe^{(-3+2i)x} = 39e^{(-3+2i)x}.$$

Dividing by the exponential and simplifying, we get the equation $4ia = 39$, whence $a = -\frac{39}{4}i$. So, the auxiliary solution is found in the form $\tilde{y} = -\frac{39}{4}ixe^{(3+2i)x}$. Therefore, the particular solution of the original equation is defined by the formula

$$y_{pn} = Im(\tilde{y}) = Im\left(-\frac{39}{4}ixe^{3x}(\cos 2x + i \sin 2x)\right) = -\frac{39}{4}xe^{3x}\cos 2x.$$

Finally, the general solution of the original equation is found in the form

$$y_{gn} = y_{gh} + y_{pn} = (C_1 \cos 2x + C_2 \sin 2x)e^{3x} - \frac{39}{4}xe^{3x}\cos 2x.$$

The particular solution of the nonhomogeneous equation can still be found directly in the real form $y_{pn} = x\,(A\cos 2x + B\sin 2x)\,e^{3x}$, but the work required is much substantial than when using the complex form. (The task for the reader: solve using the real form and compare the solutions.)

10. $y'' + 16y = 8\cos 4x$.

We start with the homogeneous part $y'' + 16y = 0$. Solving the characteristic equation $\lambda^2 + 16 = 0$, we find the complex conjugate roots $\lambda_{1,2} = \pm 4i$ and the corresponding solutions $y_{1,2} = e^{(\pm 4i)x}$, which form the general solution $y_{gh} = C_1 e^{-4ix} + C_2 e^{4ix}$. The real form of the general solution is $y_{gh} = C_1 \cos 4x + C_2 \sin 4x$. To find the particular solution of the nonhomogeneous equation, we solve the auxiliary equation $y'' + 16y = 8e^{4ix} = \tilde{f}$, the right-hand side of which is related to $f_c = 8 \cos 4x$ by the formula $f_c = Re(\tilde{f})$. The particular solution of the auxiliary equation is found in the form $\tilde{y} = axe^{4ix}$ ($s = 1$ because $\gamma = 4i$ is a simple root of the characteristic equation and the degree of the polynomial is 0 because the polynomial in $\tilde{f}$ is the constant 8). The coefficient a is found by substituting $\tilde{y}$ into the auxiliary equation: $a(-16x + 8i)e^{4ix} + 16axe^{4ix} = 8e^{4ix}$. Dividing by the exponential and simplifying, we get the equation $8ia = 8$ or $a = -i$. So, the auxiliary solution is found in the form $\tilde{y} = -ixe^{4ix}$. Therefore, the particular solution of the original equation is defined by the formula

$$y_{pn} = Re(\tilde{y}) = Re\left(-ix(\cos 4x + i \sin 4x)\right) = x \sin 4x.$$

Finally, the general solution of the original equation is found in the form

$$y_{gn} = y_{gh} + y_{pn} = C_1 \cos 4x + C_2 \sin 4x + x \sin 4x.$$

The particular solution of the nonhomogeneous equation can also be found using only the real form $y_{pn} = x\left(A \cos 4x + B \sin 4x\right)$, but the technical work is more involved. (The task for the reader: solve using the real form and compare the solutions.)

Exercises for section 2

Solve the following nonhomogeneous equations by the method of undetermined coefficients:

1. $y'' - 5y' - 6y = 3 \cos x + 19 \sin x$.
2. $y'' + 2y' - 3y = (12x^2 + 6x - 4)e^x$.
3. $y'' - 9y' + 20y = 126e^{-2x}$.
4. $y'' + 10y' + 25y = 40 + 52x - 240x^2 - 200x^3$.
5. $y'' + 36y = 36 + 66x - 36x^3$.
6. $y'' + 4y' + 20y = -4 \cos 4x - 52 \sin 4x$.
7. $y'' + 5y' = 39 \cos 3x - 105 \sin 3x$.
8. $y'' + 6y' + 9y = 72e^{3x}$.
9. $y'' + y' - 2y = 9 \cos x - 7 \sin x$.
10. $y'' + 2y' + y = 6e^{-x}$.
11. $4y'' - 4y' + y = -25 \cos x$.

12. $y'' - 2y' - 8y = 12\sin 2x - 36\cos 2x$.
13. $y'' - 7y' + 12y = 3e^{4x}$.
14. $y'' - 6y' + 10y = 51e^{-x}$.
15. $y'' - 2y' = (4x + 4)e^{2x}$.
16. $y'' - 3y' + 2y = (34 - 12x)e^{-x}$.
17. $y'' - 6y' + 34y = 18\cos 5x + 60\sin 5x$.
18. $y'' + 6y' + 10y = 74e^{3x}$.
19. $y'' - 4y' = 8 - 16x$.
20. $y'' - 2y' + 3y = x^3 + \sin x$.
21. $y''' + 2y'' - y' - 2y = e^x + x^2$.
22. $y'' - 4y' + 4y = (x^3 + x)e^{2x}$.
23. $y'' + 4y = x^2 \sin 2x$.
24. $y'' + 2y' + 2y = x^2 + \sin x$.
25. $y'' - 9y = x + e^{2x} - \sin 2x$.
26. $y''' + 3y'' + 2y' = x^2 + 4x + 8$.
27. $y'' + y = -2\sin x + 4x \cos x$.
28. $y''' - y'' - 4y' + 4y = 2x^2 - 4x - 1 + (2x^2 + 5x + 1)e^{2x}$.

3   Nonhomogeneous equations: method of variation of parameters (Lagrange method)

The Lagrange method makes it theoretically possible to find a particular solution for an arbitrary right-hand side. The idea behind the method is similar to the method of variation of parameter for the nonhomogeneous linear equation of the first order: first, we find the general solution of the corresponding homogeneous equation, and then we use this form, with unknown functions instead of arbitrary constants, to find the solution of the nonhomogeneous equation. In what follows, we explain this method for the second-order equation and then generalize it to an arbitrary order.

Second order equation

Let us illustrate the idea of this method by considering the second-order linear equation in the normalized form (with the main coefficient equal to 1):

$$y'' + ay' + by = f,$$

where a, b are constants and $f(x)$ is a function depending on x. We first find the general solution of the corresponding homogeneous equation, which has the form $y_{gh} = C_1 y_1 + C_2 y_2$, where C_1, C_2 are arbitrary constants and y_1, y_2 are two linearly independent particular solutions of the homogeneous equation. This expression of y_{gh} suggests searching for the general solution of the original equation in the form $y_{gn} = C_1(x)y_1 + C_2(x)y_2$, where C_1 and C_2 are two functions to be determined by substituting this form into the nonhomogeneous equation. Since we have two functions to find and substitution of y_{gn} into the original equation will give us only one differential equation, we can assume that an additional differential relation between C_1 and C_2 can be imposed to specify these two functions. For now, we will use this assumption without proof, but at the end of the explanation of the method, we will show its validity. In the technical calculations below, we will omit the index of the solution y_{gn} for brevity.

Calculating the first derivative of $y_{gn} \equiv y$, we have $y' = C_1'y_1 + C_2'y_2 + C_1y_1' + C_2y_2'$. Since we can supposedly impose one more condition on C_1 and C_2, let us require that $C_1'y_1 + C_2'y_2 = 0$, which simplifies the expression for the first derivative: $y' = C_1y_1' + C_2y_2'$. Proceeding with the calculation of the second derivative, we get $y'' = C_1'y_1' + C_2'y_2' + C_1y_1'' + C_2y_2''$. Therefore, the substitution into the original equation gives

$$y'' + ay' + by = (C_1'y_1' + C_2'y_2' + C_1y_1'' + C_2y_2'')$$

$$+a(C_1y_1' + C_2y_2') + b(C_1y_1 + C_2y_2) = f.$$

Regrouping the terms

$$(C_1'y_1' + C_2'y_2') + C_1(y_1'' + ay_1' + by_1) + C_2(y_2'' + ay_2' + by_2) = f,$$

notice that the terms in the second and third brackets cancel each other, because y_1 and y_2 are the solutions of the homogeneous equation. So the second relation between C_1 and C_2 (the main one) simplifies to the form $C_1'y_1' + C_2'y_2' = f$. In this way, we have to solve the system of the two equations $\begin{cases} C_1'y_1 + C_2'y_2 = 0 \\ C_1'y_1' + C_2'y_2' = f \end{cases}$. First, we decouple the system with respect to C_1' and C_2', that is, we solve the linear system for unknowns C_1' and C_2', whose determinant $W = \begin{vmatrix} y_1 & y_2 \\ y_1' & y_2' \end{vmatrix}$ is non-zero. In fact, if we assume, for contradiction, that $W = y_1y_2' - y_2y_1' = 0$, then this relation can be rewritten in the form $\frac{y_1y_2' - y_2y_1'}{y_1^2} = \left(\frac{y_2}{y_1}\right)' = 0$, from which it follows that $\frac{y_2}{y_1} = C = const$ or $y_2 = Cy_1$. The last relation contradicts the fact that y_1 and y_2 are two linearly independent solutions. Therefore, $W \neq 0$ that guarantees the existence and uniqueness of the solution C_1', C_2' of the system. This solution can be found in any known way, usually using some version of the Gauss elimination method. For instance, multiplying the first equation by y_1', the second by y_1 and subtracting the first result from the second, we eliminate C_1' and obtain $C_2'(y_1y_2' - y_2y_1') = y_1f$ or $C_2' = \frac{y_1f}{W}$. Similarly, multiplying the first equation by y_2', the second by y_2 and subtracting the first result from the second, we eliminate C_2' and obtain $C_1'(y_2 y_1' - y_1 y_2') = y_2 f$ or $C_1' = -\frac{y_2 f}{W}$. We note that in the decoupled equations for C_1' and C_2' the right-hand sides are known functions and, therefore, to find C_1 and C_2 it is sufficient to perform the integration with respect to x: $C_1 = -\int \frac{y_2 f}{W} dx$, $C_2 = \int \frac{y_1 f}{W} dx$. Substituting the specified functions C_1 and C_2 into the function $y_{gn} = C_1 y_1 + C_2 y_2$, we find the general solution of the original equation. Notice that the demonstrated possibility of carrying out this entire procedure justifies the preliminary assumption that the condition $C_1' y_1 + C_2' y_2 = 0$ can be additionally imposed on C_1 and C_2.

In the case of the non-normalized equation $cy'' + ay' + by = f$, the only difference is that the second equation of the system for C_1 and C_2 takes the form $c(C_1' y_1' + C_2' y_2') = f$ or $C_1' y_1' + C_2' y_2' = \frac{f}{c}$.

Higher order equations
For equation

$$Ly \equiv a_n y^{(n)} + \ldots + a_1 y' + a_0 y = f$$

of order n the algorithm of solution is a straightforward generalization of the method presented for the second-order equation. So, we only explain the procedure of solution, without demonstrating that the algorithm is feasible.

First, we find the general solution of the homogeneous equation $Ly = 0$ in the form $y_{gh} = C_1 y_1 + C_2 y_2 + \ldots + C_n y_n$, where $C_1, C_2, \ldots, C_n$ are arbitrary constants and $y_1, y_2, \ldots, y_n$ are linearly independent particular solutions of the homogeneous equation. Next, we look for the general solution of the original equation $Ly = f$ in the form $y_{gn} = C_1(x) y_1 + C_2(x) y_2 + \ldots + C_n(x) y_n$, where $C_1, C_2, \ldots, C_n$ are functions to be determined by substituting this form into the nonhomogeneous equation. Since we have n functions to find and substitution of y_{gn} into the original equation will give only one differential equation, we can assume that $n - 1$ additional differential relations between $C_1, C_2, \ldots, C_n$ can be imposed to specify these functions. These additional relations are used to simplify the technical calculation of the method and are typically as follows:

$$C_1' y_1 + C_2' y_2 + \ldots + C_n' y_n = 0, \ldots, C_1' y_1^{(n-2)} + C_2' y_2^{(n-2)} + \ldots + C_n' y_n^{(n-2)} = 0.$$

Using these relations in the simplification of the expressions for derivatives of y_{gn} and substituting these derivatives into the original equation we obtain the last nth relation, the main one:

$$C_1' y_1^{(n-1)} + C_2' y_2^{(n-1)} + \ldots + C_n' y_n^{(n-1)} = f/a_n.$$

Thus, we arrive at the following algebraic linear system of n equations for n unknowns $C_1', C_2', \ldots, C_n'$:

$$\begin{cases} C_1' y_1 + C_2' y_2 + \ldots + C_n' y_n = 0 \\ \cdots \\ C_1' y_1^{(n-2)} + C_2' y_2^{(n-2)} + \ldots + C_n' y_n^{(n-2)} = 0 \\ C_1' y_1^{(n-1)} + C_2' y_2^{(n-1)} + \ldots + C_n' y_n^{(n-1)} = f/a_n \end{cases}.$$

The determinant of this system

$$W = \begin{vmatrix} y_1 & y_2 & \cdots & y_n \\ & \cdots & & \\ y_1^{(n-2)} & y_2^{(n-2)} & \cdots & y_n^{(n-2)} \\ y_1^{(n-1)} & y_2^{(n-1)} & \cdots & y_n^{(n-1)} \end{vmatrix}$$

is non-zero that guarantees the existence and uniqueness of the solution. (Recall that this is the Wronski determinant, considered in section 2 of Chapter 8.) Solving the algebraic linear system, we obtain decoupled expressions for $C_1', C_2', \ldots, C_n'$, that is, we express each of the derivatives in terms of known functions. We then integrate each of these expressions and find $C_1, C_2, \ldots, C_n$. Substituting the latter expressions into the formula $y_{gn} = C_1(x)y_1 + C_2(x)y_2 + \ldots + C_n(x)y_n$, we obtain the general solution of the original equation.

Examples.

1. $y'' - 8y' + 16y = (1 - x)e^x$.
This equation was already solved by the method of undetermined coefficients (see Example 1 in section 2.1). Now we find its solution by the method of variation of parameters. We repeat in detail the main steps of the general algorithm to illustrate the entire procedure. The general solution y_{gh} of the homogeneous equation was found in the form $y_{gh} = (C_1 + C_2x)e^{4x}$, where $y_1 = e^{4x}$ and $y_2 = xe^{4x}$ are linearly independent particular solutions. Then, we look for the solution of the nonhomogeneous equation in the form $y_{gn} = (C_1(x) + C_2(x)x)e^{4x}$. The first derivative of $y_{gn} \equiv y$ is as follows:

$$y' = C_1'y_1 + C_2'y_2 + C_1y_1' + C_2y_2' = C_1'e^{4x} + C_2'xe^{4x} + C_1 \cdot 4e^{4x} + C_2(e^{4x} + 4xe^{4x})$$

and to simplify this expression we impose the complementary condition $C_1'y_1 + C_2'y_2 = (C_1' + C_2'x)e^{4x} = 0$. So, the expression for the second derivative is

$$y'' = C_1'y_1' + C_2'y_2' + C_1y_1'' + C_2y_2''$$
$$= C_1' \cdot 4e^{4x} + C_2'(e^{4x} + 4xe^{4x}) + C_1 \cdot 16e^{4x} + C_2(8e^{4x} + 16xe^{4x}).$$

Substituting y, y' and y'' into the original equation, we obtain

$$y''-8y'+16y = (C_1'y_1'+C_2'y_2'+C_1y_1''+C_2y_2'')-8(C_1y_1'+C_2y_2')+16(C_1y_1+C_2y_2)$$

$$= [4C_1' + C_2'(1 + 4x) + 16C_1 + C_2(8 + 16x)]e^{4x} - 8[4C_1 + C_2(1 + 4x)]e^{4x}$$

$$+16[C_1 + C_2x]e^{4x} = (1 - x)e^x.$$

Regrouping the terms

$$[4C_1'+C_2'(1+4x)]e^{4x}+C_1[16-8\cdot4+16]e^{4x}+C_2[8+16x-8(1+4x)+16x]e^{4x}$$

$$= (1 - x)e^x,$$

we notice that the terms in the second and third brackets cancel each other, and therefore, the main equation simplifies to the form

$$[4C_1' + C_2'(1 + 4x)]e^{4x} = (1 - x)e^x.$$

Thus, we have to solve the algebraic linear system

$$\begin{cases} (C_1' + C_2'x)e^{4x} = 0 \\ [4C_1' + C_2'(1 + 4x)]e^{4x} = (1 - x)e^x \end{cases},$$

or simplifying,

$$\begin{cases} C_1' + C_2'x = 0 \\ 4C_1' + C_2'(1 + 4x)) = (1 - x)e^{-3x} \end{cases}.$$

First, we solve this system for the unknowns C_1' and C_2'. To eliminate C_1', we multiply the first equation by 4 and subtract the result from the second, obtaining $C_2' = (1 - x)e^{-3x}$. Substituting this expression into the first equation, we get $C_1' = -xC_2' = -x(1-x)e^{-3x}$. Next, we find the proper functions C_1 and C_2. Integrating the relation for C_2', we obtain:

$$C_2 = \int (1 - x)e^{-3x}dx = (\frac{1}{3}x - \frac{2}{9})e^{-3x} + B_2.$$

Integrating the second relation, we find C_1:

$$C_1 = \int (x^2 - x)e^{-3x}dx = (-\frac{1}{3}x^2 + \frac{1}{9}x + \frac{1}{27})e^{-3x} + B_1.$$

Taking these results into the solution y_{gn}, we get:

$$y_{gn} = C_1y_1+C_2y_2 = [(-\frac{1}{3}x^2+\frac{1}{9}x+\frac{1}{27})e^{-3x}+B_1]e^{4x}+[(\frac{1}{3}x-\frac{2}{9})e^{-3x}+B_2]xe^{4x}$$

$$= (-\frac{1}{9}x + \frac{1}{27})e^x + B_1 e^{4x} + B_2 x e^{4x}.$$

Naturally, we found the same general solution obtained in Example 1 of section 2.1, where $(-\frac{1}{9}x + \frac{1}{27})e^x$ is the particular solution of the nonhomogeneous equation and $B_1 e^{4x} + B_2 x e^{4x}$ is the general solution of the homogeneous one.

2. $y'' + 4y = \tan x$.

This equation cannot be solved by the method of undetermined coefficients, since the form of the right-hand side does not correspond to those admissible in that method. However, the method of variation of parameters can be applied following the general scheme. In this example, we have shortened the steps of the solution, taking advantage of the results obtained in the derivation of the general algorithm. We start with the corresponding homogeneous equation $y'' + y = 0$ whose characteristic equation $\lambda^2 + 1 = 0$ has the roots $\lambda_{1,2} = \pm 2i$. Consequently, the general solution comes in the form $y_{gh} = C_1 \cos 2x + C_2 \sin 2x$, where $y_1 = \cos 2x$ and $y_2 = \sin 2x$ are linearly independent particular solutions. We then look for the solution of the nonhomogeneous equation in the form $y_{gh} = C_1(x) \cos 2x + C_2 \sin 2x$. Differentiation of $y_{gh} \equiv y$ gives

$$y' = C_1' y_1 + C_2' y_2 + C_1 y_1' + C_2 y_2' = C_1' \cos 2x + C_2' \sin 2x - 2C_1 \sin 2x + 2C_2 \cos 2x.$$

By imposing the complementary condition

$$C_1' y_1 + C_2' y_2 = C_1' \cos 2x + C_2' \sin 2x = 0$$

we simplify the derivative to the form $y' = -2C_1 \sin 2x + 2C_2 \cos 2x$. Then, we calculate the second derivative

$$y'' = C_1' y_1' + C_2' y_2' + C_1 y_1'' + C_2 y_2'' = -2C_1' \sin 2x + 2C_2' \cos 2x - 4C_1 \cos 2x - 4C_2 \sin 2x$$

and substitute the results of the differentiation into the original equation:

$$y'' + 4y = (-2C_1' \sin 2x + 2C_2' \cos 2x - 4C_1 \cos 2x - 4C_2 \sin 2x)$$

$$+4(C_1 \cos 2x + C_2 \sin 2x) = 8 \tan x$$

or, after simplification,

$$-C_1' \sin 2x + C_2' \cos 2x = 4 \tan x.$$

Thus, we have to solve the algebraic linear system of the two equations with respect to the unknowns C_1' and C_2':

$$\begin{cases} C_1' \cos 2x + C_2' \sin 2x = 0 \\ -C_1' \sin 2x + C_2' \cos 2x = 4 \tan x \end{cases}.$$

To eliminate C_1', we multiply the first equation by $\sin 2x$, the second by $\cos 2x$ and add the results: $C_2' = 4\tan x\cos 2x$. Similarly, we eliminate C_2' by multiplying the first equation by $\cos 2x$, the second by $\sin 2x$ and subtracting the second result from the first: $C_1' = -4\tan x\sin 2x$. Integrating the last relation, we get:

$$C_1 = -4\int \tan x\sin 2x dx = -4\int \tan x 2\sin x\cos x dx$$

$$= -2\int 2\sin^2 x dx = -4\int (1 - \cos 2x)dx = -4(x - \frac{1}{2}\sin 2x) + B_1.$$

Integrating the relation for C_2', we find:

$$C_2 = 4\int \tan x\cos 2x dx = 4\int \tan x(\cos^2 x - \sin^2 x)dx = 4\int \sin x\cos x - \frac{\sin^3 x}{\cos x}dx$$

$$= 4\int \sin x d(\sin x) + 4\int \frac{1 - \cos^2 x}{\cos x}d(\cos x) = 2\sin^2 x + 4\ln|\cos x| - 2\cos^2 x + B_2$$

$$= -2\cos 2x + 4\ln|\cos x| + B_2.$$

Taking these results into the solution y_{gn}, we obtain:

$$y_{gn} = C_1 y_1 + C_2 y_2 = (-4x + 2\sin 2x + B_1)\cos 2x + (-2\cos 2x + 4\ln|\cos x| + B_2)\sin 2x$$

$$= (-4x\cos 2x + 4\ln|\cos x|\sin 2x) + B_1\cos 2x + B_2\sin 2x.$$

Here, $y_{pn} = -4x\cos 2x + 4\ln|\cos x|\sin 2x$ is the particular solution of the original equation and $y_{gh} = B_1\cos 2x + B_2\sin 2x$ is the general solution of the homogeneous equation.

3. $y'' - 4y' + 5y = \frac{e^{2x}}{\cos x}$.
The right-hand side of this equation is not tractable by the method of undetermined coefficients. However, we can apply the method of variation of parameters. The homogeneous equation $y'' - 4y' + 5y = 0$ has the characteristic equation $\lambda^2 - 4\lambda + 5 = 0$ whose roots are $\lambda = 2 \pm i$. So, the general solution of the homogeneous equation is $y_{gh} = (C_1\cos x + C_2\sin x)e^{2x}$, where $y_1 = \cos x \cdot e^{2x}$ and $y_2 = \sin x \cdot e^{2x}$ are linearly independent particular solutions. The next step is to seek the solution of the nonhomogeneous equation in the form $y_{gn} = (C_1(x)\cos x + C_2(x)\sin x)e^{2x}$. The derivatives C_1' and C_2' are found by solving the algebraic linear system:

$$\begin{cases} C_1'y_1 + C_2'y_2 = C_1'\cos xe^{2x} + C_2'\sin xe^{2x} = 0 \\ C_1'y_1' + C_2'y_2' = C_1'(2\cos x - \sin x)e^{2x} + C_2'(2\sin x + \cos x)e^{2x} = \frac{e^{2x}}{\cos x} \end{cases},$$

or simplifying,

$$\begin{cases} C_1' \cos x + C_2' \sin x = 0 \\ C_1'(2\cos x - \sin x) + C_2'(2\sin x + \cos x) = \frac{1}{\cos x} \end{cases}.$$

Using the first equation, the second can be simplified even further:

$$C_1'(2\cos x - \sin x) + C_2'(2\sin x + \cos x)$$

$$= 2(C_1' \cos x + C_2' \sin x) - C_1' \sin x + C_2' \cos x = -C_1' \sin x + C_2' \cos x = \frac{1}{\cos x}.$$

By multiplying this equation by $\sin x$, the first by $\cos x$ and subtracting the first result from the second, we can eliminate C_2': $C_1' = -\frac{\sin x}{\cos x}$. Similarly, multiplying the second equation by $\cos x$, the first by $\sin x$ and adding the results, we eliminate C_1' and get: $C_2' = 1$. Integrating the last relation, we find $C_2 = x + B_2$. Integrating the relation for C_1', we find $C_1 = -\int \frac{\sin x}{\cos x} dx = \ln|\cos x| + B_1$. Taking these results into the solution y_{gn}, we obtain:

$$y_{gn} = C_1 y_1 + C_2 y_2 = (\ln|\cos x| + B_1) \cos x e^{2x} + (x + B_2) \sin x e^{2x}$$

$$= \ln|\cos x| \cdot \cos x e^{2x} + x \sin x e^{2x} + (B_1 \cos x + B_2 \sin x) e^{2x},$$

where $\ln|\cos x| \cdot \cos x e^{2x} + x \sin x e^{2x}$ is the particular solution of the nonhomogeneous equation and $(B_1 \cos x + B_2 \sin x) e^{2x}$ is the general solution of the homogeneous one.

4. $y'' + 4y = \cot 2x$.
It is not feasible to find solution of this equation by the method of undetermined coefficients. So, we apply the method of variation of parameters. The homogeneous equation $y'' + 4y = 0$ has the characteristic equation $\lambda^2 + 4 = 0$ whose roots are $\lambda = \pm 2i$. Then, the general solution of the homogeneous equation is $y_{gh} = C_1 \cos 2x + C_2 \sin 2x$, where $y_1 = \cos 2x$ and $y_2 = \sin 2x$ are linearly independent particular solutions. Therefore, we can seek the solution of the nonhomogeneous equation in the form $y_{gn} = C_1(x) \cos 2x + C_2(x) \sin 2x$. The derivatives C_1' and C_2' are found by solving the algebraic linear system:

$$\begin{cases} C_1' y_1 + C_2' y_2 = C_1' \cos 2x + C_2' \sin 2x = 0 \\ C_1' y_1' + C_2' y_2' = -2C_1' \sin 2x + 2C_2' \cos 2x = \cot 2x \end{cases}.$$

Multiplying the first equation by $2\sin 2x$, the second by $\cos 2x$ and adding the results, we get $2C_2' = \cos 2x \cot 2x$. Similarly, multiplying the first equation by $2\cos 2x$, the second by $\sin 2x$ and subtracting the second result from the first, we find $2C_1' = -\sin 2x \cot 2x$. Integrating the last relation, we find

$C_1 = -\frac{1}{2}\int \cos 2x dx = -\frac{1}{4}\sin 2x + B_1$. To calculate the integral of C_2', we change the variable $p = \cos 2x$ and find

$$C_2 = \frac{1}{2}\int \cos 2x \cot 2x dx = \frac{1}{2}\int \frac{\cos^2 2x}{\sin^2 2x}\sin 2x dx = -\frac{1}{4}\int \frac{\cos^2 2x}{1-\cos^2 2x}d(\cos 2x)$$

$$= -\frac{1}{4}\int \frac{p^2}{1-p^2}dp = -\frac{1}{4}\int \frac{p^2-1+1}{1-p^2}dp = -\frac{1}{4}\int -1+\frac{1}{1-p^2}dp$$

$$= -\frac{1}{4}\left[-p+\frac{1}{2}\int \frac{1}{1-p}+\frac{1}{1+p}dp\right] = \frac{p}{4}-\frac{1}{8}\left[-\ln|1-p|+\ln|1+p|\right]+B_2$$

$$= \frac{p}{4}-\frac{1}{8}\ln\left|\frac{1+p}{1-p}\right|+B_2 = \frac{\cos 2x}{4}-\frac{1}{8}\ln\left|\frac{\cos^2 x}{\sin^2 x}\right|+B_2 = \frac{\cos 2x}{4}-\frac{1}{4}\ln|\cot x|+B_2.$$

Taking these results into the solution y_{gn}, we obtain:

$$y_{gn}=C_1y_1+C_2y_2=\left(-\frac{\sin 2x}{4}+B_1\right)\cos 2x+\left(\frac{\cos 2x}{4}-\frac{1}{4}\ln|\cot x|+B_2\right)\sin 2x$$

$$= -\frac{1}{4}\ln|\cot x|\sin 2x + B_1\cos 2x + B_2\sin 2x,$$

where $-\frac{1}{4}\ln|\cot x|\sin 2x$ is the particular solution of the nonhomogeneous equation and $B_1\cos 2x+B_2\sin 2x$ is the general solution of the homogeneous one.

5. $y'' - 6y' + 8y = \frac{4e^{2x}}{1+e^{-2x}}$, $y(0) = 0, y'(0) = 0$.

The function on the right-hand side is not tractable by the method of undetermined coefficients. However, we can apply the method of variation of parameters. The homogeneous equation $y'' - 6y' + 8y = 0$ has the characteristic equation $\lambda^2 - 6\lambda + 8 = 0$, whose roots are $\lambda = 2, 4$. So the general solution of the homogeneous equation is $y_{gh} = C_1e^{2x} + C_2e^{4x}$, where $y_1 = e^{2x}$ and $y_2 = e^{4x}$ are linearly independent particular solutions. Next, we look for the solution of the nonhomogeneous equation in the form $y_{gn} = C_1(x)e^{2x} + C_2(x)e^{4x}$. The derivatives C_1' and C_2' are found from the algebraic linear system:

$$\begin{cases} C_1'y_1 + C_2'y_2 = C_1'e^{2x} + C_2'e^{4x} = 0 \\ C_1'y_1' + C_2'y_2' = 2C_1'e^{2x} + 4C_2'e^{4x} = \frac{4e^{2x}}{1+e^{-2x}} \end{cases}$$

or, in simplified form,

$$\begin{cases} C_1' + C_2'e^{2x} = 0 \\ C_1' + 2C_2'e^{2x} = \frac{2}{1+e^{-2x}} \end{cases}.$$

Subtracting the first equation from the second, we get $C'_2 e^{2x} = \frac{2}{1+e^{-2x}}$. Substituting this expression into the first equation, we find $C'_1 = -\frac{2}{1+e^{-2x}}$. The integral on the right-hand side of C'_1 can be calculated using the change of variable $p = e^x$:

$$C_1 = -\int \frac{2}{1+e^{-2x}} dx = -2 \int \frac{e^x}{e^x + e^{-x}} dx = -2 \int \frac{1}{p+1/p} dp$$

$$= -2 \int \frac{p}{p^2+1} dp = -\ln(p^2+1) + B_1 = -\ln(e^{2x}+1) + B_1.$$

The integration for C_2 can be done using the change of variable $t = e^{-2x}$:

$$C_2 = \int \frac{2e^{-2x}}{1+e^{-2x}} dx = -\int \frac{1}{1+t} dt = -\ln|t+1| + B_2 = -\ln(e^{-2x}+1) + B_2.$$

Substituting these results into the solution y_{gn}, we get:

$$y_{gn} = C_1 y_1 + C_2 y_2 = \left(-\ln(e^{2x}+1) + B_1\right) e^{2x} + \left(-\ln(e^{-2x}+1) + B_2\right) e^{4x}$$

$$= -\ln[(e^{2x}+1)(e^{-2x}+1)] + B_1 e^{2x} + B_2 e^{4x} = -\ln(e^{2x}+e^{-2x}+2) + B_1 e^{2x} + B_2 e^{4x}.$$

To find the solution of the Cauchy problem, we differentiate y_{gn}:

$$y'_{gn} = -\frac{2e^{2x} - 2e^{-2x}}{e^{2x} + e^{-2x} + 2} + 2B_1 e^{2x} + 4B_2 e^{4x}$$

and substitute the initial conditions in the solution y_{gn} and its derivative:

$$y_{gn}(0) = -\ln 4 + B_1 + B_2 = 0, \quad y'_{gn}(0) = -\frac{0}{4} + 2B_1 + 4B_2 = 0.$$

From the second equation we get $B_1 = -2B_2$, and then the first takes the form $-\ln 4 - B_2 = 0$. Therefore, $B_2 = -\ln 4$, $B_1 = 2\ln 4$ and the solution of the Cauchy problem is

$$y = -\ln(e^{2x} + e^{-2x} + 2) + \ln 4 \cdot (2e^{2x} - e^{4x}).$$

6. $y'' - 2y' + y = \frac{e^x}{x^2}$, $y(1) = 0, y'(1) = -e$.
This equations cannot be solved by the method of undetermined coefficients, but the method of variation of parameters can be employed. The homogeneous equation $y'' - 2y' + y = 0$ has the characteristic equation $\lambda^2 - 2\lambda + 1 = 0$, whose roots coincide and are equal to 1. So, the general solution of the homogeneous equation is $y_{gh} = (C_1 + C_2 x)e^x$, where $y_1 = e^x$ and $y_2 = xe^x$ are linearly independent particular solutions. Next, we look for the solution of

the nonhomogeneous equation in the form $y_{gn} = C_1(x)e^x + C_2(x)xe^x$. The derivatives C_1' and C_2' are found from the algebraic linear system:

$$\begin{cases} C_1'y_1 + C_2'y_2 = C_1'e^x + C_2'xe^x = 0 \\ C_1'y_1' + C_2'y_2' = C_1'e^x + C_2'(1+x)e^x = \frac{e^x}{x^2} \end{cases}$$

or, in a simplified form,

$$\begin{cases} C_1' + C_2'x = 0 \\ C_1' + C_2'(1+x) = \frac{1}{x^2} \end{cases}.$$

Subtracting the first equation from the second, we get $C_2' = \frac{1}{x^2}$. Substituting this expression into the first equation, we find $C_1' = -\frac{1}{x}$. Integrating both relations, we get $C_1 = -\ln|x| + B_1$ and $C_2 = -\frac{1}{x} + B_2$. Substituting these results into y_{gn}, we obtain:

$$y_{gn} = C_1y_1 + C_2y_2 = \left(-\ln|x| + B_1\right)e^x + \left(-\frac{1}{x} + B_2\right)xe^x$$

$$= -\left(\ln|x| + 1\right)e^x + B_1e^x + B_2xe^x$$

or denoting $A_1 = B_1 - 1$ and $A_2 = B_2$, we have

$$y_{gn} = \left(-\ln|x| + A_1 + A_2x\right)e^x.$$

To find the solution of the Cauchy problem, we calculate the derivative

$$y_{gn}' = \left(-\frac{1}{x} + A_2 - \ln|x| + A_1 + A_2x\right)e^x$$

and substitute the initial conditions into y_{gn} and y_{gn}':

$$y_{gn}(1) = (A_1 + A_2)e = 0, \quad y_{gn}'(1) = (-1 + A_2 + A_1 + A_2)e = -e.$$

From the first equation it follows that $A_1 = -A_2$, and from the second $-1 + A_2 = -1$, that is, $A_2 = 0$ and $A_1 = 0$. Therefore, the solution of the Cauchy problem is

$$y = -\ln|x| \cdot e^x.$$

7. $y'' - y = \frac{e^x}{e^x+1}$.

This equations cannot be solved by the method of undetermined coefficients, but the method of variation of parameters can be employed. The homogeneous equation $y'' - y = 0$ has the characteristic equation $\lambda^2 - 1 = 0$ whose roots are $\lambda = \pm 1$. So, the general solution of the homogeneous equation is

$y_{gh} = C_1 e^{-x} + C_2 e^x$, where $y_1 = e^{-x}$ and $y_2 = e^x$ are linearly independent particular solutions. Then, the solution of the nonhomogeneous equation are sought in the form $y_{gn} = C_1(x)e^{-x} + C_2(x)e^x$. The derivatives C_1' and C_2' are found from the algebraic linear system:

$$\begin{cases} C_1'y_1 + C_2'y_2 = C_1'e^{-x} + C_2'e^x = 0 \\ C_1'y_1' + C_2'y_2' = -C_1'e^{-x} + C_2'e^x = \frac{e^x}{e^x+1} \end{cases}.$$

Adding two equations together, we get $2C_2'e^x = \frac{e^x}{e^x+1}$ or $C_2' = \frac{1}{2(e^x+1)}$. Substituting this expression into the first equation, we have $C_1' = -\frac{e^{2x}}{2(e^x+1)}$. Integrating two relations, we find C_1 e C_2:

$$C_1 = -\frac{1}{2}\int\frac{e^x d(e^x)}{e^x+1} = -\frac{1}{2}\int\frac{p\,dp}{p+1} = -\frac{1}{2}\int\frac{p+1-1}{p+1}dp$$

$$= -\frac{1}{2}\int 1 - \frac{1}{p+1}dp = -\frac{1}{2}(p - \ln|p+1|) + B_1 = -\frac{1}{2}(e^x - \ln(e^x+1)) + B_1$$

and

$$C_2 = \frac{1}{2}\int\frac{dx}{e^x+1} = \frac{1}{2}\int\frac{d(e^x)}{e^{2x}+e^x} = \frac{1}{2}\int\frac{dp}{p^2+p}$$

$$= \frac{1}{2}\int\frac{1}{p} - \frac{1}{p+1}dp = \frac{1}{2}(\ln|p| - \ln|p+1|) + B_2 = \frac{1}{2}(x - \ln(e^x+1)) + B_2.$$

Substituting these results into the solution y_{gn}, we obtain:

$$y_{gn} = C_1 y_1 + C_2 y_2 = \left[-\frac{1}{2}(e^x - \ln(e^x+1)) + B_1\right]e^{-x} + \left[\frac{1}{2}(x - \ln(e^x+1)) + B_2\right]e^x.$$

$$= \left(\frac{xe^x - 1}{2} - \ln(e^x+1)\frac{e^x - e^{-x}}{2}\right) + B_1 e^{-x} + B_2 e^x.$$

8. $y'' + 4y = \frac{1}{\cos 2x}$.
The function on the right-hand side is not tractable by the method of undetermined coefficients, but we can still use the method of variation of parameters. The homogeneous equation $y'' + 4y = 0$ has the characteristic equation $\lambda^2 + 4 = 0$, whose roots are $\lambda = \pm 2i$. So, the general solution of the homogeneous equation is $y_{gh} = C_1\cos 2x + C_2\sin 2x$, where $y_1 = \cos 2x$ and $y_2 = \sin 2x$ are linearly independent particular solutions. Next, we look for the solution of the nonhomogeneous equation in the form $y_{gn} = C_1(x)\cos 2x + C_2(x)\sin 2x$. The derivatives C_1' and C_2' are found by solving the algebraic linear system:

$$\begin{cases} C_1'y_1 + C_2'y_2 = C_1'\cos 2x + C_2'\sin 2x = 0 \\ C_1'y_1' + C_2'y_2' = -2C_1'\sin 2x + 2C_2'\cos 2x = \frac{1}{\cos 2x} \end{cases}.$$

Multiplying the first equation by $2\sin 2x$, the second by $\cos 2x$ and adding the results, we get $2C_2' = 1$. Substituting this value of C_2' into the first equation, we find $2C_1' = -\tan 2x$. Integrating these two relations, we obtain

$$C_1 = -\frac{1}{2}\int \tan 2x\, dx = \frac{1}{4}\int \frac{d\cos 2x}{\cos 2x} = \frac{1}{4}\ln|\cos 2x| + B_1,\ C_2 = \frac{1}{2}\int dx = \frac{x}{2} + B_2.$$

Substituting these results into the solution y_{gn}, we get:

$$y_{gn} = C_1 y_1 + C_2 y_2 = \left(\frac{1}{4}\ln|\cos 2x| + B_1\right)\cos 2x + \left(\frac{x}{2} + B_2\right)\sin 2x$$

$$= \frac{1}{4}\ln|\cos 2x|\cdot\cos 2x + \frac{x}{2}\sin 2x + B_1\cos 2x + B_2\sin 2x,$$

where $\frac{1}{4}\ln|\cos 2x|\cdot\cos 2x + \frac{x}{2}\sin 2x$ is the particular solution of the nonhomogeneous equation and $B_1\cos 2x + B_2\sin 2x$ is the general solution of the homogeneous one.

9. $y'' + 2y' + 2y = e^{-x}\cot x$.
The right-hand side of this equation is not solvable by the method of undetermined coefficients, but we can apply the method of variation of parameters. The homogeneous equation $y'' + 2y' + 2y = 0$ has the characteristic equation $\lambda^2 + 2\lambda + 2 = 0$, whose roots are $\lambda = -1 \pm i$. So, the general solution of the homogeneous equation is $y_{gh} = (C_1\cos x + C_2\sin x)e^{-x}$, where $y_1 = e^{-x}\cos x$ and $y_2 = e^{-x}\sin x$ are linearly independent particular solutions. Therefore, we seek the solution of the nonhomogeneous equation in the form $y_{gn} = (C_1(x)\cos x + C_2(x)\sin x)e^{-x}$. The derivatives C_1' and C_2' are found from the algebraic linear system:

$$\begin{cases} C_1'y_1 + C_2'y_2 = C_1'e^{-x}\cos x + C_2'e^{-x}\sin x = 0 \\ C_1'y_1' + C_2'y_2' = C_1'(-\cos x - \sin x)e^{-x} + C_2'(-\sin x + \cos x)e^{-x} = e^{-x}\cot x \end{cases}.$$

Dividing by e^{-x}, we simplify the system to the form

$$\begin{cases} C_1'\cos x + C_2'\sin x = 0 \\ -C_1'(\cos x + \sin x) + C_2'(\cos x - \sin x) = \cot x \end{cases}.$$

Multiplying the first equation by $\cos x + \sin x$, the second by $\cos x$ and adding the results, we eliminate the unknown C_1' and get

$$C_2'\sin x(\cos x + \sin x) + C_2'\cos x(\cos x - \sin x) = \frac{\cos^2 x}{\sin x}$$

or simplifying $C_2' = \frac{\cos^2 x}{\sin x}$. Substituting this expression for C_2' into the first equation, we find $C_1' = -\cos x$. Integrating both relations, we obtain

$C_1 = -\sin x + B_1$ and

$$C_2 = \int \frac{\cos^2 x}{\sin x}\,dx = \int \frac{1 - \sin^2 x}{\sin x}\,dx = \int \frac{\sin x}{\sin^2 x} - \sin x\,dx$$

$$= \cos x - \int \frac{d(\cos x)}{1 - \cos^2 x} = \cos x - \int \frac{dp}{1 - p^2} = \cos x - \frac{1}{2}\ln\left|\frac{1+p}{1-p}\right| + B_2$$

$$= \cos x - \frac{1}{2}\ln\left|\frac{1+\cos x}{1-\cos x}\right| + B_2 = \cos x - \frac{1}{2}\ln\frac{\cos^2(x/2)}{\sin^2(x/2)} + B_2 = \cos x - \ln\left|\cot\frac{x}{2}\right| + B_2.$$

Substituting these results into the solution y_{gn}, we get:

$$y_{gn} = C_1 y_1 + C_2 y_2 = (-\sin x + B_1)\,e^{-x}\cos x + \left(\cos x - \ln\left|\cot\frac{x}{2}\right| + B_2\right)e^{-x}\sin x$$

$$= -\ln\left|\cot\frac{x}{2}\right| \cdot e^{-x}\sin x + e^{-x}(B_1\cos x + B_2\sin x),$$

where $-\ln\left|\cot\frac{x}{2}\right| \cdot e^{-x}\sin x$ is the particular solution of the nonhomogeneous equation and $e^{-x}(B_1\cos x + B_2\sin x)$ is the general solution of the homogeneous one.

10. $y'' + 2y' + y = 3e^{-x}\sqrt{x+1}$.
The right-hand side of this kind is not solvable using the method of undetermined coefficients, but we can apply the method of variation of parameters. The characteristic equation of the homogeneous part $\lambda^2 + 2\lambda + 1 = 0$ has a double real root $\lambda_{1,2} = -1$. So the general solution of the homogeneous equation is $y_{gh} = (C_1 + C_2 x)e^{-x}$, where $y_1 = e^{-x}$ and $y_2 = xe^{-x}$ are linearly independent particular solutions. Next, we look for the solution of the nonhomogeneous equation in the form $y_{gn} = C_1(x)e^{-x} + C_2(x)xe^{-x}$. The derivatives C_1' and C_2' are found from the algebraic linear system:

$$\begin{cases} C_1'y_1 + C_2'y_2 = C_1'e^{-x} + C_2'xe^{-x} = 0 \\ C_1'y_1' + C_2'y_2' = C_1' \cdot (-e^{-x}) + C_2'(1-x)e^{-x} = 3e^{-x}\sqrt{x+1} \end{cases}$$

By dividing by e^{-x}, we simplify the system to the form

$$\begin{cases} C_1' + C_2'x = 0 \\ -C_1' + C_2'(1-x) = 3\sqrt{x+1} \end{cases}$$

Adding the two equations together, we get $C_2' = 3\sqrt{x+1}$. Substituting this expression for C_2' into the first equation, we find $C_1' = -3x\sqrt{x+1}$. Integrating these two relations, we get

$$C_1 = -3\int x\sqrt{x+1}\,dx = -3\int (x+1-1)\sqrt{x+1}\,dx$$

$$= -3 \int (x+1)^{3/2} - \sqrt{x+1} dx = -3 \left(\frac{2}{5}(x+1)^{5/2} - \frac{2}{3}(x+1)^{3/2} \right) + B_1$$

and

$$C_2 = 3 \int \sqrt{x+1} dx = 2(x+1)^{3/2} + B_2.$$

Taking these results into the solution y_{gn}, we obtain:

$$y_{gn} = C_1 y_1 + C_2 y_2 = \left(-\frac{6}{5}(x+1)^{5/2} + 2(x+1)^{3/2} + B_1 \right) e^{-x}$$

$$+ \left(2(x+1)^{3/2} + B_2 \right) x e^{-x} = \left(-\frac{6}{5}(x+1)^{5/2} + 2(x+1)(x+1)^{3/2} \right) e^{-x}$$

$$+ (B_1 + B_2 x) e^{-x} = \frac{4}{5}(x+1)^{5/2} e^{-x} + (B_1 + B_2 x) e^{-x},$$

where $\frac{4}{5}(x+1)^{5/2} e^{-x}$ is the particular solution of the nonhomogeneous equation and $(B_1 + B_2 x) e^{-x}$ is the general solution of the homogeneous one.

11. $y'' + y = \tan^2 x$.

The right-hand side of this kind cannot be solved using the method of undetermined coefficients, so we use the method of variation of parameters. The homogeneous equation $y'' + y = 0$ has characteristic equation $\lambda^2 + 1 = 0$, whose roots are $\lambda = \pm i$. So, the general solution of the homogeneous equation is $y_{gh} = C_1 \cos x + C_2 \sin x$, where $y_1 = \cos x$ and $y_2 = \sin x$ are linearly independent particular solutions. Next, we seek the solution of the nonhomogeneous equation in the form $y_{gn} = C_1(x) \cos x + C_2(x) \sin x$. The derivatives C_1' and C_2' are found by solving the algebraic linear system:

$$\begin{cases} C_1' y_1 + C_2' y_2 = C_1' \cos x + C_2' \sin x = 0 \\ C_1' y_1' + C_2' y_2' = -C_1' \sin x + C_2' \cos x = \tan^2 x \end{cases}.$$

Multiplying the first equation by $\sin x$, the second by $\cos x$ and adding the results, we get $C_2' = \tan^2 x \cos x$. Substituting this expression for C_2' into the first equation, we have $C_1' \cos x = -\tan^2 x \cos x \sin x$ or $C_1' = -\tan^2 x \sin x$. Integrating, we obtain

$$C_1 = -\int \frac{\sin^3 x}{\cos^2 x} dx = \int \frac{1 - \cos^2 x}{\cos^2 x} d(\cos x) = \int \frac{1 - t^2}{t^2} dt$$

$$= -\frac{1}{t} - t + B_1 = -\frac{1}{\cos x} - \cos x + B_1$$

and

$$C_2 = \int \frac{\sin^2 x}{\cos x} dx = \int \frac{1 - \cos^2 x}{\cos x} dx = -\sin x + \int \frac{\cos x}{\cos^2 x} dx = -\sin x + \int \frac{d(\sin x)}{1 - \sin^2 x}$$

$$=-\sin x + \int \frac{dp}{1-p^2} = -\sin x + \frac{1}{2}\ln\left|\frac{1+p}{1-p}\right| + B_2 = -\sin x + \frac{1}{2}\ln\left|\frac{1+\sin x}{1-\sin x}\right| + B_2$$

$$= -\sin x + \frac{1}{2}\ln\frac{\cos^2\left(\frac{x}{2} - \frac{\pi}{4}\right)}{\sin^2\left(\frac{x}{2} - \frac{\pi}{4}\right)} + B_2 = -\sin x + \ln\left|\cot\left(\frac{x}{2} - \frac{\pi}{4}\right)\right| + B_2.$$

Taking these results into the solution y_{gn}, we get:

$$y_{gn} = C_1 y_1 + C_2 y_2$$

$$= \left(-\frac{1}{\cos x} - \cos x + B_1\right)\cos x + \left(-\sin x + \ln\left|\cot\left(\frac{x}{2} - \frac{\pi}{4}\right)\right| + B_2\right)\sin x$$

$$= -2 + \ln\left|\cot\left(\frac{x}{2} - \frac{\pi}{4}\right)\right| \cdot \sin x + B_1 \cos x + B_2 \sin x,$$

where $-2 + \ln\left|\cot\left(\frac{x}{2} - \frac{\pi}{4}\right)\right| \cdot \sin x$ is the particular solution of the nonhomogeneous equation and $B_1 \cos x + B_2 \sin x$ is the general solution of the homogeneous one.

12. $y'' + 2y' + y = xe^x + \frac{1}{xe^x}$.
The equation with the right-hand side of this kind cannot be solved using the method of undetermined coefficients, so we use the method of variation of parameters. The homogeneous equation $y'' + 2y' + y = 0$ has the characteristic equation $\lambda^2 + 2\lambda + 1 = 0$, whose double root is $\lambda = -1$. Therefore, the general solution of the homogeneous equation is $y_{gh} = (C_1 + C_2 x)e^{-x}$, where $y_1 = e^{-x}$ and $y_2 = xe^{-x}$ are linearly independent particular solutions. Next, we look for the solution of the nonhomogeneous equation in the form $y_{gn} = C_1 e^{-x} + C_2 xe^{-x}$. The derivatives C_1' and C_2' are found from the algebraic linear system:

$$\begin{cases} C_1' y_1 + C_2' y_2 = C_1' e^{-x} + C_2' xe^{-x} = 0 \\ C_1' y_1' + C_2' y_2' = -C_1' e^{-x} + C_2'(1-x)e^{-x} = xe^x + \frac{1}{xe^x} \end{cases}.$$

By dividing by e^{-x}, we simplify the system to the form

$$\begin{cases} C_1' + C_2' x = 0 \\ -C_1' + C_2'(1-x) = xe^{2x} + \frac{1}{x} \end{cases}.$$

Adding the two equations together, we get $C_2' = xe^{2x} + \frac{1}{x}$. Substituting this expression for C_2' into the first equation, we find $C_1' = -x^2 e^{2x} + 1$. Integrating these two relations, we get

$$C_1 = -\int x^2 e^{2x} + 1\, dx = -\frac{1}{2}x^2 e^{2x} + \int xe^{2x}\, dx - x$$

$$= -\frac{1}{2}x^2 e^{2x} + \left(\frac{1}{2}xe^{2x} - \frac{1}{2}\int e^{2x}dx\right) - x = -\frac{1}{2}x^2 e^{2x} + \frac{1}{2}xe^{2x} - \frac{1}{4}e^{2x} - x + B_1$$

and

$$C_2 = \int xe^{2x} + \frac{1}{x}dx = \frac{1}{2}xe^{2x} - \frac{1}{2}\int e^{2x}dx + \ln|x| = \frac{1}{2}xe^{2x} - \frac{1}{4}e^{2x} + \ln|x| + B_2.$$

Taking these results into the solution y_{gn}, we obtain:

$$y_{gn} = C_1 y_1 + C_2 y_2$$

$$= \left(-\frac{1}{2}x^2 e^{2x} + \frac{1}{2}xe^{2x} - \frac{1}{4}e^{2x} - x + B_1\right)e^{-x} + \left(\frac{1}{2}xe^{2x} - \frac{1}{4}e^{2x} + \ln|x| + B_2\right)xe^{-x}$$

$$= \left(\frac{1}{4}xe^{2x} - \frac{1}{4}e^{2x} - x + x\ln|x|\right)e^{-x} + (B_1 + B_2 x)e^{-x},$$

where $\left(\frac{1}{4}xe^{2x} - \frac{1}{4}e^{2x} - x + x\ln|x|\right)e^{-x}$ is the particular solution of the non-homogeneous equation and $(B_1 + B_2 x)e^{-x}$ is the general solution of the homogeneous one.

Exercises for section 3

Solve the following nonhomogeneous equations by the Lagrange method:

1. $y'' - 2y' + y = \frac{e^x}{x}$.
2. $y'' + 4y = \tan 2x$.
3. $y'' + 9y = \frac{1}{\sin 3x}$.
4. $y'' - 2y' + 2y = \frac{e^x}{\sin^2 x}$.
5. $y'' + 3y' + 2y = \frac{e^{-x}}{e^x + 2}$.
6. $y'' + y' = e^{2x}\cos e^x$.
7. $y'' + 4y' = \frac{1}{\sin^2 x}$.
8. $y'' - 2y' + 5y = 3e^x + e^x \tan 2x$.
9. $y'' + 4y' + 4y = e^{-2x}\ln x$.
10. $y'' - 2y' + y = \frac{e^x}{x^2 + 1}$.
11. $y''' + y' = \frac{\sin x}{\cos^2 x}$.
12. $y'' + 2y' + 2y = \frac{e^x}{\cos x}$.
13. $y'' - 2y' + y = \frac{e^x}{x^2}$.
14. $y'' + 4y' + 4y = \frac{e^{-2x}}{x^3}$.
15. $y'' + y = \frac{1}{\sin^2 x}$.
16. $y'' - 3y' + 2y = \frac{1}{1 + e^x}$.
17. $y'' - y = \frac{e^x - e^{-x}}{e^x + e^{-x}}$.
18. $y'' - 2y' = 5(3 - 4x)\sqrt{x}$.

19. $y'' - 2y' + 10y = \frac{9e^x}{\cos 3x}$.
20. $y'' - 4y' + 8y = 4(7 - 21x + 18x^2)\sqrt[3]{x}$.
21. $y'' + y = -\cot^2 x$.
22. $y'' + 3y' = \frac{3x-1}{x^2}$.
23. $y'' - 4y' + 4y = \frac{2e^{2x}}{1+x^2}$.
24. $y'' + y' = 7(4 + 3x)\sqrt[3]{x}$.
25. $y'' + 2y' + 2y = \frac{e^{-x}}{\sin x}$.
26. $y'' + 2y = 2 - 4x^2 \sin x^2$.
27. $y'' + 2y' + y = (x + 2)(\ln x + \frac{1}{x})$.
28. $y'' - y' = \frac{x+1}{x^2}$.

Exercises for sections 2 and 3

Solve the following nonhomogeneous equations and the corresponding initial value problems; if possible, use both the method of undetermined coefficients and variation of parameters and compare the obtained solutions:

1. $y'' + 2y' + y = \frac{e^{-x}}{x}$.
2. $y'' + 4y = \frac{1}{\sin 2x}$.
3. $y'' + 10y' + 34y = -9e^{-5x}$, $y(0) = 0, y'(0) = 6$.
4. $y'' + 2y' + 5y = -8e^{-x} \sin 2x$, $y(0) = 2, y'(0) = 6$.
5. $y'' - y' = e^{2x} \sin e^x$.
6. $y'' - y = \frac{e^x}{e^x+1}$.
7. $y'' - 10y' + 25y = e^{5x}$, $y(0) = 1, y'(0) = 0$.
8. $y'' - 2y' + 37y = 36e^x \cos 6x$, $y(0) = 0, y'(0) = 6$.
9. $y'' - 2y' + 2y = \frac{e^x}{\sin^2 x}$.
10. $y'' + 3y' = (40x + 58)e^{2x}$, $y(0) = 0, y'(0) = -2$.
11. $y'' + 2y' = 6x^2 + 2x + 1$, $y(0) = 2, y'(0) = 2$.
12. $y'' - 3y' + 2y = \frac{e^x}{1+e^x}$.
13. $y'' + 2y' + 2y = 2x^2 + 8x + 6$, $y(0) = 1, y'(0) = 4$.
14. $y'' - 4y = (15 - 16x^2)\sqrt{x}$.
15. $y'' + 4y' + 4y = \frac{e^{-2x}}{x+1}$.
16. $y'' + 16y = (\cos 4x - 8 \sin 4x)e^x$, $y(0) = 0, y'(0) = 5$.
17. $y'' - 14y' + 53y = 53x^3 - 42x^2 + 59x - 14$, $y(0) = 0, y'(0) = 7$.
18. $y'' + 2y' + 5y = \frac{2e^{-x}}{\cos 2x}$.
19. $y'' - 12y' + 36y = 32 \cos 2x + 24 \sin 2x$, $y(0) = 2, y'(0) = 4$.
20. $y'' - 2y = -2 - 4x^2 \cos x^2$.
21. $y'' - 2y' = \frac{1}{x} - 2\ln(ex)$.
22. $y'' + 2y' + 2y = xe^{-x}$, $y(0) = 0, y'(0) = 0$.
23. $y''' - 3y' - 2y = 9e^{2x}$, $y(0) = 0, y'(0) = -3, y''(0) = 3$.
24. $y'' + y = \frac{1}{\cos^2 x}$.

4 Cauchy problem (initial value problem)

4.1 Traditional method of solution

We have already encountered the initial value problems for various types of differential equations, and there is nothing new in solving it for linear equations using a standard approach. Some problems of this type were already proposed and solved in the previous sections. So, in the part concerned with the standard algorithm, we will just give a briefly review and solve a few more examples. The traditional method consists of finding the general solution of the equation and subsequent application of n initial conditions to this solution to specify n arbitrary coefficients $C_1, \ldots, C_n$. Usually, the way the initial conditions are applied to the solutions of linear equations does not depend on a specific case of the linear equation and whether it is homogeneous or not. Frequently this requires to solve an algebraic linear system of n equations for n unknowns $C_1, \ldots, C_n$, although sometimes this system can be split into lower order subsystems. Recall that the Cauchy theorem for linear equations guarantees the existence and uniqueness of the solution of the Cauchy problem under the condition of smoothness of the coefficients and the right-hand side of the equation. In terms of the linear system for $C_1, \ldots, C_n$, this means that the algebraic system has the only solution (its determinant is nonzero).

Examples.

1. $y'' - 2y' - 3y = 0$, $y(0) = 1, y'(0) = -5$.
The general solution of this equation was found in the form $y_{gh} = C_1 e^{-x} + C_2 e^{3x}$ (see Example 2 in Case 1 of section 1). Using the initial conditions, we find the system

$$\begin{cases} C_1 + C_2 = 1 \\ -C_1 + 3C_2 = -5 \end{cases},$$

whose solution is $C_1 = 2$, $C_2 = -1$. Therefore, the solution of the Cauchy problem is

$$y = 2e^{-x} - e^{3x}.$$

2. $y''' - y'' - y' + y = 0$, $y(0) = -1, y'(0) = 2, y''(0) = -3$.
The general solution of this equation is $y_{gh} = C_1 e^x + C_2 x e^x + C_3 e^{-x}$ (see Example 1 in Case 2 of section 1). Applying the initial conditions, we find the system

$$\begin{cases} C_1 + C_3 = -1 \\ C_1 + C_2 - C_3 = 2 \\ C_1 + 2C_2 + C_3 = -3 \end{cases},$$

whose solution is $C_1 = 1, C_2 = -1, C_3 = -2$. Therefore, the solution of the Cauchy problem is

$$y = (1 - x)e^x - 2e^{-x}.$$

3. $y'' + 9y = 0, y(\pi) = 2, y'(\pi) = -3$.
The general solution has the form $y_{gh} = C_1 \cos 3x + C_2 \sin 3x$ (see Example 1 in Case 3 of section 1). Substitution of the initial conditions into the general solution leads to the two decoupled equations

$$\begin{cases} -C_1 = 2 \\ -3C_2 = -3 \end{cases},$$

whence $C_1 = -2, C_2 = 1$. So, the solution to the problem is

$$y = -2\cos 3x + \sin 3x.$$

Note that the same solution can be obtained using the complex form of the general solution.

4. $y^{(4)} + 2y'' + y = 0, y(\frac{\pi}{2}) = -2 - 2\pi, y'(\frac{\pi}{2}) = -4 + \pi, y''(\frac{\pi}{2}) = 6 + 2\pi, y'''(\frac{\pi}{2}) = 14 - \pi$.
For a change, let us use the general solution of this equation in the complex form: $y_{gh} = C_1 e^{ix} + C_2 e^{-ix} + C_3 x e^{ix} + C_4 x e^{-ix}$ (see Example 1 in Case 4 of section 1). Using the initial conditions, we get the system

$$\begin{cases} i(C_1 - C_2) + i\frac{\pi}{2}(C_3 - C_4) = -2 - 2\pi \\ -(C_1 + C_2) + i(C_3 - c_4) - \frac{\pi}{2}(C_3 + C_4) = -4 + \pi \\ -i(C_1 - C_2) - 2(C_3 + C_4) - i\frac{\pi}{2}(C_3 - C_4) = 6 + 2\pi \\ (C_1 + C_2) - 3i(C_3 - C_4) + \frac{\pi}{2}(C_3 + C_4) = 14 - \pi \end{cases}.$$

This system is solved in the same way as a system of real coefficients (for example, using the Gauss elimination method). The coefficients that satisfy the system are $C_1 = 1 + i, C_2 = 1 - i, C_3 = -1 + 2i, C_4 = -1 - 2i$. Substituting them into the general solution, we obtain the solution of the Cauchy problem:

$$y = (1 + i)e^{ix} + (1 - i)e^{-ix} + (-1 + 2i)xe^{ix} + (-1 - 2i)xe^{-ix}.$$

Simplifying, we arrive at the form

$$y = 2\cos x - 2\sin x - 2x\cos x - 4x\sin x = 2(1 - x)\cos x - 2(1 + 2x)\sin x.$$

The same solution can be found using the real form of the general solution $y_{gh} = C_1 \cos x + C_2 \sin x + C_3 x \cos x + C_4 x \sin x$.

5. $y'' - 8y' + 16y = (1 - x)e^x, y(-1) = e^{-4} + \frac{4}{27}e^{-1}, y'(-1) = 6e^{-4} + \frac{1}{27}e^{-1}$.
The general solution of the equation was found in the form $y_{gn} = (C_1 +$

$C_2x)e^{4x} + \left(-\frac{1}{9}x + \frac{1}{27}\right)e^x$ (see Example 1 in section 2.1). Substituting the initial conditions into this formula, we get the system

$$\begin{cases} (C_1-C_2)e^{-4}+\frac{4}{27}e^{-1}=e^{-4}+\frac{4}{27}e^{-1} \\ (4C_1-3C_2)e^{-4}+\frac{1}{27}e^{-1}=6e^{-4}+\frac{1}{27}e^{-1} \end{cases}$$

whose solution is $C_1 = 3, C_2 = 2$. So, the solution of the Cauchy problem is

$$y = (3 + 2x)e^{4x} + \left(-\frac{1}{9}x + \frac{1}{27}\right)e^x.$$

6. $y'' + y = x\cos x$, $y(0) = -\frac{5}{4}, y'(0) = \frac{1}{2}$.
The general solution of the equation was found in the form $y_{gn} = C_1\cos x + C_2\sin x + \frac{1}{4}(x\cos x + x^2\sin x)$ (see Example 2 in section 2.2). Substituting the initial conditions into this expression, we get the two decoupled equations

$$\begin{cases} C_1 = -\frac{5}{4} \\ C_2 + \frac{1}{4} = \frac{1}{2} \end{cases}$$

which determine $C_1 = -\frac{5}{4}$ and $C_2 = \frac{1}{4}$. So, the solution of the Cauchy problem is

$$y=-\frac{5}{4}\cos x+\frac{1}{4}\sin x+\frac{1}{4}(x\cos x+x^2\sin x)=\frac{1}{4}\left((x-5)\cos x + (x^2+1)\sin x\right).$$

4.2   Laplace transform method

Preliminary considerations

The Laplace transform is an Operational Calculus method, that is, a technique that transforms a differential problem into an algebraic problem consisting of solving a polynomial equation. In the case of linear ordinary differential equations, the most efficient and fairly simple implementation of this method occurs when it is applied to solution of the Cauchy problem for linear equations with constant coefficients.

The definition of the Laplace transform of a function $f(x)$ is as follows.

Definition. The Laplace transform $\mathcal{L}$ applied to a function $f(x)$, $x \in [0, +\infty)$ is defined by the formula

$$F(p) \equiv \mathcal{L}[f](p) = \int_0^{+\infty} e^{-px} f(x)dx.$$

The function $f(x)$ is called the *original function* (or simply the *original*) and $F(p)$ is called the *image function* (or simply the *image*).

Recalling the properties of the improper integral involved in this definition, we can see that the Laplace transform may not exist for some functions due to the non-existence or divergence of the improper integral. For instance, the integral cannot be defined for $f(x) = \frac{1}{x-1}$, because this function is not bounded in the neighborhood of $x = 1$ and, consequently, it is not Riemann integrable on any finite interval containing point 1, which makes it impossible to define the improper integral of $f(x)$ and also of $e^{-px}f(x)$, for $\forall p$. On the other hand, the functions $f(x) = e^x$ and $f(x) = e^{x^2}$ are continuous and therefore Riemann integrable over any finite interval, but the improper integral in the definition diverges for $f(x) = e^x$ and $p \leq 1$ and also for $f(x) = e^{x^2}$ and $\forall p$. Therefore, in order to apply the Laplace transform, we have to restrict the class of functions.

Let us impose the following conditions on $f(x)$:

1. $f(x)$ is a piecewise continuous on $[0, +\infty)$, which means that on each finite segment of the semi-axis $x \geq 0$ the function $f(x)$ is continuous except, possibly, for a finite number of points of discontinuity of the removable or jump type;

2. $f(x)$ is of exponential order on $[0, +\infty)$, that is, there exist numbers $M > 0$ and c such that $|f(x)| \leq Me^{cx}$ for $\forall x \geq x_0$, where $x_0 \geq 0$ is a fixed point.

These conditions guarantee the existence of the Laplace transform, that is, the convergence of the improper integral, for all $p > c$. Furthermore, under these conditions on $f(x)$, the image $F(p)$ is infinitely differentiable and satisfies the condition $\lim_{p \to +\infty} F(p) = 0$.

According to the properties of the improper integral, for the linear combination $\alpha f + \beta g$, where α and β are two constants, we have

$$\mathcal{L}[\alpha f + \beta g](p) = \int_0^{+\infty} e^{-px}(\alpha f + \beta g)dx$$

$$= \alpha \int_0^{+\infty} e^{-px}f dx + \beta \int_0^{+\infty} e^{-px}g dx = \alpha \mathcal{L}[f](p) + \beta \mathcal{L}[g](p),$$

which means that the Laplace transform is a linear operator.

If $f'(x)$ is continuous on $x \geq 0$, then, using integration by parts, we get

$$\mathcal{L}[f'](p) = \int_0^{+\infty} e^{-px}f'(x)dx$$

$$= e^{-px}f(x)|_0^{+\infty} + p \int_0^{+\infty} e^{-px}f(x)dx = -f(0) + p\mathcal{L}[f](p).$$

Applying this result to a function $f(x)$ twice continuously differentiable in $x \geq 0$, we have

$$\mathcal{L}[f''](p) = -f'(0) + p\mathcal{L}[f'](p)$$

$$= -f'(0) + p(-f(0) + p\mathcal{L}[f](p)) = -f'(0) - pf(0) + p^2\mathcal{L}[f](p)).$$

Continuing in this way, using mathematical induction, we arrive at the differentiation formula

$$\mathcal{L}[f^{(n)}](p) = -f^{(n-1)}(0) - pf^{(n-2)}(0) - \ldots - p^{n-2}f'(0) - p^{n-1}f(0) + p^n\mathcal{L}[f](p))$$

for any function $f(x)$ continuously differentiable n times in $x \geq 0$.

The last property and the linearity property make the Laplace transform a convenient tool for solving the initial value problem for linear equations with constant coefficients:

$$\begin{cases} a_n y^{(n)} + \ldots + a_1 y' + a_0 y = h(x) \\ y(0) = y_0, y'(0) = y_1, \ldots, y^{(n-1)}(0) = y_{n-1} \end{cases}.$$

In fact, according to linearity, application of the Laplace transform to the equation results in

$$a_n \mathcal{L}[y^{(n)}] + \ldots + a_1 \mathcal{L}[y'] + a_0 \mathcal{L}[y] = \mathcal{L}[h],$$

and the differentiation formula allows us to rewrite this equation in the form

$$a_n[-y^{(n-1)}(0) - py^{(n-2)}(0) - \ldots - p^{n-2}y'(0) - p^{n-1}y(0) + p^n Y] + \ldots$$

$$+ a_1[-y(0) + pY] + a_0 Y = H,$$

where $Y = \mathcal{L}[y]$ and $H = \mathcal{L}[h]$. Thus, the differential equation for $y(x)$ is transformed into an algebraic equation for the image $Y(p)$. Collecting the terms with $Y(p)$ and adding the initial condition, we obtain

$$P(p)Y = Q(p) + H(p),$$

where

$$P(p) = a_n p^n + a_{n-1} p^{n-1} + \ldots + a_1 p + a_0,$$

and

$$Q(p) = a_n y^{(n-1)}(0) + (pa_n + a_{n-1}) y^{(n-2)}(0) + \ldots + (p^{n-2} a_n + \ldots + a_2) y'(0)$$

$$+ (p^{n-1} a_n + \ldots + pa_2 + a_1) y(0).$$

Therefore, the solution for the image is found by the formula

$$Y = \frac{Q(p)}{P(p)} + \frac{H(p)}{P(p)}.$$

It remains to apply the inverse transform $\mathcal{L}^{-1}$ to find the original y of the image Y, which will be the solution to the Cauchy problem. In general, the inverse transform involves working with functions of complex variable. However, to avoid this, there are tables of originals and images that allow us to solve the Cauchy problem for many linear equations without involving a complex integration. In Table 1 we present some results for the Laplace transform and its inverse, which are employed in the following examples.

Table 1. Laplace transform of some functions

$f(x) = \mathcal{L}^{-1}[F]$	$F(p) = \mathcal{L}[f]$
1	$\frac{1}{p}$
x	$\frac{1}{p^2}$
x^n	$\frac{n!}{p^{n+1}}$
e^{ax}	$\frac{1}{p-a}$
$x^n e^{ax}$	$\frac{n!}{(p-a)^{n+1}}$
$\sin ax$	$\frac{a}{p^2+a^2}$
$\cos ax$	$\frac{p}{p^2+a^2}$
$e^{ax}\sin bx$	$\frac{b}{(p-a)^2+b^2}$
$e^{ax}\cos bx$	$\frac{p-a}{(p-a)^2+b^2}$
$x\sin ax$	$\frac{2ap}{(p^2+a^2)^2}$
$x\cos ax$	$\frac{p^2-a^2}{(p^2+a^2)^2}$

Examples.

1. $y' + 3y = 13\sin 2x$, $y(0) = 6$.
Applying the Laplace transform to the given equation, we get $\mathcal{L}[y' + 3y] = \mathcal{L}[13\sin 2x]$ or, due to the linearity of $\mathcal{L}$:

$$\mathcal{L}[y'] + 3\mathcal{L}[y] = 13\mathcal{L}[\sin 2x].$$

Denoting $Y = \mathcal{L}[y]$ and using the differentiation formula and Table 1 of transforms, we find: $\mathcal{L}[y'] = pY - y(0) = pY - 6$, $\mathcal{L}[\sin 2x] = \frac{2}{p^2+4}$. So, we arrive at the linear algebraic equation for Y:

$$pY - 6 + 3Y = \frac{26}{p^2 + 4},$$

whose solution is

$$Y = \frac{6}{p + 3} + \frac{26}{(p + 3)(p^2 + 4)}.$$

In order to find the original y, we need to transform the last fraction into the form found between the results in Table 1. To do this, we use the representation in partial fractions, noting that the denominator $(p+3)(p^2+4)$ already has maximum factoring in the field of real numbers: $\frac{26}{(p+3)(p^2+4)} =$ $\frac{A}{p+3} + \frac{Bp+C}{p^2+4}$. (Recall that the same technique is used for integration of rational functions.) To find the constants A, B, C, we go back to the common denominator on the left-hand side

$$\frac{26}{(p+3)(p^2+4)} = \frac{A(p^2+4) + (Bp+C)(p+3)}{(p+3)(p^2+4)},$$

which implies the equality between the numerators: $A(p^2+4)+(Bp+C)(p+3) = 26$. Then, $A + B = 0, 3B + C = 0, 4A + 3C = 26$, and consequently, $A = 2, B = -2, C = 6$. So the image Y takes the form

$$Y = \frac{6}{p+3} + \frac{2}{p+3} + \frac{-2p+6}{p^2+4} = \frac{8}{p+3} + \frac{-2p+6}{p^2+4}.$$

Finding in Table 1 that $\mathcal{L}^{-1}[p+3] = e^{-3x}$, $\mathcal{L}^{-1}[\frac{p}{p^2+4}] = \cos 2x$ and $\mathcal{L}^{-1}[\frac{2}{p^2+4}] = \sin 2x$, and using the linearity of the inverse transform, we arrive at the following result:

$$y = \mathcal{L}^{-1}[Y] = \mathcal{L}^{-1}[\frac{8}{p+3} + \frac{-2p+6}{p^2+4}]$$

$$= 8\mathcal{L}^{-1}[\frac{1}{p+3}] - 2\mathcal{L}^{-1}[\frac{p}{p^2+4}] + 3\mathcal{L}^{-1}[\frac{2}{p^2+4}] = 8e^{-3x} - 2\cos 2x + 3\sin 2x.$$

This is the solution of the Cauchy problem.

2. $y'' - y = e^{2x}$, $y(0) = 0, y'(0) = 1$.
Applying the Laplace transform to the given equation, using its linearity and the differentiation formula, we get

$$p^2Y - py(0) - y'(0) - Y = \mathcal{L}[e^{2x}].$$

Using the initial conditions and Table 1 of the transforms, we find the algebraic equation for the image Y:

$$p^2Y - 1 - Y = \frac{1}{p-2}, \quad whence \quad Y = \frac{1}{(p-2)(p+1)}.$$

In order to find the original y, we need to transform the last fraction into the form found between the results in the table. To do this, we use the partial fraction representation: $\frac{1}{(p-2)(p+1)} = \frac{A}{p-2} + \frac{B}{p+1}$. To find the constants A, B, we go back to the common denominator on the left-hand side

$$\frac{1}{(p-2)(p+1)} = \frac{A(p+1) + B(p-2)}{(p-2)(p+1)},$$

whence $A + B = 0, A - 2B = 1$ and then $A = \frac{1}{3}, B = -\frac{1}{3}$. Therefore, the image Y takes the form

$$Y = \frac{1/3}{p-2} - \frac{1/3}{p+1}.$$

Finding in Table 1 that $\mathcal{L}^{-1}[\frac{1}{p-2}] = e^{2x}$ and $\mathcal{L}^{-1}[\frac{1}{p+1}] = e^{-x}$, and using the linearity of the inverse transform, we conclude that the solution to the Cauchy problem is

$$y = \mathcal{L}^{-1}[Y] = \frac{1}{3}\mathcal{L}^{-1}[\frac{1}{p-2} - \frac{1}{p+1}] = \frac{1}{3}(e^{2x} - e^{-x}).$$

3. $y'' - 2y' - 3y = 0, y(0) = 1, y'(0) = 0.$
Applying the Laplace transform to the given equation, using its linearity and the differentiation formula, we get

$$(p^2Y - py(0) - y'(0)) - 2(pY - y(0)) - 3Y = (p^2 - 2p - 3)Y - (p - 2) = 0.$$

Solving for Y, we obtain

$$Y = \frac{p-2}{p^2 - 2p - 3}.$$

In order to find the original y, we factor the denominator $p^2 - 2p - 3 = (p - 3)(p+1)$ and expand the right side in partial fractions: $\frac{p-2}{p^2-2p-3} = \frac{A}{p-3} + \frac{B}{p+1}$. For the constants A, B we have the relation $A(p + 1) + B(p - 3) = p - 2$ which leads to the system $A + B = 1, A - 3B = -2$. Therefore, $A = \frac{1}{4}$, $B = \frac{3}{4}$ and the image Y takes the form

$$Y = \frac{1/4}{p-3} + \frac{3/4}{p+1}.$$

Finding in Table 1 that $\mathcal{L}^{-1}[\frac{1}{p-3}] = e^{3x}$ and $\mathcal{L}^{-1}[\frac{1}{p+1}] = e^{-x}$, and using the linearity of the inverse transform, we conclude that the solution to the Cauchy problem is

$$y = \mathcal{L}^{-1}[Y] = \frac{1}{4}\mathcal{L}^{-1}[\frac{1}{p-3}] + \frac{3}{4}\mathcal{L}^{-1}[\frac{1}{p+1}] = \frac{1}{4}e^{3x} + \frac{3}{4}e^{-x}.$$

4. $y'' + 2y' + 2y = \cos 2x, y(0) = 0, y'(0) = 1.$
Applying the Laplace transform to the given equation, using its linearity and the differentiation formula, we get

$$(p^2Y - py(0) - y'(0)) + 2(pY - y(0)) + 2Y = \mathcal{L}[\cos 2x].$$

Since $\mathcal{L}[\cos 2x] = \frac{p}{p^2+4}$, we obtain the algebraic equation

$$(p^2 + 2p + 2)Y - 1 = \frac{p}{p^2 + 4},$$

whose solution is

$$Y = \frac{1}{p^2 + 2p + 2} + \frac{p}{(p^2 + 2p + 2)(p^2 + 4)}.$$

Representing the first fraction in the form $\frac{1}{(p+1)^2+1}$, we find its original from Table 1: $\mathcal{L}^{-1}[\frac{1}{(p+1)^2+1}] = e^{-x} \sin x$. To find the original of the second fraction, we expand it in partial fractions:

$$\frac{p}{(p^2 + 2p + 2)(p^2 + 4)} = \frac{Ap + B}{p^2 + 2p + 2} + \frac{Cp + D}{p^2 + 4}.$$

For the constants A, B, C, D we have the relation

$$(Ap + B)(p^2 + 4) + (Cp + D)(p^2 + 2p + 2) = p,$$

which leads to the system $A + C = 0, B + 2C + D = 0, 4A + 2C + 2D = 1, 4B + 2D = 0$. Solving, we find, $A = \frac{1}{10}$, $B = -\frac{1}{5}$, $C = -\frac{1}{10}$, $D = \frac{2}{5}$, that is, the second fraction is represented in the form:

$$\frac{p}{(p^2 + 2p + 2)(p^2 + 4)} = \frac{1}{10}\frac{p - 2}{p^2 + 2p + 2} - \frac{1}{10}\frac{p - 4}{p^2 + 4}.$$

According to Table 1, the original of the first term is

$$\mathcal{L}^{-1}[\frac{1}{10}\frac{p - 2}{(p + 1)^2 + 1}] = \frac{1}{10}(\mathcal{L}^{-1}[\frac{p + 1}{(p + 1)^2 + 1}] - \mathcal{L}^{-1}[\frac{3}{(p + 1)^2 + 1}])$$

$$= \frac{1}{10}(e^{-x} \cos x - 3e^{-x} \sin x).$$

For the second, we have

$$\mathcal{L}^{-1}[\frac{1}{10}\frac{p - 4}{p^2 + 4}] = \frac{1}{10}(\mathcal{L}^{-1}[\frac{p}{p^2 + 4}] - 2\mathcal{L}^{-1}[\frac{2}{p^2 + 4}]) = \frac{1}{10}(\cos 2x - 2\sin 2x).$$

Collecting all the results, we find

$$y = \mathcal{L}^{-1}[Y] = e^{-x} \sin x + \frac{1}{10}(e^{-x} \cos x - 3e^{-x} \sin x) - \frac{1}{10}(\cos 2x - 2\sin 2x).$$

5. $y'' - 3y' + 2y = 6e^{-x}$, $y(0) = 2, y'(0) = 0$.

Applying the Laplace transform to the given equation, using its linearity and the differentiation formula, we get

$$(p^2Y - py(0) - y'(0)) - 3(pY - y(0)) + 2Y = \mathcal{L}[6e^{-x}].$$

Since $\mathcal{L}[e^{-x}] = \frac{1}{p+1}$, we obtain the algebraic equation $(p^2 - 3p + 2)Y - 2p + 6 = \frac{6}{p+1}$, or simplifying,

$$(p^2 - 3p + 2)Y = \frac{2p^2 - 4p}{p+1}, \ whence \ \ Y = \frac{2p}{(p+1)(p-1)}.$$

We express the right side in partial fractions: $\frac{2p}{(p+1)(p-1)} = \frac{1}{p-1} + \frac{1}{p+1}$, and find the original functions from Table 1: $\mathcal{L}^{-1}[\frac{1}{p-1}] = e^x$ and $\mathcal{L}^{-1}[\frac{1}{p+1}] = e^{-x}$. Therefore, the solution to the Cauchy problem is

$$y = \mathcal{L}^{-1}[Y] = \mathcal{L}^{-1}[\frac{1}{p-1}] + \mathcal{L}^{-1}[\frac{1}{p+1}] = e^x + e^{-x}.$$

6. $y'' + y = x \cos 2x$, $y(0) = 0, y'(0) = 0$.

Applying the Laplace transform to the given equation, using its linearity and the differentiation formula, we get

$$(p^2Y - py(0) - y'(0)) + Y = \mathcal{L}[x \cos 2x].$$

Taking into account the initial conditions and the formula $\mathcal{L}[x \cos 2x] = \frac{p^2-4}{(p^2+4)^2}$ of Table 1, we arrive at the algebraic equation $(p^2 + 1)Y = \frac{p^2-4}{(p^2+4)^2}$, whose solution is

$$Y = \frac{p^2 - 4}{(p^2 + 1)(p^2 + 4)^2}.$$

We represent the right-hand side in partial fractions:

$$\frac{p^2 - 4}{(p^2 + 1)(p^2 + 4)^2} = \frac{A}{p^2 + 1} + \frac{B}{p^2 + 4} + \frac{C}{(p^2 + 4)^2}$$

and obtain the relation

$$A(p^2 + 4)^2 + B(p^2 + 1)(p^2 + 4) + C(p^2 + 1) = p^2 - 4$$

for the constants A, B, C. Denoting $s = p^2$ we simplify the latter to the form

$$A(s + 4)^2 + B(s + 1)(s + 4) + C(s + 1) = s - 4$$

and obtain the system $A + B = 0, 8A + 5B + C = 1, 16A + 4B + C = -4$. Solving, we find $A = -\frac{5}{9}$, $B = \frac{5}{9}$, $C = \frac{8}{3}$. Then, we have the following partial fractions:

$$\frac{p^2 - 4}{(p^2 + 1)(p^2 + 4)^2} = \frac{-5/9}{p^2 + 1} + \frac{5/9}{p^2 + 4} + \frac{8/3}{(p^2 + 4)^2}.$$

According to Table 1,

$$\mathcal{L}^{-1}[\frac{1}{p^2 + 1}] = \sin x, \ \mathcal{L}^{-1}[\frac{1}{p^2 + 4}] = \frac{1}{2}\sin 2x, \ \mathcal{L}^{-1}[\frac{1}{(p^2 + 4)^2}]$$

$$= \mathcal{L}^{-1}[\frac{1}{8}\frac{p^2 + 4 - (p^2 - 4)}{(p^2 + 4)^2}] = \frac{1}{8}\mathcal{L}^{-1}[\frac{1}{p^2 + 4}] - \mathcal{L}^{-1}[\frac{p^2 - 4}{(p^2 + 4)^2}] = \frac{1}{8}(\frac{1}{2}\sin 2x - x\cos 2x).$$

Gathering all the results, we find the solution to the Cauchy problem:

$$y = \mathcal{L}^{-1}[Y] = -\frac{5}{9}\sin x + \frac{5}{9}\cdot\frac{1}{2}\sin 2x + \frac{8}{3}\cdot\frac{1}{8}(\frac{1}{2}\sin 2x - x\cos 2x)$$

$$= -\frac{5}{9}\sin x + \frac{4}{9}\sin 2x - \frac{1}{3}x\cos 2x.$$

7. $y'' - 4y' + 3y = 2e^x + 2e^{3x}$, $y(0) = -1, y'(0) = 1$.

Applying the Laplace transform to the given equation, using its linearity and the differentiation formula, we get

$$(p^2Y - py(0) - y'(0)) - 4(pY - y(0)) + 3Y = \mathcal{L}[2e^x + 2e^{3x}].$$

Since $\mathcal{L}[e^x] = \frac{1}{p-1}$ and $\mathcal{L}[e^{3x}] = \frac{1}{p-3}$, employing the initial conditions, we arrive at the algebraic equation

$$(p^2 - 4p + 3)Y + p - 5 = \frac{2}{p-1} + \frac{2}{p-3}$$

or

$$(p - 1)(p - 3)Y = \frac{2}{p-1} + \frac{2}{p-3} - (p - 5).$$

The solution is found in the form

$$Y = \frac{2}{(p-1)^2(p-3)} + \frac{2}{(p-1)(p-3)^2} - \frac{p-5}{(p-1)(p-3)}.$$

Next, we express the right side through the partial fractions:

$$\frac{2}{(p-1)^2(p-3)} + \frac{2}{(p-1)(p-3)^2} - \frac{p-5}{(p-1)(p-3)}$$

$$= \frac{A}{p-1} + \frac{B}{(p-1)^2} + \frac{C}{p-3} + \frac{D}{(p-3)^2}.$$

For A, B, C, D we have the relation

$$A(p-1)(p-3)^2 + B(p-3)^2 + C(p-3)(p-1)^2 + D(p-1)^2$$

$$= 2(p-3) + 2(p-1) - (p-5)(p-1)(p-3),$$

whose solution is $A = -2$, $B = -1$, $C = 1$, $D = 1$. Thus,

$$Y = \frac{-2}{p-1} + \frac{-1}{(p-1)^2} + \frac{1}{p-3} + \frac{1}{(p-3)^2}.$$

Using Table 1, we find

$$\mathcal{L}^{-1}[\frac{1}{p-1}] = e^x, \mathcal{L}^{-1}[\frac{1}{(p-1)^2}] = xe^x, \mathcal{L}^{-1}[\frac{1}{p-3}] = e^{3x}, \mathcal{L}^{-1}[\frac{1}{(p-3)^2}] = xe^{3x}.$$

Therefore, the solution of the Cauchy problem is

$$y = \mathcal{L}^{-1}[Y] = -2e^x - xe^x + e^{3x} + xe^{3x}.$$

Exercises for section 4

Solve the following initial value problems using the traditional method and the Laplace transform; compare the found solutions:
1. $y'' - 3y' + 2y = e^{-x}$, $y(0) = 0, y'(0) = 1$.
2. $y'' - y' - 2y = 3xe^x$, $y(0) = 0, y'(0) = 0$.
3. $y'' - 5y' + 4y = (10x + 1)e^{-x}$, $y(0) = 0, y'(0) = 0$.
4. $y'' + 5y' + 6y = e^{-2x}$, $y(0) = -1, y'(0) = 0$.
5. $y'' - 2y' + y = 2e^x$, $y(0) = 1, y'(0) = 1$.
6. $y'' + 2y' + y = (x + 2)e^{-x}$, $y(0) = 1, y'(0) = -1$.
7. $y'' - 2y' - 3y = 4e^{3x} - 4e^{-x}$, $y(0) = 2, y'(0) = 0$.
8. $y'' + y = 4\cos x$, $y(0) = 1, y'(0) = -1$.
9. $y'' + y = 5xe^{2x}$, $y(0) = 0, y'(0) = 1$.
10. $y'' + 9y = 6\cos 3x + 9\sin 3x$, $y(0) = 1, y'(0) = 0$.
11. $y'' + 4y = 4(\cos 2x + \sin 2x)$, $y(0) = 0, y'(0) = 1$.
12. $y'' + y = 2(\cos x - \sin x)$, $y(0) = 1, y'(0) = 2$.

5   Boundary value problem for linear second order equations

The boundary value problem for second order equations was defined in section 3 of Chapter 6. Let us specify that definition for linear equations.

Definition of the boundary value problem. The linear equation

$$a_2(x)y'' + a_1(x)y' + a_0(x)y = f(x), a_2 \neq 0,$$

considered in interval $[x_1, x_2]$, together with the *boundary conditions* imposed at the endpoints of the interval

$$\alpha_1 y(x_1) + \beta_1 y'(x_1) = \gamma_1, \alpha_2 y(x_2) + \beta_2 y'(x_2) = \gamma_2,$$

where $\alpha_1^2 + \beta_1^2 \neq 0$, $\alpha_2^2 + \beta_2^2 \neq 0$, form the *boundary value problem*.

There are three specific cases of the boundary conditions that have proper names due to their importance:

1) the conditions $y(x_1) = \gamma_1, y(x_2) = \gamma_2$ (that is, $\beta_1 = \beta_2 = 0$) are called the *Dirichlet boundary conditions* (or the *conditions of the first kind*);

2) the conditions $y'(x_1) = \gamma_1, y'(x_2) = \gamma_2$ (that is, $\alpha_1 = \alpha_2 = 0$) are called the *Neumann boundary conditions* (or the *conditions of the second kind*);

3) the conditions $\alpha y(x_1) - \beta y'(x_1) = \gamma_1, \alpha y(x_2) + \beta y'(x_2) = \gamma_2$ (that is, $\alpha_1 = \alpha_2 = \alpha \neq 0, -\beta_1 = \beta_2 = \beta \neq 0$) are called the *Robin boundary conditions* (or the *conditions of the third kind*).

In the remaining cases the boundary conditions are called *mixed*.

First of all, let us show that the situation with the existence and uniqueness of the solution of a boundary value problem is quite different and more complicated than for the Cauchy problem. To do this, we consider the simple equation $y'' + y = 0$ which satisfies the conditions of Cauchy's Theorem on the entire real axis: the coefficients $a_2 = 1$, $a_1 = 0$, $a_0 = 1$ and the right-hand side $f = 0$ are continuous functions on $\mathbb{R}$. Therefore, whatever the point is chosen to set the initial conditions, the solution to the Cauchy problem exists and is unique in $\mathbb{R}$.

The situation is totally different for the boundary value problem involving this equation. First, we add the condition $y(0) = 0$ to the given equation and look for all the functions that satisfy both the equation and this complementary condition. Solving the equation, we find the general solution $y = A \cos x + B \sin x$, which contains all the particular solutions. Applying the condition $y(0) = 0$, we get $A = 0$ and, consequently, all the functions that satisfy the equation and the complementary condition are found in the form $y = B \sin x$, where B is an arbitrary constant.

Now we add the second boundary condition in different forms. The first option is $y(1) = 0$. In this case, in the family $y = B \sin x$ there is the only function $y = 0$ that satisfies this second boundary condition, and therefore, the boundary value problem $\begin{cases} y'' + y = 0, x \in [0, 1] \\ y(0) = 0, y(1) = 0 \end{cases}$ has the only solution $y = 0$. The second option of the second boundary condition is $y(\pi) = 0$. Since any function in the family $y = B \sin x$ satisfies this condition, then the boundary value problem $\begin{cases} y'' + y = 0, x \in [0, \pi] \\ y(0) = 0, y(\pi) = 0 \end{cases}$ has infinitely many solutions in the form $y = B \sin x$ with an arbitrary constant B. Finally, we choose the second boundary condition to be $y(\pi) = 1$. In this case, there is no function from the set $y = B \sin x$ that satisfies this condition. Since all the functions that satisfy both the equation and the first boundary condition are included in the family $y = B \sin x$, this means that the boundary value problem $\begin{cases} y'' + y = 0, x \in [0, \pi] \\ y(0) = 0, y(\pi) = 1 \end{cases}$ has no solution.

Thus, we have constructed three examples which show that simple boundary conditions for the same elementary equation can lead to different situations for the corresponding boundary value problems: it can happen that the solution to the problem exists and is unique, that there are infinitely many solutions, and that there is no solution to the problem.

The formulation of conditions that guarantee the existence and uniqueness of the solution of a boundary value problem is much more complicated than for the Cauchy problem and we leave these results out of the text. Nevertheless, in some specific cases the existence and uniqueness of solution can be ensured. In particular, the following theorem takes place.

Theorem on existence and uniqueness of some boundary value problems. The boundary value problem

$$\begin{cases} y'' + a_1(x)y' + a_0(x)y = f(x), a_0(x) \le 0, x \in [x_1, x_2] \\ y(x_1) = \gamma_1, y(x_2) = \gamma_2 \end{cases}$$

(the Dirichlet boundary conditions) has a unique solution.

Remark 1. The statement of the Theorem is also true for the same equation with the following boundary conditions:

$$\alpha_1 y(x_1) + \beta_1 y'(x_1) = \gamma_1, \alpha_2 y(x_2) + \beta_2 y'(x_2) = \gamma_2, \alpha_1 \beta_1 = 0, \alpha_2 \beta_2 = 0,$$

that is, when the function or its derivative (but not their linear combination) is given at each endpoint. This includes the case of the Neumann boundary conditions.

Remark 2. Even having this theorem, we need to take into account that the solution of a general boundary value problem can lead to theoretical problems of infinitely many solutions or their non-existence, which is not determined from the beginning of the solution (unlike the Cauchy problem).

In the next two sections we study the two methods of solution of boundary value problems. The first, traditional, follows the scheme of solution of the Cauchy problem, although without guaranteed existence and uniqueness of the solution, which is verified during the solving process. The second is the Green method, which has a different, more complicated approach, but allows us to find a solution for a family of equations with different right-hand sides $f(x)$.

5.1   Traditional method of solution

The algorithm of this method is standard, following the procedure of solution of the Cauchy problem, and it does not require additional explanations: first we find all the solutions of the equation and then apply the boundary conditions. The only difference is that there may arise the problems of non-existence or non-uniqueness of solutions, which is checked during the process of solution.

Examples.

1. $y'' - 2y' - 3y = 0$, $y(0) = 0, y'(1) = 1$.
The general solution of this equation was found in the form $y_{gh} = C_1 e^{-x} + C_2 e^{3x}$ (see Example 2 in Case 1 of section 1). Using the boundary conditions, we get the system $\begin{cases} C_1 + C_2 = 0 \\ -C_1 e^{-1} + 3C_2 e^3 = 1 \end{cases}$, whose solution is $C_1 = -C_2 = -\frac{1}{e^{-1}+3e^3}$. Therefore, the solution of the problem is

$$y = \frac{1}{e^{-1} + 3e^3}(e^{3x} - e^{-x}).$$

2. $y'' + 9y = 0$, $y(0) = 0, y(\frac{\pi}{2}) = 2$.
The general solution has the form $y_{gh} = C_1 \cos 3x + C_2 \sin 3x$ (see Example 1 in Case 3 of section 1). Application of the boundary conditions leads to the two decoupled equations $\begin{cases} C_1 = 0 \\ -C_2 = 2 \end{cases}$. So, the solution of the problem is

$$y = -2 \sin 3x.$$

3. $y'' + y = 1$, $y'(0) = 0, y(\pi) = 0$.
The general solution of the homogeneous equation $y'' + y = 0$ has the form $y_{gh} = C_1 \cos x + C_2 \sin x$ (the characteristic equation $\lambda^2 + 1 = 0$ has the two complex conjugate roots $\lambda_1 = i$ and $\lambda_2 = \overline{\lambda}_1 = -i$). The particular solution

of the nonhomogeneous part can be found in the form $y_{pn} = C$, and substitution into the original equation gives $y_{pn} = 1$. Thus, the general solution of the given equation is $y_{gh} = C_1 \cos x + C_2 \sin x + 1$. Application of the boundary conditions leads to the two decoupled equations $\begin{cases} C_2 = 0 \\ -C_1 + 1 = 0 \end{cases}$.

So, the solution to the problem is

$$y = \cos x + 1.$$

4. $y'' + y' = 2e^x$, $y(0) = 0, y'(1) = 0$.
The general solution of the homogeneous equation $y'' + y' = 0$ has the form $y_{gh} = C_1 + C_2 e^{-x}$ (the characteristic equation $\lambda^2 + \lambda = 0$ has the simple real roots $\lambda_1 = 0$ and $\lambda_2 = -1$). The particular solution of the nonhomogeneous part can be found in the form $y_{pn} = Ce^x$, and substitution into the original equation specifies the constant $C = 1$. Thus, the general solution of the given equation is $y_{gh} = C_1 + C_2 e^{-x} + e^x$. Using the boundary conditions, we get the system $\begin{cases} C_1 + C_2 + 1 = 0 \\ -C_2 e^{-1} + e = 0 \end{cases}$, whence $C_2 = e^2$, $C_1 = -e^2 - 1$. Thus, the solution of the problem is

$$y = -e^2 - 1 + e^{2-x} + e^x.$$

5.2 Green method

The *Green method* is usually applied to the problems with nonhomogeneous equations and homogeneous boundary conditions.

Green method for boundary value problem. The linear equation

$$a_2(x)y'' + a_1(x)y' + a_0(x)y = f(x), \ a_2 \neq 0, \ x \in [x_1, x_2]$$

is considered with the boundary conditions

$$\alpha_1 y(x_1) + \beta_1 y'(x_1) = 0, \alpha_2 y(x_2) + \beta_2 y'(x_2) = 0,$$

where $\alpha_1^2 + \beta_1^2 \neq 0$, $\alpha_2^2 + \beta_2^2 \neq 0$.

The essential part of the Green method is the construction of *Green's function* $G(x, s)$ which is a function of two variables defined on the square $[x_1, x_2] \times [x_1, x_2]$, such that for any fixed $s \in [x_1, x_2]$, as a function of a single variable x, $G(x, s)$ satisfies the following conditions (see Fig.9.1):
1) $G(x, s)$ is a solution of the homogeneous differential equation $a_2(x)y'' + a_1(x)y' + a_0(x)y = 0$ on $[x_1, x_2]$, except at the point $x = s$;
2) $G(x, s)$ satisfies the boundary conditions;

3) at the point $x = s$, the function $G(x, s)$ is continuous and its derivative (with respect to x) has a jump discontinuity of the value $\frac{1}{a_2(s)}$.

Usually, the construction of Green's function follows the following algorithm. First, we find two nonzero solutions $y_1(x)$ and $y_2(x)$ of the homogeneous equation such that $y_1(x)$ satisfies the first boundary condition and $y_2(x)$ the second. If the functions $y_1(x)$ and $y_2(x)$ are linearly independent (or equivalently, $y_1(x)$ does not satisfy the second boundary condition and $y_2(x)$ does not satisfy the first), then Green's function exists and can be found in the form

$$G(x, s) = \begin{cases} a(s)y_1(x), x_1 \leq x \leq s \\ b(s)y_2(x), s \leq x \leq x_2 \end{cases}.$$

Obviously, this function satisfies the first two conditions of the definition of $G(x, s)$, and it remains to determine the functions $a(s)$ and $b(s)$ in such a way that the third condition is also satisfied. This means that $a(s)$ and $b(s)$ must satisfy the relations $a(s)y_1(s) = b(s)y_2(s)$ (continuity of $G(x, s)$ at $x = s$) and $a(s)y_1'(s) = b(s)y_2'(s) - \frac{1}{a_2(s)}$ (jump of the value $\frac{1}{a_2(s)}$ at $x = s$), for any $s \in [x_1, x_2]$ (see Fig.9.1).

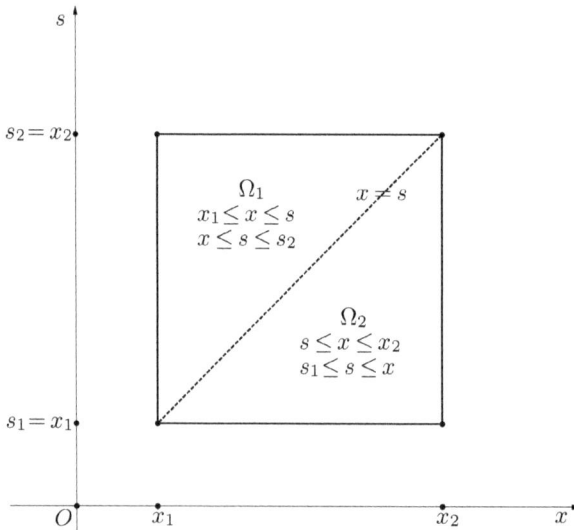

Figure 9.1: Domain of a Green's function

Therefore, we have the following system for unknowns $a(s)$ and $b(s)$:

$$\begin{cases} a(s)y_1(s) - b(s)y_2(s) = 0 \\ a(s)y_1'(s) - b(s)y_2'(s) = -\frac{1}{a_2(s)} \end{cases}.$$

This system has a unique solution if its determinant is nonzero: $D = \begin{vmatrix} y_1(s) & y_2(s) \\ y_1'(s) & y_2'(s) \end{vmatrix} \neq 0$. This happens if and only if the columns of D are linearly independent, which is equivalent to the condition of linear independence of the functions $y_1(s)$ and $y_2(s)$, which was indicated earlier as the condition of existence of Green's function.

After finding Green's function (if it exists), the solution to the boundary condition problem with the arbitrary (but continuous) right-hand side $f(x)$ is found via the formula $y(x) = \int_{x_1}^{x_2} G(x,s)f(s)ds$. Due to the definition of $G(x,s)$ by different sentences in the regions Ω_1 and Ω_2 (see Fig.9.1), the latter formula in practice is expressed as follows:

$$y(x) = \int_{x_1}^{x_2} G(x,s)f(s)ds = \int_{x_1}^{x} b(s)y_2(x)f(s)ds + \int_{x}^{x_2} a(s)y_1(x)f(s)ds.$$

To check that $y(x) = \int_{x_1}^{x_2} G(x,s)f(s)ds$ is indeed the desired solution to the problem, we first calculate the two derivatives of this function:

$$y'(x) = \left(\int_{x_1}^{x_2} G(x,s)f(s)ds \right)_x = \left(\int_{x_1}^{x} b(s)y_2(x)f(s)ds + \int_{x}^{x_2} a(s)y_1(x)f(s)ds \right)_x$$

$$= \left(\int_{x_1}^{x} b(s)y_2(x)f(s)ds \right)_x + \left(\int_{x}^{x_2} a(s)y_1(x)f(s)ds \right)_x$$

$$= \int_{x_1}^{x} b(s)y_2'(x)f(s)ds + b(x)y_2(x)f(x) + \int_{x}^{x_2} a(s)y_1'(x)f(s)ds - a(x)y_1(x)f(x)$$

$$= \int_{x_1}^{x} b(s)y_2'(x)f(s)ds + \int_{x}^{x_2} a(s)y_1'(x)f(s)ds$$

(note that $b(x)y_2(x) - a(x)y_1(x) = 0$ due to the continuity of $G(x,s)$ at $x = s$);

$$y''(x) = \left(\int_{x_1}^{x} b(s)y_2'(x)f(s)ds + \int_{x}^{x_2} a(s)y_1'(x)f(s)ds \right)_x$$

$$= \left(\int_{x_1}^{x} b(s)y_2'(x)f(s)ds \right)_x + \left(\int_{x}^{x_2} a(s)y_1'(x)f(s)ds \right)_x$$

$$= \int_{x_1}^{x} b(s)y_2''(x)f(s)ds + b(x)y_2'(x)f(x) + \int_{x}^{x_2} a(s)y_1''(x)f(s)ds - a(x)y_1'(x)f(x)$$

$$= \int_{x_1}^{x} b(s)y_2''(x)f(s)ds + \int_{x}^{x_2} a(s)y_1''(x)f(s)ds + \frac{1}{a_2(x)}f(x)$$

(note that $b(x)y_2'(x) - a(x)y_1'(x) = \frac{1}{a_2(x)}f(x)$ due to the jump condition at the point $x = s$).

Now we check the boundary conditions. At the point x_1 we have

$$\alpha_1 y(x_1) + \beta_1 y'(x_1) = \alpha_1 \left(\int_{x_1}^{x} b(s)y_2(x)f(s)ds + \int_{x}^{x_2} a(s)y_1(x)f(s)ds \right) \Big|_{x=x_1}$$

$$+ \beta_1 \left(\int_{x_1}^{x} b(s)y_2'(x)f(s)ds + \int_{x}^{x_2} a(s)y_1'(x)f(s)ds \right) \Big|_{x=x_1}$$

$$= \alpha_1 \left(\int_{x_1}^{x_1} b(s)y_2(x_1)f(s)ds + \int_{x_1}^{x_2} a(s)y_1(x_1)f(s)ds \right)$$

$$+ \beta_1 \left(\int_{x_1}^{x_1} b(s)y_2'(x_1)f(s)ds + \int_{x_1}^{x_2} a(s)y_1'(x_1)f(s)ds \right)$$

$$= \alpha_1 \int_{x_1}^{x_2} a(s)y_1(x_1)f(s)ds + \beta_1 \int_{x_1}^{x_2} a(s)y_1'(x_1)f(s)ds$$

$$= \int_{x_1}^{x_2} (\alpha_1 y_1(x_1) + \beta_1 y_1'(x_1))a(s)f(s)ds = \int_{x_1}^{x_2} 0 \cdot a(s)f(s)ds = 0$$

(note that $\alpha_1 y_1(x_1) + \beta_1 y_1'(x_1) = 0$ because $y_1(x)$ satisfies the first boundary condition). At the point x_2 we get the same result.

Finally, we substitute $y(x)$ and its derivatives into the equation:

$$a_2(x)y'' + a_1(x)y' + a_0(x)y$$

$$= a_2(x) \left(\int_{x_1}^{x} b(s)y_2''(x)f(s)ds + \int_{x}^{x_2} a(s)y_1''(x)f(s)ds + \frac{1}{a_2(x)}f(x) \right)$$

$$+ a_1(x) \left(\int_{x_1}^{x} b(s)y_2'(x)f(s)ds + \int_{x}^{x_2} a(s)y_1'(x)f(s)ds \right)$$

$$+ a_0(x) \left(\int_{x_1}^{x} b(s)y_2(x)f(s)ds + \int_{x}^{x_2} a(s)y_1(x)f(s)ds \right)$$

$$= \int_{x_1}^{x} (a_2(x)y_2''(x) + a_1(x)y_2'(x) + a_0(x)y_2(x))\, b(s)f(s)ds$$

$$+ \int_x^{x_2} (a_2(x)y_1''(x) + a_1(x)y_1'(x) + a_0(x)y_1(x)) a(s)f(s)ds + f(x)$$

$$= \int_{x_1}^x 0 \cdot b(s)f(s)ds + \int_x^{x_2} 0 \cdot a(s)f(s)ds + f(x) = f(x)$$

(we use $a_2(x)y_2''(x) + a_1(x)y_2'(x) + a_0(x)y_2(x) = 0$ and $a_2(x)y_1''(x) + a_1(x)y_1'(x) + a_0(x)y_1(x) = 0$ because y_1 and y_2 are solutions of the homogeneous equation). Thus, the original equation is also satisfied.

Examples.

1. $y'' + y = 1$, $y'(0) = 0, y(\pi) = 0$.
This problem has already been solved using the traditional method (see Example 3 in section 5.1). Let us see how the Green method works for this problem. The first step is to construct Green's function following the general algorithm presented above. Recall that the general solution of the homogeneous equation $y'' + y = 0$ has the form $y_{gh} = C_1 \cos x + C_2 \sin x$ and its derivative is $y_{gh}' = -C_1 \sin x + C_2 \cos x$. From this solution we can pick the function y_1 which satisfies the first boundary condition: $-C_1 \sin 0 + C_2 \cos 0 = C_2 = 0$. So, we can take $y_1 = \cos x$ (any constant $C_1 \neq 0$ can be chosen). From the same general solution, we now select the function y_2 which satisfies the second boundary condition: $-C_1 \cos 0 + C_2 \sin 0 = -C_1 = 0$. Now, we can take $y_2 = \sin x$ (again, any constant $C_2 \neq 0$ can be chosen). Since y_1 and y_2 are linearly independent, Green's function exists and the Green method is applicable to this problem. According to the general formula, we look for Green's function in the form $G(x,s) = \begin{cases} a(s)\cos x, 0 \leq x \leq s \\ b(s)\sin x, s \leq x \leq \pi \end{cases}$, where $a(s)$ and $b(s)$ are found from the system $\begin{cases} a(s)\cos s - b(s)\sin s = 0 \\ -a(s)\sin s - b(s)\cos s = -1 \end{cases}$.
To solve the last one, we multiply the first equation by $\sin s$, the second by $\cos s$ and add the two results together. In this way, we eliminate $a(s)$ and obtain the following expression for $b(s)$: $b(s) = \cos s$. Substituting this $b(s)$ into the first equation, we find $a(s) = \sin s$. Thus, Green's function has the form

$$G(x,s) = \begin{cases} \sin s \cos x, 0 \leq x \leq s \\ \cos s \sin x, s \leq x \leq \pi \end{cases}.$$

At the second step, we find the solution of the problem using the formula

$$y(x) = \int_0^\pi G(x,s)f(s)ds = \int_0^x \cos s \sin x \cdot 1 ds + \int_x^\pi \sin s \cos x \cdot 1 ds$$

$$= \sin x \int_0^x \cos s ds + \cos x \int_x^\pi \sin s ds = \sin x \sin s|_0^x - \cos x \cos s|_x^\pi$$

$$= \sin^2 x + \cos^2 x + \cos x = 1 + \cos x.$$

We note that the same Green's function can be used to solve any problem of the type $y'' + y = f(x)$, $y'(0) = 0, y(\pi) = 0$. For example, if $f(x) = x$, then the solution is found by the formula

$$y(x) = \int_0^{\pi} G(x,s)f(s)ds = \int_0^x \cos s \sin x \cdot s ds + \int_x^{\pi} \sin s \cos x \cdot s ds$$

$$= \sin x \int_0^x s \cos s ds + \cos x \int_x^{\pi} s \sin s ds$$

$$= \sin x (s \sin s + \cos s)|_0^x + \cos x(-s \cos s + \sin s)|_x^{\pi} = x + \pi \cos x - \sin x.$$

The reader can check the obtained solution employing the traditional method.

2. $y'' + y' = 2e^x$, $y(0) = 0, y'(1) = 0$.

This problem has already been solved using the traditional method (see Example 4 in section 5.1). To employ the Green method, we first construct Green's function in the form $G(x,s) = \begin{cases} a(s)y_1(x), 0 \le x \le s \\ b(s)y_2(x), s \le x \le 1 \end{cases}$, where the functions $y_1(x)$ and $y_2(x)$ are solutions of the homogeneous equation $y'' + y' = 0$ with the first and second boundary conditions, respectively. Recalling that the general solution of the homogeneous equation is $y_{gh} = C_1 + C_2 e^{-x}$ and applying the first boundary condition, we have $C_1 + C_2 = 0$, that is, $y_1 = 1 - e^{-x}$ (or any multiple of this function). Calculating the derivative $y'_{gh} = -C_2 e^{-x}$ and using the second boundary condition, we get $C_2 = 0$ and therefore $y_2 = 1$ (or any non-zero constant). Since y_1 and y_2 are linearly independent, Green's function exists and we specify it in the form $G(x,s) = \begin{cases} a(s)(1 - e^{-x}), 0 \le x \le s \\ b(s) \cdot 1, s \le x \le 1 \end{cases}$, where $a(s)$ and $b(s)$ are still to be determined from the smoothness conditions $\begin{cases} a(s)(1 - e^{-s}) - b(s) = 0 \\ a(s)e^{-s} - 0 = -1 \end{cases}$. So, the second equation gives $a(s) = -e^s$ and from the first we find $b(s) = -e^s(1 - e^{-s}) = 1 - e^s$. Thus, Green's function is

$$G(x,s) = \begin{cases} -e^s(1 - e^{-x}), 0 \le x \le s \\ 1 - e^s, s \le x \le 1 \end{cases}.$$

At the second step, we find the solution to the problem using the formula

$$y(x) = \int_0^1 G(x,s)f(s)ds = \int_0^x (1 - e^s)2e^s ds + \int_x^1 -e^s(1 - e^{-x})2e^s ds$$

$$= 2\int_0^x e^s - e^{2s} ds - 2(1 - e^{-x})\int_x^1 e^{2s} ds = 2(e^s - \frac{1}{2}e^{2s})|_0^x - 2(1 - e^{-x})\frac{1}{2}e^{2s}|_x^1$$

$$= 2(e^x - \frac{1}{2}e^{2x}) - 1 - (1 - e^{-x})(e^2 - e^{2x}) = e^x + e^{2-x} - 1 - e^2.$$

We note that the same Green's function can be used to solve the problem $y'' + y' = f(x)$, $y'(0) = 0, y(\pi) = 0$ with any right-hand side $f(x)$. For example, if $f(x) = 1$, then we have

$$y(x) = \int_0^1 G(x, s)f(s)ds = \int_0^x (1 - e^s) \cdot 1 ds + \int_x^1 -e^s(1 - e^{-x}) \cdot 1 ds$$

$$= (s - e^s)|_0^x - (1 - e^{-x})e^s|_x^1 = x - e^x + 1 - (1 - e^{-x})(e - e^x) = x + e^{1-x} - e.$$

The reader can verify the obtained solution using the traditional method.

3. $y'' - 2y' - 3y = 0$, $y(0) = 0, y'(1) = 1$.
This problem was solved using the traditional method in Example 1 of section 5.1. The problem includes nonhomogeneous boundary conditions (the second condition), so the Green method is not directly applicable. However, there is a simple way to transform the given conditions into homogeneous ones, which works for any type of nonhomogeneous conditions. It is sufficient to construct a line that satisfies the nonhomogeneous conditions and subtract this line from y, introducing a new function that satisfies the homogeneous conditions (and the equation with the new right-hand side). In the case of the given problem, the line that satisfies the boundary conditions (but not the equation) is $y_r = x$ and the new function is $z = y - x$. For this function, the conditions become homogeneous and the equation becomes $z'' - 2z' - 3z = 2 + 3x$.
The auxiliary problem

$$z'' - 2z' - 3z = 2 + 3x, z(0) = 0, z'(1) = 0$$

can be solved using the Green method. Green's function can be found in the form $G(x, s) = \begin{cases} a(s)z_1(x), 0 \le x \le s \\ b(s)z_2(x), s \le x \le 1 \end{cases}$, where the functions $z_1(x)$ and $z_2(x)$ are solutions of the homogeneous equation $z'' - 2z' - 3z = 0$ with the first and second boundary conditions, respectively. Recalling that the general solution of the homogeneous equation is $z_{gh} = C_1 e^{-x} + C_2 e^{3x}$ and applying the first boundary condition, we have $C_1 + C_2 = 0$, that is, $z_1 = e^{-x} - e^{3x}$. Calculating the derivative $z'_{gh} = -C_1 e^{-x} + 3C_2 e^{3x}$ and using the second boundary condition, we get $C_1 = 3e^4 C_2$ and, therefore, $z_2 = 3e^4 e^{-x} + e^{3x}$. Since z_1 and z_2 are linearly independent, Green's function exists and we specify it in the form $G(x, s) = \begin{cases} a(s)(e^{-x} - e^{3x}), 0 \le x \le s \\ b(s)(3e^{4-x} + e^{3x}), s \le x \le 1 \end{cases}$.
The functions $a(s)$ and $b(s)$ are determined from the smoothness conditions

$$\begin{cases} a(s)(e^{-s} - e^{3s}) - b(s)(3e^{4-s} + e^{3s}) = 0 \\ a(s)(-e^{-s} - 3e^{3s}) - b(s)(-3e^{4-s} + 3e^{3s}) = -1 \end{cases}.$$ Adding the two equa-

tions together, we find the following relationship $-4ae^{3s} = 4be^{3s} - 1$ or $b = -a + \frac{1}{4}e^{-3s}$. Substituting this expression into the first equation of the system and simplifying, we get $a = \frac{3e^{4-3s}+e^s}{4(1+3e^4)}$. Returning to b we get $b = \frac{e^{-3s}-e^s}{4(1+3e^4)}$. This result can also be written as $a = C(3e^{4-3s} + e^s)$ and $b = C(e^{-3s} - e^s)$, where $C = \frac{1}{4(1+3e^4)}$ is a constant. Therefore,

$$G(x,s) = C \begin{cases} (3e^{4-3s} + e^s)(e^{-x} - e^{3x}), 0 \le x \le s \\ (e^{-3s} - e^s)(3e^{4-x} + e^{3x}), s \le x \le 1 \end{cases}.$$

At the second step, we find the solution of the auxiliary problem for z using the formula

$$z(x) = \int_0^1 G(x,s)f(s)ds = \int_0^x C(e^{-3s} - e^s)(3e^{4-x} + e^{3x})(2 + 3s)ds$$

$$+ \int_x^1 C(3e^{4-3s}+e^s)(e^{-x}-e^{3x})(2+3s)ds = C(3e^{4-x}+e^{3x})\int_0^x(e^{-3s}-e^s)(2+3s)ds$$

$$+ C(e^{-x}-e^{3x})\int_x^1(3e^{4-3s}+e^s)(2+3s)ds = C(3e^{4-x}+e^{3x})(e^x-e^{-3x}-x(e^{-3x}+3e^x))$$

$$+ C(e^{-x}-e^{3x})(3e^{4-3x}+e^x+3x(e^{4-3x}-e^x)-4e) = C\left(4e(e^{3x} - e^{-x}) - x(12e^4+4)\right)$$

$$= \frac{4e(e^{3x} - e^{-x})}{4(1+3e^4)} - x\frac{12e^4 + 4}{4(1+3e^4)} = \frac{e^{3x} - e^{-x}}{e^{-1} + 3e^3} - x.$$

Finally, the solution of the original problem is

$$y = z + x = \frac{e^{3x} - e^{-x}}{e^{-1} + 3e^3}.$$

The same solution was obtained using the traditional method.

4. $x^2y'' + xy' - y = 6x^2, y(1) = 0, y'(2) = 0$.
In order to construct Green's function, we have to find a general solution of the corresponding homogeneous equation $x^2y'' + xy' - y = 0$. This equation is linear, but it has variable coefficients, so the methods of solution known so far are not applicable. Nevertheless, there is a simple way to transform this equation into a linear one with constant coefficients. It can be made by changing the independent variable by the formula $x = e^t$. Then, $y_x = y_t t_x = y_t \frac{1}{x}$ and $y_{xx} = (y_t \frac{1}{x})_x = y_{tt} \frac{1}{x^2} - y_t \frac{1}{x^2}$. Consequently, $xy_x = y_t$ and $x^2y_{xx} = y_{tt} - y_t$, and, substituting into the homogeneous equation, we have $y_{tt} - y = 0$.

The general solution of this equation is $y_{gh}(t) = C_1 e^t + C_2 e^{-t}$, or returning to x, $y_{gh}(x) = C_1 x + C_2 \frac{1}{x}$. Next, among solutions $y_{gh}(x)$ we should select such that satisfy the first and second boundary conditions. The first condition leads to the relation $y_{gh}(1) = C_1 + C_2 = 0$, that is, $C_2 = -C_1$, and consequently, we can choose $y_1(x) = x - \frac{1}{x}$. Calculating the derivative $y'_{gh}(x) = C_1 - C_2 \frac{1}{x^2}$ and applying the second condition, we have $y'_{gh}(2) = C_1 - C_2 \frac{1}{4} = 0$, whence $C_2 = 4C_1$ and $y_2(x) = x + \frac{4}{x}$. Since y_1 and y_2 are linearly independent, Green's function exists and can be found in the form $G(x,s) = \begin{cases} a(s)(x - \frac{1}{x}), 1 \leq x \leq s \\ b(s)(x + \frac{4}{x}), s \leq x \leq 2 \end{cases}$, where $a(s)$ and $b(s)$ are the function determined

by the smoothness conditions $\begin{cases} a(s)(s - \frac{1}{s}) - b(s)(s + \frac{4}{s}) = 0 \\ b(s)(1 - \frac{4}{s^2}) - a(s)(1 + \frac{1}{s^2}) = \frac{1}{s^2} \end{cases}$. Multiplying the first equation by s and the second by s^2, we can rewrite this system as follows: $\begin{cases} a(s)(s^2 - 1) = b(s)(s^2 + 4) \\ b(s)(s^2 - 4) - a(s)(s^2 + 1) = 1 \end{cases}$. Substituting the expression for $a(s)$ from the first equation $a(s) = \frac{s^2+4}{s^2-1} b(s)$ into the second one, we obtain $b(s)(s^2 - 4) - b(s)\frac{s^2+4}{s^2-1}(s^2 + 1) = 1$, or simplifying, $b(s) \cdot (-10s^2) = s^2 - 1$, whence $b(s) = \frac{1-s^2}{10s^2}$. Consequently, $a(s) = -\frac{s^2+4}{10s^2}$. Thus, Green's function is

$$G(x,s) = \begin{cases} -\frac{s^2+4}{10s^2}(x - \frac{1}{x}), 1 \leq x \leq s \\ \frac{1-s^2}{10s^2}(x + \frac{4}{x}), s \leq x \leq 2 \end{cases}.$$

At the second step, we find the solution to the problem using the formula

$$y(x) = \int_1^2 G(x,s)f(s)ds = \int_1^x \frac{1-s^2}{10s^2}(x + \frac{4}{x})6s^2 ds + \int_x^2 -\frac{s^2+4}{10s^2}(x - \frac{1}{x})6s^2 ds$$

$$= \frac{3}{5}(x + \frac{4}{x})\int_1^x 1 - s^2 ds - \frac{3}{5}(x - \frac{1}{x})\int_x^2 s^2 + 4ds$$

$$= \frac{3}{5}(x + \frac{4}{x})(s - \frac{s^3}{3})|_1^x - \frac{3}{5}(x - \frac{1}{x})(\frac{s^3}{3} + 4s)|_x^2$$

$$= \frac{3}{5}(x + \frac{4}{x})(x - \frac{x^3}{3} - \frac{2}{3}) + \frac{3}{5}(x - \frac{1}{x})(\frac{x^3}{3} + 4x - \frac{32}{3}) = 2x^2 - \frac{34}{5}x + \frac{24}{5}\frac{1}{x}.$$

The reader can verify that the same solution of the problem can be obtained by applying the boundary conditions to the general solution of the original equation $y_{gn} = C_1 x + C_2 \frac{1}{x} + 2x^2$ (here, $y_{gh} = C_1 x + C_2 \frac{1}{x}$ is the general solution of the corresponding homogeneous equation and $y_{pn} = 2x^2$ is a particular solution of the original equation).

Remark. The presented algorithm of solution of the homogeneous equation $x^2 y'' + xy' - y = 0$ can be generalized to the case of the second-order equation $x^2 y'' + pxy' + qy = 0, x > 0$, where p and q are constants. By changing the independent variable $x = e^t$, this equation is transformed into the linear equation with constant coefficients for $y(t)$: $y_{tt} + (p-1)y_t + qy = 0$, whose solutions can be found by traditional method (see section 1 of this chapter). The equations of this type (called the Euler equations) will be considered in detail in section 5 of Chapter 10.

Remark about application of the Green method. We can see that the Green method requires considerably more technical work than the traditional one. This is a common situation when using the Green method and this is why it is not very popular in practical applications.

Exercises for section 5

Solve the following boundary value problems using the traditional method and the Green method; compare the found solutions:

1. $y'' + 4y = 2x$, $y(0) = 0, y(\frac{\pi}{8}) = 0$.
2. $y'' = e^{3x}$, $y(0) = 0, 3y(1) + y'(1) = 0$.
3. $y'' - y = \sin x$, $y'(0) = 0, y'(2) + y(2) = 0$.
4. $y'' - y = \cos x$, $y'(0) = 0, y'(2) + y(2) = 0$.
5. $y'' + y = x^2 + 2x$, $y(0) = 0, y'(1) = 0$.
6. $y'' + 4y = 2^{-x}$, $y'(0) = 0, y(1) = 0$.
7. $y'' - 4y = xe^x$, $y'(0) = 0, 2y(1) - y'(1) = 0$.
8. $y'' - y' = 2e^x - e^x$, $y(0) = 0, y(1) - y'(1) = 0$.
9. $y'' - y = 2x$, $y(0) = 0, y(1) = 0$.
10. $x^2 y'' + 3xy' - 3y = 2x^3 - 3x^4$, $y(1) = 0, y(2) - 2y'(2) = 0$.

6   Sturm-Liouville problem for second order equations

Definition of the Sturm-Liouville problem. The problem of finding a function $y \not\equiv 0$ and a parameter λ that satisfy the equation

$$y'' + a(x)y' = \lambda y$$

on an interval $[x_1, x_2]$ and the boundary conditions at the endpoints of the interval

$$\alpha_1 y(x_1) + \beta_1 y'(x_1) = 0, \alpha_2 y(x_2) + \beta_2 y'(x_2) = 0,$$

where $\alpha_1^2 + \beta_1^2 \neq 0$, $\alpha_2^2 + \beta_2^2 \neq 0$, is called the *Sturm-Liouville problem*. The unknowns λ and y are called *eigenvalue* and *eigenfunction*, respectively, and, for this reason, the problem is also called the *eigenvalue and eigenfunction problem*.

This type of problem is even more complicated than the boundary value problem. Therefore, we will only consider two simple examples.

Examples.

1. $y'' = \lambda y$, $x \in [0, d]$, $y(0) = y(d) = 0$.
We start with the general solution of the given equation. The characteristic equation for unknown μ has the form $\mu^2 = \lambda$ with the roots $\mu_{1,2} = \pm\sqrt{\lambda}$. The roots depend on the value of the parameter λ, which has not yet been defined, which means that we have to consider all the options of $\lambda \in \mathbb{R}$. We divide the study into three cases.

1) If $\lambda > 0$, then $\mu_{1,2} = \pm\sqrt{\lambda}$ are two different real roots that generate the general solution $y_{gh} = C_1 e^{-\sqrt{\lambda}x} + C_2 e^{\sqrt{\lambda}x}$ (which contains all the particular solutions). Substituting this solution into the boundary conditions, we get $\begin{cases} C_1 + C_2 = 0 \\ C_1 e^{-\sqrt{\lambda}d} + C_2 e^{\sqrt{\lambda}d} = 0 \end{cases}$. From the first equation follows $C_2 = -C_1$ and, substituting this relation into the second, we get $C_1 e^{-\sqrt{\lambda}d} - C_1 e^{\sqrt{\lambda}d} = C_1(e^{-\sqrt{\lambda}d} - e^{\sqrt{\lambda}d}) = 0$. Since $\sqrt{\lambda} > 0$ and $d > 0$, it follows that $e^{-\sqrt{\lambda}d} \neq e^{\sqrt{\lambda}d}$. So, the only option to satisfy this equation is to set $C_1 = 0$. But in this case $C_2 = 0$ and therefore $y \equiv 0$, that is, there is no eigenfunction.

2) If $\lambda = 0$, then $\mu_{1,2} = 0$ is the double root that leads to the general solution $y_{gh} = C_1 + C_2 x$ (which contains all the particular solutions). Substituting this solution into the boundary conditions, we get $\begin{cases} C_1 = 0 \\ C_1 + C_2 d = 0 \end{cases}$, whence $C_1 = C_2 = 0$. This again gives only the trivial function $y = 0$ and, since there are no other options to satisfy the equation and boundary conditions, we conclude that there are no eigenfunctions in this case.

3) If $\lambda < 0$, then $\mu_{1,2} = \pm\sqrt{\lambda}$ are the two complex conjugate roots and the two fundamental solutions in complex form are $y_1 = e^{-\sqrt{\lambda}x}$ and $y_2 = e^{\sqrt{\lambda}x}$. We write the latter in the form $y_2 = e^{i\sqrt{-\lambda}x}$, where $-\lambda > 0$ and $\sqrt{-\lambda}$ is a real number, and generate from it the two fundamental solutions in real form: $u = Re y_2 = \cos\sqrt{-\lambda}x$ and $v = Im y_2 = \sin\sqrt{-\lambda}x$. Then, the general solution in real form (which contains all particular solutions) is $y_{gh} = C_1 \cos\sqrt{-\lambda}x + C_2 \sin\sqrt{-\lambda}x$. Now, applying the boundary conditions, we have $\begin{cases} C_1 = 0 \\ C_1 \cos\sqrt{-\lambda}d + C_2 \sin\sqrt{-\lambda}d = 0 \end{cases}$. So, $C_1 = 0$ and $C_2 \sin\sqrt{-\lambda}d = 0$. If $C_2 = 0$, then $y \equiv 0$, which is not an eigenfunction. It remains to try the condition $\sin\sqrt{-\lambda}d = 0$, which gives the solutions $\sqrt{-\lambda_k}d = k\pi$, $k \in \mathbb{N}$ (we

can only take positive solutions, because $\sqrt{-\lambda}d > 0$). Since there are several solutions, it is convenient to use the index k together with λ to number these solutions. Isolating λ_k, we get $\lambda_k = -\left(\frac{k\pi}{d}\right)^2$, $k \in \mathbb{N}$. Returning to y, we find the corresponding eigenfunctions $y_k = \sin \sqrt{-\lambda_k}x = \sin \frac{k\pi}{d}x$. It can be seen that the eigenfunctions are always determined up to a multiplicative factor (different from 0) and usually the most convenient constant is chosen. Thus, we find an infinite number of pairs of eigenvalues and eigenfunctions

$$(\lambda_k, y_k) = \left(-\left(\frac{k\pi}{d}\right)^2, y_k = \sin \frac{k\pi}{d}x\right), \ k \in \mathbb{N}.$$

2. $y'' = \lambda y$, $x \in [0, d]$, $y'(0) = y'(d) = 0$.
The procedure for solving this problem follows the steps of the previous example. The equation is the same and, therefore, its general solution has the same form determined by the roots of the characteristic equation $\mu_{1,2} = \pm\sqrt{\lambda}$. Again, we divide the study into three cases.

1) If $\lambda > 0$, then the general solution has the real form $y_{gh} = C_1 e^{-\sqrt{\lambda}x} + C_2 e^{\sqrt{\lambda}x}$ (which contains all the particular solutions). Substituting its derivative $y'_{gh} = -\sqrt{\lambda}C_1 e^{-\sqrt{\lambda}x} + \sqrt{\lambda}C_2 e^{\sqrt{\lambda}x}$ in the boundary conditions, we get

$\begin{cases} -\sqrt{\lambda}C_1 + \sqrt{\lambda}C_2 = 0 \\ -\sqrt{\lambda}C_1 e^{-\sqrt{\lambda}d} + \sqrt{\lambda}C_2 e^{\sqrt{\lambda}d} = 0 \end{cases}$. Since $\sqrt{\lambda} > 0$, it follows from the

first equation that $C_2 = C_1$. Using this relationship in the second, we get $-\sqrt{\lambda}C_1(e^{-\sqrt{\lambda}d} - e^{\sqrt{\lambda}d}) = 0$. For the positive parameters $\sqrt{\lambda} > 0$ and $d > 0$, we get $e^{-\sqrt{\lambda}d} \neq e^{\sqrt{\lambda}d}$. Therefore, the only solution to this equation is $C_1 = 0$. But in this case $C_2 = 0$ and therefore $y \equiv 0$, that is, there is no eigenfunction.

2) If $\lambda = 0$, then the general solution takes the form $y_{gh} = C_1 + C_2 x$ (which contains all the particular solutions). By substituting its derivative $y'_{gh} = C_2$ into the boundary conditions, we get the same constraint twice: $\begin{cases} C_2 = 0 \\ C_2 = 0 \end{cases}$. With parameter C_1 free, we have the eigenfunction $y_0 = 1$. Thus, we obtain a pair of the eigenvalue and eigenfunction:

$$(\lambda_0, y_0) = (0, y_0 = 1).$$

3) If $\lambda < 0$, then the general solution in real form (which contains all the particular solutions) is $y_{gh} = C_1 \cos \sqrt{-\lambda}x + C_2 \sin \sqrt{-\lambda}x$. Substituting y_{gh} and its derivative $y'_{gh} = -\sqrt{-\lambda}C_1 \sin \sqrt{-\lambda}x + \sqrt{-\lambda}C_2 \cos \sqrt{-\lambda}x$ into the boundary conditions, we get $\begin{cases} \sqrt{-\lambda}C_2 = 0 \\ -\sqrt{-\lambda}C_1 \sin \sqrt{-\lambda}d + \sqrt{-\lambda}C_2 \cos \sqrt{-\lambda}d = 0 \end{cases}$.

From the first relation it follows that $C_2 = 0$ (because $\sqrt{-\lambda} > 0$) and then the second takes the form $\sqrt{-\lambda} C_1 \sin \sqrt{-\lambda} d = 0$. The first factor is positive ($\sqrt{-\lambda} > 0$), so we can try to vanish the second or third. If $C_1 = 0$, then $y \equiv 0$, which is not eigenfunction. This leaves the only option $\sqrt{-\lambda} d = 0$, which has a number of solutions $\sqrt{-\lambda_k} d = k\pi$, $k \in \mathbb{N}$ (since $\sqrt{-\lambda} d > 0$, we can only take positive solutions). Isolating λ_k, we get $\lambda_k = - \left(\frac{k\pi}{d} \right)^2$, $k \in \mathbb{N}$. Returning to y, we find the corresponding eigenfunctions $y_k = \cos \sqrt{-\lambda_k} x = \cos \frac{k\pi}{d} x$. Thus, for $\lambda < 0$, we find an infinite number of pairs of the eigenvalues and eigenfunctions

$$(\lambda_k, y_k) = \left(- \left(\frac{k\pi}{d} \right)^2, y_k = \cos \frac{k\pi}{d} x \right), \quad k \in \mathbb{N}.$$

Together with the pair $(\lambda_0, y_0) = (0, y_0 = 1)$ this forms the complete set of the eigenvalues and eigenfunctions for this problem.

Exercises for section 6

Solve the following Sturm-Liouville problems:
1. $y'' = \lambda y$, $x \in [0, d]$, $y(0) = y'(d) = 0$;
2. $y'' = \lambda y$, $x \in [0, d]$, $y'(0) = y(d) = 0$;
3. $y'' = \lambda y$, $x \in [0, d]$, $y(0) = y'(d) + y(d) = 0$;
4. $y'' = \lambda y$, $x \in [0, d]$, $y'(0) - y(0) = y(1) = 0$;
5. $y'' + 2y' + (1 - \lambda)y = 0$, $x \in [0, 1]$, $y(0) = y(1) = 0$ (hint: multiply the equation by e^x and change the function by the formula $z = e^x y$);
6. $x^2 y'' - xy' + y = \lambda y$, $x \in [1, 2]$, $y(1) = y(2) = 0$ (hint: use the method of solving the Euler equation).

Chapter 10

Linear equations with variable coefficients: power series method

1 Review of power series

Definition (power series). A *power series* has the form

$$\sum_{n=0}^{\infty} c_n (x-a)^n,$$

where c_n are *coefficients* (constants) and a is *center point*.

A frequently used particular case is the power series centered at $a=0$:

$$\sum_{n=0}^{\infty} c_n x^n.$$

Definition (convergence at a point). For a specific value of x a power series is a series of numbers. If this series converges, the power series is said to converge at x. Otherwise, the series is said to diverge at x. Obviously, any power series converges at its central point.

Definition (set of convergence of a power series). A *set of convergence* of a power series $\sum_{n=0}^{\infty} c_n (x-a)^n$ is the set of all points where this series converges. We denote this set by E and define $R = \sup_{\forall x \in E} |x-a|$ (if E is not limited, we consider $R = +\infty$).

Typical examples of a power series are the series $\sum_{n=0}^{\infty} x^n$, $\sum_{n=0}^{\infty} \frac{x^n}{n!}$, $\sum_{n=1}^{\infty} (xn)^n$. The first converges on $(-1, 1)$ and diverges on $\mathbb{R}\backslash[-1, 1]$, the second converges on $\mathbb{R}$ and the last converges only at the point 0. These are examples of all kinds of sets of convergence that a power series can have: a power series centered at a can converge only at the central point, or converge on an interval centered at a, or converge on the entire real axis. This result is formalized in the following theorem.

Theorem (set of convergence of a power series). If E is a set of convergence of a power series $\sum_{n=0}^{\infty} c_n(x - a)^n$ and $R = \sup_{\forall x \in E} |x - a|$, then only one of the following three options can occur:
1) if $R = +\infty$, then the series converges on $\mathbb{R}$;
2) if $R = 0$, then the series converges only at the point $x = a$ and diverges at all other real points;
3) if $0 < R < +\infty$, then the series converges on the interval $(a - R, a + R)$ and diverges outside the interval $[a - R, a + R]$.

Remark. The first two cases can be included in the third (which is the main case), using the convention that the interval $(a - \infty, a + \infty) = (-\infty, +\infty) = \mathbb{R}$ when $R = +\infty$ and $(a - 0, a + 0) = \{a\}$, $[a - 0, a + 0] = \{a\}$ when $R = 0$.

Definition. The interval $(a - R, a + R)$, where $R = \sup_{\forall x \in E} |x - a|$ and E is the set of convergence of a series $\sum_{n=0}^{\infty} c_n(x - a)^n$, is called the *interval of convergence* of this series and R the *radius of convergence*.

Meeting the radius of convergence.

The radius of convergence R can frequently be found using D'Alembert's (the ratio) test or Cauchy's (the root) test. Recall that the applicability of D'Alembert's test to the study of (absolute) convergence of the series of numbers $\sum_{n=0}^{\infty} a_n$ depends on the existence of the limit $A = \lim\limits_{n \to \infty} \left| \frac{a_{n+1}}{a_n} \right|$. In the case of a power series $\sum_{n=0}^{\infty} c_n(x - a)^n$, for each fixed x we have $a_n = c_n(x - a)^n$, and consequently,

$$A = \lim_{n \to \infty} \left| \frac{a_{n+1}}{a_n} \right| = \lim_{n \to \infty} \left| \frac{c_{n+1}}{c_n} \right| |x - a| = |x - a| \lim_{n \to \infty} \left| \frac{c_{n+1}}{c_n} \right|.$$

Let us assume that the last limit exists. Then D'Alembert's test states that for $A < 1$ we have convergence and for $A > 1$ divergence. This means that the series $\sum_{n=0}^{\infty} c_n(x - a)^n$ converges on the set $|x - a| < \frac{1}{\lim\limits_{n \to \infty} |c_{n+1}/c_n|}$ and diverges on the set $|x - a| > \frac{1}{\lim\limits_{n \to \infty} |c_{n+1}/c_n|}$. Therefore, the radius of convergence can be found by D'Alembert's formula: $R = \frac{1}{\lim\limits_{n \to \infty} |c_{n+1}/c_n|}$.

Similarly, Cauchy's test involves the calculation of the limit

$$C = \lim_{n \to \infty} \sqrt[n]{|a_n|} = \lim_{n \to \infty} \sqrt[n]{|c_n (x-a)^n|} = |x-a| \cdot \lim_{n \to \infty} \sqrt[n]{|c_n|}.$$

If the last limit exists, then for $C < 1$ we have convergence and for $C > 1$ divergence. Therefore, the series $\sum_{n=0}^{\infty} c_n(x-a)^n$ converges for $|x-a| < \dfrac{1}{\lim\limits_{n \to \infty} \sqrt[n]{|c_n|}}$ and diverges for $|x-a| > \dfrac{1}{\lim\limits_{n \to \infty} \sqrt[n]{|c_n|}}$. Hence, the radius of convergence is determined by Cauchy's formula $R = \dfrac{1}{\lim\limits_{n \to \infty} \sqrt[n]{|c_n|}}$.

Recall that the existence of the limit $\lim\limits_{n \to \infty} \left| \dfrac{c_{n+1}}{c_n} \right|$ guarantees the existence of the limit $\lim\limits_{n \to \infty} \sqrt[n]{|c_n|}$ equal to the first one (that is why Cauchy's test is stronger than D'Alembert's test), but in practice, calculation of the first limit (if it exists) is frequently simpler.

It may happen that even the second (Cauchy) limit does not exist. In this case, we have to look for alternative, finer options for finding the radius R. The most general formula used to determine the radius of convergence in problematic situations (when the simpler Cauchy and D'Alembert limits do not exist), is the Cauchy-Hadamard formula which comes from the same Cauchy's test, albeit in a finer form. Let us recall that Cauchy's test can be formulated using the upper limit $\overline{C} = \limsup\limits_{n \to \infty} \sqrt[n]{|a_n|}$ as follows: if $\overline{C} < 1$ then the series $\sum_{n=0}^{\infty} a_n$ converges, and if $\overline{C} > 1$ then it diverges. Formally, this formulation is a replica of the formulation with the general limit $C = \lim\limits_{n \to \infty} \sqrt[n]{|a_n|}$, but the important detail is that the upper limit always exists (if we include infinite limits), while the general limit may not exist. In the case of power series, we have $a_n = c_n(x-a)^n$ for each fixed x, which leads to the following form of Cauchy's test with the upper limit: if $|x-a| < \dfrac{1}{\limsup\limits_{n \to \infty} \sqrt[n]{|c_n|}}$, then the series $\sum_{n=0}^{\infty} c_n(x-a)^n$ converges, and if $|x-a| > \dfrac{1}{\limsup\limits_{n \to \infty} \sqrt[n]{|c_n|}}$ then it diverges. Therefore, the radius of convergence is defined as $R = \dfrac{1}{\limsup\limits_{n \to \infty} \sqrt[n]{|c_n|}}$. This is the Cauchy-Hadamard formula and its advantage is that the limit $H = \limsup\limits_{n \to \infty} \sqrt[n]{|c_n|}$ can always be found (at least theoretically), while the limits in the D'Alembert and Cauchy formulas may not exist.

In view of the determination of the radius of convergence by the Cauchy-Hadamard formula, the Theorem on the set of convergence can be reformulated as follows.

Cauchy-Hadamard Theorem. A power series $\sum_{n=0}^{\infty} c_n(x-a)^n$ admits only one of the three options which can be defined as a function of the limit

$H = \limsup\limits_{n\to\infty} \sqrt[n]{|c_n|}$:

1) if $R = \dfrac{1}{\limsup\limits_{n\to\infty} \sqrt[n]{|c_n|}} = +\infty$, then the series converges on $\mathbb{R}$;

2) if $R = \dfrac{1}{\limsup\limits_{n\to\infty} \sqrt[n]{|c_n|}} = 0$, then the series converges only at the central point $x = a$ and diverges at all other real points;

3) if $0 < R = \dfrac{1}{\limsup\limits_{n\to\infty} \sqrt[n]{|c_n|}} < +\infty$, then the series converges on the interval $(a - R, a + R)$ and diverges outside the interval $[a - R, a + R]$. Recall that the number $R = \frac{1}{H}$ is the radius of convergence and $(a - R, a + R)$ is the interval of convergence.

Theorem (convergence of the series of derivatives). If R is the radius of convergence of the series $\sum_{n=0}^{\infty} c_n(x-a)^n$, then the series of derivatives of any order

$$\sum_{n=0}^{\infty} c_n((x-a)^n)^{(k)} = \sum_{n=k}^{\infty} c_n n(n-1)\cdot \ldots \cdot (n-k+1)(x-a)^{n-k}$$

converges on the interval $(a - R, a + R)$ and diverges outside the interval $[a - R, a + R]$.

Properties of power series and their sums

Uniqueness theorem. A representation of a function $f(x)$ in a power series $\sum_{n=0}^{\infty} c_n(x - a)^n$ is unique on any interval $(a - c, a + c), c > 0$ where the series converges to $f(x)$.

Property 1. Sum of power series. If the power series $\sum_{n=0}^{\infty} c_n(x-a)^n$ and $\sum_{n=0}^{\infty} d_n(x - a)^n$ have the intervals of convergence $(a - R_c, a + R_c)$ and $(a - R_d, a + R_d)$, respectively, then the series $\sum_{n=0}^{\infty}(c_n + d_n)(x - a)^n$ converges on the interval $(a - R, a + R)$, $R = \min\{R_c, R_d\}$.

Property 2. If the power series $\sum_{n=0}^{\infty} c_n(x - a)^n$ has the interval of convergence $(a - R, a + R)$ and the function $g(x)$ is bounded on $(a - R, a + R)$, then the series $\sum_{n=0}^{\infty} g(x)c_n(x - a)^n$ converges on the interval $(a - R, a + R)$.

Definition of the Cauchy product. For two series $\sum_{n=0}^{\infty} c_n x^n$ and $\sum_{n=0}^{\infty} d_n x^n$ centered at the point 0, their *Cauchy product* is the new power series (centered at the same point) obtained by the formula

$$\sum_{n=0}^{\infty} c_n x^n \cdot \sum_{n=0}^{\infty} d_n x^n = (c_0 + c_1 x + c_2 x^2 + \ldots)(d_0 + d_1 x + d_2 x^2 + \ldots)$$

$$= c_0 d_0 + (c_0 d_1 + c_1 d_0)x + (c_0 d_2 + c_1 d_1 + c_2 d_0)x^2 + \ldots = \sum_{n=0}^{\infty} e_n x^n, \ e_n = \sum_{k=0}^{n} c_k d_{n-k}.$$

Similarly, the Cauchy product is defined for series centered at a point a.

We note that this type of product of series is especially suitable for power series since it uses grouping of terms of the original series according to the power of x and represents the generalization of the multiplication of two polynomials in the form distributed by powers.

Property 3. Product of power series. If the series $f(x) = \sum_{n=0}^{\infty} c_n(x - a)^n$ and $g(x) = \sum_{n=0}^{\infty} d_n(x - a)^n$ have convergence intervals $(a - R_c, a + R_c)$ and $(a - R_d, a + R_d)$, respectively, then their Cauchy product $\sum_{n=0}^{\infty} e_n(x - a)^n$, $e_n = \sum_{k=0}^{n} c_k d_{n-k}$ converges to the function $f(x)g(x)$ on the interval $(a - R, a + R)$, $R = \min\{R_c, R_d\}$.

Property 4. Change of variable. Let us assume that the series $f(x) = \sum_{n=0}^{\infty} c_n(x - a)^n$ converges on $(a - R, a + R)$, $R > 0$ and that $x = g(t) = a + \alpha(t - b)^k$, $k \in \mathbb{N}$, $\alpha \neq 0$. In this case, the power series $h(t) = \sum_{n=0}^{\infty} c_n \alpha^n (t - b)^{kn}$ converges to the function $f(g(t))$ on $(b - R_1, b + R_1)$, $R_1 = \left(\frac{R}{|\alpha|}\right)^{1/k}$.

Property 5. Change of center point. If the series $f(x) = \sum_{n=0}^{\infty} c_n(x - a)^n$ converges on $|x - a| < R$, then for any $b \in (a - R, a + R)$, the same function can be represented as a power series centered at b: $f(x) = \sum_{n=0}^{\infty} d_n(x - b)^n$, which converges on $|x - b| < R - |b - a|$.

Property 6. Differentiability. The sum $f(x)$ of a power series $\sum_{n=0}^{\infty} c_n(x - a)^n$ is infinitely differentiable over the entire interval of convergence $(a - R, a + R)$ and its derivative of order m is equal to the sum of the series of derivatives of order m:

$$\left(\sum_{n=0}^{\infty} c_n(x - a)^n\right)^{(m)} = \sum_{n=0}^{\infty} (c_n(x - a)^n)^{(m)}$$

$$= \sum_{n=m}^{\infty} c_n n(n - 1) \cdot \ldots \cdot (n - m + 1)(x - a)^{n-m} \ , \forall x \in (a - R, a + R) \ .$$

Thus, the power series can be infinitely differentiated term by term.

Property 7. Parity. If the series $\sum_{n=0}^{\infty} c_n x^n$ represents an even function $f(x)$ in a neighborhood of the origin, then this series contains only even powers. Similarly, if the series $\sum_{n=0}^{\infty} c_n x^n$ represents an odd function $f(x)$ in a neighborhood of the origin, then this series contains only odd powers.

Definition (analytic function). The function $f(x)$ that represents the sum of a power series $f(x) = \sum_{n=0}^{\infty} c_n(x - a)^n$ converging on the interval $(a - R, a + R)$, $R > 0$ is called *analytic* on this interval. The function $f(x)$ is analytic at a point a if it is analytic in a neighborhood of this point.

Notice that Property 5 can be reformulated as follows: if a function $f(x)$ is analytic on the interval $(a - R, a + R)$, then it is analytic on any interval within that interval, including $f(x)$ being analytic at any point of the interval $(a - R, a + R)$.

Taylor coefficients and Taylor series

Theorem (Taylor coefficients). The coefficients of the power series $f(x) = \sum_{n=0}^{\infty} c_n (x-a)^n$ can be calculated using the formula $c_n = \frac{f^{(n)}(a)}{n!}$, $\forall n$. (As always, the derivative of order 0 is understood as the function itself).

Definition. The coefficients $\frac{f^{(n)}(a)}{n!}$ are called *Taylor coefficients* and the power series $\sum_{n=0}^{\infty} c_n (x - a)^n$ written in the form

$$f(x) = \sum_{n=0}^{\infty} \frac{f^{(n)}(a)}{n!} (x - a)^n$$

is called the *Taylor series* of the function $f(x)$ at a point a.

Power series of some elementary functions

$$\frac{1}{1 - x} = \sum_{n=0}^{\infty} x^n = 1 + x + x^2 + x^3 + \dots, \ \forall x \in (-1, 1).$$

$$\frac{1}{(1-x)^p} = \sum_{m=0}^{\infty} \frac{(m+p-1)(m+p-2) \cdot \dots \cdot (m+1)}{(p - 1)!} x^m, \forall p \in \mathbb{N}, \ \forall x \in (-1, 1).$$

$$\ln(1 - x) = - \sum_{n=1}^{\infty} \frac{x^n}{n} = -x - \frac{x^2}{2} - \frac{x^3}{3} - \frac{x^4}{4} - \dots, \ \forall x \in [-1, 1).$$

$$\frac{1}{1 + x} = \sum_{n=0}^{\infty} (-1)^n x^n = 1 - x + x^2 - x^3 + \dots, \ \forall x \in (-1, 1).$$

$$\frac{1}{(1+x)^p} = \sum_{n=0}^{\infty} (-1)^n \frac{(n+p-1)(n+p-2) \cdot \dots \cdot (n+1)}{(p - 1)!} x^n, \forall p \in \mathbb{N}, \forall x \in (-1, 1).$$

$$\ln(1 + x) = \sum_{n=1}^{\infty} (-1)^{n-1} \frac{x^n}{n} = x - \frac{x^2}{2} + \frac{x^3}{3} - \frac{x^4}{4} + \dots, \ \forall x \in (-1, 1].$$

$$(1 + x)^p = \sum_{n=0}^{\infty} \frac{p(p - 1) \cdot \dots \cdot (p - n + 1)}{n!} x^n$$

$$= 1 + \frac{p}{1!} x + \frac{p(p - 1)}{2!} x^2 + \frac{p(p - 1)(p - 2)}{3!} x^3 + \dots, \forall p \notin \mathbb{N}, p \neq 0, \ \forall x \in (-1, 1).$$

$$e^x = \sum_{n=0}^{\infty} \frac{x^n}{n!} = 1 + \frac{x}{1!} + \frac{x^2}{2!} + \frac{x^3}{3!} + \dots, \ \forall x \in \mathbb{R}.$$

$$b^x = \sum_{n=0}^{\infty} \frac{\ln^n b}{n!} x^n = 1 + \frac{\ln b}{1!} x + \frac{\ln^2 b}{2!} x^2 + \frac{\ln^3 b}{3!} x^3 + \dots, \ b > 0, b \neq 1, \forall x \in \mathbb{R}.$$

$$\sin x = \sum_{n=0}^{\infty} (-1)^n \frac{x^{2n+1}}{(2n+1)!} = x - \frac{x^3}{3!} + \frac{x^5}{5!} - \frac{x^7}{7!} + \dots, \ \forall x \in \mathbb{R}.$$

$$\cos x = \sum_{n=0}^{\infty} (-1)^n \frac{x^{2n}}{(2n)!} = 1 - \frac{x^2}{2!} + \frac{x^4}{4!} - \frac{x^6}{6!} + \dots, \ \forall x \in \mathbb{R}.$$

$$\tan x = x + \frac{x^3}{3} + \frac{2x^5}{15} + \frac{17x^7}{315} + \dots, \ \forall x \in \left(-\frac{\pi}{2}, \frac{\pi}{2}\right).$$

$$\cot x - \frac{1}{x} = -\frac{1}{3}x - \frac{1}{45}x^3 - \frac{2}{945}x^5 - \frac{1}{4725}x^7 + \dots, \forall x \in \left(-\frac{\pi}{2}, \frac{\pi}{2}\right).$$

$$\arcsin x = \sum_{n=0}^{\infty} \frac{(2n-1)!!}{(2n+1)2^n n!} x^{2n+1} = x + \frac{1}{6}x^3 + \frac{3}{40}x^5 + \frac{5}{112}x^7 + \dots, \forall x \in (-1,1).$$

$$\arccos x = \frac{\pi}{2} - \sum_{n=0}^{\infty} \frac{(2n-1)!!}{(2n+1)2^n n!} x^{2n+1} = \frac{\pi}{2} - x - \frac{1}{6}x^3 - \frac{3}{40}x^5 - \dots, \forall x \in (-1,1).$$

$$\arctan x = \sum_{n=0}^{\infty} (-1)^n \frac{x^{2n+1}}{2n+1} = x - \frac{x^3}{3} + \frac{x^5}{5} - \frac{x^7}{7} + \dots, \ \forall x \in (-1,1).$$

$$\operatorname{arccot} x = \frac{\pi}{2} - \sum_{n=0}^{\infty} (-1)^n \frac{x^{2n+1}}{2n+1} = \frac{\pi}{2} - x + \frac{x^3}{3} - \frac{x^5}{5} + \frac{x^7}{7} + \dots, \ \forall x \in (-1,1).$$

2 Power series solutions: theoretical results

Solutions of linear ordinary differential equations can be found in power series form even in cases when methods desidned for solution of special types of equations, such as linear with constant coefficients and of reducible order, do not work (we note again that there is no universal method of solution even for equations of the first order). Let us recall some results from the theory of linear differential equations.

A *homogeneous linear equation of order* n

$$y^{(n)} + a_{n-1}(x)y^{(n-1)} + \dots + a_1(x)y' + a_0(x)y = 0, \tag{2.1}$$

with coefficients $a_0(x), a_1(x), \ldots, a_{n-1}(x)$ continuous over an interval I, has the *general solution* $y_g(x)$ in I which contains all particular solutions and can be found as a *linear combination* :

$$y_g(x) = C_1 y_1(x) + C_2 y_2(x) + \ldots + C_n y_n(x),$$

where $y_1(x), y_2(x), \ldots, y_n(x)$ are *linearly independent particular solutions* and $C_1, C_2, \ldots, C_n$ are arbitrary coefficients.

A *(nonhomogeneous) linear equation of order* n

$$y^{(n)} + a_{n-1}(x)y^{(n-1)} + \ldots + a_1(x)y' + a_0(x)y = b(x), \qquad (2.2)$$

with coefficients $a_0(x), a_1(x), \ldots, a_{n-1}(x)$ and right-hand side $b(x)$ continuous on an interval I, has a *general solution* $y_n(x)$ on I which contains all particular solutions and can be represented in the form

$$y_n(x) = y_g(x) + y_p(x),$$

where $y_g(x)$ is the *general solution of the homogeneous equation* (2.1) and $y_p(x)$ is any *particular solution of the nonhomogeneous equation* (2.2).

The forms (2.1) and (2.2) are called *canonical* (or *normalized*) for homogeneous and nonhomogeneous linear equations, respectively.

A *continuous Cauchy (initial value) problem* consists of the equation (2.2) in interval I, which has coefficients $a_0(x), a_1(x), \ldots, a_{n-1}(x)$ and right-hand side $b(x)$ continuous on an interval I, together with initial conditions $y(x_0) = b_0, y'(x_0) = b_1, \ldots, y^{(n-1)}(x_0) = b_{n-1}$, where $x_0 \in I$. The continuous Cauchy problem has a unique solution on I.

Under certain conditions on the coefficients and right-hand side of the linear equation, all particular solutions can be expressed in power series, that is, as analytic functions. We formulate the corresponding results in the following two theorems.

Theorem E1 (power series solution for homogeneous equation). If the coefficients $a_0(x), a_1(x), \ldots, a_{n-1}(x)$ of homogeneous linear equation (2.1) are analytic functions of the radii of convergence $R_0, R_1, \ldots, R_{n-1}$, respectively (all with respect to the center point x_0), then any particular solution of (2.1) can be represented as a power series with the center point x_0 and radius of convergence $R \geq \min\{R_0, R_1, \ldots, R_{n-1}\}$.

Theorem E2 (power series solution for nonhomogeneous equation). Any particular solution of nonhomogeneous linear equation (2.2) with analytic coefficients $a_0(x), a_1(x), \ldots, a_{n-1}(x)$ of the radii of convergence $R_0, R_1, \ldots, R_{n-1}$, respectively, and the analytic right-hand side $b(x)$ of the radius of convergence R_b (all with respect to the center point x_0), can be represented as a power series with the center point x_0 and the radius of convergence $R \geq \min\{R_0, R_1, \ldots, R_{n-1}, R_b\}$.

If the conditions of Theorem E1 are satisfied in a neighborhood of the point x_0, then this point is called regular. We formalize this in the following definition.

Definition of ordinary and singular point. A point x_0 is called an *ordinary (or non-singular) point* of the equation (2.1) if the coefficients $a_0(x), a_1(x), \ldots, a_{n-1}(x)$ are analytic functions on x_0. If x_0 is not an ordinary point, it is called *singular point.*

3   Solution about ordinary points

In the case of an ordinary point, the solution procedure consists of searching for particular solutions in the form of a formal power series, whose coefficients must be specified by substituting the series into the original equation, with subsequent analysis of the convergence of the obtained formal series. Notice that it is frequently difficult or impossible to find a general form of the coefficients when the relationships in which they are involved are complicated. In such cases, we are able to specify only a few first coefficients of the series, and we use the first terms of the series to represent an approximation of the exact solution that has a required accuracy at least in a small neighborhood of the center point.

Let us consider below several examples, starting with some elementary equations, even of the first order, whose solutions can be easily found using the methods simpler than power series expansion, and then these solutions can be used to check how the series solutions works.

Examples.

1. The first order homogeneous equation $y' - y = 0$.
This is a trivial equation of separable variables (that is, of the type $y' = f(x)g(y)$), whose solution can be found immediately by separating y and x and integrating both sides of the equation:

$$\int \frac{dy}{y} = \int 1 dx \;\Rightarrow\; \ln y = x + A; \; y \equiv 0 \;\Rightarrow\; y = Ce^x, \forall C \in \mathbb{R}$$

(see section 1 in Chapter 3).

For illustration purposes, let us find the solution of this equation using the formal power series $y(x) = \sum_{n=0}^{\infty} c_n x^n$ whose coefficients are determined by substituting this series into the original equation:

$$\left(\sum_{n=0}^{\infty} c_n x^n\right)' - \sum_{n=0}^{\infty} c_n x^n = 0.$$

Assuming that the series has a non-zero radius of convergence, we apply the term-by-term differentiation within the interval of convergence and obtain

$$\sum_{n=1}^{\infty} n c_n x^{n-1} - \sum_{n=0}^{\infty} c_n x^n = 0$$

or, changing the index variation in the second series and joining the two series,

$$\sum_{n=1}^{\infty} (n c_n - c_{n-1}) x^{n-1} = 0.$$

Then, from the uniqueness of a power series, we obtain the recurrence relation $n c_n = c_{n-1}, \forall n \in \mathbb{N}$ and the coefficient c_0 is arbitrary. The recurrence relation can be solved for c_n in terms of c_0:

$$c_n = \frac{1}{n} c_{n-1} = \frac{1}{n(n-1)} c_{n-2} = \ldots = \frac{1}{n!} c_0.$$

Therefore, we obtain the general solution in the form

$$y(x) = C \sum_{n=0}^{\infty} \frac{1}{n!} x^n = C e^x, \forall C = c_0 \in \mathbb{R},$$

that is, we arrive at the solution obtained before in a simpler way. Since the generated series has already been recognized as the Taylor series of e^x, there is no need to check its convergence: we know that it converges on $\mathbb{R}$, which justifies the operations performed formally for any $x \in \mathbb{R}$. (Actually, the general result on the existence of solutions in the form of power series, presented in Theorem E1 at the end of section 2, already indicates that the considered series has the radius of convergence $R = \infty$, since the coefficient $a_0 \equiv -1$ is the analytic function in $\mathbb{R}$).

2. The second order homogeneous equation $y'' + y = 0$.
This is a homogeneous linear equation with constant coefficients, whose solution is usually found by solving the characteristic equation $\lambda^2 + 1 = 0$ and forming two corresponding particular solutions in the form $y_1(x) = \cos x$ and $y_2(x) = \sin x$, which lead to the general solution $y_g(x) = C_1 \cos x + C_2 \sin x$ (see section 1 in Chapter 9).

Now we solve the same equation using the formal power series $y(x) = \sum_{n=0}^{\infty} c_n x^n$, whose coefficients are determined by substituting this series into the original equation:

$$\left(\sum_{n=0}^{\infty} c_n x^n\right)'' + \sum_{n=0}^{\infty} c_n x^n = 0.$$

Assuming that the series has a non-zero radius of convergence, we apply the term-by-term differentiation twice within the interval of convergence and obtain

$$\sum_{n=2}^{\infty} n(n-1)c_n x^{n-2} + \sum_{n=0}^{\infty} c_n x^n = 0$$

or, changing the index in the second series from n to $n-2$ and joining the two series,

$$\sum_{n=2}^{\infty} [n(n-1)c_n + c_{n-2}] x^{n-2} = 0.$$

The uniqueness of the power series leads to the recurrence relation $n(n-1)c_n = -c_{n-2}, \forall n = 2, 3, \ldots$ and the parameters c_0 and c_1 are arbitrary. The recurrence relation can be divided into two independent groups: the first starts with c_0 and determines all the even-indexed coefficients as a function of c_0, and the second starts with c_1 and determines all the odd-indexed coefficients as a function of c_1. Solving the relations of the first group, we have for $\forall k \in \mathbb{N}$

$$c_n = c_{2k} = -\frac{1}{2k(2k-1)} c_{2k-2} = (-1)^2 \frac{1}{2k(2k-1)(2k-2)(2k-3)} c_{2k-4} = \ldots$$

$$= (-1)^k \frac{1}{(2k)!} c_0,$$

and similarly for the second group

$$c_n = c_{2k+1} = -\frac{1}{(2k+1)(2k)} c_{2k-1} = (-1)^2 \frac{1}{(2k+1)2k(2k-1)(2k-2)} c_{2k-3} = \ldots$$

$$= (-1)^k \frac{1}{(2k+1)!} c_1.$$

Therefore, we obtain the two linearly independent solutions in the form of power series (we fix here $c_0 = 1$ and $c_1 = 1$)

$$y_1(x) = \sum_{n=0}^{\infty} (-1)^n \frac{1}{(2n)!} x^{2n}, \quad y_2(x) = \sum_{n=0}^{\infty} (-1)^n \frac{1}{(2n+1)!} x^{2n+1}.$$

We easily recognize in these two functions the series for $y_1(x) = \cos x$ and $y_2(x) = \sin x$ (convergent on $\mathbb{R}$), and therefore, we arrive at the same general solution found by the traditional method in a much simpler way.

3. The second order homogeneous equation $y'' - xy = 0$ (Airy equation). This is a homogeneous linear equation with variable coefficients of the type (2.1), whose solution requires the use of power series: $y(x) = \sum_{n=0}^{\infty} c_n x^n$. To find the values of c_n we substitute the series into the original equation

$$\sum_{n=2}^{\infty} n(n-1)c_n x^{n-2} - \sum_{n=0}^{\infty} c_n x^{n+1} = 0.$$

We change the index in the second series from $n+1$ to $n-2$ and add it to the first one:

$$2c_2 + \sum_{n=3}^{\infty} [n(n-1)c_n - c_{n-3}]x^{n-2} = 0.$$

So, we determine $c_2 = 0$ and obtain the following recurrence relation: $n(n-1)c_n = c_{n-3}, \forall n = 3, 4, \ldots$ (this follows from the uniqueness of the power series). Obviously, this relation can be split into three groups: the first contains indices $3k$, $k \in \mathbb{N}$ and is determined by the coefficient c_0, the second, with indices $3k+1$, is determined by c_1, and the third, with indices $3k+2$, is determined by c_2. Since $c_2 = 0$, all the coefficients of the third group vanish. But the two other parameters - c_0 and c_1 - can be chosen arbitrarily, that gives the two groups of coefficients:

$$c_{3k} = \frac{1}{3k(3k-1)}c_{3k-3} = \frac{1}{3k(3k-1)(3k-3)(3k-4)}c_{3k-6} = \ldots$$

$$= \frac{1}{2 \cdot 3 \cdot \ldots \cdot (3k-4)(3k-3)(3k-1)3k}c_0$$

and

$$c_{3k+1} = \frac{1}{(3k+1)3k}c_{3k-2} = \frac{1}{(3k+1)3k(3k-2)(3k-3)}c_{3k-5} = \ldots$$

$$= \frac{1}{3 \cdot 4 \cdot \ldots \cdot (3k-3)(3k-2)3k(3k+1)}c_1.$$

Consequently, the two linearly independent solutions can be found in the form (we choose here $c_0 = 1$ and $c_1 = 1$)

$$y_1(x) = \sum_{n=0}^{\infty} \frac{x^{3n}}{2 \cdot 3 \cdot \ldots \cdot (3n-4)(3n-3)(3n-1)3n}$$

and

$$y_2(x) = \sum_{n=0}^{\infty} \frac{x^{3n+1}}{3 \cdot 4 \cdot \ldots \cdot (3n-3)(3n-2)3n(3n+1)}.$$

Finally, the general solution has the common form $y_g(x) = C_1 y_1(x) + C_2 y_2(x)$.

To justify the applied procedures, we have to check where the two obtained series converge. This can be done in two ways. First, observing that the coefficient of the original equation $a_0 = -x$ is an analytic function in $\mathbb{R}$, we can use the general statement that guarantees that the series of solutions have the same radius of convergence and, therefore, converge in $\mathbb{R}$ (see Theorem E1 in section 2). Another way is to check the convergence of the series directly by applying one of the tests. For example, using D'Alembert's test, we have for the first series

$$\frac{|x^{3n+3}|}{2 \cdot 3 \cdot \ldots \cdot (3n-1)3n(3n+2)(3n+3)} \cdot \frac{2 \cdot 3 \cdot \ldots \cdot (3n-1)3n}{|x^{3n}|}$$

$$= |x^3| \frac{1}{(3n+2)(3n+3)} \xrightarrow[n \to \infty]{} 0, \forall x \in \mathbb{R},$$

which means that this series converges on $\mathbb{R}$. The second series has the same behavior. This justifies all the steps taken to derive these two series.

4. The second order homogeneous equation $(x^2 + 1)y'' + xy' - y = 0$. This is a homogeneous linear equation with variable coefficients, whose solution requires the use of power series: $y(x) = \sum_{n=0}^{\infty} c_n x^n$. To find the values of c_n we substitute the series into the original equation

$$(x^2 + 1) \sum_{n=2}^{\infty} n(n-1)c_n x^{n-2} + x \sum_{n=1}^{\infty} nc_n x^{n-1} - \sum_{n=0}^{\infty} c_n x^n = 0$$

or, separating terms with equal powers,

$$\sum_{n=2}^{\infty} n(n-1)c_n x^n + \left[2c_2 x^0 + 3 \cdot 2c_3 x^1 + \sum_{n=2}^{\infty} (n+2)(n+1)c_{n+2} x^n \right]$$

$$+ \left[c_1 x^1 + \sum_{n=2}^{\infty} nc_n x^n \right] - \left[c_0 x^0 + c_1 x^1 + \sum_{n=2}^{\infty} c_n x^n \right] = 0.$$

Regrouping the terms, we get

$$(2c_2 - c_0) + 6c_3 x + \sum_{n=2}^{\infty} \left[n(n-1)c_n + (n+2)(n+1)c_{n+2} + nc_n - c_n \right] x^n = 0.$$

Consequently, we have the following formulas for the coefficients:

$$2c_2 - c_0 = 0, c_3 = 0; (n+1)(n-1)c_n + (n+2)(n+1)c_{n+2} = 0, \forall n = 2, 3, \ldots .$$

The last set represents the recurrence relation that can be simplified to the form $c_{n+2} = -\frac{n-1}{n+2}c_n, \forall n = 2, 3, \ldots$ and divided into two groups – the even-indixed $n = 2k$ and odd-indixed $n = 2k + 1$. In the first group we have

$$c_{2k} = -\frac{2k-3}{2k}c_{2k-2} = (-1)^2 \frac{2k-3}{2k}\frac{2k-5}{2k-2}c_{2k-4} = \ldots$$

$$= (-1)^{k-1}\frac{(2k-3)(2k-5)\cdot \ldots \cdot 3 \cdot 1}{2k(2k-2)\cdot \ldots \cdot 6 \cdot 4}c_2, \forall k = 2, 3, \ldots; \ c_2 = \frac{1}{2}c_0.$$

In the second group, all coefficients starting from the index 3 are zero, because $c_3 = 0$, and the only arbitrary coefficient is c_1:

$$c_{2k+1} = 0, \forall k = 1, 2, \ldots; \ \forall c_1.$$

Thus, we have the two linearly independent solutions: $y_1 = x$ and the series

$$y_2(x) = 2 + x^2 + \sum_{n=2}^{\infty}(-1)^{k-1}\frac{1 \cdot 3 \cdot \ldots \cdot (2n-3)}{4 \cdot 6 \cdot \ldots \cdot 2n}x^{2n}$$

(in the last we chose $c_2 = 1$).

To justify the applied procedure, we have to analyze the convergence of the obtained series. This can be done in two ways. First, noting that the coefficients of the original normalized equation (written in the canonical form) are $a_1 = \frac{x}{x^2+1}$ and $a_0 = -\frac{1}{x^2+1}$, we see that these two functions are analytic on $(-1, 1)$, and then, we can use the general statement (Theorem E1 in section 2), which guarantees that the solution series converges on $(-1, 1)$. Another way is to check the convergence of the series directly by applying D'Alembert's test:

$$\frac{|x^{2n+2}| \cdot 1 \cdot 3 \cdot \ldots \cdot (2n-3)(2n-1)}{4 \cdot 6 \cdot \ldots \cdot 2n(2n+2)} \cdot \frac{4 \cdot 6 \cdot \ldots \cdot 2n}{|x^{2n}| \cdot 1 \cdot 3 \cdot \ldots \cdot (2n-3)}$$

$$= x^2\frac{2n-1}{2n+2} \xrightarrow[n\to\infty]{} x^2, \forall x \in \mathbb{R}.$$

Thus, the series converges for $|x| < 1$ and diverges for $|x| > 1$.

We can also see that the series $y_2(x)$ represents the expansion of the function $2(1 + x^2)^{1/2}$ in power series, and, therefore, the general solution can also be expressed in elementary functions: $y_g(x) = C_1 y_1 + C_2 y_2 = C_1 x + C_2 (1 + x)^{1/2}$.

5. The second order homogeneous equation $(x^2 - 4)y'' + 3xy' + y = 0$. This is a homogeneous linear equation with variable coefficients that can be solved by the power series method. Since there are only two singular points $x = \pm 2$, the expansion of the solution in series $y(x) = \sum_{n=0}^{\infty} c_n x^n$ has the radius of convergence greater than or equal to 2. Substituting this power series into the equation, we get:

$$(x^2 - 4)\sum_{n=2}^{\infty} n(n-1)c_n x^{n-2} + 3x\sum_{n=1}^{\infty} nc_n x^{n-1} + \sum_{n=0}^{\infty} c_n x^n$$

$$= \sum_{n=0}^{\infty} n(n-1)c_n x^n - 4\sum_{n=0}^{\infty}(n+2)(n+1)c_{n+2}x^n + 3\sum_{n=0}^{\infty} nc_n x^n + \sum_{n=0}^{\infty} c_n x^n$$

$$= \sum_{n=0}^{\infty}[(n^2 + 2n + 1)c_n - 4(n+2)(n+1)c_{n+2}]x^n = 0.$$

Therefore, we obtain the following recurrence relation for the coefficients:

$$c_{n+2} = \frac{n+1}{4(n+2)}c_n, \forall n \geq 0.$$

This set of relations can be decoupled into two groups – the even-indexed terms and the odd-indexed ones. For the first group we have

$$c_{2n} = \frac{1}{4}\frac{2n-1}{2n}c_{2n-2} = \frac{1}{4^2}\frac{(2n-1)(2n-3)}{2n \cdot (2n-2)}c_{2n-4} = \ldots = \frac{1}{4^n}\frac{(2n-1)!!}{(2n)!!}c_0, \forall n \geq 1.$$

Similarly, for the second group we get

$$c_{2n+1} = \frac{1}{4^n}\frac{(2n)!!}{(2n+1)!!}c_1, \forall n \geq 1.$$

Hence, all the even-indexed coefficients can be expressed in terms of c_0 and all the odd-indexed coefficients in terms of c_1. Consequently, choosing $c_0 = 1$ and $c_1 = 1$, we construct two linearly independent solutions in the power series:

$$y_1 = \sum_{n=0}^{\infty} \frac{1}{4^n}\frac{(2n-1)!!}{(2n)!!}x^{2n} \ , \ y_2 = \sum_{n=0}^{\infty} \frac{1}{4^n}\frac{(2n)!!}{(2n+1)!!}x^{2n+1}.$$

(As usual, it is defined that $(-1)!! = 1$ and $0!! = 1$, which generates for $n = 0$ the first term 1 in the first series and the first term x in the second).

Application of D'Alembert's test shows that both series converge on the interval $(-2, 2)$:

$$\frac{|c_{2n+2}x^{2n+2}|}{|c_{2n}x^{2n}|} = \frac{2n+1}{2n+2} \cdot \frac{x^2}{4} \underset{n\to\infty}{\to} \frac{x^2}{4} < 1$$

and

$$\frac{|c_{2n+1}x^{2n+1}|}{|c_{2n-1}x^{2n-1}|} = \frac{2n}{2n+1} \cdot \frac{x^2}{4} \underset{n\to\infty}{\to} \frac{x^2}{4} < 1.$$

Therefore, the linear combination of these two series, which represents the general solution of the equation, also converges on $(-2, 2)$.

Thus, the general solution can be represented in the form

$$y = C_1 y_1 + C_2 y_2 = C_1 \sum_{n=0}^{\infty} \frac{1}{4^n} \frac{(2n-1)!!}{(2n)!!} x^{2n} + C_2 \sum_{n=0}^{\infty} \frac{1}{4^n} \frac{(2n)!!}{(2n+1)!!} x^{2n+1},$$

where C_1, C_2 are arbitrary real constants.

Remark. As was noted, it may be difficult or impossible to find the general form of coefficients of a power series. This may happen when the recurrence relation becomes complicated because of involvement of coefficients with distant indices and because of the sophisticated formulas for coefficients. In the next example, this happens because the indices of the coefficients are far apart. In this case, as an alternative, we find a few first coefficients of the series, and use the first terms of the series to represent an approximation of the exact solution near the development point.

Lacking a general formula for the coefficients, we cannot verify the convergence of found series by a convergence test. However, we can still use Theorem E1 or Theorem E2 (section 2) to establish the convergence.

6. The second order homogeneous equation $y'' - (1+x)y = 0$.

This is a homogeneous linear equation with variable coefficients, whose solution requires the use of a power series: $y(x) = \sum_{n=0}^{\infty} c_n x^n$. To find the values of c_n we substitute the series into the original equation

$$\sum_{n=2}^{\infty} n(n-1)c_n x^{n-2} - (1+x) \sum_{n=0}^{\infty} c_n x^n = 0$$

or, standardizing the powers in the three series,

$$\sum_{m=0}^{\infty} (m+2)(m+1)c_{m+2}x^m - \sum_{n=0}^{\infty} c_n x^n - \sum_{k=1}^{\infty} c_{k-1}x^k = 0.$$

So, we get

$$2c_2 - c_0 = 0; \quad (n+2)(n+1)c_{n+2} - c_n - c_{n-1} = 0, \forall n = 1, 2, \ldots$$

or

$$c_2 = \frac{c_0}{2}; \quad c_{n+2} = \frac{c_n + c_{n-1}}{(n+2)(n+1)}, \forall n = 1, 2, \ldots.$$

In this case, the recurrence relation involves the distant indices $n+2$ and $n-1$ and cannot be decoupled, which makes it difficult to find general formulas for c_n.

What we can do is to specify the first coefficients of two linearly independent solutions. For the first solution, we choose $c_0 = 1$ and $c_1 = 0$, which will give $c_2 = \frac{1}{2}, c_3 = \frac{c_1+c_0}{2\cdot3} = \frac{1}{6}, c_4 = \frac{c_2+c_1}{3\cdot4} = \frac{1}{24}$, etc. Therefore, the first solution has the form

$$y_1 = 1 + \frac{2}{2}x^2 + \frac{1}{6}x^3 + \frac{1}{24}x^4 + \cdots .$$

For the second one, choosing $c_0 = 0$ and $c_1 = 1$, we get $c_2 = 0, c_3 = \frac{c_1+c_0}{2\cdot3} = \frac{1}{6}, c_4 = \frac{c_2+c_1}{3\cdot4} = \frac{1}{12}$, etc. Therefore, the second solution takes the form

$$y_2 = x + \frac{1}{6}x^3 + \frac{1}{12}x^4 + \cdots .$$

Lacking a general formula for the coefficients, we cannot verify the convergence of the two obtained series by a convergence test (such as the ratio test). However, to justify the procedures applied we can still use Theorem E1 (section 2). Notice that the coefficients of the original equation (which is already in canonical form) are analytic functions in $\mathbb{R}$, and therefore, both series also converge in $\mathbb{R}$. The general solution, as usual, is found via a linear combination of linearly independent solutions: $y_g(x) = C_1y_1 + C_2y_2$.

7. The second order homogeneous equation $(1-x^2)y'' - 2xy' + p(p+1)y = 0$, $p \in \mathbb{R}$ (the Legendre equation).
This is a homogeneous linear equation with variable coefficients that can be solved by the power series method. Since the only singular points are $x = \pm 1$, the power series solution has the radius of convergence (about $x = 0$) greater than or equal to 1. We note that it is sufficient to consider only the case $p > -1$, since for $p \le -1$ we can perform the substitution $q = -(1+p)$, which leads to the Legendre equation $(1-x^2)y'' - 2xy' + q(q+1)y = 0$ with $q \ge 0$. Substituting the power series $y(x) = \sum_{n=0}^{\infty} c_n x^n$ into the differential equation, we get:

$$(1-x^2) \sum_{n=2}^{\infty} n(n-1)c_n x^{n-2} - 2x \sum_{n=1}^{\infty} nc_n x^{n-1} + p(p+1) \sum_{n=0}^{\infty} c_n x^n$$

$$= \sum_{n=0}^{\infty}(n+2)(n+1)c_{n+2}x^n - \sum_{n=2}^{\infty} n(n-1)c_n x^n - 2\sum_{n=1}^{\infty} nc_n x^n + p(p+1)\sum_{n=0}^{\infty} c_n x^n$$

$$= \sum_{n=0}^{\infty}[(n+2)(n+1)c_{n+2} - (n^2 + n - p(p+1))c_n]x^n = 0.$$

(The second and third series in the middle row can be used with the index of the sum starting from 0, because $n(n-1)c_n = 0$ for $n = 0, 1$ and $nc_n = 0$ for $n = 0$.) So, we have the following recurrence relation for the coefficients c_n:

$$(n+2)(n+1)c_{n+2} - (n(n+1) - p(p+1))c_n = 0, \forall n \geq 0,$$

or equivalently,

$$c_{n+2} = \frac{(n(n+1) - p(p+1))}{(n+2)(n+1)}c_n = \frac{(n-p)(n+p+1)}{(n+2)(n+1)}c_n, \forall n \geq 0.$$

This set of relations can be decoupled into two groups – with the even-numbered coefficients and with the odd-numbered ones. The formula for the first group can be written as follows: $c_{2n} = \frac{(2n-2-p)(2n-1+p)}{2n(2n-1)}c_{2n-2}, \forall n \geq 1$. These coefficients are determined by a choice of c_0 according to the following formula:

$$c_{2n} = \frac{(2n-2-p)(2n-1+p)}{2n(2n-1)}c_{2n-2}$$

$$= \frac{(2n-2-p)(2n-1+p)(2n-4-p)(2n-3+p)}{2n(2n-1)(2n-2)(2n-3)}c_{2n-4} = \dots$$

$$= \frac{(2n-2-p)(2n-4-p)\cdot \dots \cdot (2-p)(-p) \cdot (2n-1+p)\cdot \dots \cdot (3+p)(1+p)}{2n(2n-1)\cdot \dots \cdot 2 \cdot 1}c_0.$$

Similarly, the coefficients of the second group satisfy the relation $c_{2n+1} = \frac{(2n-1-p)(2n+p)}{(2n+1)2n}c_{2n-1}, \forall n \geq 1$ and are uniquely defined by a choice of c_1:

$$c_{2n+1}=\frac{(2n-1-p)(2n-3-p)\cdot \dots \cdot (3-p)(1-p) \cdot (2n+p)\cdot \dots \cdot (4+p)(2+p)}{(2n+1)(2n)\cdot \dots \cdot 2 \cdot 1}c_1.$$

Therefore, the two linearly independent solutions of the Legendre equation are as follows:

$$y_1 = \sum_{n=0}^{\infty} c_{2n}x^{2n}, \quad y_2 = \sum_{n=0}^{\infty} c_{2n+1}x^{2n+1}.$$

According to the theory of analytic solutions (Theorem E1 in section 2), both series converge at least on the interval $(-1, 1)$, which justifies the transformations performed in this interval. The same convergence result can be obtained using D'Alembert's test:

$$\frac{|c_{2n+2}x^{2n+2}|}{|c_{2n}x^{2n}|} = \frac{(2n-p)(2n+1+p)}{(2n+2)(2n+1)} \cdot x^2 \underset{n\to\infty}{\to} x^2 < 1$$

and

$$\frac{|c_{2n+1}x^{2n+1}|}{|c_{2n-1}x^{2n-1}|} = \frac{(2n-1-p)(2n+p)}{(2n+1)2n} \cdot x^2 \underset{n\to\infty}{\to} x^2 < 1.$$

Therefore, the linear combination of these two series $y = C_1 y_1 + C_2 y_2$, which represents the general solution of the Legendre equation, also converges on $(-1, 1)$. Here, the arbitrary constants C_1 and C_2 represent the first two coefficients c_0 and c_1 in the power series solution.

We note that if $p = 0$ or $p \in \mathbb{N}$, then one of the two particular solutions y_1 and y_2 are polynomials of degree p. Indeed, if p is even, then $c_{2n} = 0, \forall 2n > p$, and consequently, y_1 is a polynomial of degree p (while y_2 is an infinite series). If p is odd, then $c_{2n+1} = 0, \forall (2n+1) > p$, and consequently, y_2 is a polynomial of degree p (while y_1 is an infinite series). Finally, if $p = 0$, then the recurrence relation for the even-numbered coefficients simplifies to the form $c_{2n} = \frac{n-1}{n} c_{2n-2}, \forall n \geq 1$. For $n = 1$ we have $c_2 = 0$, which vanishes all the even-numbered coefficients, leaving only the arbitrary value c_0. For the odd-numbered coefficients we have $c_{2n+1} = \frac{2n-1}{2n+1} c_{2n-1} = \ldots = \frac{1}{2n+1} c_1, \forall n \geq 1$. Therefore, the general solution has the form

$$y = c_0 + c_1 \sum_{n=1}^{\infty} \frac{x^{2n+1}}{2n+1} = c_0 + c_1 \cdot \frac{1}{2} \ln \frac{1-x}{1+x}.$$

This solution can also be obtained using the order reduction method. In fact, by introducing the function $z = y'$, we transform the equation $(1 - x^2)y'' - 2xy' = 0$ into the first order equation $(1 - x^2)z' - 2xz = 0$. The last equation is a separable one, whose solution is obtained by integrating $\int \frac{dz}{z} = \int \frac{2x\,dx}{1-x^2}$, whence $\ln z = -\ln(1-x^2) + C$ or, eliminating the logarithm, $z = \frac{C}{1-x^2}$. Then, $y' = \frac{C}{1-x^2}$ and, integrating once more, we find

$$y = C \int \frac{dx}{1-x^2} = \frac{C}{2} \ln \frac{1+x}{1-x} + B.$$

8. The second order homogeneous equation $(1 - x^2)y'' - xy' + p^2 y = 0$, $p \in \mathbb{R}$ (the Chebyshev equation).
This is a homogeneous linear equation with variable coefficients that can be solved by applying the power series method. Since the only singular

points are $x = \pm 1$, the power series solution has the radius of convergence (about $x = 0$) greater than or equal to 1. Substituting the power series $y(x) = \sum_{n=0}^{\infty} c_n x^n$ into the differential equation, we get:

$$(1 - x^2) \sum_{n=2}^{\infty} n(n-1)c_n x^{n-2} - x \sum_{n=1}^{\infty} nc_n x^{n-1} + p^2 \sum_{n=0}^{\infty} c_n x^n$$

$$= \sum_{n=0}^{\infty} (n+2)(n+1)c_{n+2} x^n - \sum_{n=2}^{\infty} n(n-1)c_n x^n - \sum_{n=1}^{\infty} nc_n x^n + p^2 \sum_{n=0}^{\infty} c_n x^n$$

$$= \sum_{n=0}^{\infty} [(n+2)(n+1)c_{n+2} - (n^2 - p^2)c_n] x^n = 0.$$

(The second and third series in the middle row can be used with the index of the sum starting from 0, because $n(n-1)c_n = 0$ for $n = 0, 1$ and $nc_n = 0$ for $n = 0$.) So, we have the following recurrence relation for the coefficients:

$$(n+2)(n+1)c_{n+2} - (n^2 - p^2))c_n = 0, \forall n \geq 0,$$

or equivalently,

$$c_{n+2} = \frac{n^2 - p^2}{(n+2)(n+1)} c_n, \forall n \geq 0.$$

This set of relations can be decoupled into two groups – with the even-indexed and with odd-indexed terms. The coefficients of the first group are determined by choosing c_0 as follows:

$$c_{2n} = \frac{(2n-2)^2 - p^2}{2n(2n-1)} c_{2n-2} = \ldots = \frac{((2n-2)^2 - p^2) \cdot \ldots \cdot (2^2 - p^2) \cdot (-p^2)}{(2n)!} c_0.$$

Similarly, the coefficients of the second group are defined by choosing c_1:

$$c_{2n+1} = \frac{(2n-1)^2 - p^2}{(2n+1)2n} c_{2n-1} = \ldots = \frac{((2n-1)^2 - p^2) \cdot \ldots \cdot (3^2 - p^2) \cdot (1^2 - p^2)}{(2n+1)!} c_1.$$

Therefore, the two linearly independent solutions of the Chebyshev equation are as follows:

$$y_1 = \sum_{n=0}^{\infty} c_{2n} x^{2n}, \quad y_2 = \sum_{n=0}^{\infty} c_{2n+1} x^{2n+1}.$$

According to the theory of analytic solutions (Theorem E1 in section 2), both series converge at least on the interval $(-1, 1)$, which justifies the transformations performed in this interval. The same convergence result can be obtained by applying D'Alembert's test:

$$\frac{|c_{2n+2}x^{2n+2}|}{|c_{2n}x^{2n}|} = \frac{(2n)^2 - p^2}{(2n+2)(2n+1)} \cdot x^2 \underset{n \to \infty}{\to} x^2 < 1$$

and

$$\frac{|c_{2n+1}x^{2n+1}|}{|c_{2n-1}x^{2n-1}|} = \frac{(2n-1)^2 - p^2}{(2n+1)2n} \cdot x^2 \underset{n \to \infty}{\to} x^2 < 1.$$

Therefore, the linear combination of these two series $y = C_1 y_1 + C_2 y_2$, which represents the general solution of the Chebyshev equation, also converges on $(-1, 1)$. Here, the arbitrary constants C_1 and C_2 represent the first two coefficients c_0 and c_1 in the power series solution.

We note that if $|p| \in \mathbb{N}$, one of the particular solutions y_1 or y_2 is a polynomial of degree $|p|$. Indeed, if $|p|$ is even, then $c_{2n} = 0, \forall 2n > |p|$, and consequently, y_1 is a polynomial of degree $|p|$ (while y_2 is an infinite series). If $|p|$ is odd, then $c_{2n+1} = 0, \forall (2n+1) > |p|$ and y_2 is a polynomial of degree $|p|$ (while y_1 is an infinite series). Finally, if $p = 0$, then the recurrence relation for the even indices simplifies to the form

$$c_{2n} = \frac{2(n-1)^2}{n(2n-1)} c_{2n-2}, \forall n \geq 1.$$

For $n = 1$ we have $c_2 = 0$, which vanishes all the even-indexed coefficients, leaving only the arbitrary value c_0. For the odd indices we have

$$c_{2n+1} = \frac{(2n-1)^2}{(2n+1)2n} c_{2n-1} = \ldots = \frac{((2n-1)!!)^2}{(2n+1)!} c_1 = \frac{(2n-1)!!}{(2n+1) \cdot 2^n \cdot n!} c_1, \forall n \geq 1.$$

Therefore, the general solution takes the form

$$y = c_0 + c_1 \sum_{n=0}^{\infty} \frac{(2n-1)!!}{(2n+1) \cdot 2^n \cdot n!} x^{2n+1} = c_0 + c_1 \arcsin x.$$

This solution can also be obtained using the order reduction method. In fact, by introducing the function $z = y'$, we transform the equation $(1 - x^2)y'' - xy' = 0$ into the first-order equation $(1 - x^2)z' - xz = 0$. The last equation is a separable one, whose solution is obtained by integrating $\int \frac{dz}{z} = \int \frac{x dx}{1-x^2}$, whence $\ln z = -\frac{1}{2} \ln(1-x^2) + C$ or, eliminating the logarithm, $z = \frac{C}{\sqrt{1-x^2}}$. Then $y' = \frac{C}{\sqrt{1-x^2}}$ and, integrating once more, we find

$$y = C \int \frac{dx}{\sqrt{1 - x^2}} = C \arcsin x + B.$$

9. The second order nonhomogeneous equation $y'' - xy' = 12x^3$.
This equation is linear with variable coefficients of the type (2.2). Both the coefficients of the equation and the right-hand side are analytic functions in $\mathbb{R}$. We look for the solution of the problem in the form of a power series $y(x) = \sum_{n=0}^{\infty} c_n x^n$. Substituting this series into the original equation, we get

$$\sum_{n=2}^{\infty} n(n-1)c_n x^{n-2} - x\sum_{n=1}^{\infty} nc_n x^{n-1} = 12x^3$$

or

$$2c_2 + \sum_{n=1}^{\infty} ((n+2)(n+1)c_{n+2} - nc_n)\, x^n = 12x^3.$$

Equating the coefficients with the same powers of x, we have:

$$2c_2 = 0,\ \ 6c_3 - c_1 = 0,\ \ 12c_4 - 2c_2 = 0,\ \ 20c_5 - 3c_3 = 12$$

for the first indices and

$$(n+2)(n+1)c_{n+2} - nc_n = 0$$

for the remaining indices. The recurrence relation can be decoupled into two groups – the odd-indexed and even-indexed terms. Since $c_2 = 0$, the recurrence relation implies that all the even-indexed coefficients are zero, except for c_0 which has an arbitrary value. For the odd-indexed coefficients we have $c_3 = \frac{1}{6}c_1$, $c_5 = \frac{1}{20}(\frac{1}{2}c_1 + 12)$ and

$$c_{2k+1} = \frac{2k-1}{(2k+1)2k}c_{2k-1} = \frac{(2k-1)(2k-3)}{(2k+1)2k(2k-1)(2k-2)}c_{2k-3} = \ldots$$

$$= \frac{(2k-1)(2k-3)\cdot\ldots\cdot 7\cdot 5}{(2k+1)2k(2k-1)(2k-2)\cdot\ldots\cdot 7\cdot 6}c_5$$

$$= \frac{(2k-1)(2k-3)\cdot\ldots\cdot 7\cdot 5}{(2k+1)2k(2k-1)(2k-2)\cdot\ldots\cdot 7\cdot 6}\cdot\frac{3}{5}\cdot\frac{1}{4\cdot 3\cdot 2}\cdot 24$$

$$+ \frac{(2k-1)(2k-3)\cdot\ldots\cdot 7\cdot 5}{(2k+1)2k(2k-1)(2k-2)\cdot\ldots\cdot 7\cdot 6}\cdot\frac{3}{5\cdot 4\cdot 3\cdot 2}c_1$$

$$= \frac{(2k-1)!!}{(2k+1)!}\cdot 24 + \frac{(2k-1)!!}{(2k+1)!}c_1 = \frac{24}{(2k+1)2^k k!} + \frac{c_1}{(2k+1)2^k k!}, \forall k > 2.$$

Therefore, the general solution has the form

$$y(x) = c_0 + c_1\left[x + \frac{1}{6}x^3 + \sum_{n=2}^{\infty}\frac{1}{(2n+1)2^n n!}x^{2n+1}\right] + 24\sum_{n=2}^{\infty}\frac{x^{2n+1}}{(2n+1)2^n n!}.$$

Notice that the term c_0 is the first particular solution and the series with the multiplier c_1 is the second particular solution of the corresponding homogeneous equation, and the last term (the second series) is the particular solution of the nonhomogeneous equation. It is straightforward to show that both series converge in $\mathbb{R}$.

10. The second order nonhomogeneous equation $y'' - xy' = e^x$.
This equation is linear with variable coefficients of the type (2.2). Both the coefficients of the equation and the right-hand side are analytic functions in $\mathbb{R}$. We look for the solution of the problem in the form of a power series $y(x) = \sum_{n=0}^{\infty} c_n x^n$. Substituting this series into the original equation and using the power series of e^x, we get

$$\sum_{n=2}^{\infty} n(n-1)c_n x^{n-2} - x \sum_{n=1}^{\infty} n c_n x^{n-1} = \sum_{n=0}^{\infty} \frac{1}{n!} x^n$$

or

$$2c_2 + \sum_{n=1}^{\infty} \left((n+2)(n+1)c_{n+2} - nc_n\right) x^n = 1 + \sum_{n=1}^{\infty} \frac{1}{n!} x^n.$$

Equating the coefficients with the same powers of x, we have the first relation $2c_2 = 1$ and the others in the form

$$(n+2)(n+1)c_{n+2} - nc_n = \frac{1}{n!}$$

or

$$c_{n+2} = \frac{1}{(n+2)!} + \frac{n}{(n+2)(n+1)} c_n.$$

Developing this relation, we obtain

$$c_{n+2} = \frac{1}{(n+2)!} + \frac{n}{(n+2)(n+1)} \left[\frac{1}{n!} + \frac{n-2}{n(n-1)} c_{n-2} \right]$$

$$= \frac{1+n}{(n+2)!} + \frac{n(n-2)}{(n+2)(n+1)n(n-1)} c_{n-2}$$

$$= \frac{1+n}{(n+2)!} + \frac{n(n-2)}{(n+2)(n+1)n(n-1)} \left[\frac{1}{(n-2)!} + \frac{n-4}{(n-2)(n-3)} c_{n-4} \right]$$

$$= \frac{1+n+n(n-2)}{(n+2)!} + \frac{n(n-2)(n-4)}{(n+2)\cdot\ldots\cdot(n-3)} c_{n-4}$$

$$= \frac{1+n+n(n-2)+n(n-2)(n-4)}{(n+2)!} + \frac{n(n-2)(n-4)(n-6)}{(n+2)\cdot\ldots\cdot(n-5)} c_{n-6} = \ldots.$$

Now it is clear that for the even indices we have

$$c_{2k+2} = \frac{1 + \sum_{i=1}^{k-1} 2^i \prod_{j=1}^{i}(k-i+j)}{(2k+2)!} + \frac{(2k)!!}{(2k+2)!}c_2 \equiv p_{2k+2} + \frac{(2k)!!}{(2k+2)!}c_2$$

and for the odd indices

$$c_{2k+1} = \frac{1 + \sum_{i=1}^{k-1} 2^i \prod_{j=i}^{i}(k-i+j)}{(2k+1)!} + \frac{(2k-1)!!}{(2k+1)!}c_1 \equiv p_{2k+1} + \frac{(2k-1)!!}{(2k+1)!}c_1.$$

The coefficient c_0 is arbitrary (there is no condition for it), as well as the coefficient c_1. Given that $c_2 = \frac{1}{2}$, we obtain the general solution of the equation in the form

$$y(x) = c_0 + c_1 x + \frac{1}{2}x^2 + \sum_{k=1}^{\infty} c_{2k+2}x^{2k+2} + \sum_{k=1}^{\infty} c_{2k+1}x^{2k+1}$$

$$= c_0 + c_1 x + \frac{1}{2}x^2 + \sum_{k=1}^{\infty}\left(p_{2k+2} + \frac{(2k)!!}{2(2k+2)!}\right)x^{2k+2} + \sum_{k=1}^{\infty}\left(p_{2k+1} + \frac{(2k-1)!!}{(2k+1)!}c_1\right)x^{2k+1}$$

$$= c_0 + c_1\left[x + \sum_{k=1}^{\infty}\frac{(2k-1)!!}{(2k+1)!}x^{2k+1}\right]$$

$$+ \left[\frac{1}{2}x^2 + \sum_{k=1}^{\infty}\left(p_{2k+2} + \frac{(2k)!!}{2(2k+2)!}\right)x^{2k+2} + \sum_{k=1}^{\infty}p_{2k+1}x^{2k+1}\right],$$

where p_{2k+2} and p_{2k+1} are found by the above specified formulas. Notice that the term c_0 is the first particular solution of the respective homogeneous equation, the term together with c_1 is the second particular solution of the homogeneous equation, and the remaining terms represent the particular solution of the nonhomogeneous equation. It is straightforward to show that all the series converge in $\mathbb{R}$.

11. Cauchy problem: the homogeneous equation $(1 - x^2)y'' - xy' = 0$ of the second order with the initial conditions $y(0) = 0, y'(0) = 1$.
The equation of this problem is linear with variable coefficients, which indicates that it can be solved using the power series method. Looking for the solution in the form $y(x) = \sum_{n=0}^{\infty} c_n x^n$ with undetermined coefficients c_n, we first substitute this series into the initial conditions to specify the coefficients $c_0 = 0$ and $c_1 = 1$, and then substitute it into the equation:

$$(1 - x^2)\sum_{n=2}^{\infty} n(n-1)c_n x^{n-2} - x\sum_{n=1}^{\infty} nc_n x^{n-1}$$

$$= \sum_{n=2}^{\infty} n(n-1)c_n x^{n-2} - \sum_{n=2}^{\infty} n(n-1)c_n x^n - \sum_{n=1}^{\infty} nc_n x^n$$

$$= 2c_2 + 6c_3 x - c_1 x + \sum_{n=2}^{\infty} [(n+2)(n+1)c_{n+2} - n(n-1)c_n - nc_n]x^n = 0.$$

Therefore, we get the following relations between the coefficients c_n:

$$c_2 = 0; \quad 2 \cdot 3 \cdot c_3 - c_1 = 0; \quad (n+2)(n+1)c_{n+2} - n^2 c_n = 0, \forall n \geq 2.$$

The recurrence relation can be divided into two groups – with the even-numbered coefficients $n = 2k$, $k \geq 1$ and with the odd-numbered coefficients $n = 2k+1$, $k \geq 1$. Since $c_2 = 0$, all the coefficients in the first group are zero. For the second group, we start with $c_3 = \frac{1}{2 \cdot 3} c_1 = \frac{1}{2 \cdot 3}$ and apply the recurrence relation:

$$c_5 = \frac{3^2}{4 \cdot 5} c_3 = \frac{3}{2 \cdot 4 \cdot 5}, \quad c_7 = \frac{5^2}{6 \cdot 7} c_5 = \frac{3 \cdot 5}{2 \cdot 4 \cdot 6} \cdot \frac{1}{7}, \ldots,$$

$$c_{2n-1} = \frac{3 \cdot 5 \cdot \ldots \cdot (2n-3)}{2 \cdot 4 \cdot \ldots \cdot (2n-2)} \cdot \frac{1}{2n-1} = \frac{(2n-3)!!}{(2n-2)!!} \cdot \frac{1}{2n-1}.$$

Thus, we construct the solution in the form

$$y = \sum_{n=0}^{\infty} \frac{(2n-1)!!}{(2n)!!} \frac{x^{2n+1}}{2n+1},$$

which represents the expansion of $\arcsin x$ in the power series convergent on $(-1, 1)$.

In this specific case, we can derive the same solution without involving power series. In fact, by replacing the unknown function with the formula $z = y'$, we reduce the original equation to the first order separable equation $(1 - x^2)z' - xz = 0$. Its solution is immediately found by direct integration $\int \frac{dz}{z} = \int \frac{x}{1-x^2} dx$, that gives the solution $z = \frac{C}{\sqrt{1-x^2}}$. The constant C is defined from the second initial condition: $z(0) = y'(0) = \frac{C}{1} = 1$, whence $C = 1$. Integrating the relation $y' = z = \frac{1}{\sqrt{1-x^2}}$ once again, we arrive at the solution of the original equation: $y = \int \frac{dx}{\sqrt{1-x^2}} = \arcsin x + B$. The constant B is found by applying the first initial condition: $y(0) = \arcsin 0 + B = 0$, whence $B = 0$. Thus, the solution of the problem is $y = \arcsin x$, the same as obtained by the power series method.

12. Cauchy problem: the homogeneous equation $(1-x^2)y'' - 5xy' - 4y = 0$ of the second order with the initial conditions $y(0) = 1, y'(0) = 0$.

The equation of this problem is linear with variable coefficients, which indicates that it can be solved using the power series method. Since there are only two singular points $x = \pm 1$, the expansion of the solution in the series $y(x) = \sum_{n=0}^{\infty} c_n x^n$ has the radius of convergence at least 1. To find the coefficients c_n, we first substitute this series into the initial conditions to specify the coefficients $c_0 = 1$ and $c_1 = 0$, and then substitute it into the equation:

$$(1 - x^2) \sum_{n=2}^{\infty} n(n-1)c_n x^{n-2} - 5x \sum_{n=1}^{\infty} n c_n x^{n-1} - 4 \sum_{n=0}^{\infty} c_n x^n$$

$$= 2c_2 + 2 \cdot 3 c_3 x - 4c_0 - 4c_1 x - 5c_1 x + \sum_{n=2}^{\infty}(n+2)(n+1)c_{n+2} x^n - \sum_{n=2}^{\infty}(n^2 - n + 5n + 4)c_n x^n$$

$$= 2c_2 + 6c_3 x - 4 + \sum_{n=2}^{\infty}[(n+2)(n+1)c_{n+2} - (n+2)^2 c_n] x^n = 0.$$

Consequently, we have the following relations for the coefficients c_n:

$$2c_2 - 4 = 0, c_3 = 0, c_{n+2} = \frac{n+2}{n+1} c_n, \forall n \geq 2.$$

The recurrence relation can be divided into two groups – with the even-numbered coefficients and with the odd-numbered ones. Since $c_3 = 0$, all the odd-numbered coefficients are zero. For the even-numbered ones, we start with $c_2 = 2$ and get:

$$c_4 = \frac{2 \cdot 4}{3}, c_6 = \frac{2 \cdot 4 \cdot 6}{1 \cdot 3 \cdot 5}, \ldots, c_{2n} = \frac{2 \cdot 4 \cdot \ldots \cdot 2n}{1 \cdot 3 \cdot \ldots \cdot (2n-1)} = \frac{(2n)!!}{(2n-1)!!}.$$

Thus, we find the series in the form

$$y = \sum_{n=0}^{\infty} \frac{(2n)!!}{(2n-1)!!} x^{2n}.$$

According to the theory of analytic solutions, this series converges at least on the interval $(-1, 1)$, which justifies the performed transformations in this interval. The interval of convergence can also be determined by D'Alembert's test:

$$\frac{|c_{2n+2} x^{2n+2}|}{|c_{2n} x^{2n}|} = \frac{2n+2}{2n+1} x^2 \xrightarrow[n \to \infty]{} x^2 < 1,$$

which shows that the series converges on $(-1, 1)$.

13. Cauchy problem: the homogeneous equation $y'' + \cos x \cdot y = 0$ of second order together with the initial conditions $y(0) = 1$, $y'(0) = 0$.
The equation of the problem is linear with variable coefficients of the type (2.1) with the coefficient $a_0(x) = \cos x$ which is the analytic function in $\mathbb{R}$. Since the coefficient $\cos x$ is not a polynomial, in order to find the solution in the form of a power series $y(x) = \sum_{n=0}^{\infty} c_n x^n$ we need to use the Cauchy product of series. Substituting the series

$$\cos x = \sum_{n=0}^{\infty} (-1)^n \frac{x^{2n}}{(2n)!} = 1 - \frac{1}{2!} x^2 + \frac{1}{4!} x^4 - \frac{1}{6!} x^6 + \dots$$

and $y(x) = \sum_{n=0}^{\infty} c_n x^n$ in the original equation, we get

$$\sum_{n=2}^{\infty} n(n-1) c_n x^{n-2} + \sum_{n=0}^{\infty} (-1)^n \frac{x^{2n}}{(2n)!} \cdot \sum_{n=0}^{\infty} c_n x^n = 0$$

or

$$\sum_{n=0}^{\infty} (n+2)(n+1) c_{n+2} x^n + \sum_{n=0}^{\infty} e_n x^n = 0,$$

where e_n are the coefficients of the Cauchy product. It follows that

$$(n+2)(n+1) c_{n+2} + e_n = 0, \quad n = 0, 1, \dots.$$

To find the coefficients c_n we need to know expressions for e_n. Since the odd-indexed coefficients of the series for $\cos x$ are zero, the expression for e_n is convenient to separate into two cases - even-indexed and odd-indexed. If $n = 2k$, $k \in \mathbb{N}$, then

$$e_{2k} = c_{2k} - \frac{1}{2!} c_{2k-2} + \frac{1}{4!} c_{2k-4} - \dots + \frac{(-1)^{k-1}}{(2k-2)!} c_2 + \frac{(-1)^k}{(2k)!} c_0.$$

If $n = 2k + 1$, $k \in \mathbb{N}$, then

$$e_{2k+1} = c_{2k+1} - \frac{1}{2!} c_{2k-1} + \frac{1}{4!} c_{2k-3} - \dots + \frac{(-1)^{k-1}}{(2k-2)!} c_3 + \frac{(-1)^k}{(2k)!} c_1.$$

Correspondingly, we separate the relations for c_n into even-indexed and odd-indexed:

$$n = 2k: \quad (2k+2)(2k+1) c_{2k+2} + e_{2k} = 0$$

$$\Rightarrow (2k+2)(2k+1) c_{2k+2} = -\left(c_{2k} - \frac{1}{2!} c_{2k-2} + \dots + \frac{(-1)^{k-1}}{(2k-2)!} c_2 + \frac{(-1)^k}{(2k)!} c_0 \right);$$

$$n = 2k + 1 : \quad (2k + 3)(2k + 2)c_{2k+3} + e_{2k+1} = 0$$

$$\Rightarrow (2k+3)(2k+2)c_{2k+3} = -\left(c_{2k+1} - \frac{1}{2!}c_{2k-1} + \ldots + \frac{(-1)^{k-1}}{(2k-2)!}c_3 + \frac{(-1)^k}{(2k)!}c_1 \right).$$

Thus, we have two separate sets of recurrence relations, for the even- and odd-indexed coefficients. To better see the type of obtained relations, we specify the first formulas for even and odd indices:

$$n = 0 : 2c_2 = -c_0, \quad n = 2 : 12c_4 = -\left(c_2 - \frac{1}{2!}c_0 \right),$$

$$n = 4 : 30c_6 = -\left(c_4 - \frac{1}{2!}c_2 + \frac{1}{4!}c_0 \right), \quad \ldots \; .$$

$$n = 1 : \; 6c_3 = -c_1, \quad n = 3 : \; 20c_5 = -\left(c_3 - \frac{1}{2!}c_1 \right),$$

$$n = 5 : 42c_7 = -\left(c_5 - \frac{1}{2!}c_3 + \frac{1}{4!}c_1 \right), \quad \ldots \; .$$

Starting with the coefficient c_0 and using the first set, we find all the even-indexed coefficients one by one; likewise, knowing c_1 and using the second set, we find all the odd-indexed coefficients. The arbitrary choice of two coefficients corresponds to the determination of two initial conditions.

Now we can apply the two initial conditions. From the condition $y(0) = 1$ it follows that $c_0 = 1$ and from the condition $y'(0) = 0$ we have $c_1 = 0$. Consequently, all the odd-indexed coefficients are zero and the solution of the Cauchy problem can be expressed in the form $y(x) = \sum_{k=0}^{\infty} c_{2k}x^{2k}$, where $c_0 = 1$ and the other coefficients are found successively for $k = 0, 1, \ldots$ employing the recurrence relation

$$c_{2k+2} = -\frac{1}{(2k+2)(2k+1)}\left(c_{2k} - \frac{1}{2!}c_{2k-2} + \ldots + \frac{(-1)^{k-1}}{(2k-2)!}c_2 + \frac{(-1)^k}{(2k)!}c_0 \right).$$

Finally, we note that the coefficient $a_0 = \cos x$ of the original equation is the analytic function in $\mathbb{R}$. So, by the general result (Theorem E1 in section 2), the power series representing the solution of the problem converges on $\mathbb{R}$.

14. Cauchy problem: the nonhomogeneous equation $y'' - xy' + y = 1$ of the second order with the initial conditions $y(0) = 0$, $y'(0) = 0$.
The equation of the problem is linear with variable coefficients of the type (2.2). Both the coefficients of the equation and the right-hand side are analytic functions in $\mathbb{R}$. We look for the solution of the problem in the form

of a power series $y(x) = \sum_{n=0}^{\infty} c_n x^n$. Substituting this series into the original equation, we get

$$\sum_{n=2}^{\infty} n(n-1)c_n x^{n-2} - x \sum_{n=1}^{\infty} n c_n x^{n-1} + \sum_{n=0}^{\infty} c_n x^n = 1$$

or

$$2c_2 + c_0 + \sum_{n=1}^{\infty} \left((n+2)(n+1)c_{n+2} - n c_n + c_n \right) x^n = 1.$$

Equating coefficients with the same powers of x, we have:

$$2c_2 + c_0 = 1; \ (n+2)(n+1)c_{n+2} + (1-n)c_n = 0, \forall n \in \mathbb{N}.$$

Using the initial conditions, we find $c_0 = 0$ and $c_1 = 0$. So, the first equation of the system of coefficients gives $c_2 = \frac{1}{2}$ and the others form the recurrence relation

$$c_{n+2} = \frac{n-1}{(n+2)(n+1)} c_n, \forall n \in \mathbb{N},$$

which can be decoupled into two groups – the odd-indexed starting from c_1 and the even-indexed starting from c_2. Since c_1 and c_2 have already been found, all the coefficients c_n are determined: for the odd-indexed we have $c_{2k-1} = 0, \forall k \in \mathbb{N}$ (since $c_1 = 0$), and for the even-indexed we have

$$c_{2k} = \frac{2k-3}{2k(2k-1)} c_{2k-2} = \frac{(2k-3)(2k-5)}{2k(2k-1)(2k-2)(2k-3)} c_{2k-4} = \dots$$

$$= \frac{(2k-3)(2k-5)\dots 1}{2k(2k-1)(2k-2)(2k-3)\cdot \dots \cdot 4\cdot 3} c_2, \forall k \geq 2.$$

Since $c_2 = \frac{1}{2}$, the last formula becomes

$$c_{2k} = \frac{1\cdot 3\cdot \dots \cdot (2k-5)(2k-3)}{3\cdot 4\cdot \dots \cdot (2k-1)2k} \cdot \frac{1}{2}, \forall k \geq 2.$$

Therefore,

$$y(x) = \frac{x^2}{2} + \sum_{n=2}^{\infty} \frac{1\cdot 3\cdot \dots \cdot (2n-5)(2n-3)}{(2n)!} x^{2n}.$$

Finally, we note that the coefficients $a_1 = -x$ and $a_0 = 1$, as well as the right-hand part $b(x) = 1$ of the original equation are analytic functions in $\mathbb{R}$. So, by the general result (Theorem E2 in section 2), the power series of the obtained solution converges on $\mathbb{R}$. The reader can verify that the same result follows from application of one of the convergence tests.

Exercises for section 3

Verify whether the indicated center point is ordinary and solve in the corresponding power series; find the interval of validity of the obtained solution:
1. Solve $xy' - y - x - 1 = 0$ about the center point 1.
2. Solve $y'' + xy' + y = 0$ about the center point 0.
3. Solve $y'' - xy' - y = 0$ about the center point 0.
4. Solve $y'' + xy' + 2y = 0$, $y(0) = 3, y'(0) = -2$.
5. Solve $y'' + x^2y' + xy = 0$ about the center point 0.
6. Solve $(x - 1)y'' + y' = 0$ about the center point 0.
7. Solve $(x^2 - 1)y'' + 4xy' + 2y = 0$ about the center point 0.
8. Solve $(x - 1)y'' - xy' + y = 0$, $y(0) = -2, y'(0) = 6$.
9. Solve $y'' - 2xy' + 8y = 0$, $y(0) = 3, y'(0) = 0$.
10. Solve $y'' - xy = 1$ about the center point 0.
11. Solve $y'' - xy' + 2y = 0$ about the center point 0.
12. Solve $y'' + 2xy' + 2y = 0$ about the center point 0.
13. Solve $(x^2 + 1)y'' - 6y = 0$ about the center point 0.
14. Solve $(x^2 + 1)y'' + 2xy' = 0$, $y(0) = 0, y'(0) = 1$.

4 Solution about singular points: introductory examples

In the case of a singular point, the algorithm described above does not work. We illustrate the problems that can arise with the two examples.

Example 1. The second order homogeneous equation $x^2y'' + 3y' - xy = 0$. Obviously, $x = 0$ is the singular point of this equation, because in its canonical form (2.1) the coefficients $a_1 = \frac{3}{x^2}$ and $a_0 = -\frac{1}{x}$ are not defined at 0 (without talking about differentiability or analyticity at this point). Even so, let us try to use the formal power series $y(x) = \sum_{n=0}^{\infty} c_n x^n$ (centered at 0). Substituting this series into the equation, we get:

$$x^2 \left(\sum_{n=0}^{\infty} c_n x^n \right)'' + 3 \left(\sum_{n=0}^{\infty} c_n x^n \right)' - x \sum_{n=0}^{\infty} c_n x^n = 0.$$

Assuming that the series has a non-zero radius of convergence, we apply term-by-term differentiation in the interval of convergence to find

$$x^2 \sum_{n=2}^{\infty} n(n-1)c_n x^{n-2} + 3 \sum_{n=1}^{\infty} nc_n x^{n-1} - x \sum_{n=0}^{\infty} c_n x^n = 0$$

or

$$\sum_{n=2}^{\infty} n(n-1)c_n x^n + \sum_{n=0}^{\infty} 3(n+1)c_{n+1}x^n - \sum_{n=1}^{\infty} c_{n-1}x^n = 0.$$

Regrouping the terms, we get

$$3c_1 + (6c_2 - c_0)x + \sum_{n=2}^{\infty} [n(n-1)c_n + 3(n+1)c_{n+1} - c_{n-1}]x^n = 0.$$

From the uniqueness of a power series, the following relations follow:

$$3c_1 = 0, \quad c_2 = \frac{c_0}{6}, \quad c_{n+1} = -\frac{n(n-1)c_n - c_{n-1}}{3(n+1)}, \forall n = 2, 3, \ldots.$$

The only arbitrary parameter here is c_0, that is, all the coefficients c_n are determined uniquely by choosing c_0. Thus, the power series found has only one arbitrary constant, while the second-order equation must have two arbitrary constants in its general solution. This means that not all solutions of the given equation can be represented as a power series.

Example 2. The second order homogeneous equation $x^2 y'' + y = 0$. This is the Euler equation with the singular point $x = 0$. Trying to find the power series solution $y(x) = \sum_{n=0}^{\infty} c_n x^n$ (centered at 0), we substitute this series into the equation and, assuming that the series has a non-zero radius of convergence, we get:

$$x^2 \sum_{n=2}^{\infty} n(n-1)c_n x^{n-2} + \sum_{n=0}^{\infty} c_n x^n = 0$$

or

$$\sum_{n=2}^{\infty} n(n-1)c_n x^n + \sum_{n=0}^{\infty} c_n x^n = 0.$$

Regrouping the terms, we have

$$c_0 + c_1 x + \sum_{n=2}^{\infty} [n(n-1) + 1]c_n x^n = 0.$$

Then, the following relations follow from the uniqueness of a power series:

$$c_0 = 0, \quad c_1 = 0, \quad [n(n-1) + 1]c_n = 0, \forall n = 2, 3, \ldots.$$

Therefore, the only solution obtained in power series is zero, which does not reveal any information about the solutions, because the existence of the zero solution of a homogeneous linear equation is known from the beginning.

However, the general solution of this equation can be found in a simple way by substituting the independent variable $x = e^t$ for $x > 0$ or $x = -e^t$ for $x < 0$. Considering the first option, we have

$$y_x = y_t t_x = y_t \frac{1}{x} = y_t e^{-t}, \ y_{xx} = (y_t e^{-t})_t t_x = (y_{tt} e^{-t} - y_t e^{-t}) e^{-t} = (y_{tt} - y_t) e^{-2t}.$$

So, the original equation is reduced to the linear equation with constant coefficients

$$y_{tt} - y_t + y = 0,$$

whose characteristic equation $\lambda^2 - \lambda + 1 = 0$ has the roots $\lambda_{1,2} = \frac{1 \pm \sqrt{3}i}{2}$. Therefore, the general solution of the reduced equation is

$$y(t) = e^{t/2}(C_1 \cos \sqrt{3}t + C_2 \sin \sqrt{3}t),$$

and consequently, the general solution of the original equation is

$$y(x) = \sqrt{|x|} \left(C_1 \cos(\sqrt{3} \ln |x|) + C_2 \sin(\sqrt{3} \ln |x|) \right).$$

Hence, as we have seen in these examples, the algorithm used for ordinary points does not work in the case of singular points. However, a variation of this method, called the Frobenius method, can be used in the case of regular singular points. A presentation of this method will be given in section 6. For now, we just provide a classification of singular points required in this method.

Definition (classification of singular points). A singular point x_0 is called a *regular singular point* of the equation (2.1) if the functions $(x - x_0)^n a_0(x)$, $(x - x_0)^{n-1} a_1(x)$, ..., $(x - x_0) a_{n-1}(x)$ are analytic at x_0. If x_0 is not a regular singular point, it is called an *irregular singular point*.

Remark. If x_0 is a regular singular point, then the coefficients of (2.1) can be represented in the form $a_0(x) = \frac{1}{(x-x_0)^n} b_0(x)$, $a_1(x) = \frac{1}{(x-x_0)^{n-1}} b_1(x)$, ..., $a_{n-1}(x) = \frac{1}{x-x_0} b_{n-1}(x)$, where $b_0(x), b_1(x), \ldots, b_{n-1}(x)$ are analytic functions at x_0. Notice that, in this definition, if the function $(x-x_0)^{n-k} a_k(x)$ is not defined at x_0, as in the case of $a_k(x) = \frac{1}{(x-x_0)^{n-k}}$, then it is considered the analyticity of the function $b_k(x) = \begin{cases} (x - x_0)^{n-k} a_k(x), & x \neq x_0 \\ b_{k0}, & x = x_0 \end{cases}$, where $b_{k0} = \lim_{x \to x_0} (x - x_0)^{n-k} a_k(x)$.

5   Euler equation

Let us start our study of the Frobenius method of series solution about a regular singular point by considering the specific case of the Euler equation

$$x^n y^{(n)} + p_{n-1} x^{n-1} y^{(n-1)} + \ldots + p_1 x y' + p_0 y = 0, \qquad (5.1)$$

where the coefficients $p_0, p_1, \ldots, p_{n-1}$ are constants. Rewriting the equation in the canonical form (2.1)

$$y^{(n)} + \frac{1}{x} p_{n-1} y^{(n-1)} + \ldots + \frac{1}{x^{n-1}} p_1 y' + \frac{1}{x^n} p_0 y = 0,$$

we immediately notice that the functions $x^n a_0(x) = p_0$, $x^{n-1} a_1(x) = p_1, \ldots,$ $x a_{n-1}(x) = p_{n-1}$ are constant and therefore analytic at $x_0 = 0$. This means that the Euler equation is a special case of equations with a regular singular point $x_0 = 0$.

First, we show how to solve the Euler equation without use of power series. The equation (5.1) can be reduced to an equation with constant coefficients by substituting the independent variable by the formula $x = e^t$ for $x > 0$ or $x = -e^t$ for $x < 0$. Considering the first option, we have $y_x = y_t t_x = y_t \frac{1}{x} = Dy \cdot e^{-t}$. For convenience, we denote the derivative with respect to t by D. Therefore,

$$y_{xx} = D(Dy \cdot e^{-t}) t_x = (D^2 y \cdot e^{-t} - Dy \cdot e^{-t}) e^{-t} = D(D-1) y \cdot e^{-2t},$$

$$y_{3x} = D\left[D(D-1) y \cdot e^{-2t}\right] t_x = \left[D^2(D-1) y \cdot e^{-2t} - 2D(D-1) y \cdot e^{-2t}\right] e^{-t}$$

$$= D(D-1)(D-2) y \cdot e^{-3t},$$

and so on. From this pattern of the form of the first derivatives, we can suggest that $y_{kx} = D(D-1) \cdot \ldots \cdot (D-k+1) y \cdot e^{-kt}$. Let us prove this formula by induction, deriving the formula for $y_{(k+1)x}$:

$$y_{(k+1)x} = D\left[D(D-1) \cdot \ldots \cdot (D-k+1) y \cdot e^{-kt}\right] t_x$$

$$= \left[D^2(D-1) \cdot \ldots \cdot (D-k+1) y \cdot e^{-kt} - kD(D-1) \cdot \ldots \cdot (D-k+1) y \cdot e^{-kt}\right] e^{-t}$$

$$= D(D-1) \cdot \ldots \cdot (D-k+1)(D-k) y \cdot e^{-(k+1)t}.$$

Thus, the assumption is proved. Therefore, the Euler equation can be reduced to the equation with constant coefficients for the function $y(t)$, whose characteristic equation is

$$\lambda(\lambda-1) \cdot \ldots \cdot (\lambda-n+1) + p_{n-1} \lambda(\lambda-1) \cdot \ldots \cdot (\lambda-n+2) + \ldots + p_2 \lambda(\lambda-1) + p_1 \lambda + p_0 = 0.$$

Finding the roots of this equation, we construct n linearly independent particular solutions and form the general solution as their linear combination. Returning from t to x in this solution, we find the general solution of the Euler equation. In this way, the Euler equation can always be solved completely (at least theoretically).

Remark. The case of the Euler equation with the singular point x_0

$$(x - x_0)^n y^{(n)} + p_{n-1}(x - x_0)^{n-1} y^{(n-1)} + \ldots + p_1(x - x_0)y' + p_0 y = 0,$$

can be reduced to the equation (5.1) for the new function $y(s)$ using the change of the independent variable $s = x - x_0$.

Let us specify the presented algorithm of solution in the case of the second order Euler equation

$$x^2 y'' + pxy' + qy = 0, x > 0 \tag{5.2}$$

where p and q are constants. By changing the independent variable $x = e^t$, we obtain the equation for $y(t)$ in the form $y_{tt} + (p-1)y_t + qy = 0$. Its characteristic equation $\lambda^2 + (p-1)\lambda + q = 0$ has the roots $\lambda_{1,2} = \frac{1-p \pm \sqrt{(1-p)^2 - 4q}}{2}$. As we know from the theory of equations with constant coefficients, the three situations can occur:

1) if $\lambda_{1,2}$ are real and distinct (when $(1-p)^2 > 4q$), then

$$y = C_1 e^{\lambda_1 t} + C_2 e^{\lambda_2 t} = C_1 x^{\lambda_1} + C_2 x^{\lambda_2};$$

2) if $\lambda_{1,2}$ are complex conjugate (when $(1-p)^2 < 4q$), that is, $\lambda_{1,2} = \alpha \pm i\beta$, then

$$y = e^{\alpha t}(C_1 \cos \beta t + C_2 \sin \beta t) = x^\alpha (C_1 \cos(\beta \ln x) + C_2 \sin(\beta \ln x));$$

3) if $\lambda_{1,2} = \lambda$ are real and equal (when $(1-p)^2 = 4q$), then

$$y = e^{\lambda t}(C_1 + C_2 t) = x^\lambda(C_1 + C_2 \ln x).$$

From this result we can clearly see that, unless both roots $\lambda_{1,2}$ are natural numbers, there is no way to find the solution of the equation (5.2) in the form of a power series $y(x) = \sum_{n=0}^{\infty} c_n x^n$ (centered at the singular point $x = 0$), since the solutions of the equation are not analytic at the origin. However, in the first case, if $\lambda_{1,2}$ are not natural, we can look for the solution in the form of a generalized series $\sum_{n=0}^{\infty} c_n x^{n+\lambda} = x^\lambda \sum_{n=0}^{\infty} c_n x^n$, assuming that the factor x^λ will absorb the non-analytic part. In the second case, the appropriate forms of solutions could be $x^\alpha \cos(\beta \ln x) \sum_{n=0}^{\infty} c_n x^n$ and $x^\alpha \sin(\beta \ln x) \sum_{n=0}^{\infty} c_n x^n$. Considering the factors $x^\alpha \cos(\beta \ln x)$ and $x^\alpha \sin(\beta \ln x)$ as the real and imaginary parts of $x^{\alpha+i\beta}$, we can again look for the solution in the form $x^\lambda \sum_{n=0}^{\infty} c_n x^n$, where $\lambda = \alpha + i\beta$. In the third case, one of the particular solutions can be found in the form $x^\lambda \sum_{n=0}^{\infty} c_n x^n$, but the second solution contains the term $\ln x$ which does not fit into this form. Nevertheless, this second solution can be obtained from the first by differentiating it with respect to λ: $(x^\lambda)_\lambda = x^\lambda \ln x$. In this way, in all cases, we can find a solution in the form of a generalized series $x^\lambda \sum_{n=0}^{\infty} c_n x^n$, with different subsequent treatment depending on the case considered.

Let us apply this procedure to equation (5.2) and compare the series solutions with those obtained by reducing (5.2) to the equation with constant coefficients. Substituting

$$y(x) = x^\lambda \sum_{n=0}^{\infty} c_n x^n = \sum_{n=0}^{\infty} c_n x^{n+\lambda}$$

into the equation, we get

$$x^2 \sum_{n=0}^{\infty} c_n (n+\lambda)(n+\lambda-1) x^{n+\lambda-2} + px \sum_{n=0}^{\infty} c_n (n+\lambda) x^{n+\lambda-1} + q \sum_{n=0}^{\infty} c_n x^{n+\lambda}$$

$$= \sum_{n=0}^{\infty} c_n [(n+\lambda)(n+\lambda-1) + p(n+\lambda) + q] x^{n+\lambda}$$

$$= x^\lambda \sum_{n=0}^{\infty} c_n [(n+\lambda)(n+\lambda-1) + p(n+\lambda) + q] x^n = 0.$$

Due to the uniqueness of a power series, we obtain the following relations:

$$c_n [(n+\lambda)(n+\lambda-1) + p(n+\lambda) + q] = 0, n = 0, 1, \ldots.$$

For $n = 0$ we have (under the condition $c_0 \neq 0$) the so-called *indicial equation* $\lambda(\lambda-1) + p\lambda + q = 0$ or $\lambda^2 + (p-1)\lambda + q = 0$, which determines the values $\lambda_{1,2} = \frac{1-p\pm\sqrt{(1-p)^2-4q}}{2}$ (note that the indicial equation coincides with the characteristic one found for $y(t)$). From the remaining relations for $n = 1, 2, \ldots$, it follows that $c_1 = c_2 = \ldots = 0$, since the expression inside the brackets does not vanish. In fact, for any n, this expression can be written as $\mu^2 + (p-1)\mu + q = 0$ with $\mu = n+\lambda$. The two (unique) roots of this equation are $\lambda_{1,2}$ and, therefore, $\mu_{1,2} = n + \lambda_{1,2}$ cannot satisfy this equation for any $n > 0$. Thus, if $\lambda_1 \neq \lambda_2$, then we arrive at the same linearly independent solutions x^{λ_1} and x^{λ_2} found before in the cases 1) and 2). If $\lambda_1 = \lambda_2$, then we find only one solution, without logarithm, x^{λ_1} from the case 3).

As noted, the Euler equation (5.2) is a very special case of equations with a regular singular point $x = 0$, when the functions $x^2 a_0(x) = q$ and $x a_1(x) = p$ are constant. This results in the singular form of the generalized series solution $y(x) = x^\lambda \sum_{n=0}^{\infty} c_n x^n$, in which all the terms vanish except for a single term of index 0 (then the series actually disappears). Even so, the three different cases of solutions that occur with the Euler equation (5.2), are also observed for a general second order linear equation at a regular singular point x_0.

Examples.

We solve below the Euler equations by a simpler method, reducing them to linear equations with constant coefficients.

1. $x^2y'' + 4xy' + 2y = 0$.
This is the Euler equation with the regular singular point $x = 0$. By changing the independent variable $x = e^t$, we get the equation for $y(t)$ in the form $y_{tt} + 3y_t + 2y = 0$. Its characteristic equation $\lambda^2 + 3\lambda + 2 = 0$ has two distinct real roots $\lambda_1 = -1$ and $\lambda_2 = -2$. Then, the two linearly independent solutions are $y_1(t) = e^{-t}$ and $y_2(t) = e^{-2t}$. Therefore, for variable x, we have $y_1(x) = x^{-1}$ and $y_2(x) = x^{-2}$ and, consequently, the general solution is

$$y(x) = C_1 x^{-1} + C_2 x^{-2}.$$

2. $(x-1)^2 y'' - (x-1)y' + 5y = 0$.
This is the Euler equation with the regular singular point $x = 1$. First, we reduce this equation to the form with the singular point at the origin by changing the variable $s = x - 1$: $s^2 y_{ss} - s y_s + 5y = 0$. Now we follow the standard algorithm and perform the substitution of the variable $s = e^t$, obtaining the equation for $y(t)$: $y_{tt} - 2y_t + 5y = 0$. The characteristic equation $\lambda^2 - 2\lambda + 5 = 0$ has the complex roots $\lambda_{1,2} = 1 \pm 2i$. Therefore, the two linearly independent solutions are $y_1(t) = e^t \cos 2t$ and $y_2(t) = e^t \sin 2t$. Consequently, for variable s, we have $y_1(s) = s \cos(2\ln s)$ and $y_2(s) = s \sin(2\ln s)$. Finally, returning to the original variable x, we have $y_1(x) = (x-1)\cos(2\ln(x-1))$ and $y_2(x) = (x-1)\sin(2\ln(x-1))$. Hence, the general solution is

$$y(x) = (x-1)[C_1 \cos(2\ln(x-1)) + C_2 \sin(2\ln(x-1))].$$

3. $x^2 y'' + 3xy' + y = 0$.
This is the Euler equation with the regular singular point $x = 0$. By changing the independent variable $x = e^t$, we get the equation for $y(t)$ in the form $y_{tt} + 2y_t + y = 0$. The characteristic equation $\lambda^2 + 2\lambda + 1 = 0$ has the two equal roots $\lambda_{1,2} = -1$. Therefore, the two linearly independent solutions are $y_1(t) = e^{-t}$ and $y_2(t) = te^{-t}$. Returning to the variable x, we have $y_1(x) = x^{-1}$ and $y_2(x) = x^{-1}\ln x$ and, consequently, the general solution is

$$y(x) = x^{-1}(C_1 + C_2 \ln x).$$

Exercises for section 5

Solve the following Euler equations and the corresponding initial value problems:

1. $x^2 y'' - 3xy' - 5y = 0$.
2. $x^2 y'' + 5xy' + 3y = 0$.
3. $x^2 y'' - xy' + y = 0$.
4. $x^2 y'' + 9xy' + 16y = 0$.
5. $x^2 y'' + 2xy' - 6y = 0$, $y(1) = 3, y'(1) = 1$.
6. $(x - 3)^2 y'' + 5(x - 3)y' + 4y = 0$, $y(4) = 1, y'(4) = 1$.

6 Solution about singular points: Frobenius method

Let us consider a general linear equation of the second order in the canonical form

$$y'' + a(x)y' + b(x)y = 0. \tag{6.1}$$

with the regular singular point x_0, that is, with analytic functions $(x - x_0)a(x) = \sum_{n=0}^{\infty} a_n (x - x_0)^n$ and $(x - x_0)^2 b(x) = \sum_{n=0}^{\infty} b_n (x - x_0)^n$. In this case, the simple procedure used to solve the Euler equation (5.2) does not apply and a more general approach, called the Frobenius method, must be used. However, as we will see, some important points in the search for solutions of the Euler equation can also be observed in the application of the Frobenius method.

The Frobenius method is based on the following theorem.

Frobenius theorem. Let x_0 be a regular singular point of equation (6.1). Let $\lambda_{1,2}$ be the roots of the *indicial equation* $\lambda(\lambda - 1) + a_0\lambda + b_0 = 0$ such that $\text{Re}\lambda_1 \geq \text{Re}\lambda_2$. Then, the *generalized series*

$$y_1(x) = |x - x_0|^{\lambda_1} \sum_{n=0}^{\infty} c_n (x - x_0)^n$$

is the particular solution of equation (6.1), whose coefficients c_n are determined by substituting this solution into equation (6.1), which leads to the following recurrence relation for $n = 0, 1, 2, \ldots$:

$$c_n[(n+\lambda_1)(n+\lambda_1-1)+a_0(n+\lambda_1)+b_0] + \sum_{m=0}^{n-1} c_m[(m+\lambda_1)a_{n-m}+b_{n-m}] = 0.$$

Finding the second solution, linearly independent of the first, depends on the relationship between the roots λ_1 and λ_2, and can be specified as follows:

1) if $\lambda_1 \neq \lambda_2$ and $\lambda_1 - \lambda_2$ is not an integer, then the second solution can be found in the form

$$y_2(x) = |x - x_0|^{\lambda_2} \sum_{n=0}^{\infty} d_n (x - x_0)^n,$$

where the coefficients d_n, $n = 0, 1, \dots$ are found from the recurrence relation

$$d_n[(n+\lambda_2)(n+\lambda_2-1)+a_0(n+\lambda_2)+b_0] + \sum_{m=0}^{n-1} d_m[(m+\lambda_d)a_{n-m}+b_{n-m}] = 0 \, ;$$

2) if $\lambda_1 = \lambda_2$, then the second solution can be found in the form

$$y_2(x) = y_1(x) \ln|x - x_0| + |x - x_0|^{\lambda_1} \sum_{n=1}^{\infty} d_n (x - x_0)^n,$$

where the coefficients d_n are found by substituting y_2 in (6.1);

3) if $\lambda_1 - \lambda_2$ is an integer (nonzero), then the second solution can be found in the form

$$y_2(x) = cy_1(x) \ln|x - x_0| + |x - x_0|^{\lambda_2} \sum_{n=0}^{\infty} d_n (x - x_0)^n$$

(note that the coefficient c can be zero).

All the above solutions converge on $0 < |x - x_0| < R$, $R \geq \min\{R_a, R_b\}$, where R_a and R_b are the radii of convergence of the series $(x - x_0)a(x) = \sum_{n=0}^{\infty} a_n(x - x_0)^n$ and $(x - x_0)^2 b(x) = \sum_{n=0}^{\infty} b_n(x - x_0)^n$.

Examples.

1. $3xy'' + y' - y = 0$. Find generalized series solutions at the singular points.

This is a homogeneous linear equation with variable coefficients, which has the regular singular point $x = 0$, because the functions

$$xa(x) = x \cdot \frac{1}{3x} = \frac{1}{3} \quad \text{and} \quad x^2 b(x) = x^2 \cdot \frac{-1}{3x} = -\frac{x}{3}$$

are analytic at 0. There are no other singular points of this equation. Therefore, we look for the generalized series solution $y(x) = \sum_{n=0}^{\infty} c_n x^{n+\lambda}$. Substituting into the equation, we get

$$3x \sum_{n=0}^{\infty} c_n(n+\lambda)(n+\lambda-1)x^{n+\lambda-2} + \sum_{n=0}^{\infty} c_n(n+\lambda)x^{n+\lambda-1} - \sum_{n=0}^{\infty} c_n x^{n+\lambda}$$

$$= \sum_{n=0}^{\infty} c_n[3(n+\lambda)(n+\lambda-1)+(n+\lambda)]x^{n+\lambda-1} - \sum_{m=1}^{\infty} c_{m-1}x^{m-1+\lambda}$$

$$= x^{\lambda}\left[c_0\lambda(3\lambda-2)x^{-1} + \sum_{n=1}^{\infty}\{c_n(n+\lambda)(3n+3\lambda-2)-c_{n-1}\}x^{n-1}\right] = 0.$$

The choice $c_0 = 0$ gives the zero solution, which is of no interest. So, for $n = 0$ we have the indicial equation $\lambda(3\lambda-2) = 0$ which determines λ: $\lambda_1 = 0$, $\lambda_2 = \frac{2}{3}$. For others n we have the recurrence relation: $c_n = \frac{c_{n-1}}{(n+\lambda)(3n+3\lambda-2)}$, $n = 1, 2, \ldots$. Since the difference between λ_1 and λ_2 is not an integer, by the Frobenius Theorem, the two solutions can be found in the form of the proposed series. For the first series, we substitute $\lambda_1 = 0$ into the recurrence relations and get

$$c_n = \frac{c_{n-1}}{n(3n-2)}, \quad \text{whence} \quad c_n = \frac{c_0}{n! \cdot 1 \cdot 4 \cdot \ldots \cdot (3n-2)}, \quad n = 1, 2, \ldots.$$

For the second, we find

$$c_n = \frac{c_{n-1}}{(n+\frac{2}{3})3n} \text{ or } c_n = \frac{c_{n-1}}{(3n+2)n}, \quad \text{whence } c_n = \frac{c_0}{n! \cdot 5 \cdot 8 \cdot \ldots \cdot (3n+2)}, n = 1, 2, \ldots.$$

Choosing $c_0 = 1$, we obtain the two linearly independent solutions in the form

$$y_1(x) = 1 + \sum_{n=1}^{\infty} \frac{1}{n! \cdot 1 \cdot 4 \cdot \ldots \cdot (3n-2)}x^n$$

and

$$y_2(x) = |x|^{2/3}\left[1 + \sum_{n=1}^{\infty} \frac{1}{n! \cdot 5 \cdot 8 \cdot \ldots \cdot (3n+2)}x^n\right].$$

From the Frobenius Theorem it follows that the series found converge in $\mathbb{R}$, since the functions $xa(x) = x \cdot \frac{1}{3x}$ and $x^2b(x) = x^2 \cdot \frac{-1}{3x}$ are analytic in $\mathbb{R}$. We can also check this using D'Alembert's test:

$$\frac{|x^{n+1}| \cdot n! \cdot 1 \cdot 4 \cdot \ldots \cdot (3n-2)}{|x^n| \cdot (n+1)! \cdot 1 \cdot 4 \cdot \ldots \cdot (3n-2)(3n+1)} = \frac{|x|}{(n+1)(3n+1)} \xrightarrow[n\to\infty]{} 0, \forall x \in \mathbb{R}$$

for first series and

$$\frac{|x^{n+1}| \cdot n! \cdot 5 \cdot 8 \cdot \ldots \cdot (3n+2)}{|x^n| \cdot (n+1)! \cdot 5 \cdot 8 \cdot \ldots \cdot (3n+2)(3n+5)} = \frac{|x|}{(n+1)(3n+5)} \xrightarrow[n\to\infty]{} 0, \forall x \in \mathbb{R}$$

for the second one. Finally, the general solution of the original equation is found as a linear combination of the two obtained particular solutions: $y = C_1 y_1 + C_2 y_2$.

2. $x^2 y'' + x(x - \frac{1}{2})y' + \frac{1}{2}y = 0$. Find generalized series solutions at the singular points.

This is a homogeneous linear equation with variable coefficients, which has the regular singular point $x = 0$, because the functions

$$xa(x) = x \cdot \frac{x(x - \frac{1}{2})}{x^2} = x - \frac{1}{2} \quad \text{and} \quad x^2 b(x) = x^2 \cdot \frac{\frac{1}{2}}{x^2} = \frac{1}{2}$$

are analytic at 0. There are no other singular points. Therefore, we look for the generalized series solution $y(x) = \sum_{n=0}^{\infty} c_n x^{n+\lambda}$. Substituting into the equation, we get

$$x^2 \sum_{n=0}^{\infty} c_n (n+\lambda)(n+\lambda-1)x^{n+\lambda-2} + x(x - \frac{1}{2})\sum_{n=0}^{\infty} c_n (n+\lambda)x^{n+\lambda-1} + \frac{1}{2}\sum_{n=0}^{\infty} c_n x^{n+\lambda}$$

$$= \sum_{n=0}^{\infty} c_n (n+\lambda)(n+\lambda-1)x^{n+\lambda} + \sum_{n=0}^{\infty} c_n (n+\lambda)x^{n+\lambda+1} - \sum_{n=0}^{\infty} \frac{c_n}{2}(n+\lambda)x^{n+\lambda} + \sum_{n=0}^{\infty} \frac{c_n}{2}x^{n+\lambda}$$

$$= x^\lambda \left[\sum_{n=0}^{\infty} c_n (n+\lambda)(n+\lambda-1)x^n + \sum_{m=1}^{\infty} c_{m-1}(m-1+\lambda)x^m - \sum_{n=0}^{\infty} \frac{c_n}{2}(n+\lambda-1)x^n \right]$$

$$= x^\lambda \left[c_0[\lambda(\lambda-1) - \frac{1}{2}(\lambda-1)] + \sum_{n=1}^{\infty} [c_n (n+\lambda-\frac{1}{2})(n+\lambda-1) + c_{n-1}(n+\lambda-1)]x^n \right] = 0.$$

For $c_0 = 0$ we have the zero solution, which is of no interest. So, for $n = 0$ we get the indicial equation $(\lambda - 1)(\lambda - \frac{1}{2}) = 0$, which determines λ: $\lambda_1 = 1$, $\lambda_2 = \frac{1}{2}$. For others n we have the recurrence relation: $c_n = -\frac{c_{n-1}}{n+\lambda-\frac{1}{2}}$, $n = 1, 2, \ldots$. Since the difference between λ_1 and λ_2 is not an integer, by the Frobenius Theorem, the two solutions can be found in the proposed form of series. For the first series, we substitute $\lambda_1 = 1$ into the recurrence relation and get

$$c_n = -\frac{2c_{n-1}}{2n+1}, \quad \text{whence} \quad c_n = (-1)^n \frac{2^n}{3 \cdot 5 \cdot \ldots \cdot (2n+1)}c_0, n = 1, 2, \ldots.$$

For the second we find

$$c_n = -\frac{c_{n-1}}{n}, \quad \text{whence} \quad c_n = (-1)^n \frac{1}{n!}c_0, n = 1, 2, \ldots.$$

By choosing $c_0 = 1$, we get the two linearly independent solutions in the form

$$y_1(x) = |x| \left[1 + \sum_{n=1}^{\infty} (-1)^n \frac{2^n}{3 \cdot 5 \cdot \ldots \cdot (2n+1)} x^n \right]$$

and

$$y_2(x) = |x|^{1/2} \left[1 + \sum_{n=1}^{\infty} \frac{(-1)^n}{n!} x^n \right] = |x|^{1/2} e^{-x}.$$

From the Frobenius Theorem it follows that the series found converge in $\mathbb{R}$, since the functions $xa(x) = x - \frac{1}{2}$ and $x^2 b(x) = \frac{1}{2}$ are analytic in $\mathbb{R}$. We can also check this using D'Alembert's test:

$$\frac{|x^{n+1}| \cdot 2^{n+1} \cdot 3 \cdot 5 \cdot \ldots \cdot (2n+1)}{|x^n| \cdot 2^n \cdot 3 \cdot 5 \cdot \ldots \cdot (2n+1)(2n+3)} = |x| \frac{2}{2n+3} \xrightarrow[n \to \infty]{} 0, \forall x \in \mathbb{R}$$

for first series and

$$\frac{|x^{n+1}| \cdot n!}{|x^n| \cdot (n+1)!} = |x| \frac{1}{n+1} \xrightarrow[n \to \infty]{} 0, \forall x \in \mathbb{R}$$

for the second one. Finally, the general solution of the original equation can be found as a linear combination of the two obtained particular solutions: $y = C_1 y_1 + C_2 y_2$.

3. $2x^2(1 - x^2)y'' - xy' + y = 0$. Find generalized series solutions at the origin.

The origin is the regular singular point of this homogeneous linear equation, since the functions $xa(x) = x \cdot \frac{-x}{2x^2(1-x^2)} = -\frac{1}{2(1-x^2)}$ and $x^2 b(x) = x^2 \cdot \frac{1}{2x^2(1-x^2)} = \frac{1}{2(1-x^2)}$ are analytic at 0. Therefore, we search for the solution in the form $y(x) = \sum_{n=0}^{\infty} c_n x^{n+\lambda}$. Substituting into the equation, we get

$$2x^2(1-x^2)\sum_{n=0}^{\infty} c_n(n+\lambda)(n+\lambda-1)x^{n+\lambda-2} - x\sum_{n=0}^{\infty} c_n(n+\lambda)x^{n+\lambda-1} + \sum_{n=0}^{\infty} c_n x^{n+\lambda}$$

$$= \sum_{n=0}^{\infty} 2c_n(n+\lambda)(n+\lambda-1)x^{n+\lambda} - \sum_{n=0}^{\infty} 2c_n(n+\lambda)(n+\lambda-1)x^{n+\lambda+2} - \sum_{n=0}^{\infty} c_n(n+\lambda)x^{n+\lambda}$$

$$+ \sum_{n=0}^{\infty} c_n x^{n+\lambda} = x^{\lambda} \left\{ \sum_{n=0}^{\infty} 2c_n(n+\lambda)(n+\lambda-1)x^n - \sum_{m=2}^{\infty} 2c_{m-2}(m-2+\lambda)(m-3+\lambda)x^m \right.$$

$$\left. - \sum_{n=0}^{\infty} c_n(n+\lambda)x^n + \sum_{n=0}^{\infty} c_n x^n \right\} = x^{\lambda} \left\{ c_0(2\lambda - 1)(\lambda - 1) + c_1\lambda(2\lambda + 1) + \right.$$

$$\sum_{n=2}^{\infty} [c_n(n + \lambda - 1)(2n + 2\lambda - 1) - 2c_{n-2}(n - 2 + \lambda)(n - 3 + \lambda)]x^n \bigg\} = 0.$$

For $n = 0$ we have the indicial equation $(2\lambda - 1)(\lambda - 1) = 0$, which has the two roots: $\lambda_1 = 1$, $\lambda_2 = \frac{1}{2}$. For $n = 1$ we have the condition $c_1\lambda(2\lambda+1) = 0$. Since $\lambda(2\lambda+1) \neq 0$ for $\lambda_{1,2}$, it follows that $c_1 = 0$. For others n we have the recurrence relation:

$$c_n = \frac{2(n - 2 + \lambda)(n - 3 + \lambda)}{(n + \lambda - 1)(2n + 2\lambda - 1)} c_{n-2}, \quad n = 2, \ldots .$$

Since the difference between λ_1 and λ_2 is not an integer, by the Frobenius Theorem, the two solutions can be found in the form of generalized series.

For both series, the condition $c_1 = 0$ results in vanishing all the odd-indexed coefficients due to the recurrence relation. To determine the even-indexed coefficients, we consider the two roots separately. Substituting $\lambda_1 = 1$ into the recurrence relation, we get $c_n = \frac{2(n-1)(n-2)}{n(2n+1)} c_{n-2}$, $n = 2, 4, \ldots .$ Note that for $n = 2$ the quotient is zero, and therefore, $c_2 = 0$, whence $c_n = 0$, $n = 2, 4, \ldots .$ Therefore, for the first solution, the only remaining term is the term with c_0, that is, $y_1 = c_0 x^{\lambda_1} = c_0 x$.

For $\lambda_2 = \frac{1}{2}$ the recurrence relation takes the form

$$c_n = \frac{(2n-3)(2n-5)}{2n(2n-1)} c_{n-2}, n = 2, 4, \ldots \text{ or } c_{2k} = \frac{(4k-5)(4k-3)}{(4k-1)4k} c_{2k-2}, k = 1, 2, \ldots .$$

These coefficients can be expressed in the form

$$c_{2k} = -\frac{1 \cdot 3 \cdot 5 \cdot 7 \cdot 9 \cdot \ldots \cdot (4k - 5)(4k - 3)}{3 \cdot 7 \cdot \ldots \cdot (4k - 1) \cdot 4^k \cdot k!} c_0,$$

or simplifying,

$$c_{2k} = -\frac{5 \cdot 9 \cdot \ldots \cdot (4k - 3)}{(4k - 1) \cdot 4^k \cdot k!} c_0, \quad k = 1, 2, \ldots .$$

Therefore, the second solution is

$$y_2(x) = |x|^{1/2} \sum_{n=0}^{\infty} c_{2n} x^{2n}.$$

It is easy to show that this series converges on $(-1, 1)$. Finally, we find the general solution of the original equation as a linear combination of the two particular solutions: $y = C_1 y_1 + C_2 y_2$.

4. $4(1 + x)^2 y'' - 3(1 - x^2)y' + 4y = 0$. Find generalized series solutions at the singular points.

The only singular point of this linear equation is $x = -1$. This is the regular singular point, because the functions

$$(x+1)a(x)=(x+1)\cdot\frac{-3(1-x^2)}{4(1+x)^2}=-\frac{3}{4}(1-x) \text{ and } (x+1)^2 b(x)=(x+1)^2\cdot\frac{4}{4(1+x)^2}=1$$

are analytic at -1. Therefore, we can search for the solution in the form $y(x) = \sum_{n=0}^{\infty} c_n(x+1)^{n+\lambda}$. Substituting into the equation, we get

$$4(1+x)^2\sum_{n=0}^{\infty}c_n(n+\lambda)(n+\lambda-1)(x+1)^{n+\lambda-2}+3(x+1)(x+1-2)\sum_{n=0}^{\infty}c_n(n+\lambda)(x+1)^{n+\lambda-1}$$

$$+4\sum_{n=0}^{\infty}c_n(x+1)^{n+\lambda}=\sum_{n=0}^{\infty}4c_n(n+\lambda)(n+\lambda-1)(x+1)^{n+\lambda}+\sum_{n=0}^{\infty}3c_n(n+\lambda)(x+1)^{n+\lambda+1}$$

$$-\sum_{n=0}^{\infty}6c_n(n+\lambda)(x+1)^{n+\lambda}+4\sum_{n=0}^{\infty}c_n(x+1)^{n+\lambda}$$

$$=(x+1)^{\lambda}\Big\{\sum_{n=0}^{\infty}[4c_n(n+\lambda)(n+\lambda-1)-6c_n(n+\lambda)+4c_n](x+1)^n$$

$$+\sum_{m=1}^{\infty}3c_{m-1}(m-1+\lambda)(x+1)^m\Big\}=(x+1)^{\lambda}\Big\{c_0[4\lambda(\lambda-1)-6\lambda+4]$$

$$+\sum_{n=1}^{\infty}[c_n(4(n+\lambda)(n+\lambda-1)-6(n+\lambda)+4)+3c_{n-1}(n+\lambda-1)](x+1)^n\Big\}=0.$$

For $n=0$ the indicial equation is $4\lambda(\lambda-1)-6\lambda+4=2(2\lambda^2-5\lambda+2)=0$, whose roots are $\lambda_1=2$, $\lambda_2=\frac{1}{2}$. For others n we have the recurrence relation:

$$c_n=-\frac{3(n+\lambda-1)}{4(n+\lambda)(n+\lambda-1)-6(n+\lambda)+4}c_{n-1}, n=1,2,\ldots.$$

Since the difference between λ_1 and λ_2 is not an integer, by the Frobenius Theorem, the two solutions can be found in the form of the proposed generalized series. For the first series, we substitute $\lambda_1=2$ into the recurrence relation and obtain

$$c_n=-\frac{3(n+1)}{4(n+2)(n+1)-6(n+2)+4}c_{n-1}, n=1,2,\ldots,$$

whence

$$c_n=-\frac{3(n+1)}{2n(2n+3)}c_{n-1}=\frac{3(n+1)3n}{2n(2n+3)2(n-1)(2n+1)}c_{n-2}=\ldots$$

$$=(-1)^n \frac{3(n+1)3n\cdot\ldots\cdot 3\cdot 2}{2n(2n+3)2(n-1)(2n+1)\cdot\ldots\cdot 2\cdot 1\cdot 5}c_0=(-1)^n\left(\frac{3}{2}\right)^n\frac{3(n+1)}{(2n+3)!!}c_0.$$

For the second series we find

$$c_n=-\frac{3(n-\frac{1}{2})}{4(n+\frac{1}{2})(n-\frac{1}{2})-6(n+\frac{1}{2})+4}c_{n-1}, n=1,2,\ldots,$$

whence

$$c_n=-\frac{3(2n-1)}{4n(2n-3)}c_{n-1}=\frac{3(2n-1)3(2n-3)}{4n(2n-3)4(n-1)(2n-5)}c_{n-2}=\ldots$$

$$=(-1)^n\frac{3(2n-1)3(2n-3)\cdot\ldots\cdot 3\cdot 1}{4n(2n-3)4(n-1)(2n-5)\cdot\ldots\cdot 4\cdot 1\cdot(-1)}c_0=(-1)^{n+1}\left(\frac{3}{4}\right)^n\frac{2n-1}{n!}c_0.$$

Choosing $c_0=1$, we get the two linearly independent solutions in the form

$$y_1(x)=(x+1)^2\left[1+\sum_{n=1}^{\infty}(-1)^n\left(\frac{3}{2}\right)^n\frac{3(n+1)}{(2n+3)!!}(x+1)^n\right].$$

and

$$y_2(x)=|x+1|^{1/2}\left[1-\sum_{n=1}^{\infty}(-1)^n\left(\frac{3}{4}\right)^n\frac{2n-1}{n!}(x+1)^n\right].$$

From the Frobenius Theorem it follows that the found series converge in $\mathbb{R}$, since the functions $(x+1)a(x)=-\frac{3}{4}(1-x)$ and $(x+1)^2b(x)=1$ are analytic in $\mathbb{R}$. Finally, the general solution of the original equation is found as a linear combination of the two particular solutions: $y=C_1y_1+C_2y_2$.

5. $x(1-x)y''+(1-x)y'-y=0$. Find generalized series solutions at the origin.

The point $x=0$ is a regular singular point of this equation, because the functions

$$xa(x)=x\cdot\frac{1-x}{x(1-x)}=1 \text{ and } x^2b(x)=x^2\cdot\frac{-1}{x(1-x)}=-\frac{x}{1-x}$$

are analytic at 0. Therefore, we can search for the solution in the form $y(x)=\sum_{n=0}^{\infty}c_nx^{n+\lambda}$. Substituting into the equation, we get

$$x(1-x)\sum_{n=0}^{\infty}c_n(n+\lambda)(n+\lambda-1)x^{n+\lambda-2}+(1-x)\sum_{n=0}^{\infty}c_n(n+\lambda)x^{n+\lambda-1}-\sum_{n=0}^{\infty}c_nx^{n+\lambda}$$

$$=\sum_{n=0}^{\infty}c_n(n+\lambda)(n+\lambda-1)(x^{n+\lambda-1}-x^{n+\lambda})+\sum_{n=0}^{\infty}c_n(n+\lambda)(x^{n+\lambda-1}-x^{n+\lambda})+\sum_{n=0}^{\infty}c_nx^{n+\lambda}$$

$$= x^{\lambda-1} \left\{ \sum_{n=0}^{\infty} c_n[(n+\lambda)(n+\lambda-1)+(n+\lambda)]x^n \right.$$

$$\left. - \sum_{n=0}^{\infty} c_n[(n+\lambda)(n+\lambda-1)+(n+\lambda)+1]x^{n+1} \right\}$$

$$= x^{\lambda-1} \left\{ \sum_{n=0}^{\infty} c_n(n+\lambda)^2 x^n - \sum_{m=1}^{\infty} c_{m-1}((m+\lambda-1)^2+1)x^m \right\}$$

$$= x^{\lambda-1} \left\{ c_0\lambda^2 + \sum_{n=1}^{\infty} [c_n(n+\lambda)^2 - c_{n-1}((n+\lambda-1)^2+1)]x^n \right\} = 0.$$

From this formula we obtain the indicial equation $\lambda^2 = 0$, which has the two equal roots $\lambda_{1,2} = 0$, and the recurrence relation $c_n = \frac{(n+\lambda-1)^2+1}{(n+\lambda)^2} c_{n-1}$, $n = 1, 2, \ldots$. Since $\lambda_1 = \lambda_2$, by the Frobenius Theorem, one of the solutions can be found in the form of generalized series and the other should be found in a different form. For the first solution, we substitute $\lambda = 0$ into the recurrence relations and get

$$c_n = \frac{(n-1)^2+1}{n^2} c_{n-1}, \text{ whence } c_n = \frac{1 \cdot 2 \cdot 5 \cdot \ldots \cdot ((n-1)^2+1)}{(n!)^2} c_0, n = 1, 2, \ldots.$$

By choosing $c_0 = 1$, we have the first solution in the form $y_1(x) = \sum_{n=0}^{\infty} c_n x^n$.

To find the second solution, we need to keep the dependence of the coefficients c_n on λ: $c_n = \frac{(n+\lambda-1)^2+1}{(n+\lambda)^2} c_{n-1}$, whence

$$c_n = \frac{(\lambda^2+1)((1+\lambda)^2+1) \cdot \ldots \cdot ((n+\lambda-1)^2+1)}{(1+\lambda)^2(2+\lambda)^2 \cdot \ldots \cdot (n+\lambda)^2} c_0, n = 1, 2, \ldots.$$

The coefficients d_n of the second solution series

$$y_2(x) = y_1(x) \ln|x - x_0| + |x - x_0|^{\lambda_1} \sum_{n=1}^{\infty} d_n(x - x_0)^n$$

can be found by differentiating $c_n(\lambda)$. The simplest way to do this is to use the logarithmic form:

$$\frac{c_{n\lambda}}{c_n} = (\ln c_n)_\lambda = \left[\sum_{k=1}^{n} \ln((k+\lambda-1)^2+1) - 2\ln(k+\lambda) \right]_\lambda$$

$$= 2 \sum_{k=1}^{n} \left[\frac{k+\lambda-1}{(k+\lambda-1)^2+1} - \frac{1}{k+\lambda} \right].$$

Then, for $\lambda = 0$ we have $(\ln c_n)_\lambda(0) = 2\sum_{k=1}^{n}\frac{k-2}{k((k-1)^2+1)}$. Consequently,

$$d_n = c_{n\lambda}(0) = c_n(0)\cdot 2\sum_{k=1}^{n}\frac{k-2}{k((k-1)^2+1)}$$

$$= 2\frac{1\cdot 2\cdot 5\cdot\ldots\cdot((n-1)^2+1)}{(n!)^2}\cdot\sum_{k=1}^{n}\frac{k-2}{k((k-1)^2+1)}.$$

In this way, the series in y_2 is specified. Finally, the general solution is found as a linear combination of the two linearly independent solutions: $y = C_1y_1 + C_2y_2$. It can be shown that the series in y_1 and y_2 have the radius of convergence equal to 1.

Remark. Several cases of the Frobenius method, including the most complicated ones, can be illustrated by solving the Bessel equation $x^2y'' + xy' + (x^2 - \nu^2)y = 0$, where $\nu \in \mathbb{R}$ is called the order of the Bessel equation. The only singular point in this equation is $x = 0$. This is the regular singular point, because the functions

$$xa(x) = x\cdot\frac{x}{x^2} = 1 \text{ and } x^2b(x) = x^2\cdot\frac{x^2-\nu^2}{x^2} = x^2-\nu^2$$

are analytic at 0.

6. $x^2y'' + xy' + x^2y = 0$ (Bessel equation of order 0). Find generalized series solutions at the singular point.
We search for generalized series solutions $y(x) = \sum_{n=0}^{\infty}c_nx^{n+\lambda}$. Substituting into the equation, we get

$$x^2\sum_{n=0}^{\infty}c_n(n+\lambda)(n+\lambda-1)x^{n+\lambda-2} + x\sum_{n=0}^{\infty}c_n(n+\lambda)x^{n+\lambda-1} + x^2\sum_{n=0}^{\infty}c_nx^{n+\lambda}$$

$$= x^\lambda\left\{\sum_{n=0}^{\infty}c_n(n+\lambda)(n+\lambda-1)x^n + \sum_{n=0}^{\infty}c_n(n+\lambda)x^n + \sum_{m=2}^{\infty}c_{m-2}x^m\right\}$$

$$= x^\lambda\left\{c_0\lambda^2 + c_1(\lambda+1)^2 + \sum_{n=2}^{\infty}[c_n(n+\lambda)^2 + c_{n-2}]x^n\right\} = 0.$$

From this formula we obtain the indicial equation $\lambda^2 = 0$, the equation $c_1(\lambda + 1)^2 = 0$ and the recurrence relation $c_n = -\frac{1}{(n+\lambda)^2}c_{n-2}$, $n = 2, 3, \ldots$. From the indicial equation it follows that $\lambda_{1,2} = 0$ and, therefore, from the second equation it follows that $c_1 = 0$. Consequently, due to the recurrence relation, all the odd-indexed coefficients vanish. For the even-indexed coefficients we have

$$c_{2n} = -\frac{c_{2n-2}}{(2n+\lambda)^2}, \text{ whence } c_{2n} = (-1)^n \frac{c_0}{(2+\lambda)^2 \cdot \ldots \cdot (2n+\lambda)^2}, n = 1, 2, \ldots.$$

For $\lambda = 0$ we get $c_{2n} = (-1)^n \frac{c_0}{2^{2n}(n!)^2}$, $n = 1, 2, \ldots$. Choosing $c_0 = 1$, we find the first series solution

$$y_1(x) = \sum_{n=0}^{\infty} (-1)^n \frac{1}{2^{2n}(n!)^2} x^{2n}.$$

To find the second solution, we need to maintain the dependence of coefficients c_{2n} on λ: $c_{2n} = (-1)^n \frac{c_0}{(2+\lambda)^2 \cdot \ldots \cdot (2n+\lambda)^2}$, $n = 1, 2, \ldots$. The coefficients d_n of the series in the second solution

$$y_2(x) = y_1(x) \ln|x - x_0| + |x - x_0|^{\lambda_1} \sum_{n=1}^{\infty} d_n (x - x_0)^n$$

can be found by differentiating $c_n(\lambda)$. Since $c_{2n-1} = 0$, it follows that $d_{2n-1} = 0$. For the even-indexed coefficients, we perform the differentiation in the logarithmic form:

$$\frac{c_{2n\lambda}}{c_{2n}} = (\ln c_{2n})_\lambda = -2 \left[\sum_{k=1}^{n} \ln(2k + \lambda) \right]_\lambda = -2 \sum_{k=1}^{n} \frac{1}{2k + \lambda}.$$

Evaluating at $\lambda = 0$ we have $(\ln c_{2n})_\lambda(0) = -2 \sum_{k=1}^{n} \frac{1}{2k}$. Therefore,

$$d_{2n} = c_{2n\lambda}(0) = c_{2n}(0) \cdot (-2) \sum_{k=1}^{n} \frac{1}{2k} = (-1)^{n+1} \frac{2}{2^{2n}(n!)^2} \sum_{k=1}^{n} \frac{1}{2k}.$$

In this way, the series in y_2 is specified. Finally, the general solution is found as a linear combination of the two linearly independent solutions: $y = C_1 y_1 + C_2 y_2$. It can be shown that the series in y_1 and y_2 converge in $\mathbb{R}$.

7. $x^2 y'' + x y' + (x^2 - \frac{1}{4})y = 0$ (Bessel equation of order $\frac{1}{2}$). Find generalized series solutions at the singular point.
We can search for the solution in the form $y(x) = \sum_{n=0}^{\infty} c_n x^{n+\lambda}$. Substituting into the equation, we get

$$x^2 \sum_{n=0}^{\infty} c_n(n+\lambda)(n+\lambda-1)x^{n+\lambda-2} + x \sum_{n=0}^{\infty} c_n(n+\lambda)x^{n+\lambda-1} + (x^2 - \frac{1}{4}) \sum_{n=0}^{\infty} c_n x^{n+\lambda}$$

$$= x^\lambda \left\{ \sum_{n=0}^{\infty} c_n (n+\lambda)(n+\lambda-1)x^n + \sum_{n=0}^{\infty} c_n (n+\lambda)x^n + \sum_{m=2}^{\infty} c_{m-2}x^m - \frac{1}{4}\sum_{n=0}^{\infty} c_n x^n \right\}$$

$$= x^\lambda \left\{ c_0(\lambda^2 - \frac{1}{4}) + c_1((\lambda+1)^2 - \frac{1}{4}) + \sum_{n=2}^{\infty} [c_n((n+\lambda)^2 - \frac{1}{4}) + c_{n-2}]x^n \right\} = 0.$$

From this formula we obtain the indicial equation $\lambda^2 - \frac{1}{4} = 0$, the equation $c_1((\lambda + 1)^2 - \frac{1}{4}) = 0$ and the recurrence relation $c_n = -\frac{1}{(n+\lambda)^2 - \frac{1}{4}}c_{n-2}$, $n = 2, 3, \ldots$. The roots of the indicial equation are $\lambda_{1,2} = \pm\frac{1}{2}$. Let us first consider $\lambda_1 = -\frac{1}{2}$. In this case, the second equation takes the form $c_1 \cdot 0 = 0$ and, therefore, c_1 is an arbitrary parameter. The recurrence relation takes the form $c_n = -\frac{1}{(n-\frac{1}{2})^2 - \frac{1}{4}}c_{n-2} = -\frac{1}{n(n-1)}c_{n-2}$, $n = 2, 3, \ldots$ and can be divided into the two separate groups, with the odd and even indices. For the former we have

$$c_{2k} = -\frac{1}{2k(2k-1)}c_{2k-2} = (-1)^k \frac{1}{(2k)!}c_0, k = 1, 2, \ldots$$

and for the latter

$$c_{2k+1} = -\frac{1}{(2k+1)2k}c_{2k-1} = (-1)^k \frac{1}{(2k+1)!}c_1, k = 1, 2, \ldots.$$

In this way, we find the two linearly independent solutions, which correspond to the same $\lambda_1 = -\frac{1}{2}$:

$$y_1(x) = |x|^{-1/2} \sum_{k=0}^{\infty} (-1)^k \frac{1}{(2k)!} x^{2k}$$

and

$$y_2(x) = |x|^{-1/2} \sum_{k=0}^{\infty} (-1)^k \frac{1}{(2k+1)!} x^{2k+1}.$$

We can immediately realize that the two power series found are the power series for $\cos x$ and $\sin x$, which converge on the entire real axis. Then, we can rewrite the solutions in the form

$$y_1(x) = |x|^{-1/2} \cos x, \quad y_2(x) = |x|^{-1/2} \sin x,$$

and the general solution is $y = C_1 y_1 + C_2 y_2 = |x|^{-1/2}(C_1 \cos x + C_2 \sin x)$.

Notice that in this case the difference between λ_1 and λ_2 is a non-zero integer, but even so, it was possible to find the two independent solutions in the form of generalized series due to the arbitrariness of the choice of c_1.

This is the case 3) of the Frobenius Theorem with the coefficient $c = 0$ in the solution y_2. Such cases are called false exceptions.

8. $x^2 y'' + xy' + (x^2 - 1)y = 0$ (Bessel equation of order 1). Find generalized series solutions at the singular point.
We search for the solution in the form $y(x) = \sum_{n=0}^{\infty} c_n x^{n+\lambda}$. Substituting into the equation, we get

$$x^2 \sum_{n=0}^{\infty} c_n (n+\lambda)(n+\lambda-1) x^{n+\lambda-2} + x \sum_{n=0}^{\infty} c_n (n+\lambda) x^{n+\lambda-1} + (x^2-1) \sum_{n=0}^{\infty} c_n x^{n+\lambda}$$

$$= x^\lambda \left\{ \sum_{n=0}^{\infty} c_n (n+\lambda)(n+\lambda-1) x^n + \sum_{n=0}^{\infty} c_n (n+\lambda) x^n + \sum_{m=2}^{\infty} c_{m-2} x^m - \sum_{n=0}^{\infty} c_n x^n \right\}$$

$$= x^\lambda \left\{ c_0(\lambda^2 - 1) + c_1((\lambda+1)^2 - 1) + \sum_{n=2}^{\infty} [c_n((n+\lambda)^2 - 1) + c_{n-2}] x^n \right\} = 0.$$

From this formula we obtain the indicial equation $\lambda^2 - 1 = 0$, the equation $c_1((\lambda+1)^2 - 1) = 0$ and the recurrence relation $c_n = -\frac{1}{(n+\lambda)^2 - 1} c_{n-2}$, $n = 2, 3, \ldots$. The roots of the initial equation are $\lambda_{1,2} = \pm 1$. Let us first consider $\lambda_2 = 1$. In this case, the second equation takes the form $c_1 \cdot 3 = 0$, that is, $c_1 = 0$, which implies that all the odd-indexed coefficients are zero (due to the recurrence relation). For the even-indexed coefficients, we have the recurrence relation $c_{2k} = -\frac{c_{2k-2}}{(2k+2)2k} = -\frac{c_{2k-2}}{4(k+1)k}$, $k = 1, 2, \ldots$. Expressing in terms of c_0, we have $c_{2k} = (-1)^k \frac{c_0}{4^k (k+1)!k!}$, $k = 1, 2, \ldots$. Therefore, the first solution is found in the form (we put $c_0 = 1$):

$$y_1(x) = x \sum_{k=0}^{\infty} (-1)^k \frac{1}{4^k (k+1)!k!} x^{2k}.$$

The second solution can be found by substituting the formula proposed in the Frobenius Theorem (the case 3))

$$y_2(x) = c y_1(x) \ln|x| + x^{-1} \sum_{n=0}^{\infty} d_n x^n$$

in the original equation. Taking into account that y_1 is the solution of the equation under consideration, we obtain

$$2cx y_1' + x^2 \sum_{n=0}^{\infty} d_n (n-1)(n-2) x^{n-3} + x \sum_{n=0}^{\infty} d_n (n-1) x^{n-2} + (x^2-1) \sum_{n=0}^{\infty} d_n x^{n-1}$$

$$= 2cxy_1' + x^{-1}\left\{\sum_{n=0}^{\infty}d_n(n-1)(n-2)x^n + \sum_{n=0}^{\infty}d_n(n-1)x^n + \sum_{m=2}^{\infty}d_{m-2}x^m - \sum_{n=0}^{\infty}d_nx^n\right\}$$

$$= 2cxy_1' + x^{-1}\left\{d_0(2-1-1) - d_1x + \sum_{n=2}^{\infty}(d_n(n-2)n + d_{n-2})x^n\right\} = 0.$$

Substituting

$$xy_1' = \sum_{n=0}^{\infty}(-1)^n\frac{2n+1}{4^n(n+1)!n!}x^{2n+1}$$

in the last equation, we get

$$2c\sum_{n=0}^{\infty}(-1)^n\frac{2n+1}{4^n(n+1)!n!}x^{2n+1} - d_1 + \sum_{n=1}^{\infty}(d_{n+1}(n-1)(n+1) + d_{n-1})x^n$$

$$= 2cx + 2c\sum_{n=1}^{\infty}\frac{(-1)^n(2n+1)}{4^n(n+1)!n!}x^{2n+1} - d_1 + d_0x + d_2\cdot 0 + \sum_{n=2}^{\infty}(d_{n+1}(n-1)(n+1) + d_{n-1})x^n$$

$$= -d_1 + (2c+d_0)x + 2c\sum_{n=1}^{\infty}\frac{(-1)^n(2n+1)}{4^n(n+1)!n!}x^{2n+1} + \sum_{n=2}^{\infty}(d_{n+1}(n-1)(n+1) + d_{n-1})x^n = 0.$$

It follows that $d_1 = 0$, $2c + d_0 = 0$ and d_2 is arbitrary, and that the relations for the even and odd exponents are decoupled. For the even ones (x^{2k}) we have

$$d_{2k+1}(2k-1)(2k+1) + d_{2k-1} = 0, \forall k \geq 1.$$

Taking into account that $d_1 = 0$, we see that $d_{2k+1} = 0, \forall k$. For the odd exponents (x^{2k+1}) we have

$$d_{2k+2}2k(2k+2) + d_{2k} + 2c\cdot(-1)^k\frac{2k+1}{4^k(k+1)!k!} = 0, \forall k \geq 1.$$

Then,

$$d_{2k+2} = -\frac{d_{2k}}{4k(k+1)} - 2c\frac{(-1)^k}{4k(k+1)}\cdot\frac{2k+1}{4^k(k+1)!k!}, \forall k \geq 1.$$

Using mathematical induction, we can show that

$$d_{2k+2} = (-1)^k\frac{d_2}{4^k k!(k+1)!} - 2c\frac{(-1)^k}{4^{k+1}k!(k+1)!}\cdot\sum_{j=1}^{k}\left(\frac{1}{j} + \frac{1}{j+1}\right), \forall k \geq 1.$$

Therefore, the second solution is (we set $d_2 = 1$):

$$y_2(x) = cy_1(x)\ln|x| - \frac{2c}{x} + x + \sum_{k=2}^{\infty} \frac{(-1)^k}{4^{k+1}k!(k+1)!}\left\{4 - 2c\sum_{j=1}^{k}\left(\frac{1}{j} + \frac{1}{j+1}\right)\right\}x^{2k+1}.$$

Finally, the general solution is expressed as the linear combination $y = C_1 y_1 + C_2 y_2$.

Notice that in this case the difference between λ_1 and λ_2 is a nonzero integer, and it was not possible to find the two independent solutions in the form of generalized series. Consequently, it was necessary to search for the second solution including an additional logarithmic term. Such cases are called real exceptions.

9. $x^2 y'' + xy' + (x^2 - p^2)y = 0$, $p \in \mathbb{R}$ (general Bessel equation of order p). Find generalized series solutions at the singular point.
Finally, we solve the general Bessel equation with arbitrary real parameter p. Bearing in mind that the case $p = 0$ has already been considered, without loss of generality we can assume that $p > 0$. We look for the generalized series solution $y(x) = \sum_{n=0}^{\infty} c_n x^{n+\lambda}$. Substituting in the given equation, we get

$$x^2 \sum_{n=0}^{\infty} c_n(n+\lambda)(n+\lambda-1)x^{n+\lambda-2} + x\sum_{n=0}^{\infty} c_n(n+\lambda)x^{n+\lambda-1} + (x^2 - p^2)\sum_{n=0}^{\infty} c_n x^{n+\lambda}$$

$$= x^\lambda\left\{\sum_{n=0}^{\infty} c_n(n+\lambda)(n+\lambda-1)x^n + \sum_{n=0}^{\infty} c_n(n+\lambda)x^n + \sum_{m=2}^{\infty} c_{m-2}x^m - p^2\sum_{n=0}^{\infty} c_n x^n\right\}$$

$$= x^\lambda\left\{c_0(\lambda^2 - p^2) + c_1((\lambda+1)^2 - p^2) + \sum_{n=2}^{\infty}[c_n(n+\lambda)^2 + c_{n-2} - p^2 c_n]x^n\right\} = 0.$$

From this formula we obtain the indicial equation $\lambda^2 = p^2$, the equation $c_1((\lambda+1)^2 - p^2) = 0$ and the recurrence relation $c_n = -\frac{1}{(n+\lambda)^2 - p^2}c_{n-2}$, $n = 2, 3, \dots$. The roots of the indicial equation are $\lambda_{1,2} = \pm p$. For the root $\lambda_1 = p$ we have $c_1((p+1)^2 - p^2) = 0$, whence $c_1 = 0$, and the recurrence relation takes the form $c_n = -\frac{1}{(n+p)^2 - p^2}c_{n-2} = -\frac{1}{n(n+2p)}c_{n-2}$, $n = 2, 3, \dots$. Therefore, all the odd-indexed coefficients vanish and for the even-indexed we have

$$c_{2n} = -\frac{c_{2n-2}}{4n(n+p)}, \text{ whence } c_{2n} = (-1)^n \frac{c_0}{4^n n!(n+p)(n-1+p)\cdot\ldots\cdot(1+p)},$$

where $n = 1, 2, \dots$. Traditionally, the constant c_0 is chosen in the form $c_0 = \frac{1}{2^p \Gamma(p+1)}$, where $\Gamma(p)$ is the Euler gamma-function defined by the formula

$\Gamma(p) = \int_0^{+\infty} e^{-x} x^{p-1} dx$, $\forall p > 0$. It can be shown that this function satisfies the following properties:

1. $\Gamma(p + 1) = p\Gamma(p)$;
2. $\Gamma(1) = 1$;
3. $\Gamma(p + n + 1) = (n + p)(n - 1 + p) \cdot \ldots \cdot (1 + p)\Gamma(p + 1)$;
4. $\Gamma(n + 1) = n!$.

Using these properties, the formula for c_{2n}, $n = 1, 2, \ldots$ can be written in the form

$$c_{2n} = \frac{(-1)^n}{4^n n!(n+p)(n-1+p) \cdot \ldots \cdot (1+p) \cdot 2^p\,\Gamma(p+1)} = \frac{(-1)^n}{4^n 2^p n!\,\Gamma(p+n+1)}.$$

The particular solution found is usually called the Bessel function of the first kind of order p and is denoted $J_p(x)$:

$$J_p(x) = \sum_{n=0}^{\infty} \frac{(-1)^n}{n!\,\Gamma(p+n+1)} \left(\frac{x}{2}\right)^{2n+p}.$$

The second solution, which corresponds to the root $\lambda_2 = -p$, can be obtained from the function $J_p(x)$ by replacing p by $-p$, since the original equation contains only p^2 and does not change its form when p is replaced by $-p$. So,

$$J_{-p}(x) = \sum_{n=0}^{\infty} \frac{(-1)^n}{n!\,\Gamma(-p+n+1)} \left(\frac{x}{2}\right)^{2n-p}.$$

This function is called the Bessel function of the first kind of order $-p$.

If p is not a natural number, then the solutions $J_p(x)$ and $J_{-p}(x)$ are linearly independent, because their series start from different exponents of x and, therefore, the linear combination $\alpha_1 J_p(x) + \alpha_2 J_{-p}(x)$ is zero only when $\alpha_1 = \alpha_2 = 0$.

If p is a natural number, then the solutions $J_p(x)$ and $J_{-p}(x)$ are linearly dependent, namely: $J_{-p}(x) = (-1)^p J_p(x)$, $p \in \mathbb{N}$. Therefore, in this case, it is necessary to find another particular solution, which is linearly independent of $J_p(x)$. To this end, we introduce the new function

$$Y_p(x) = \frac{J_p(x)\cos p\pi - J_{-p}(x)}{\sin p\pi},$$

initially considering that p is not a natural number. Obviously, the function $Y_p(x)$ is a solution of the Bessel equation, since it represents the linear combination of the two particular solutions of this equation. Taking the limit in the definition of $Y_p(x)$ when p approaches a natural number, we find the particular solution defined for $p \in \mathbb{N}$, which is linearly independent of $J_p(x)$.

The function $Y_p(x)$ is called the Bessel function of the second kind of order p.

Finally, the general solution is found as a linear combination of two particular linearly independent solutions. For any $p > 0$ the general solution has the form $y_g = C_1 J_p(x) + C_2 Y_p(x)$. In the case $p \notin \mathbb{N}$, the general solution can also be given by the formula $y_g = C_1 J_p(x) + C_2 J_{-p}(x)$.

Exercises for section 6

Classify the indicated center point and solve using the corresponding series; find the set of validity of the obtained solution:
1. Solve $2x^2 y'' + (3x - x^2)y' - (x+1)y = 0$ about the center point 0.
2. Solve $4xy'' + 2y' + y = 0$ about the center point 0.
3. Solve $(1 + x)y' - py = 0$ about the center point 0.
4. Solve $9x(1 - x)y'' - 12y' + 4y = 0$ about the center point 0.
5. Solve $x^2 y'' + xy' + (4x^2 - \frac{1}{9})y = 0$ about the center point 0.
6. Solve $x^2 y'' + xy' + (x^2 - \frac{1}{9})y = 0$ about the center point 0.
7. Solve $x^2 y'' + xy' + (4x^2 - \frac{1}{9})y = 0$ about the center point 0.
8. Solve $2xy'' + (1 + x)y' + y = 0$ about the center point 0.
9. Solve $xy'' + (x - 6)y' - 3y = 0$ about the center point 0.
10. Solve $2xy'' - y' + 2y = 0$ about the center point 0.
11. Solve $2xy'' - y' + 2y = 0$ about the center point 0.
12. Solve $2x^2 y'' - x(x - 1)y' - y = 0$ about the center point 0.
13. Solve $xy'' + 2y' - xy = 0$ about the center point 0.
14. Solve $x(x - 1)y'' + 3y' - 2y = 0$ about the center point 0.
15. Solve $xy'' + (1 - x)y' - y = 0$ about the center point 0.
16. Solve $2x^2 y'' + xy' - (x + 1)y = 0$ about the center point 0.
17. Solve $xy'' - (2x - 1)y' + (x - 1)y = 0$ about the center point 0.
18. Solve $x^2 y'' + x(2 + 3x)y' - 2y = 0$ about the center point 0.
19. Solve $x^2 y'' + xy' - xy = 0$ about the center point 0.
20. Solve $4x^2 y'' + (1 - 2x)y = 0$ about the center point 0.

Exercises for sections 3 and 6

Classify the indicated center point and solve using the corresponding series; find the set of validity of the obtained solution:
1. Solve $(1 - x^2)y'' - 6xy' - 4y = 0$ about the center point 0.
2. Solve $2xy'' + (1 + x)y' - 2y = 0$ about the center point 0.
3. Solve $xy'' + y = 0$ about the center point 0.
4. Solve $y'' + (x - 1)^2 y' - 4(x - 1)y = 0$ about the center point 1.

5. Solve $y'' + 3xy' + 3y = 0$ about the center point 0.
6. Solve $x^2y'' + 3xy' + (1 - 2x)y = 0$ about the center point 0.
7. Solve $2x(x + 1)y'' + 3(x + 1)y' - y = 0$ about the center point 0.
8. Solve $(x^2 + 4)y'' + 2xy' - 12y = 0$ about the center point 0.
9. Solve $(1 + 4x^2)y'' - 8y = 0$ about the center point 0.
10. Solve $x(x - 2)y'' + 2(x - 1)y' - 2y = 0$ about the center point 2.
11. Solve $4xy'' + 3y' + 3y = 0$ about the center point 0.
12. Solve $y'' - 2(x + 3)y' - 3y = 0$ about the center point -3.
13. Solve $x(1 + x)y'' + (1 + 5x)y' + 3y = 0$ about the center point 0.
14. Solve $(1 + x^2)y'' + 10xy' + 20y = 0$ about the center point 0.
15. Solve $2x^2(1 - x)y'' - x(1 + 7x)y' + y = 0$ about the center point 0.
16. Solve $x(1 + x)y'' + (x + 5)y' - 4y = 0$ about the center point -1.
17. Solve $y'' + (x - 2)y = 0$ about the center point 2.
18. Solve $x^2(1 + 2x)y'' + 2x(1 + 6x)y' - 2y = 0$ about the center point 0.
19. Solve $x^2y'' + x(1 - x)y' - (1 + 3x)y = 0$ about the center point 0.
20. Solve $x^2y'' + x(x - 1)y' + (1 - x)y = 0$ about the center point 0.

Chapter 11

Higher order equations: applications

In this part of the text, we consider some applications of higher order equations, namely spring-mass models, electrical circuit models, pendulum models and pursuit models.

1 Spring-mass problem

Consider a flexible spring hanging vertically and attached at the top to a rigid support. If a weight of mass m is attached to the free lower end of this spring, then it will elongate, extending downwards from its original position, and this elongation will depend on the mass of the body. In the new static (equilibrium) position of the spring, there are two balanced forces acting on the point where the mass is attached to the spring: the gravitational force $F_g = mg$ which pulls downwards ($g \approx 10m/s^2$ is the gravitational acceleration) and the force of the spring resistance F_r which acts upwards. If the elongation L from the original position is quite small, then experiments show that the force F_r is proportional to L: $F_r = -kL$, where the coefficient of proportionality $k > 0$ is called the spring constant, and the negative sign is due to the fact that the spring force pulls upwards, in the opposite direction to the gravitational force (we chose the positive direction downwards). This relationship is called Hooke's law. In the new position, the spring is in equilibrium, which means that $F_g + F_r = mg - kL = 0$ (see Fig.11.1). If we know the mass and elongation measurements, then from this relationship we can find the coefficient k, specific to each type of spring.

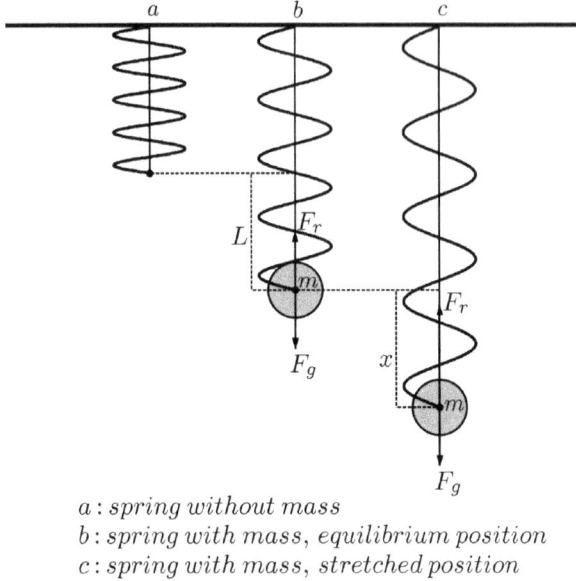

a : *spring without mass*
b : *spring with mass, equilibrium position*
c : *spring with mass, stretched position*

Figure 11.1: Spring-mass system

Let us study the motion of the spring with a mass attached to the lower end when it is displaced from the equilibrium position due to the initial impulse or in the presence of an external force. We denote by x the position of the mass (the lower end of the spring), measured from the equilibrium position L, and we use the vertical axis x with a positive downward direction and the origin at the equilibrium point (see Fig.11.1). We denote the velocity of the mass by $v(t)$ and its acceleration by $a(t)$. According to the definition, the relationships between these quantities are $x_t = v$ and $x_{tt} = v_t = a$.

1.1   Undamped free oscillations

First, let us consider the situation when there are no external forces. In this case, according to Newton's second law, $ma = F_g + F_r$ or $ma = mg - k(x+L)$. Taking into account that $x_{tt} = a$ and $mg - kL = 0$, we get $mx_{tt} = -kx$. The negative sign, together with the positive coefficient k, indicates that the spring resistance force acts in the opposite direction to the direction of displacement from equilibrium. Dividing the obtained equation by m and denoting $\omega^2 = \frac{k}{m}$, we arrive at the linear equation of the second order

$$x_{tt} + \omega^2 x = 0,$$

which describes undamped free oscillations (also called simple harmonic oscillations).

Equations of this type have already been solved in several previous examples (see Chapter 9, especially section 1) and their general solution is

$$x = A \cos \omega t + B \sin \omega t,$$

where A and B are two arbitrary constants. These constants are determined by the initial conditions: the initial position of the spring $x(t_0) = x_0$ and its initial velocity $x_t(t_0) = v_0$. To analyze properties of this solution, it is convenient to use the following representation:

$$x = C \cos(\omega t - \alpha),$$

where $C = \sqrt{A^2 + B^2}$ and the angle α is determined by the relations $\cos \alpha = \frac{A}{C}$ and $\sin \alpha = \frac{B}{C}$. (This expression can be deduced by writing the solution in the form $x = C(\frac{A}{C} \cos \omega t + \frac{B}{C} \sin \omega t)$, introducing the angle α by the relations $\cos \alpha = \frac{A}{C}$ and $\sin \alpha = \frac{B}{C}$ and using the trigonometric formula for the cosine of the difference of two angles). From the last representation of the solution it follows that the maximum value of x is $C = \sqrt{A^2 + B^2}$, which physically corresponds to the greatest downward displacement (in the positive direction of the x-axis) from the equilibrium position, while the minimum value of x is $-C = -\sqrt{A^2 + B^2}$, which corresponds to the greatest upward displacement from the equilibrium point.

The period of the functions involved in the solution is $T = \frac{2\pi}{\omega}$, which physically represents the time it takes the mass to perform one cycle of oscillations, that is, the minimum time needed for the mass to return to the position from which it starts moving. For example, if at time t_0 the mass was in its lowest position (below the equilibrium point), then it will take T seconds to rise to the highest position (above the equilibrium point) and then return to the same lowest position at time $t_0 + T$. Another characteristic of this motion is the frequency $f = \frac{1}{T} = \frac{\omega}{2\pi}$ which shows the number of cycles (oscillations) completed in one second. The parameter $\omega = 2\pi f = \sqrt{km}$ itself represents the number of radians traveled in one second and is called the circular frequency. For example, if $x = \cos 4\pi t - 3 \sin 4\pi t$, then the circular frequency is $\omega = 4\pi$, the frequency is $f = 2$ and the period is $T = \frac{2\pi}{4\pi} = \frac{1}{2}$, which means that within 1 second the mass passes the distance of 4π radians or, equivalently, makes two complete oscillations (cycles), and also that the mass makes one complete oscillation within $\frac{1}{2}$ seconds. To define the extreme displacements (the minimum x_{min} and maximum x_{max} of $x(t)$), we use the formula $x = C\cos(\omega t - \alpha)$, where $C = \sqrt{1+9} = \sqrt{10}$ and the angle α is determined by the relations $\cos\alpha = \frac{1}{\sqrt{10}}$ and $\sin\alpha = \frac{-3}{\sqrt{10}}$. It is clear from this that the largest downward displacement is $x_{max} = \sqrt{10}$ and the highest point the mass reaches is $x_{min} = -\sqrt{10}$.

Problem 1. A mass of 2 kilograms causes the spring to stretch 40 centimeters. Knowing that the spring with the mass was released 24 centimeters below the equilibrium point with an upward velocity of $20cm/s$, find the position of the spring at instant t.

Solution.

The conditions of the problem correspond to those of undamped free oscillations. Therefore, the governing differential equation is $x_{tt} + \omega^2 x = 0$, whose general solution is $x = A\cos\omega t + B\sin\omega t$. To specify the parameter ω, we note that equilibrium with a hanging mass of $2kg$ occurs when $L = 40cm = 40 \cdot 10^{-2}m$. So, from the relation $mg = kL$ it follows that $k = \frac{mg}{L} = \frac{2 \cdot 10}{40 \cdot 10^{-2}} = 50$. Therefore, $\omega^2 = \frac{50}{2} = 25$ and $\omega = 5$. (Note that this value can be found more directly, without calculating k, by the formula $\omega^2 = \frac{k}{m} = \frac{mg}{mL} = \frac{g}{L}$.) To find the constants A and B, we apply the initial conditions $x(0) = 24 \cdot 10^{-2}$ and $x_t(0) = -20 \cdot 10^{-2}$, where $x(0) = A = 24 \cdot 10^{-2}$ and $x_t(0) = B\omega = -20 \cdot 10^{-2}$, $B = \frac{-20 \cdot 10^{-2}}{5} = -4 \cdot 10^{-2}$. Thus, the position of the spring at time t is determined by the formula $x = 10^{-2} \cdot (24\cos 5t - 4\sin 5t)$, or in the alternative form, $x = C\cos(5t+\alpha)$ with $C = 4\sqrt{37} \cdot 10^{-2} \approx 0.2433$ and $\alpha \approx -0.1651$. In addition, we can find the extrema, period and frequency of these oscillations: $x_{max} = 4\sqrt{37} \cdot 10^{-2}$, $x_{min} = -4\sqrt{37} \cdot 10^{-2}$, $T = \frac{2\pi}{5}$ and $f = \frac{5}{2\pi}$. Fig.11.2 shows the oscillations of constant amplitude (simple harmonic oscillations) with the maximum values on the line $x = C$ and the minimum values on the line $x = -C$.

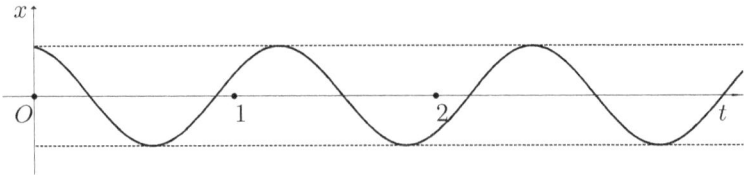

Figure 11.2: Spring-mass problem: undamped free oscillations

1.2 Damped free oscillations

Now let us consider a more realistic situation when there are retarding forces, for example, the resistance forces of the environment, gaseous or liquid, in which the spring and mass are located. Usually, these forces are considered to be proportional to some power of the velocity of the mass, in the simplest case, to the velocity itself. In this case, Newton's second law leads to the relation $ma = mg - \kappa v - k(x + L)$, or recalling that $mg = kL$, we obtain $mx_{tt} = -kx - \kappa x_t$, where $\kappa > 0$ is the damping coefficient and the negative sign is a consequence of the fact that the retarding force acts in the opposite direction to the motion. Dividing the last equation by m and using the parameters $\omega^2 = \frac{k}{m}$ and $2\nu = \frac{\kappa}{m}$, we get the second order linear equation

$$x_{tt} + 2\nu x_t + \omega^2 x = 0,$$

which describes damped free oscillations.

Equations of this type were considered in Chapter 9, section 1. To find the general solution of this equation, we first solve its characteristic equation $\lambda^2 + 2\nu\lambda + \omega^2 = 0$, whose roots are $\lambda_{1,2} = -\nu \pm \sqrt{\nu^2 - \omega^2}$. If the roots are different, then the complex form of the general solution is $x = Ae^{\lambda_1 t} + Be^{\lambda_2 t}$, but the physical interpretation requires the representation in real form, which depends on the sign of the expression $\nu^2 - \omega^2$. Let us consider the three possible cases.

Case 1, overdamping. If $\nu^2 - \omega^2 > 0$ (that is, $\nu > \omega$), then $\lambda_{1,2}$ are real and the real form of the solution is

$$x = e^{-\nu t}\left(Ae^{\sqrt{\nu^2 - \omega^2}\,t} + Be^{-\sqrt{\nu^2 - \omega^2}\,t}\right).$$

Since the damping coefficient is quite large compared to the spring constant (the case of overdamping), the movement does not represent oscillations and the mass quickly approaches the equilibrium point (see Fig.11.3).

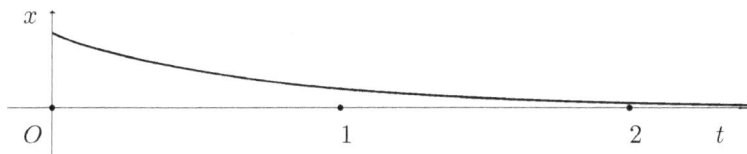

Figure 11.3: Spring-mass problem: overdamped free oscillations.

Case 2, critical damping. If $\nu^2 - \omega^2 = 0$ (that is, $\nu = \omega$), then the roots are real and equal to each other: $\lambda_1 = \lambda_2 = -\nu$, and therefore, the

real solution takes the form

$$x = e^{-\nu t}(A + Bt).$$

Although the expression in the brackets has linear growth, the exponential term dominates in this solution and causes the mass to rapidly approach the equilibrium point (see Fig.11.4).

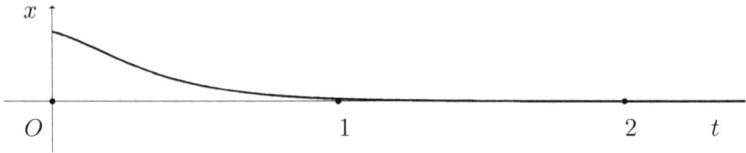

Figure 11.4: Spring-mass problem: critically damped free oscillations

Case 3, underdamping. If $\nu^2 - \omega^2 < 0$ (that is, $\nu < \omega$), then the roots $\lambda_{1,2}$ are complex and the real form of the solution is

$$x = e^{-\nu t}(A \cos \sqrt{\omega^2 - \nu^2}t + B \sin \sqrt{\omega^2 - \nu^2}t).$$

Since the damping coefficient is quite small compared to the spring constant (the underdamping case), mass oscillations are observed throughout the motion, but unlike undamped oscillations, the amplitude of these oscillations is falling with time, getting closer and closer to the equilibrium point (see Fig.11.5).

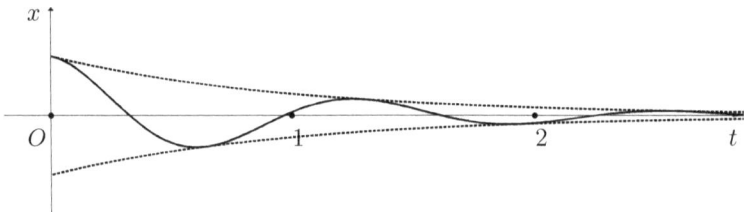

Figure 11.5: Spring-mass problem: underdamped free oscillations

Problem 2. A spring with a mass of 2 kilograms is released 24 centimeters below the equilibrium point with an upward velocity of $20cm/s$. Knowing that the spring constant is $k = 50kg/s^2$ and the force of air resistance is proportional to the speed of motion with the damping coefficient $\kappa = 40s^{-1}$, find the position of the mass at instant t.

Solution.

The conditions of the problem correspond to those of damped free oscillations. Therefore, the governing differential equation is $x_{tt} + 2\nu x_t + \omega^2 x = 0$, where $\omega = \sqrt{\frac{k}{m}} = \sqrt{\frac{50}{2}} = 5$ and $\nu = \frac{\kappa}{2m} = \frac{40}{4} = 10$. Since $\nu > \omega$, we have the overdamping case and the general solution in real form is

$$x = e^{-\nu t}(Ae^{\sqrt{\nu^2 - \omega^2}\,t} + Be^{-\sqrt{\nu^2 - \omega^2}\,t}) = e^{-10t}(Ae^{5\sqrt{3}t} + Be^{-5\sqrt{3}t}).$$

Application of the initial conditions $x(0) = 0.24$ and $x_t(0) = -0.2$ results in the system for A and B: $A + B = 0.24$, $A(-10 + 5\sqrt{3}) + B(-10 - 5\sqrt{3}) = -0.2$. The solution of this system is $A = \frac{-0.1 + 0.6(2 + \sqrt{3})}{5\sqrt{3}} \approx 0.2470$, $B = \frac{0.1 - 0.6(2 - \sqrt{3})}{5\sqrt{3}} \approx 0.0070$. The mass quickly approaches the equilibrium point (see Fig.11.3).

Problem 3. A spring with a mass of 2 kilograms is released 24 centimeters below the equilibrium point with an upward velocity of $20cm/s$. Knowing that the spring constant is $k = 50kg/s^2$ and the force of air resistance is proportional to the speed of motion with the damping coefficient $\kappa = 20s^{-1}$, find the position of the spring at instant t.

Solution.

The conditions of the problem correspond to those of damped free oscillations. Therefore, the governing differential equation is $x_{tt} + 2\nu x_t + \omega^2 x = 0$, where $\omega = \sqrt{\frac{k}{m}} = \sqrt{\frac{50}{2}} = 5$ and $\nu = \frac{\kappa}{2m} = \frac{20}{4} = 5$. Since $\nu = \omega$, we have the case of critical damping and the general solution has the real form

$$x = e^{-\nu t}(A + Bt) = e^{-5t}(A + Bt).$$

Application of the initial conditions $x(0) = 0.24$ and $x_t(0) = -0.2$ results in the following relationships: $A = 0.24$, $-5A + B = -0.2$, whence $B = 1$. The mass quickly approaches the equilibrium point (see Fig.11.4).

Problem 4. A spring with a mass of 2 kilograms is released 24 centimeters below the equilibrium point with an upward velocity of $20cm/s$. Knowing that the spring constant is $k = 50kg/s^2$ and the force of air resistance is proportional to the speed of motion with the damping coefficient $\kappa = 4s^{-1}$, find the position of the spring at instant t.

Solution.

The conditions of the problem correspond to those of damped free oscillations. Therefore, the governing differential equation is $x_{tt} + 2\nu x_t + \omega^2 x = 0$, where $\omega = \sqrt{\frac{k}{m}} = \sqrt{\frac{50}{2}} = 5$ and $\nu = \frac{\kappa}{2m} = \frac{4}{4} = 1$. Since $\nu < \omega$, we have the underdamping case and the general solution has the following real form:

$$x = e^{-\nu t}(A\cos\sqrt{\omega^2 - \nu^2}\,t + B\sin\sqrt{\omega^2 - \nu^2}\,t) = e^{-t}(A\cos 2\sqrt{6}t + B\sin 2\sqrt{6}t).$$

Application of the initial conditions $x(0) = 0.24$ and $x_t(0) = 0.2$ results in the following equations: $A = 0.24$ and $-A + 2\sqrt{6}B = -0.2$, whence $B = \frac{0.02}{\sqrt{6}} \approx$ 0.0082. The alternative form of the solution is $x = e^{-t}C\cos(2\sqrt{6}t + \alpha)$ with $C \approx 0.2401$ and $\alpha \approx 0.0340$. The mass is oscillating around the equilibrium point, with the amplitude of the oscillations decreasing with time. Fig.11.5 shows that the extreme values are contained between the curves $x = \pm Ce^{-t}$.

1.3 Forced oscillations

Let us consider the motion of the spring in the presence of an external force $F(t)$. According to Newton's second law, we have $ma = mg - \kappa v - k(x + L) + F$. Using the relation $mg = kL$, dividing by m and denoting $f = \frac{F}{m}$, we get the second order linear equation

$$x_{tt} + 2\nu x_t + \omega^2 x = f$$

which describes forced damped oscillations. Equations of this type with different right-hand parts f were considered in Chapter 9, sections 2 and 3. A phenomenon of special interest occurs when the motion of a spring is carried out in the presence of a vibrating force (for example, the spring is hanging from an oscillating support). In this case, we can model the external force by the function $f = D\sin\gamma t$, where constants D and γ represent the amplitude and circular frequency of external oscillations. Let us consider two cases – undamped and underdamped.

    Case 1, undamping. This situation is described by the equation

$$x_{tt} + \omega^2 x = D\sin\gamma t.$$

Remember that the general solution of this equation can be represented as the sum of the general solution of the corresponding homogeneous equation and a particular solution of the original equation. The general solution of the homogeneous part is $x_{gh} = A\cos\omega t + B\sin\omega t$, where A and B are two arbitrary constants. If $\omega \neq \gamma$, then a particular solution can be found in the form $x_{pn} = a\cos\gamma t + b\sin\gamma t$. Substituting x_{pn} into the original equation, we get

$$(-a\gamma^2\cos\gamma t - b\gamma^2\sin\gamma t) + \omega^2(a\cos\gamma t + b\sin\gamma t) = D\sin\gamma t.$$

Comparing the coefficients on both sides, we find $a(-\gamma^2 + \omega^2) = 0$ and $b(-\gamma^2 + \omega^2) = D$, whence $a = 0$ and $b = \frac{D}{\omega^2 - \gamma^2}$. Therefore, the general solution has the form

$$x_{gn} = x_{gh} + x_{pn} = A\cos\omega t + B\sin\omega t + \frac{D}{\omega^2 - \gamma^2}\sin\gamma t.$$

If $\omega = \gamma$, then a particular solution can be sought in the form $x_{pn} = at \cos \gamma t + bt \sin \gamma t$. Substituting x_{pn} into the original equation, we get

$$(-at\gamma^2 \cos\gamma t - bt\gamma^2 \sin\gamma t - 2a\gamma \sin\gamma t + 2b\gamma\cos\gamma t) + \omega^2(at\cos\gamma t + bt \sin\gamma t) = D\sin\gamma t.$$

Taking into account that $\omega = \gamma$, we simplify this relation to the form $-2a\gamma \sin \gamma t + 2b\gamma \cos \gamma t = D \sin \gamma t$, whence $a = -\frac{D}{2\gamma}$ and $b = 0$. Thus, we find the particular solution $x_{pn} = -\frac{D}{2\omega} t \cos \omega t$ and the general one

$$x_{gn} = x_{gh} + x_{pn} = A \cos \omega t + B \sin \omega t - \frac{D}{2\omega} t \cos \omega t.$$

An important specificity of this solution is that the amplitude of its oscillations grows with time, going to infinity. This is a situation of resonance, which occurs when the frequency of the spring free oscillations coincides with the frequency of the external force. Fig.11.6 shows that in the initial period, when the contribution of the linear term is small, the oscillations are similar to the case of free motion, but after this period the linear term begins to dominate and, with increasing time, the amplitude of the oscillations almost reach the values on the lines $x = \pm\frac{D}{2\omega} t$.

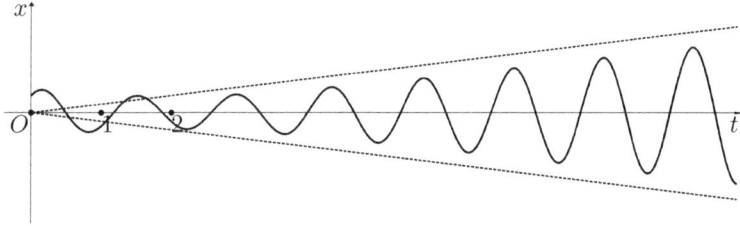

Figure 11.6: Spring-mass problem: forced undamped oscillations

Case 2, underdamping. In this case, the equation takes the form

$$x_{tt} + 2\nu x_t + \omega^2 x = D \sin \gamma t$$

with $\nu < \omega$. The roots of the characteristic equation are complex $\lambda_{1,2} = -\nu \pm i\sqrt{\omega^2 - \nu^2}$ and the general solution of the homogeneous part has the real form

$$x_{gh} = e^{-\nu t}(A \cos \sqrt{\omega^2 - \nu^2}t + B \sin \sqrt{\omega^2 - \nu^2}t),$$

where A and B are two arbitrary constants. A particular solution can be found in the form $x_{pn} = a \cos \gamma t + b \sin \gamma t$. Substituting x_{pn} into the original equation, we get

$$(-a\gamma^2\cos\gamma t - b\gamma^2\sin\gamma t) - 2\nu(a\gamma\sin\gamma t - b\gamma\cos\gamma t) + \omega^2(a\cos\gamma t + b\sin\gamma t) = D\sin\gamma t.$$

Comparing the coefficients on both sides, we find the system of two equations $(\omega^2 - \gamma^2)a + 2\nu\gamma b = 0$ and $-2\nu\gamma a + (\omega^2 - \gamma^2)b = D$, whence $a = \frac{-2\nu\gamma D}{(\omega^2-\gamma^2)^2+4\nu^2\gamma^2}$ and $b = \frac{(\omega^2-\gamma^2)D}{(\omega^2-\gamma^2)^2+4\nu^2\gamma^2}$. Therefore, the general solution has the form

$$x_{gn} = x_{gh} + x_{pn} = e^{-\nu t}(A\cos\sqrt{\omega^2 - \nu^2}t + B\sin\sqrt{\omega^2 - \nu^2}t)$$

$$+\frac{-2\nu\gamma D}{(\omega^2 - \gamma^2)^2 + 4\nu^2\gamma^2}\cos\gamma t + \frac{(\omega^2 - \gamma^2)D}{(\omega^2 - \gamma^2)^2 + 4\nu^2\gamma^2}\sin\gamma t.$$

Although in this case we do not observe the growth of oscillations, as in the case of resonance, but if $\omega = \gamma$, then the particular solution takes the form $x_{pn} = \frac{-D}{2\nu\gamma}\cos\gamma t$ and for small values of ν the forced oscillations can still have a large amplitude. As time progresses, the homogeneous part of the solution becomes smaller and smaller due to the exponential multiplier $e^{-\nu t}$ (even for small values of ν), while the particular part of the solution keeps the same amplitude $\frac{D}{2\nu\gamma}$ (which can be quite large for small ν) and becomes the main part of the solution. Some situations with different values of ν are shown in Fig.11.7 and 11.8.

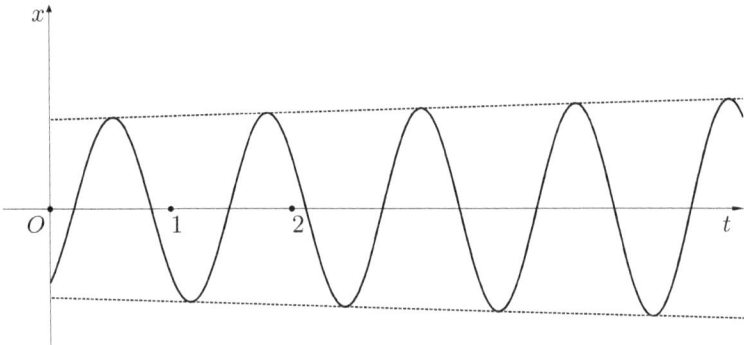

Figure 11.7: Spring-mass problem: forced underdamped oscillations, $\nu{=}0.1$.

2   Electrical circuit problem

A mathematical model similar to the spring-mass problem occurs when considering the flow of electric current $I(t)$ in a closed series circuit. In this

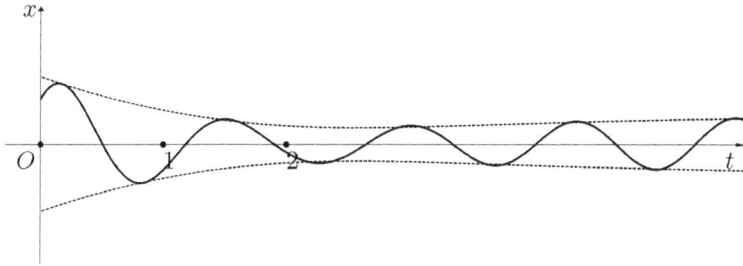

Figure 11.8: Spring-mass problem: forced underdamped oscillations, $\nu=0.5$.

circuit, the following elements are included: the inductor L of a constant inductance, which causes the voltage drop $L\frac{dI}{dt}$; the constant resistance R, which causes the voltage drop IR; and the capacitor C of a constant capacitance, which causes the voltage drop $\frac{q}{C}$, where q is the charge on the capacitor, related to the current by the formula $I = \frac{dq}{dt}$. This type of electric current is called the $L-R-C$ circuit. According to Kirchoff's law, in a closed circuit, the voltage $E(t)$ applied in the circuit is equal to the sum of the voltage drops in the rest of the circuit, that is, $L\frac{dI}{dt}+RI+\frac{q}{C} = E(t)$. Substituting the relation $I = \frac{dq}{dt}$ into the last formula, we get the second-order linear equation:

$$Lq_{tt} + Rq_t + \frac{1}{C}q = E(t).$$

Mathematically, the form of the last equation coincides with that obtained in the spring-mass problem in the case of forced oscillations with damping. Therefore, the nomenclature used in circuit analysis is similar to that of the spring-mass problem: if $E = 0$, the electrical vibrations of the circuit are said to be free. If additionally $R = 0$ (there is no resistance), then the oscillations are undamped, maintaining the same amplitude throughout the period (free oscillations or simple harmonic oscillations). When $R > 0$, there are three forms of the current, depending on the discriminant $D = R^2 - \frac{4L}{C}$. If $D > 0$ then we have the overdamping regime; if $D = 0$ then we have critical damping; and if $D < 0$ then we have underdamping. In all three cases the current tends to 0 as time progresses. When there is a voltage $E(t)$ induced in the circuit, the electrical vibrations are called forced. In particular, in a circuit without resistance, there will be resonance if the frequency of the voltage $E(t)$ is equal to the frequency of free oscillations.

Problem 1. Determine the charge on the capacitor in a circuit with no resistor and no induced voltage, if initially there is no current in the circuit.

Solution.

In this case, we have the free vibration equation $Lq_{tt} + \frac{1}{C}q = 0$, whose general solution is $q = C_1 \cos \frac{1}{LC}t + C_2 \sin \frac{1}{LC}t$. Assuming that the initial charge on the capacitor was $q(0) = q_0$ and applying the condition of no current at the initial instant $q_t(0) = 0$, we specify the constants C_1 and C_2: $C_1 = q_0$ and $C_2 = 0$. Then, $q = q_0 \cos \frac{1}{LC}t$.

Problem 2. Find the charge on the capacitor in a circuit with no induced voltage and parameters $L = 0.25 henry$, $R = 10 ohms$, $C = 0.001 farad$, if the initial charge is q_0 and the initial current is zero.

Solution.

In this case, the governing equation is $\frac{1}{4}q_{tt} + 10q_t + 1000q = 0$. Since $D = R^2 - \frac{4L}{C} = 100 - 1000 = -900 < 0$, we have the case of underdamped vibrations. The roots of the characteristic equation are $\lambda_{1,2} = -20 \pm 60i$ and the general solution is $q = e^{-20t}(C_1 \cos 60t + C_2 \sin 60t)$. Applying the initial conditions $q(0) = q_0$ and $q_t(0) = 0$, we specify $C_1 = q_0$ and $C_2 = \frac{q_0}{3}$. So the solution of the problem is $q = q_0 e^{-20t}(\cos 60t + \frac{1}{3}\sin 60t)$.

3   Pendulum problem

A simple pendulum is a body suspended on a rigid weightless rod, which is attached at the top to a rigid support. This body swings back and forth, that is, moves around the vertical position of equilibrium along the arc of the circle centered at the point of attachment of the rod. Let us assume that the entire mass m of a pendulum is concentrated at the end of the rod (the arm of a pendulum) at a constant distance l from the point of attachment (the length of a pendulum), and that there are only two forces acting on the body: the gravitational force and the counter-force of the rod acting along the rod. So, at each moment, the component of the gravitational force perpendicular to the motion is compensated by the counter-force of the rod. Therefore, the only force causing the pendulum to move is F_a component of the gravitational force $F_g = mg$ acting along the arc of the circle. When a pendulum is displaced from its equilibrium position, it is subject to the force F_a that will accelerate a pendulum back towards the equilibrium position. We denote by s the displacement of the pendulum along the arc and by θ the angle that the rod (the arm of a pendulum) forms with the vertical axis, setting $s = 0$ and $\theta = 0$ when the pendulum is in the vertical position (see Fig.11.9). Choosing the displacement to the right as positive, we can write Newton's second law in the form $ms_{tt} = -mg \sin \theta \equiv F_a$, or $s_{tt} = -g \sin \theta$, where the negative sign means that the force acts in the negative direction when $\theta > 0$ and in the positive direction when $\theta < 0$. The relationship between the displacement s and the angle θ is $s = l\theta$, whence $s_{tt} = l\theta_{tt}$. Therefore, in terms of the angle θ, the pendulum equation can be written in the form

$$\theta_{tt} + \omega^2 \sin\theta = 0,$$

where $\omega = \sqrt{g/l}$.

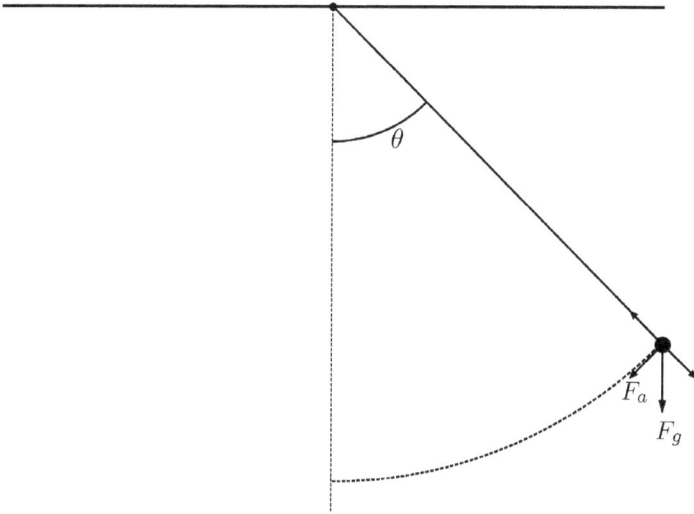

Figure 11.9: Pendulum problem: forces and motion

The standard small-angle approximation $\sin\theta \approx \theta$ results in the simple harmonic oscillation equation

$$\theta_{tt} + \omega^2\theta = 0,$$

whose solution has already been analyzed in the case of the spring-mass problem.

The complete equation is non-linear and therefore its solution is more complicated. First, we note that t is not present in the equation, which allows us to reduce its order by changing the unknown function: $y(\theta) = \theta_t$ (see Chapter 7, section 3). With the relation $y_\theta \cdot \theta_t = y_\theta \cdot y = \theta_{tt}$, this leads to the following first-order equation for $y(\theta)$: $y_\theta \cdot y + \omega^2 \sin\theta = 0$. This is an equation of separable variables, whose solution is $\int y\,dy = -\omega^2 \int \sin\theta\,d\theta$ or $y^2 = 2\omega^2 \cos\theta + C$. Considering that at the initial moment the angle of displacement was $\theta(0) = \theta_0$ and the initial angular velocity was zero $\theta_t(0) = 0$ (we use this condition to simplify later analysis), we find the solution in the form $y^2 = 2\omega^2(\cos\theta - \cos\theta_0)$. Returning to the original unknown function $\theta(t)$, we have the two separable equations $\theta_t = \pm\sqrt{2}\omega\sqrt{\cos\theta - \cos\theta_0}$, whose solution in integral form is

$$\int \frac{d\theta}{\sqrt{\cos\theta - \cos\theta_0}} = \pm\sqrt{2}\omega t.$$

The problem with the found solution is that the integral on the left-hand side cannot be calculated in terms of elementary functions. However, we can reduce this integral to a tabulated integral, whose properties are known, called the incomplete elliptic integral of the first kind. To do this, we first use the trigonometric formulas $\cos\theta = 1 - 2\sin^2\frac{\theta}{2}$ and $\cos\theta_0 = 1 - 2\sin^2\frac{\theta_0}{2}$ to rewrite the root of the denominator in the form $\sqrt{2\sin^2\frac{\theta_0}{2} - 2\sin^2\frac{\theta}{2}}$. Next, we change the variable in the integral: $\sin\phi = \frac{\sin(\theta/2)}{\sin(\theta_0/2)}$ (according to the physical meaning of the problem and the initial conditions, we have that $|\theta| \leq |\theta_0| \leq \pi$, where θ_0 is the initial displacement, and therefore, the function $\sin\phi$ is defined for the values $\theta \in [-\theta_0, \theta_0]$, and consequently, $\phi \in [-\frac{\pi}{2}, \frac{\pi}{2}]$). Denoting $k = \sin\frac{\theta_0}{2}$ and calculating the differential $\cos\phi d\phi = \frac{1}{2a}\cos\frac{\theta}{2}d\theta$, whence

$$d\theta = 2k\frac{\cos\phi}{\cos\frac{\theta}{2}}d\phi = 2k\frac{\cos\phi}{\sqrt{1 - \sin^2\frac{\theta}{2}}}d\phi = 2k\frac{\cos\phi}{\sqrt{1 - k^2\sin^2\phi}}d\phi,$$

we get

$$\int \frac{d\theta}{\sqrt{\cos\theta - \cos\theta_0}} = \frac{1}{\sqrt{2}}\int \frac{1}{\sqrt{k^2 - k^2\sin^2\phi}} \cdot \frac{2k\cos\phi}{\sqrt{1 - k^2\sin^2\phi}}d\phi$$

$$= \sqrt{2}\int \frac{d\phi}{\sqrt{1 - k^2\sin^2\phi}}.$$

Thus, the integral solution for the function ϕ can be written in the form $\int\frac{d\phi}{\sqrt{1-k^2\sin^2\phi}} = \pm\omega t$. Finally, on the left-hand side, instead of the indefinite integral, we can use one of the associated antiderivatives in the form of the integral with variable upper limit and move an arbitrary constant to the right-hand side:

$$\int_0^\phi \frac{dp}{\sqrt{1 - k^2\sin^2 p}} = \pm\omega t + C.$$

The integral on the left-hand side is the incomplete elliptic integral of the first kind. To specify the constant C, we apply the initial condition $\theta(0) = \theta_0$

which, in terms of the variable ϕ, takes the form $\phi(0) = \frac{\pi}{2}$. Therefore, $\int_0^{\pi/2} \frac{dp}{\sqrt{1-k^2 \sin^2 p}} = 0 + C$ and the solution is

$$\pm \omega t = \int_0^{\phi} \frac{dp}{\sqrt{1-k^2 \sin^2 p}} - \int_0^{\pi/2} \frac{dp}{\sqrt{1-k^2 \sin^2 p}},$$

where the second integral is the full elliptic integral of the first kind. The two integrals on the right-hand side are tabulated and, using their values, we can find the solution $\theta(t)$. The two cases of the solution with the parameter $l = 2.5$ (that is, $\omega = 2$), are shown in Fig.11.10: the graph of the first solution, with the black lines and annotations, corresponds to the case $\theta_0 = \frac{2\pi}{3}$ (that is, $k = \frac{\sqrt{3}}{2}$), the graph of the second solution, with the blue lines and annotations, corresponds to the case $\theta_0 = \frac{\pi}{3}$ (that is, $k = \frac{1}{2}$).

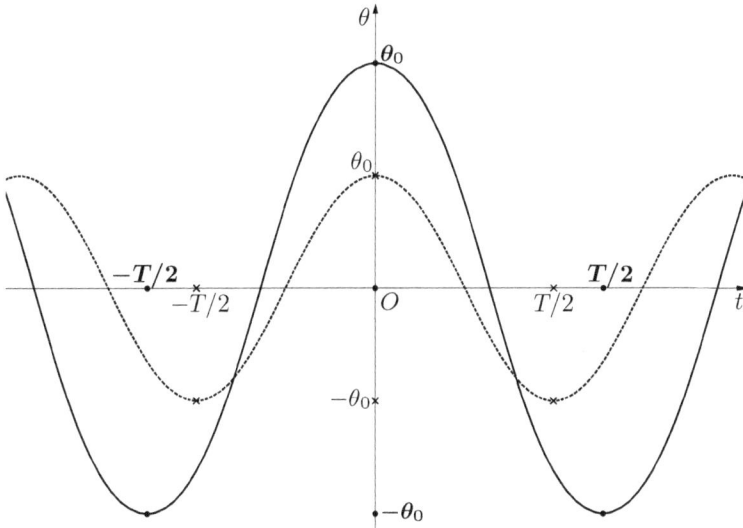

Figure 11.10: Pendulum problem: graph of the solution for $\omega = 2$ and the two initial conditions: $\theta_0 = \frac{2\pi}{3}$ (solid line, bold font) and $\theta_0 = \frac{\pi}{3}$ (dashed line, regular font).

It is interesting to note that, unlike simple harmonic oscillations, the period T of pendulum oscillations is not equal to $\frac{2\pi}{\omega}$, although in the small-angle approximation it is very close to this value. In general, the period is determined by the formula $\frac{T}{4} = \frac{1}{\omega} \int_0^{\pi/2} \frac{dp}{\sqrt{1-k^2 \sin^2 p}}$. It can be seen that T depends on k, that is, on the value of the initial position of pendulum (which is equal to the amplitude of the oscillations). For the two examples shown in Fig.11.10, the period is equal to $T \approx 4.3129$ in the case $\theta_0 = \frac{2\pi}{3}$ and $T \approx 3.3715$ in the case $\theta_0 = \frac{\pi}{3}$, while the period of harmonic oscillations would be $\frac{2\pi}{\omega} = \frac{2\pi}{2} = \pi \approx 3.1415$.

4   Pursuit problem

Let us look at the classic pursuit problem, analyzed by French mathematician Pierre Bourguer in the 18th century, and therefore, frequently called Bourguer's problem. Bourguer had studied the problem of chasing of a cargo ship by pirates under the following simplifying conditions: the two ships are sailing at a constant speed, the cargo ship is traveling along a straight line, while the pirate ship is always sailing directly towards the current position of the cargo ship. (Remember that the speed is the absolute value of the velocity vector.) Under these assumptions, denoting the initial distance between the two ships by d, we can choose the planar coordinates in such a way that the path of the cargo ship is along the line $x = x_0$ with the initial position at $(x_0, 0)$, $x_0 = d > 0$ and the initial position of the pirates is at the origin of the coordinates.

Let us denote the path of the pirates by $y(x)$ with $y(0) = 0$. According to the condition of the problem, the pirate ship always moves toward the current position of the merchant. In mathematical terms, this means that at any moment t the tangent to the trajectory of the pirates must pass through the current position of the merchant $(x_0, v_c t)$, where v_c is the constant speed of the merchant. Since the current position of the pirates is (x, y), the slope of the tangent line must coincide with the slope $\frac{y - v_c t}{x - x_0}$ of the line passing through the points $(x_0, v_c t)$ and (x, y). Recalling that the slope of the tangent (and of the path curve) is determined by the derivative y_x, we express this condition using the formula $y_x = \frac{y - v_c t}{x - x_0}$.

We can derive one more relationship involving the derivative y_x. Since the pirate ship is sailing with a constant speed v_p, we conclude that during the time t it will travel the distance $v_p t$. On the other hand, according to the arc length formula, the distance the pirate will travel along its path $y(x)$ is equal to $\int_0^x \sqrt{1 + (y_x)^2} du$ (here u is simply the variable of integration). Therefore, $v_p t = \int_0^x \sqrt{1 + (y_x)^2} du$.

In both found relations, there is the time variable t, which is unrelated to the problem under consideration (since we want to know the path y as a function of x). To eliminate t, we substitute its expression from the second formula into the first and obtain

$$y_x(x - x_0) = y - v_c \frac{1}{v_p} \int_0^x \sqrt{1 + (y_x)^2} du.$$

Finally, by differentiating the last formula with respect to x, we arrive at the second order equation for the unknown function $y(x)$:

$$y_{xx}(x - x_0) + y_x = y_x - \frac{v_c}{v_p} \sqrt{1 + (y_x)^2}$$

or, simplifying and denoting $k = \frac{v_c}{v_p}$, we get

$$y_{xx}(x - x_0) = -k\sqrt{1 + (y_x)^2}.$$

Since this equation does not contain y explicitly, it allows the order reduction by substitution of the function $y_x = p$: $p_x(x - x_0) = -k\sqrt{1 + p^2}$ (equations of this type were considered in Chapter 7, section 2). The last is a separable equation, whose solution is found using the traditional technique, separating the variables and integrating each side with respect to its variable: $\int \frac{dp}{\sqrt{1+p^2}} = -k \int \frac{dx}{x - x_0}$. Therefore,

$$\ln(p + \sqrt{1+p^2}) = -k \ln|x - x_0| + A \text{ or } p + \sqrt{1+p^2} = \frac{A}{(x_0 - x)^k}, A = const.$$

(Note that, by the meaning of the problem, $x < x_0$ at any instant t before the pirates reach the cargo ship, and therefore $|x - x_0| = x_0 - x$.)

Applying the assumption that the pirate ship is always heading toward the merchant, at the initial moment we get the condition $y_x(0) = p(0) = 0$, since the two ships are initially positioned on the x-axis. Using this condition in the formula for p, we specify the constant A: $1 = \frac{A}{x_0^k}$, whence $A = x_0^k$.

So, $p + \sqrt{1 + p^2} = \frac{1}{(1 - x/x_0)^k} \equiv q$, where the notation q has been introduced to simplify the presentation of the subsequent transformations. Let us now simplify the equation for p. First we rewrite it in the form $\sqrt{1 + p^2} = q - p$ and eliminate the root by squaring both sides: $1 + p^2 = (q - p)^2$. Opening the square on the right-hand side and cutting p^2, we get $p = \frac{q^2 - 1}{2q} = \frac{1}{2}(q - \frac{1}{q})$. Returning to the expression for q and recalling that $p = y_x$, we get

$$y_x = \frac{1}{2}\left((1 - \frac{x}{x_0})^{-k} - (1 - \frac{x}{x_0})^k\right).$$

To find $y(x)$, it remains to integrate the last equation. Changing the variable in the integral $u = 1 - \frac{x}{x_0}$, we find:

$$\int (1 - \frac{x}{x_0})^{-k} - (1 - \frac{x}{x_0})^k)dx = \int (u^{-k} - u^k) \cdot (-x_0)du$$

$$= -x_0 \left(\frac{u^{-k+1}}{-k+1} - \frac{u^{k+1}}{k+1} \right) + C = -x_0 \left(\frac{(1-x/x_0)^{-k+1}}{-k+1} - \frac{(1-x/x_0)^{k+1}}{k+1} \right) + C.$$

This formula is valid when $k \neq 1$. Substituting this result into the equation for y, we get

$$y = -\frac{x_0}{2} \left(\frac{(1-x/x_0)^{-k+1}}{-k+1} - \frac{(1-x/x_0)^{k+1}}{k+1} \right) + C.$$

Using the initial position of the pirates $y(0) = 0$, we determine the constant C: $C = \frac{x_0}{2} \left(\frac{1}{-k+1} - \frac{1}{k+1} \right) = -x_0 \frac{k}{k^2-1}$. So the final solution takes the form

$$y = -\frac{x_0}{2} \left(\frac{(1-x/x_0)^{-k+1}}{-k+1} - \frac{(1-x/x_0)^{k+1}}{k+1} \right) - x_0 \frac{k}{k^2-1}$$

or

$$y = -\frac{x_0 - x}{2} \left(\frac{(1-x/x_0)^{-k}}{-k+1} - \frac{(1-x/x_0)^{k}}{k+1} \right) - x_0 \frac{k}{k^2-1}.$$

The capture occurs when the pirate ship catch up the merchant, that is, when $x = x_0$. At this point, $y_0 = x_0 \frac{k}{1-k^2}$. This can only happen when the speed of pirates is greater than the speed of merchant, that is, when $k < 1$. For example, if $k = \frac{1}{2}$ the pirates reaches the merchant at the point $P_0 = (x_0, \frac{2}{3}x_0)$; if $k = \frac{3}{4}$, then $P_0 = (x_0, \frac{12}{7}x_0)$. Naturally, as k increases (for values less than 1) the point P_0 moves further and further away from the x-axis. Knowing the point of capture P_0, we conclude that the merchant sailed the distance $d = x_0 \frac{k}{1-k^2}$ before being captured and this occurred at the instant $\frac{d}{v_c}$. Since the speed of the pirates is $\frac{1}{k}$ times greater, the distance they will travel is $x_0 \frac{1}{1-k^2}$.

If $k > 1$, then the pirates will never capture the merchant. Even so, their path is still represented by the found solution.

If $k = 1$, the pirates will not capture the merchant either, and furthermore, the integration performed to find y is not valid. In this case, we have $y_x = \frac{1}{2} \left((1 - \frac{x}{x_0})^{-1} - (1 - \frac{x}{x_0}) \right)$. The integration of the function on the right-hand side can be done using the same change of the variable $u = 1 - \frac{x}{x_0}$:

$$\int (1 - \frac{x}{x_0})^{-1} - (1 - \frac{x}{x_0}) dx = \int (u^{-1} - u) \cdot (-x_0) du$$

$$= -x_0 (\ln|u| - \frac{u^2}{2}) + C = -x_0 \left(\ln(1 - \frac{x}{x_0}) - \frac{1}{2}(1 - \frac{x}{x_0})^2 \right) + C.$$

Then,

$$y = -\frac{x_0}{2} \left(\ln(1 - \frac{x}{x_0}) - \frac{1}{2}(1 - \frac{x}{x_0})^2 \right) + C.$$

From the initial condition $y(0) = 0$, we find $C = -\frac{x_0}{4}$ and therefore,

$$y = -\frac{x_0}{2} \left(\ln(1 - \frac{x}{x_0}) - \frac{1}{2}(1 - \frac{x}{x_0})^2 \right) - \frac{x_0}{4}.$$

This is the trajectory of the pirates in the case $k = 1$.

Fig.11.11 shows the path of the pirate ship for the values $k = \frac{1}{2}$, $k = \frac{3}{4}$, $k = 1$ and $k = 2$, and the capture points P_0 for $k = \frac{1}{2}$ and $k = \frac{3}{4}$.

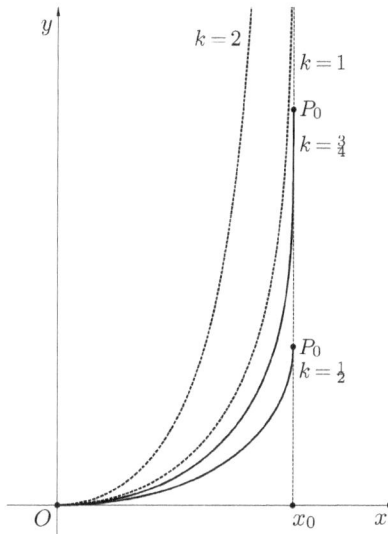

Figure 11.11: Pursuit problem: different relations between ship speeds.

Problems for Chapter 11

1. A ball of mass $1kg$ is attached to a spring hanging from a fixed upper support. In equilibrium, the spring is stretched $70cm$. The ball is pushed upwards with a speed of $1m/s$ from the equilibrium position. Deduce the equation of motion and find its solution in the absence of damping and external forces. Consider that the gravitational acceleration is $g = 9.8m/s^2$.

Solve the problem in the case of the existence of damping with the coefficient $\kappa = 10kg/s$. Solve the same problem if the damping coefficient is $\kappa = 40kg/s$.

Solve the problem when the external force $F = 5\sin t \, kg/s^2$ is present in addition to the damping.

2. A ball of mass $100g$ is attached to a spring hanging from a fixed upper support. In equilibrium, the spring is stretched $25cm$. The ball is pulled down an additional $50cm$ and released with zero initial velocity. If the damping coefficient is $\kappa = 4kg/s$, find the equation of motion and its solution. Solve the same problem if the damping coefficient is $\kappa = 8kg/s$.

Solve the problem in the presence of the external force $F = 2\sin t kg/s^2$.

3. A ball of mass $20g$ is attached to a spring hanging from a fixed upper support. In the equilibrium state, the spring stretches $4cm$. The ball has been pulled up $1cm$ from the equilibrium position and pushed down with a speed of $1cm/s$. Deduce the equation of motion and find its solution in the absence of damping and external forces. Consider that the gravitational acceleration is $g = 9.8m/s^2$.

Solve the problem in the case of the existence of damping with the coefficient $\kappa = 10g/s$ and the external force $F = \cos t kg/s^2$.

4. Find the charge on the capacitor in a circuit with no induced voltage if $L = 0.04henry$, $R = 20ohms$, $C = 4 \cdot 10^{-4} farad$. The initial charge is $q_0 = 2 \cdot 10^{-3} farad$ and the initial current is zero.

Solve the problem in the case of the constant voltage $E = 100$ induced in the circuit.

Solve the problem in the case of the voltage $E = 100\sin t$ induced in the circuit.

5. For an electrical circuit with no induced voltage, determine whether the equation of the charge on the capacitor represents simple harmonic oscillations, overdamped oscillations, underdamped oscillations or critically damped oscillations in the following cases:

1) $L = 0.5$, $R = 0$, $C = 2 \cdot 10^{-5}$;
2) $L = 0.1$, $R = 20$, $C = 1 \cdot 10^{-3}$;
3) $L = 0.1$, $R = 20$, $C = 5 \cdot 10^{-4}$;
4) $L = 0.2$, $R = 10$, $C = 2 \cdot 10^{-2}$.

In all cases L is measured in *henry*, R in *ohm* and C in *faraday*. Find the solutions to the given problems.

6. A child weighing $40kg$ is swinging on a swing whose length is $5m$. Disregarding air resistance and using the small angle approximation, determine the period and frequency of this movement. Assuming that the child was pushed with a speed of $10m/s$ from the vertical position, find the law and maximum height of these oscillations. What will change if an adult weighing $80kg$ sits instead of a child? What will change if the length of the swing becomes $8m$?

7. Consider two identical pendulums, one on the surface of the Earth, where the gravitational acceleration is $g_T = 10m/s^2$, and the other on the surface of the Moon, where the gravitational acceleration is $g_L = 1.6m/s^2$, both

subjected to the same initial displacement with the same initial velocity. Disregarding air resistance and using the small angle approximation, determine which pendulum has the greater frequency of oscillations and which pendulum has the greater amplitude.

Now suppose that the two pendulums have the same initial displacement and zero initial velocity. Disregarding air resistance, but using the complete pendulum equation, determine which pendulum has the greater frequency of oscillations and which pendulum has the greater amplitude. (Use the differential formulas of the solution and the period integrals deduced at the end of section 3.)

Suppose that the two pendulums have zero initial displacement and zero initial velocity. Disregarding air resistance, but using the complete pendulum equation, determine which pendulum has the greater frequency of oscillations and which pendulum has the greater amplitude. (Use the differential formulas of the solution and the period integrals deduced at the end of section 3.)

8. Find the trajectory of the pirate ship in the pursuit problem if it is three times faster than the merchant and the latter departs from the point $(10, 0)$, where the measurements are given in km. Also find the point of capture and the distance traveled. How long does this chase take, if the speed of the merchant is $10km/h$.

9. A rabbit runs with constant speed along the y-axis in the positive direction, staring at the point $(0, a)$, $a > 0$. A dog chases this rabbit with constant speed, starting at the point $(b, 0)$, $b > 0$, always running toward the current position of the rabbit. Deduce the differential equation that describes the path of the dog. Find the solution to this equation, analyze when the dog catches the rabbit, the point of capture and the distance covered.

Give the specific solution and geometric representation of the path of the dog in the case when $a = 300m$, $b = 100m$, $v_c = 8m/s$ (v_c is the speed of the rabbit) and the speed of the dog is

1) $v_d = 8m/s$;
2) $v_d = 10m/s$;
3) $v_d = 16m/s$.

Chapter 12

Systems of equations: theoretical properties

1 General concepts and basic results

To begin with, we introduce the basic concepts of systems of ordinary differential equations.

1.1 Definitions of systems and their solutions

Definition of a general system. A *general system of m equations for n unknowns* in general form can be written as follows

$$
\begin{cases}
F_1(x, y_1, y_1', \ldots, y_1^{(k_{11})}, y_2, y_2', \ldots, y_2^{(k_{12})}, \ldots, y_n, y_n', \ldots, y_n^{(k_{1n})}) = 0 \\
F_2(x, y_1, y_1', \ldots, y_1^{(k_{21})}, y_2, y_2', \ldots, y_2^{(k_{22})}, \ldots, y_n, y_n', \ldots, y_n^{(k_{2n})}) = 0 \\
\qquad \cdots \\
F_m(x, y_1, y_1', \ldots, y_1^{(k_{m1})}, y_2, y_2', \ldots, y_2^{(k_{m2})}, \ldots, y_n, y_n', \ldots, y_n^{(k_{mn})}) = 0
\end{cases}.
$$

Here, $F_1, F_2, \ldots, F_m$ are given functions, x is the independent variable, $y_1, y_2, \ldots, y_n$ are unknown functions and the order k of the system is defined as the highest order of the derivative found in the system: $k = \max\limits_{i,j} k_{ij}$.

Remark. An equation of order k for an unknown function is a particular case of this system with $m = 1$, $n = 1$.

Definition of a normal system. A *normal system of n equations for* *n unknowns* can be written as follows

$$\begin{cases} y_1' = f_1(x, y_1, y_2 \ldots, y_n) \\ y_2' = f_2(x, y_1, y_2 \ldots, y_n) \\ \quad \cdots \\ y_n' = f_n(x, y_1, y_2 \ldots, y_n) \end{cases}$$

Here, $f_1, f_2, \ldots, f_n$ are given functions, x is the independent variable and $y_1, y_2, \ldots, y_n$ are unknown functions. It is important to point out the following characteristics of the normal system compared to a general one:
1) the number of equations is equal to the number of unknown functions;
2) each equation in the system is of the first order;
3) the ith equation of the system contains only the derivative of the ith function y_i in explicit form.
Because of these features, the order of the normal system is usually defined as the number of equations in the system (or, equivalently, the number of unknowns), that is, n.

Remark. Introducing the vector functions $\mathbf{y} = (y_1, \ldots, y_n)^T$ and $\mathbf{f} = (f_1, \ldots, f_n)^T$, where T is the transpose symbol, we can rewrite the *normal system in vector form* $\mathbf{y}' = \mathbf{f}(x, \mathbf{y})$. This resembles the form of the first order normal (explicit) equation, which is a particular case of the normal system when $n = 1$. Due to its compactness, the vector form is convenient for representation of systems of equations.

A normal (explicit) equation of order n can be reduced to a normal system of the same order. In fact, starting with equation $y^{(n)} = f(x, y, y', \ldots, y^{(n-1)})$ and introducing the new functions by the formulas $z_1 = y, z_2 = y', \ldots, z_n = y^{(n-1)}$, we transform the given equation into the following equivalent system:

$$\begin{cases} z_1' = z_2 \\ z_2' = z_3 \\ \quad \cdots \\ z_{n-1}' = z_n \\ z_n' = f(x, z_1, z_2, \ldots, z_n) \end{cases}$$

In what follows, we will only consider normal systems, which is why we will often omit the term "normal".

Definition of particular solution. A vector function $\mathbf{y}(x)$ is a *particular solution* of a normal system if, substituted into this system, it transforms the system into an identity.

Remark 1. As in the case of an equation, the solution of a system is considered over a set of values of x, which are defined explicitly or implicitly by the form of the function $\mathbf{y}(x)$ and the original equation.

Remark 2. As in the case of an equation, the definition itself indicates a simple way to verify whether a function is a solution of the given system or not on the set X: first, we have to check whether the function is differentiable on X and, if so, then substitute it, together with its derivatives, into the original system and check whether each of its equations becomes an identity or not.

Definition of the general solution. A function $\mathbf{y}(x, \mathbf{C})$ of independent variable x and n independent parameters $\mathbf{C} = (C_1, \ldots, C_n)^T$ is a general solution of a system of order n if for any specific choice of $C_1, \ldots, C_n$ it represents a particular solution of the same system. Note that the number of parameters of a general solution must coincide with the order of the system.

Remark. As in the case of an equation, the general solution of the system may or may not contain all the particular solutions.

1.2 Cauchy problem

In the same way as an individual equation, a system generates infinitely many solutions represented by a general solution containing arbitrary parameters, and possibly by special solutions. There are different types of complementary conditions that make it possible to choose a single solution from the set of solutions of the system. One of the most common type are initial conditions, whose physical interpretation is an extension of the case of an equation: considering that each function $y_1, \ldots, y_n$ represents the position of one of n particles in motion, the system of equations gives the relationships between the velocities of these particles and their positions, while the initial conditions represent the positions of these particles at a given instant.

Formulated analytically, this leads to the following definitions.

Definition of initial conditions. For a set of functions $y_1(x), \ldots, y_n(x)$, the *initial conditions* at a point x_0 are $y_1(x_0) = b_1, \ldots, y_n(x_0) = b_n$, where $b_1, \ldots, b_n$ are the given values. In vector form we have $\mathbf{y}(x_0) = \mathbf{b}$, where $\mathbf{b} = (b_1, \ldots, b_n)^T$.

Definition of the Cauchy problem. The system of equations together with the initial conditions form the *Cauchy problem:* $\begin{cases} \mathbf{y}' = \mathbf{f}(x, \mathbf{y}) \\ \mathbf{y}(x_0) = \mathbf{b} \end{cases}$. This problem is also called the *initial value problem* or the *initial conditions problem*. Note that the number of initial conditions is always equal to the order of the system.

Similar to the case of a separate equation, there exists the *Cauchy theo-rem* (also called the *Picard* theorem or *Picard-Lindelöf* theorem) for normal systems, which guarantees the existence and uniqueness of a solution of the Cauchy problem.

Cauchy theorem (existence and uniqueness of a solution). If there exists a neighborhood of the point $(x_0, \mathbf{b})$ in $\mathbb{R}^{n+1}$ where the function $\mathbf{f}$ and its partial derivative $\mathbf{f_y}$ are continuous functions, then there exists a neighborhood of x_0 in $\mathbb{R}$ where the solution of the Cauchy problem (for a normal system) exists and is unique.

Remark. Recall that in the case of vector functions with vector argu-ments (also called vector fields), the partial derivative $\mathbf{f_y}$ means, in compo-nent form, the matrix of partial derivatives (called the Jacobi matrix)

$$\mathbf{f_y} \equiv \begin{pmatrix} f_{1y_1} & f_{1y_2} & \cdots & f_{1y_n} \\ f_{2y_1} & f_{2y_2} & \cdots & f_{2y_n} \\ & & \cdots & \\ f_{ny_1} & f_{ny_2} & \cdots & f_{ny_n} \end{pmatrix}.$$

Naturally, the continuity condition of $\mathbf{f_y}$ means the continuity of each element of this matrix, that is, all the partial derivatives f_{iy_j}, $i = 1,\ldots,n$, $j = 1,\ldots,n$ must be continuous in a neighborhood of the point $(x_0, b_1,\ldots,b_n) \in \mathbb{R}^{n+1}$.

Exercises for section 1.2

1. Verify whether the given vector functions are general solutions of the indicated systems:

1) $\begin{pmatrix} u \\ v \end{pmatrix} = \begin{pmatrix} C_1 \\ C_2 \end{pmatrix} \cos x + \begin{pmatrix} C_2 \\ -C_1 \end{pmatrix} \sin x, \quad \begin{cases} u' = v \\ v' = -u \end{cases}.$

2) $\begin{cases} y = C_1 + C_2 x + 2 \sin x \\ z = -2C_1 - C_2(2x+1) - 3\sin x - 2\cos x \end{cases}, \quad \begin{cases} y' + 2y + z = \sin x \\ z' - 4y - 2z = \cos x \end{cases}.$

2. Determine the regions where the Cauchy conditions are satisfied for the given system:

1) $\begin{cases} y' = \sqrt{z} \\ z' = y^2 + t \end{cases}.$

2) $\begin{cases} zy' = \sqrt{y - t} \\ z' = y^3 + \ln(t+1) \end{cases}.$

2 Linear systems. Definitions and basic results

2.1 Definition of a linear system

Definition of linear system of order n. The system of differential equations in the form

$$\mathbf{y}' = A\mathbf{y} + \mathbf{f},$$

where $\mathbf{y}(x) = (y_1(x), \ldots, y_n(x))^T$ is the unknown vector function, $\mathbf{f}(x) = (f_1(x), \ldots, f_n(x))^T$ is the given vector function and

$$A(x) \equiv (a_{i,j}(x))_{i,j=1}^n = \begin{pmatrix} a_{11}(x) & a_{12}(x) & \cdots & a_{1n}(x) \\ a_{21}(x) & a_{22}(x) & \cdots & a_{2n}(x) \\ & & \cdots & \\ a_{n1}(x) & a_{n2}(x) & \cdots & a_{nn}(x) \end{pmatrix}$$

is the matrix of given entries, is called the *linear system of order* n. If $\mathbf{f} \equiv 0$, the linear system is called *homogeneous*, otherwise – *nonhomogeneous*. According to operations between matrices and vectors, in open component form, this system is written as follows:

$$\begin{cases} y_1' = a_{11}(x)y_1 + a_{12}(x)y_2 + \ldots + a_{1n}(x)y_n + f_1(x) \\ y_2' = a_{21}(x)y_1 + a_{22}(x)y_2 + \ldots + a_{2n}(x)y_n + f_2(x) \\ \quad \cdots \\ y_n' = a_{n1}(x)y_1 + a_{n2}(x)y_2 + \ldots + a_{nn}(x)y_n + f_n(x) \end{cases}.$$

2.2 Relationship between linear system and linear equation

As we have already seen, a normal equation of order n can be transformed into a normal system of the same order. Specifying this result for linear systems (which is a particular case of normal systems) we have the following statement: a linear equation of order n can be transformed into a linear system of the same order. The conversion procedure follows the same scheme: given a linear equation $y^{(n)} + a_{n-1}y^{(n-1)} + \ldots + a_1y' + a_0y = f$, we introduce the functions $z_1 = y, z_2 = y', \ldots, z_n = y^{(n-1)}$ and obtain the following system equivalent to the equation:

$$\begin{cases} z_1' = z_2 \\ z_2' = z_3 \\ \quad \cdots \\ z_{n-1}' = z_n \\ z_n' = -a_{n-1}z_n - \ldots - a_1 z_2 - a_0 z_1 + f \end{cases} .$$

It turns out that, unlike normal systems in general form, any linear system of order n can be reduced to a linear equation of order at most n for one of the system's unknown functions. Let us show how a linear equation can be obtained for the function y_1. In the linear system

$$\begin{cases} y_1' = a_{11}(x)y_1 + a_{12}(x)y_2 + \ldots + a_{1n}(x)y_n + f_1(x) \\ y_2' = a_{21}(x)y_1 + a_{22}(x)y_2 + \ldots + a_{2n}(x)y_n + f_2(x) \\ \quad \cdots \\ y_n' = a_{n1}(x)y_1 + a_{n2}(x)y_2 + \ldots + a_{nn}(x)y_n + f_n(x) \end{cases} ,$$

we differentiate the first equation

$$y_1'' = a_{11}(x)y_1' + a_{12}(x)y_2' + \ldots + a_{1n}(x)y_n' + a_{11}'(x)y_1 + a_{12}'(x)y_2 + \ldots + a_{1n}'(x)y_n + f_1'(x)$$

and, using the equations of the original system, we eliminate from the last relation all the first order derivatives on the right-hand side:

$$y_1'' = b_{21}(x)y_1 + b_{22}(x)y_2 + \ldots + b_{2n}(x)y_n + g_2(x).$$

Here, b_{2j}, $j = 1, \ldots, n$ are coefficients (functions of x) expressed in terms of a_{ij}, and $g_2(x)$ is a combination of a_{ij} and f_j, but their specific representation will not interest us. Next, we differentiate the equation obtained for y_1'' once more and again substitute all the derivatives of the unknown functions using the equations of the original system:

$$y_1''' = b_{31}(x)y_1 + b_{32}(x)y_2 + \ldots + b_{3n}(x)y_n + g_3(x).$$

We continue in this way until we obtain the equation for the nth derivative

$$y_1^{(n)} = b_{n1}(x)y_1 + b_{n2}(x)y_2 + \ldots + b_{nn}(x)y_n + g_n(x).$$

We now set up the system of the derived equations

$$\begin{cases} y_1' - f_1 = a_{11}y_1 + a_{12}y_2 + \ldots + a_{1n}y_n \\ y_1'' - g_2 = b_{21}y_1 + b_{22}y_2 + \ldots + b_{2n}y_n, \\ \quad \cdots \\ y_1^{(n)} - g_n = b_{n1}y_1 + b_{n2}y_2 + \ldots + b_{nn}y_n \end{cases} .$$

For each fixed x, we can consider this system as an algebraic linear system for unknowns $y_1, y_2, \ldots, y_n$ with the matrix of coefficients

$$M = \begin{pmatrix} a_{11} & a_{12} & \cdots & a_{1n} \\ b_{21} & b_{22} & \cdots & b_{2n} \\ & & \cdots & \\ b_{n1} & b_{n2} & \cdots & b_{nn} \end{pmatrix}$$

on the right-hand side of the system. Finally, we eliminate all the unknowns except y_1 using some familiar elimination algorithm (for example, the Gauss elimination method). If the rows of the matrix M are linearly independent, then elimination of the unknowns $y_2, \ldots, y_n$ leads to a linear equation of order n for the unknown function y_1. Otherwise (if the lines of M are linearly dependent), the order of the equation for y_1 will be less than n and will be equal to the number of linearly independent lines.

We summarize this result in the form of a theorem.

Theorem 1. Linear equations and linear systems are equivalent in the following sense: an equation of order n is reducible to a system of order n, and a system of order n can be reduced to an equation of order at most n.

2.3 Properties of solutions of a linear system

Due to the close connection between linear systems and linear equations, we can transfer all the properties of linear equations to linear systems. Notice that the same properties can be proved directly for linear systems (we recommend to the reader to show this as an additional exercise).

Theorem 2. Properties of the linear combination of solutions. If $\mathbf{y}_1$ is a solution of the system $\mathbf{y}' = A\mathbf{y} + \mathbf{f}_1$ and $\mathbf{y}_2$ is a solution of the system $\mathbf{y}' = A\mathbf{y} + \mathbf{f}_2$, then $\alpha_1\mathbf{y}_1 + \alpha_2\mathbf{y}_2$ is the solution of the system $\mathbf{y}' = A\mathbf{y} + \mathbf{f}$ with $\mathbf{f} = \alpha_1\mathbf{f}_1 + \alpha_2\mathbf{f}_2$, where α_1, α_2 are arbitrary constants. Expressing this result in formulas we have:

$$\mathbf{y}_1' = A\mathbf{y}_1 + \mathbf{f}_1, \mathbf{y}_2' = A\mathbf{y}_2 + \mathbf{f}_2 \Rightarrow \mathbf{y}' = A\mathbf{y} + \mathbf{f}, \mathbf{y} = \alpha_1\mathbf{y}_1 + \alpha_2\mathbf{y}_2, \mathbf{f} = \alpha_1\mathbf{f}_1 + \alpha_2\mathbf{f}_2.$$

Corollary 1. If $\mathbf{y}_1$ and $\mathbf{y}_2$ are solutions of the homogeneous system $\mathbf{y}' = A\mathbf{y}$, then $\alpha_1\mathbf{y}_1 + \alpha_2\mathbf{y}_2$ is the solution of the same system, whatever constants α_1, α_2 are chosen, that is,

$$\mathbf{y}_1' = A\mathbf{y}_1, \mathbf{y}_2' = A\mathbf{y}_2 \Rightarrow \mathbf{y}' = A\mathbf{y}, \mathbf{y} = \alpha_1\mathbf{y}_1 + \alpha_2\mathbf{y}_2.$$

Corollary 2. If $\mathbf{y}_h$ is a solution of the homogeneous system $\mathbf{y}' = A\mathbf{y}$ and $\mathbf{y}_n$ is a solution of the nonhomogeneous system $\mathbf{y}' = A\mathbf{y} + \mathbf{f}$, then $\mathbf{y}_h + \mathbf{y}_n$

is the solution of the same nonhomogeneous system. Using formulas:

$$\mathbf{y}'_h = A\mathbf{y}_h, \mathbf{y}'_n = A\mathbf{y}_n + \mathbf{f} \;\Rightarrow\; \mathbf{y}' = A\mathbf{y} + \mathbf{f}, \mathbf{y} = \mathbf{y}_h + \mathbf{y}_n.$$

Corollary 3. If $\mathbf{y}_1$ and $\mathbf{y}_2$ are solutions of the nonhomogeneous system $\mathbf{y}' = A\mathbf{y}+\mathbf{f}$, then $\mathbf{y}_1-\mathbf{y}_2$ is the solution of the homogeneous system $\mathbf{y}' = A\mathbf{y}$. Using formulas:

$$\mathbf{y}'_1 = A\mathbf{y}_1 + \mathbf{f}, \mathbf{y}'_2 = A\mathbf{y}_2 + \mathbf{f} \;\Rightarrow\; \mathbf{y}' = A\mathbf{y}, \; \mathbf{y} = \mathbf{y}_1 - \mathbf{y}_2.$$

Theorem 3. The function $\mathbf{y} = \mathbf{u} + i\mathbf{v}$ is a solution of the system $\mathbf{y}' = A\mathbf{y}+\mathbf{h}$, $\mathbf{h} = \mathbf{f}+i\mathbf{g}$ if and only if the real part $\mathbf{u} = Re(\mathbf{y})$ is the solution of the system $\mathbf{y}' = A\mathbf{y} + \mathbf{f}$, $\mathbf{f} = Re(\mathbf{h})$, and the imaginary part $\mathbf{v} = Im(\mathbf{y})$ is the solution of the system $\mathbf{y}' = A\mathbf{y} + \mathbf{g}$, $\mathbf{g} = Im(\mathbf{h})$. Using formulas:

$$(\mathbf{u} + i\mathbf{v})' = A(\mathbf{u} + i\mathbf{v}) + \mathbf{f} + i\mathbf{g} \;\Leftrightarrow\; \mathbf{u}' = A\mathbf{u} + \mathbf{f}, \mathbf{v}' = A\mathbf{v} + \mathbf{g}.$$

Corollary. The function $\mathbf{y} = \mathbf{u} + i\mathbf{v}$ is a solution of the homogeneous system $\mathbf{y}' = A\mathbf{y}$ if and only if its real part $\mathbf{u} = Re(\mathbf{y})$ and imaginary part $\mathbf{v} = Im(\mathbf{y})$ are solutions of the same homogeneous system, that is,

$$(\mathbf{u} + i\mathbf{v})' = A(\mathbf{u} + i\mathbf{v}) \;\Leftrightarrow\; \mathbf{u}' = A\mathbf{u}, \mathbf{v}' = A\mathbf{v}.$$

Theorem 4. Cauchy theorem (existence and uniqueness theorem) for linear systems. The Cauchy problem for the linear system

$$\begin{cases} \mathbf{y}' = A\mathbf{y} + \mathbf{f} \\ \mathbf{y}(x_0) = \mathbf{b} \end{cases}$$

has a unique solution in a neighborhood of x_0 provided that the matrix of coefficients A and the right-hand side $\mathbf{f}$ are continuous in a neighborhood of x_0.

If A and $\mathbf{f}$ are continuous functions on an interval I, then the Cauchy problem has a unique solution on I for any $x_0 \in I$.

Remark. Recall that the continuity of the matrix $A(x)$ means the continuity of all its elements $a_{ij}(x)$, and the continuity of the vector function $\mathbf{f}$ means the continuity of all its components $f_j(x)$.

Usually the conditions of the Cauchy Theorem will be satisfied for Cauchy systems/problems we will encounter in this the text, so we will not specially verify the conditions of the Cauchy Theorem for each system.

2.4 Structure of solutions of linear systems

Definition. Linear independence of functions.

Functions $\mathbf{y}_1, \ldots, \mathbf{y}_n$ are *linearly independent* on a set I if their linear combination $C_1\mathbf{y}_1 + \ldots C_n\mathbf{y}_n$ is zero on I only when all the constants $C_1, \ldots, C_n$ are zero. In other words, the equation $C_1\mathbf{y}_1 + \ldots C_n\mathbf{y}_n = 0$ with respect to the unknowns $C_1, \ldots, C_n$ has the only solution $C_1 = \ldots = C_n = 0$. Otherwise, the functions $\mathbf{y}_1, \ldots, \mathbf{y}_n$ are *linearly dependent*.

Remark 1. If the set I is not explicitly indicated, then the domain of the set of functions $\mathbf{y}_1, \ldots, \mathbf{y}_n$ is considered to be I.

Remark 2. Obviously, the zero function together with any others always forms a set of linearly dependent functions.

Theorem 5. General solution of homogeneous linear system. The *general solution of homogeneous linear system*

$$\mathbf{y}' = A\mathbf{y}$$

can be found the form

$$\mathbf{y}_{gh} = C_1\mathbf{y}_1 + \ldots C_n\mathbf{y}_n,$$

where $\mathbf{y}_1, \ldots, \mathbf{y}_n$ are linearly independent particular solutions of this system and $C_1, \ldots, C_n$ are arbitrary constants. The general solution $\mathbf{y}_{gh}$ contains all the particular solutions.

Definition. The set $\mathbf{y}_1, \ldots, \mathbf{y}_n$ of n linearly independent particular solutions of the homogeneous linear system of order n is called the *fundamental set of solutions*.

Remark. With the concept of the fundamental set of solutions, Theorem 5 can be reformulated as follows: the general solution of a homogeneous linear system is the linear combination of solutions of the fundamental set. This linear combination contains all the particular solutions.

Theorem 6. General solution of the nonhomogeneous linear system. The *general solution of non-homogeneous linear system*

$$\mathbf{y}' = A\mathbf{y} + \mathbf{f}$$

can be found in the form

$$\mathbf{y}_{gn} = \mathbf{y}_{gh} + \mathbf{y}_{pn} = C_1\mathbf{y}_1 + \ldots C_n\mathbf{y}_n + \mathbf{y}_{pn},$$

where $\mathbf{y}_{gh}$ is the general solution of the corresponding homogeneous system and $\mathbf{y}_{pn}$ is a particular solution of the nonhomogeneous one. The general solution $\mathbf{y}_{gn}$ contains all the particular solutions.

Chapter 13

Linear systems: methods of solution

1 Homogeneous systems with constant coefficients

Definition. A homogeneous linear system with constant coefficients has the form

$$\mathbf{y}' = A\mathbf{y},$$

where matrix A is constant (that is, all its elements a_{ij} are constant).

1.1 Method of reduction

The relationship itself between linear systems and linear equations, established in the previous section, suggests a *method of solution of systems by means of their reduction to a linear equation*. We also note that systems with constant coefficients are transformed into equations with constant coefficients, the methods of solution of which have already been studied.

In practice, the algorithm used in the previous section to show the equivalence between linear systems and equations is not used due to its complexity. It is much simpler to follow the idea of Gaussian elimination of unknowns reformulated for differential systems. We will show how this works in the following examples, using two practical versions of this method.

Examples.

1. $\begin{cases} y' = 2z - y \\ z' = z - 5y \end{cases}$.

In this system, either of the two unknown functions can be easily eliminated. For instance, let us eliminate y, using the second equation of the system, which we rewrite in the form $y = \frac{1}{5}(z - z')$. Substituting this expression into the first equation, we find the second order equation for a single unknown function z: $\frac{1}{5}(z' - z'') = 2z - \frac{1}{5}(z - z')$ or, after simplification, $z'' + 9z = 0$. The characteristic equation $\lambda^2 + 9 = 0$ has the complex conjugate roots $\lambda_{1,2} = \pm 3i$, and consequently, the two linearly independent solutions are $z_1 = \cos 3x$ and $z_2 = \sin 3x$, which form the general solution $z = C_1 \cos 3x + C_2 \sin 3x$. So, the z-component of the system solution is already known. To find y, we return to its expression in terms of z:

$$y = \frac{1}{5}(z - z') = \frac{1}{5}(C_1 \cos 3x + C_2 \sin 3x) - \frac{1}{5}(-3C_1 \sin 3x + 3C_2 \cos 3x)$$

$$= \frac{1}{5}(C_1 - 3C_2)\cos 3x + \frac{1}{5}(3C_1 + C_2)\sin 3x.$$

Thus, the general solution of the system is

$$\begin{cases} y = \frac{1}{5}(C_1 - 3C_2)\cos 3x + \frac{1}{5}(3C_1 + C_2)\sin 3x \\ z = C_1 \cos 3x + C_2 \sin 3x \end{cases}$$

or in the vector form

$$\begin{pmatrix} y \\ z \end{pmatrix} = C_1 \begin{pmatrix} \frac{1}{5}\cos 3x + \frac{3}{5}\sin 3x \\ \cos 3x \end{pmatrix} + C_2 \begin{pmatrix} -\frac{3}{5}\cos 3x + \frac{1}{5}\sin 3x \\ \sin 3x \end{pmatrix}.$$

As a homework exercise, we propose to the reader to solve this system by eliminating the function z and compare the solution obtained with the one derived here.

Another version of the same method performs the elimination process differently. The operator of differentiation is formally replaced by the letter D, which is treated during the elimination stage as a "coefficient", and the elimination of unknowns in the (formally algebraic) system follows the Gauss algorithm, with the only exception of not being allowed to divide by expressions involving the "coefficient" D. After reducing the system to an equation for only one unknown function, the meaning of the differentiation of the letter D is "recovered" and the obtained differential equation is solved using familiar techniques. Although this approach may seem a bit artificial (and for the second-order systems it is), its advantage lies in our ability to solve algebraic linear systems, which usually makes it possible to find a simpler way of eliminating the unknowns, the way that can be difficult to see when working with the original form of differential equations. This simplification becomes apparent when solving systems of order 3 and higher.

Following this scheme, we can write the original system in the form $\begin{cases} Dy = 2z - y \\ Dz = z - 5y \end{cases}$ or, regrouping the terms, $\begin{cases} (D+1)y - 2z = 0 \\ 5y + (D-1)z = 0 \end{cases}$. Multiplying the second equation by $\frac{1}{5}(D+1)$ and subtracting the result from the first equation, we get $\frac{1}{5}(D+1)(D-1)z + 2z = 0$. Simplifying the last equation to the form $D^2z + 9z = 0$ and returning to the meaning of the letter D, we obtain the same differential equation as in the first version $z'' + 9z = 0$, and we finalize the algorithm of solution by following the same steps as in the version without D.

2. $\begin{cases} y' = 4y - z \\ z' = 2z + y \end{cases}$.

In this system, we can easily eliminate either of the two functions. For a change, let us eliminate the function z, using the first equation of the system, which we rewrite in the form $z = 4y - y'$. Substituting this expression into the second equation, we find the second order equation for a single unknown function y: $(4y - y')' = 2(4y - y') + y$, or simplifying, $y'' - 6y' + 9y = 0$. The characteristic equation $\lambda^2 - 6\lambda + 9 = 0$ has the double root $\lambda_{1,2} = 3$, and consequently, the two linearly independent solutions are $y_1 = e^{3x}$ and $y_2 = xe^{3x}$, which form the general solution $y = (C_1 + C_2x)e^{3x}$. To find the function z, we return to its expression in terms of y: $z = 4y - y' = (C_1 - C_2 + C_2x)e^{3x}$. Thus, the general solution of the system is

$$\begin{cases} y = (C_1 + C_2x)e^{3x} \\ z = (C_1 - C_2 + C_2x)e^{3x} \end{cases}$$

or in the vector form

$$\begin{pmatrix} y \\ z \end{pmatrix} = C_1 \begin{pmatrix} 1 \\ 1 \end{pmatrix} e^{3x} + C_2 \begin{pmatrix} x \\ x-1 \end{pmatrix} e^{3x}.$$

Although the second version, with the letter D, has no advantages for second order systems, we will apply it for training purposes. We write the original system in the form $\begin{cases} (D-4)y + z = 0 \\ -y + (D-2)z = 0 \end{cases}$. Multiplying the first equation by $D - 2$ and subtracting the second from the result, we get $(D -$

$2)(D-4)y+y=0$. Simplifying the last one and using again differentiation instead of D, we get the same differential equation as in the first version $y''-6y'+9y=0$. From this point on we finish with the same steps as in the first version.

As a homework exercise for the reader, we recommend to solve this system by eliminating the function y and using both versions of the elimination method.

$$3. \quad \begin{cases} u' = v + w \\ v' = 3u + w \\ w' = 3u + v \end{cases} .$$

Let us use the second version of eliminating the unknowns. We rewrite the system in the form

$$\begin{cases} Du - v - w = 0 \\ -3u + Dv - w = 0 \\ -3u - v + Dw = 0 \end{cases} .$$

The order of elimination doesn't matter, we can start, for instance, by eliminating w. First, we subtract the second equation from the first to get $(D+3)u - (D+1)v = 0$. Then, we multiply the second equation by D and add the result to the third: $-3(D+1)u + (D^2-1)v = 0$. Now we have the following system of two equations for two unknowns:

$$\begin{cases} (D+3)u - (D+1)v = 0 \\ -3(D+1)u + (D^2-1)v = 0 \end{cases} .$$

Although the second equation can be written in the form $-3(D+1)u + (D-1)(D+1)v = 0$, it should be recalled that it cannot be simplified by dividing by $D+1$, because this factor does not merely represent a numerical coefficient, but rather the differential operator $\frac{d}{dx}+1$ in the symbolic form. Simplifications of this kind lead to the loss of solutions of the original system.

Now, multiplying the first equation by $D-1$ and adding the result to the second, we arrive at the equation for the only unknown u: $(D^2-D-6)u = 0$. This means that we have the second order differential equation for u: $u'' - u' - 6u = 0$. The general solution of this equation is $u = C_1 e^{-2x} + C_2 e^{3x}$. Returning to v, we have to solve one more first order differential equation $v' + v = u' + 3u = C_1 e^{-2x} + 6C_2 e^{3x}$, whose general solution is $v = -C_1 e^{-2x} + \frac{3}{2}C_2 e^{3x} + C_3 e^{-x}$. Finally, we find w from the first equation of the original system: $w = u' - v = -C_1 e^{-2x} + \frac{3}{2}C_2 e^{3x} - C_3 e^{-x}$. Therefore, the general solution of the original system is

$$\begin{cases} u = C_1 e^{-2x} + C_2 e^{3x} \\ v = -C_1 e^{-2x} + \frac{3}{2} C_2 e^{3x} + C_3 e^{-x} \\ w = -C_1 e^{-2x} + \frac{3}{2} C_2 e^{3x} - C_3 e^{-x} \end{cases},$$

or in the vector form

$$\begin{pmatrix} u \\ v \\ w \end{pmatrix} = C_1 \begin{pmatrix} 1 \\ -1 \\ -1 \end{pmatrix} e^{-2x} + C_2 \begin{pmatrix} 1 \\ 3/2 \\ 3/2 \end{pmatrix} e^{3x} + C_3 \begin{pmatrix} 0 \\ 1 \\ -1 \end{pmatrix} e^{-x}.$$

Let us show in this example what will happen if we cut the multiplier $D + 1$ in the second equation of the system

$$\begin{cases} (D + 3)u - (D + 1)v = 0 \\ -3(D + 1)u + (D^2 - 1)v = 0 \end{cases}.$$

In this case, the system takes the form

$$\begin{cases} (D + 3)u - (D + 1)v = 0 \\ -3u + (D - 1)v = 0 \end{cases}.$$

Then, using the second equation, we substitute $\frac{1}{3}(D - 1)v$ for u in the first and get

$$\frac{1}{3}(D + 3)(D - 1)v - (D + 1)v = 0 \text{ or } (D^2 - D - 6)v = 0.$$

Therefore, we have to solve the following equation for v: $v'' - v' - 6v = 0$. The general solution of this equation is $v = A_1 e^{-2x} + A_2 e^{3x}$. Returning to u, we get $u = \frac{1}{3}(v' - v) = -A_1 e^{-2x} + \frac{2}{3} A_2 e^{3x}$. Finally, we find w from the first equation of the original system: $w = u' - v = -A_1 e^{-2x} + A_2 e^{3x}$. Thus, we arrive at the solution

$$\begin{cases} u = -A_1 e^{-2x} + \frac{2}{3} A_2 e^{3x} \\ v = -A_1 e^{-2x} + A_2 e^{3x} \\ w = A_1 e^{-2x} + A_2 e^{3x} \end{cases}.$$

Compared to the general solution, we note that this group of solutions represents only a part of the general solution, with no terms involving the function e^{-x} (the coefficients with e^{-2x} and e^{3x} in the two solutions are related by the formulas $A_1 = -C_1$ and $A_2 = \frac{3}{2} C_2$). Therefore, the obtained solution is not a general one; it is of an intermediate type, included in the general solution. The loss of the terms with e^{-x} happened because of the division by $D + 1$, which is not a numerical coefficient.

4. $\begin{cases} u' = v + w \\ v' = u + w \\ w' = u + v \end{cases}$.

Let us use the second version of eliminating the unknowns. We rewrite the system in the form

$$\begin{cases} Du - v - w = 0 \\ -u + Dv - w = 0 \\ -u - v + Dw = 0 \end{cases} .$$

Due to the symmetry of the unknowns, the order of elimination is not important. For instance, we can eliminate u first. We multiply the third equation by D and add the result to the first, obtaining $-(D + 1)v + (D^2 - 1)w = 0$. Then, subtracting the third equation from the second, we get $(D+1)v-(D+1)w = 0$. This brings us to the system of two equations for two unknowns:

$$\begin{cases} -(D + 1)v + (D^2 - 1)w = 0 \\ (D + 1)v - (D + 1)w = 0 \end{cases} .$$

It should be remembered that it is not possible to divide by expressions involving D, so the obtained equations cannot be simplified. Now adding the two equations, we eliminate v: $(D^2 - 1)w - (D + 1)w = 0$, and then we obtain the second order equation for w: $w'' - w' - 2w = 0$. The general solution of this equation is $w = C_1 e^{-x} + C_2 e^{2x}$. Returning to v, we have to solve one more first-order equation: $v' + v = w' + w = 3C_2 e^{2x}$. Its general solution is $v = C_2 e^{2x} + C_3 e^{-x}$. Finally, to find u we use the third equation $u = w' - v = -(C_1 + C_3)e^{-x} + C_2 e^{2x}$. Thus, the general solution of the original system is

$$\begin{cases} u = (C_3 - C_1)e^{-x} + C_2 e^{2x} \\ v = C_2 e^{2x} + C_3 e^{-x} \\ w = C_1 e^{-x} + C_2 e^{2x} \end{cases} ,$$

or in the vector form

$$\begin{pmatrix} u \\ v \\ w \end{pmatrix} = C_1 \begin{pmatrix} -1 \\ 0 \\ 1 \end{pmatrix} e^{-x} + C_2 \begin{pmatrix} 1 \\ 1 \\ 1 \end{pmatrix} e^{2x} + C_3 \begin{pmatrix} -1 \\ 1 \\ 0 \end{pmatrix} e^{-x}.$$

5. $\begin{cases} u' = 8v \\ v' = -2w \\ w' = 2u + 8v - 2w \end{cases}$, $\begin{cases} u(0) = -4 \\ v(0) = 0 \\ w(0) = 1 \end{cases}$.

First, we find the general solution of the system. To use the second version of the elimination of unknowns, we rewrite the system in the form

$$\begin{cases} Du - 8v = 0 \\ Dv + 2w = 0 \\ -2u - 8v + (D+2)w = 0 \end{cases}.$$

Adding the first equation multiplied by 2 with the third multiplied by D, we get $8(D+2)v - D(D+2)w = 0$. Together with the second equation of the system, this gives the subsystem of two equations for two unknowns:

$$\begin{cases} Dv + 2w = 0 \\ 8(D+2)v - D(D+2)w = 0 \end{cases}.$$

Remember that it is not possible to divide by expressions involving D, so the obtained equations cannot be simplified. Multiplying the first equation of the subsystem by $D(D+2)$ and adding it to the third multiplied by 2, we find $D^2(D+2)v + 16(D+2)v = 0$. Simplifying and returning to the differential form, we get $v''' + 2v'' + 16v' + 32v = 0$. The characteristic equation can be factored into the form $(\lambda + 2)(\lambda^2 + 16) = 0$, which gives the roots $\lambda_1 = -2$ and $\lambda_{2,3} = \pm 4i$. So the general solution for v has the following real form $v = C_1 e^{-2x} + C_2 \cos 4x + C_3 \sin 4x$. Knowing v, we find w from the first equation of the subsystem: $w = -\frac{1}{2}v' = C_1 e^{-2x} + 2C_2 \sin 4x - 2C_3 \sin 4x$. Finally, the u-component is calculated from the third equation of the original system: $u = \frac{1}{2}w' + w - 4u = -4C_1 e^{-2x} + 2C_2 \sin 4x - 2C_3 \cos 4x$. Combining the results, we get the following general solution of the original system

$$\begin{cases} u = -4C_1 e^{-2x} + 2C_2 \sin 4x - 2C_3 \cos 4x \\ v = C_1 e^{-2x} + C_2 \cos 4x + C_3 \sin 4x \\ w = C_1 e^{-2x} + 2C_2 \sin 4x - 2C_3 \cos 4x \end{cases},$$

or in the vector form

$$\begin{pmatrix} u \\ v \\ w \end{pmatrix} = C_1 \begin{pmatrix} -4 \\ 1 \\ 1 \end{pmatrix} e^{-2x} + C_2 \begin{pmatrix} 2\sin 4x \\ \cos 4x \\ 2\sin 4x \end{pmatrix} + C_3 \begin{pmatrix} -2\cos 4x \\ \sin 4x \\ -2\cos 4x \end{pmatrix}.$$

Now, we find the solution of the Cauchy problem. To do this, we substitute the initial conditions into the expressions of the three functions:

$$\begin{cases} u(0) = -4C_1 - 2C_3 = -4 \\ v(0) = C_1 + C_2 = 0 \\ w(0) = C_1 - 2C_3 = 1 \end{cases}.$$

The solution of this linear algebraic system is $C_1 = 1, C_2 = -1, C_3 = 0$. Therefore, the solution of the problem is

$$\begin{cases} u = -4e^{-2x} - 2\sin 4x \\ v = e^{-2x} - \cos 4x \\ w = e^{-2x} - 2\sin 4x \end{cases}.$$

6. $\begin{cases} u' = -u + v + w \\ v' = u - v + w \\ w' = u + v + w \end{cases}.$

Let us use the second version of eliminating the unknowns. We rewrite the system as follows:

$$\begin{cases} (D+1)u - v - w = 0 \\ -u + (D+1)v - w = 0 \\ -u - v + (D-1)w = 0 \end{cases}.$$

The order of elimination doesn't matter. For instance, we can start with the elimination of u. First, we subtract the third equation from the second to get $(D+2)v - Dw = 0$. Second, we multiply the second equation by $(D+1)$ and add the result to the first: $(D+1)^2 v - v - (D+1)w - w = 0$ or, simplifying, $D(D+2)v - (D+2)w = 0$. Note that we cannot simply eliminate the factor $D+2$ and rewrite the equation in the form $Dv - w = 0$, because the symbol $D+2$ represents the action of the differential operator $\frac{d}{dx} + 2$ and is not merely a fixed coefficient. Simplifications of this kind lead to the loss of solutions of the original system. Hence, we have the following system of two equations for two unknowns:

$$\begin{cases} (D+2)v - Dw = 0 \\ D(D+2)v - (D+2)w = 0 \end{cases}.$$

Using the first equation, we substitute Dw for $(D+2)v$ in the second and obtain a single equation for the single unknown function z: $D(Dw) - (D+2)w = 0$ or $D^2 w - Dw - 2w = 0$. This corresponds to the second order differential equation: $w'' - w' - 2w = 0$. The general solution of this equation is $w = C_1 e^{-x} + C_2 e^{2x}$. Returning to v, we have to solve one more first-order differential equation $v' + 2v = w' = -C_1 e^{-x} + 2C_2 e^{2x}$ whose general solution is $v = -C_1 e^{-x} + \frac{1}{2}C_2 e^{2x} + C_3 e^{-2x}$. Finally, we find u from the third equation of the original system:

$$u = w' - w - v = -C_1 e^{-x} + \frac{1}{2}C_2 e^{2x} - C_3 e^{-2x}.$$

Therefore, the general solution of the original system is

$$\begin{cases} u = C_1 e^{-x} + \frac{1}{2}C_2 e^{2x} - C_3 e^{-2x} \\ v = -C_1 e^{-x} + \frac{1}{2}C_2 e^{2x} + C_3 e^{-2x} \\ w = C_1 e^{-x} + C_2 e^{2x} \end{cases},$$

or in the vector form

$$\begin{pmatrix} u \\ v \\ w \end{pmatrix} = C_1 \begin{pmatrix} -1 \\ -1 \\ 1 \end{pmatrix} e^{-x} + C_2 \begin{pmatrix} 1/2 \\ 1/2 \\ 1 \end{pmatrix} e^{2x} + C_3 \begin{pmatrix} -1 \\ 1 \\ 0 \end{pmatrix} e^{-2x}.$$

7. $\begin{cases} u' = 4u - v \\ v' = 3u + v - w \\ w' = u + w \end{cases}$.

Let us use the second version of eliminating the unknowns. We rewrite the system as follows:

$$\begin{cases} (D-4)u + v = 0 \\ -3u + (D-1)v + w = 0 \\ -u + (D-1)w = 0 \end{cases}.$$

Since w does not appear in the first equation, we derive one more equation without w using the second and third equations. Multiplying the second equation by $D - 1$ and subtracting the third equation from the result, we get $-3(D-1)u + u + (D-1)^2 v = 0$. Together with the first equation, we get the following subsystem:

$$\begin{cases} (D-4)u + v = 0 \\ (4-3D)u + (D-1)^2 v = 0 \end{cases}.$$

Substituting the expression for v from the first equation into the second, we find an equation for the single unknown function u: $(4-3D)u - (D-1)^2(D-4)u = 0$ or, after simplification, $(D^3 - 6D^2 + 12D - 8)u = 0$. Therefore, we have to solve the equation $u''' - 6u'' + 12u' - 8u = 0$, whose characteristic equation $\lambda^2 - 6\lambda^2 + 12\lambda - 8 = 0$ has the only triple root $\lambda_{1,2,3} = 2$. This implies that there exist three linearly independent solutions $u_1 = e^{2x}$, $u_2 = xe^{2x}$ and $u_3 = x^2 e^{2x}$, and the general solution is $u = (C_1 + C_2 x + C_3 x^2)e^{2x}$. Next, the function v is found from the first equation of the subsystem:

$$v = 4u - u' = [(2C_1 - C_2) + (2C_2 - 2C_3)x + 2C_3 x^2]e^{2x}.$$

Finally w is found from the second equation of the original system:

$$w = 3u + v - v' = [(C_1 - C_2 + 2C_3) + (C_2 - 2C_3)x + C_3 x^2]e^{2x}.$$

Therefore, the general solution of the original system is

$$\begin{cases} u = (C_1 + C_2 x + C_3 x^2)e^{2x} \\ v = [(2C_1 - C_2) + (2C_2 - 2C_3)x + 2C_3 x^2]e^{2x} \\ w = [(C_1 - C_2 + 2C_3) + (C_2 - 2C_3)x + C_3 x^2]e^{2x} \end{cases},$$

or in the vector form:

$$\begin{pmatrix} u \\ v \\ w \end{pmatrix} = C_1 \begin{pmatrix} 1 \\ 2 \\ 1 \end{pmatrix} e^{2x} + C_2 \begin{pmatrix} x \\ 2x - 1 \\ x - 1 \end{pmatrix} e^{2x} + C_3 \begin{pmatrix} x^2 \\ 2x^2 - 2x \\ x^2 - 2x + 2 \end{pmatrix} e^{2x}.$$

1.2   Euler method

The Euler method is a different technique for solving systems with constant coefficients. Although it has no direct reference to higher order equations associated with the systems, the close relationship between systems and equations is reflected in this method as well, since the construction of its algorithm of solution is very similar to the algorithm for solving equations with constant coefficients.

The source of this method is the following observation: generalizing the search for solutions in the form $e^{\lambda x}$ used for linear equations, we can try to find the solutions of the system $\mathbf{y}' = A\mathbf{y}$ in the form $\mathbf{y} = \mathbf{s}e^{\lambda x}$, where constant vector $\mathbf{s}$ and exponent λ are parameters to be determined. Substituting this type of solution into the original system we arrive at the following relationship: $\lambda \mathbf{s}e^{\lambda x} = A\mathbf{s}e^{\lambda x}$. After eliminating $e^{\lambda x}$, we recognize the relation that defines eigenvalues λ and eigenvectors $\mathbf{s}$ of the matrix A: $A\mathbf{s} = \lambda \mathbf{s}$. As we know from Linear Algebra, any matrix has at least one pair of eigenvalue and eigenvector that offers a particular solution to the system $\mathbf{y}' = A\mathbf{y}$. It may happen that the matrix A has the complete set of eigenvectors (those that are linearly independent and form the basis in the space $\mathbb{R}^n$), in which case the fundamental set of solutions can be found in the suggested exponential form. If some of these pairs are complex, then we can always transform them into real solutions, if necessary. If the number of linearly independent eigenvectors is less than n, then the complementary solutions must be found in a different form. Below, we consider in detail all the cases that may arise.

Case 1: There is a complete system of eigenvectors.
In this case, we have n linearly independent solutions in the form

$$\mathbf{y}_1 = \mathbf{s}_1 e^{\lambda_1 x}, \ldots, \mathbf{y}_n = \mathbf{s}_n e^{\lambda_n x}$$

and the general solution is their linear combination:

$$\mathbf{y}_{gh} = C_1 \mathbf{y}_1 + \ldots + C_n \mathbf{y}_n = C_1 \mathbf{s}_1 e^{\lambda_1 x} + \ldots + C_n \mathbf{s}_n e^{\lambda_n x}.$$

It does not matter whether the eigenvalues are repeated or not, whether they are real or complex. The only relevant condition for this form of the general solution is the linear independence of the eigenvectors. If we need to transform the complex form into a real one, recall that complex eigenvalues always appear in complex conjugate pairs and the corresponding eigenvectors are also complex conjugate. So, we can choose one of the solutions generated by this pair of complex eigenvalues and eigenvectors and find its real and imaginary parts which form two solutions in real form of the same homogeneous system (according to the Corollary to Theorem 3).

Examples.

1. $\begin{cases} y' = 2z - y \\ z' = z - 5y \end{cases}$.

We rewrite this system in the vector form $\mathbf{y}' = A\mathbf{y}$, where $\mathbf{y} = \begin{pmatrix} y \\ z \end{pmatrix}$ and $A = \begin{pmatrix} -1 & 2 \\ -5 & 1 \end{pmatrix}$. To find the eigenvalues of the matrix A, we solve the characteristic equation

$$det(A - \lambda I) = det \begin{pmatrix} -1 - \lambda & 2 \\ -5 & 1 - \lambda \end{pmatrix} = 0,$$

where det is the determinant symbol and I is the identity matrix of order 2. Calculating the determinant, we get the following equation $(\lambda + 1)(\lambda - 1) + 10 = \lambda^2 + 9 = 0$. So the two eigenvalues are complex conjugate $\lambda_{1,2} = \pm 3i$, and substituting one of them, for example $\lambda_1 = 3i$, into the eigenvector equation $(A - \lambda_1 I)\mathbf{s}_1 = 0$, we obtain

$$\begin{cases} (-1 - 3i)a + 2b = 0 \\ -5a + (1 - 3i)b = 0 \end{cases},$$

where a and b are the components of the eigenvector $\mathbf{s}_1 = \begin{pmatrix} a \\ b \end{pmatrix}$. Since these two equations are linearly dependent, it is sufficient to use either of them, for example, the first one, to obtain $b = \frac{1+3i}{2}a$. Recall that eigenvectors are always found with a precision upto a non-zero factor, so we can choose some convenient value of a, for example, $a = 2$. Therefore, $\mathbf{s}_1 = \begin{pmatrix} 2 \\ 1 + 3i \end{pmatrix}$ and the corresponding solution is $\mathbf{y}_1 = \mathbf{s}_1 e^{\lambda_1 x} = \begin{pmatrix} 2 \\ 1 + 3i \end{pmatrix} e^{3ix}$. We know from Linear Algebra that another pair of the eigenvalue and eigenvector will be complex conjugate, that is, $\lambda_2 = -3i$, $\mathbf{s}_2 = \begin{pmatrix} 2 \\ 1 - 3i \end{pmatrix}$ and, consequently, the solution $\mathbf{y}_2 = \mathbf{s}_2 e^{\lambda_2 x} = \begin{pmatrix} 2 \\ 1 - 3i \end{pmatrix} e^{-3ix}$ is complex conjugate of $\mathbf{y}_1$. Therefore, the two complex solutions will generate the same pair of real solutions, and we can save our time and do not consider at all the second pair of the eigenvalue and eigenvector. Returning to $\mathbf{y}_1$, to transform it into real form, we extract the real and imaginary parts of $\mathbf{y}_1$:

$$\mathbf{u} = Re(\mathbf{y}_1) = Re(\begin{pmatrix} 2 \\ 1 + 3i \end{pmatrix}(\cos 3x + i \sin 3x))) = \begin{pmatrix} 2\cos 3x \\ \cos 3x - 3\sin 3x \end{pmatrix},$$

$$\mathbf{v} = Im(\mathbf{y}_1) = Im(\begin{pmatrix} 2 \\ 1 + 3i \end{pmatrix}(\cos 3x + i \sin 3x))) = \begin{pmatrix} 2\sin 3x \\ 3\cos 3x + \sin 3x \end{pmatrix}.$$

Thus, the general solution is represented in the form

$$\mathbf{y}_{gh} = \begin{pmatrix} y \\ z \end{pmatrix} = C_1 \mathbf{u} + C_2 \mathbf{v} = C_1 \begin{pmatrix} 2\cos 3x \\ \cos 3x - 3\sin 3x \end{pmatrix} + C_2 \begin{pmatrix} 2\sin 3x \\ 3\cos 3x + \sin 3x \end{pmatrix}.$$

This system has been solved before using the method of reduction (Example 1 in section 1.1) and the general solution found there has the form

$$\begin{pmatrix} y \\ z \end{pmatrix} = A_1 \begin{pmatrix} \frac{1}{5}\cos 3x + \frac{3}{5}\sin 3x \\ \cos 3x \end{pmatrix} + A_2 \begin{pmatrix} -\frac{3}{5}\cos 3x + \frac{1}{5}\sin 3x \\ \sin 3x \end{pmatrix}$$

(the constants of that solution are denoted here by A_1 and A_2). Although, at first glance, the solutions seem to be different, but by specifying the relationship between constants C_1, C_2 and A_1, A_2 we can check that the two are the same (as they should be). In fact, by setting $C_1 - 3C_2 = 10A_1$, $3C_1 + C_2 = 10A_2$ (with the inverse relations $C_1 = A_1 + 3A_2$, $C_2 = A_2 - 3A_1$), we can see that the two forms represent the same general solution.

2. $\begin{cases} y' = y + 5z \\ z' = -y - 3z \end{cases}$, $\begin{cases} y(0) = -2 \\ z(0) = 1 \end{cases}$.

We start by finding the general solution of the system. We rewrite the system in the vector form $\mathbf{y}' = A\mathbf{y}$, where $\mathbf{y} = \begin{pmatrix} y \\ z \end{pmatrix}$ and $A = \begin{pmatrix} 1 & 5 \\ -1 & -3 \end{pmatrix}$. To find the eigenvalues of the matrix A, we solve the characteristic equation

$$det(A - \lambda I) = det \begin{pmatrix} 1 - \lambda & 5 \\ -1 & -3 - \lambda \end{pmatrix} = \lambda^2 + 2\lambda + 2 = 0,$$

which has the two complex conjugate roots $\lambda_{1,2} = -1 \pm i$. Substituting one of them, for example $\lambda_1 = -1 - i$, into the eigenvector equation $(A - \lambda_1 I)\mathbf{s}_1 = 0$, we get

$$\begin{cases} (2 + i)a + 5b = 0 \\ -a + (-4 + i)b = 0 \end{cases},$$

where a and b are the components of the eigenvector $\mathbf{s}_1 = \begin{pmatrix} a \\ b \end{pmatrix}$. Since these two equations are linearly dependent, it is sufficient to use only one of them, for example the first one, to obtain $b = -\frac{2 \pm i}{5}a$. Choosing $a = 5$, we get $b = -2 - i$ and, consequently, $\mathbf{s}_1 = \begin{pmatrix} 5 \\ -2 - i \end{pmatrix}$. Therefore, the first complex solution has the form $\mathbf{y}_1 = \mathbf{s}_1 e^{\lambda_1 x} = \begin{pmatrix} 5 \\ -2 - i \end{pmatrix} e^{(-1-i)x}$, and the second has the complex conjugate form $\mathbf{y}_2 = \mathbf{s}_2 e^{\lambda_2 x} = \begin{pmatrix} 5 \\ -2 + i \end{pmatrix} e^{(-1+i)x}$. To form the pair of real solutions, it is sufficient to consider the real and imaginary parts of one of these complex solutions, for example, the first one. Then, we have

$$\mathbf{u} = Re(\mathbf{y}_1) = Re\left(\begin{pmatrix} 5 \\ -2 - i \end{pmatrix} (\cos x - i \sin x)e^{-x}\right) = \begin{pmatrix} 5\cos x \\ -2\cos x - \sin x \end{pmatrix} e^{-x},$$

$$\mathbf{v} = Im(\mathbf{y}_1) = Im\left(\begin{pmatrix} 5 \\ -2 - i \end{pmatrix} (\cos x - i \sin x)e^{-x}\right) = \begin{pmatrix} -5\sin x \\ -\cos x + 2\sin x \end{pmatrix} e^{-x}.$$

Therefore, the general solution in real form is as follows:

$$\mathbf{y}_{gh} = \begin{pmatrix} y \\ z \end{pmatrix} = C_1\mathbf{u} + C_2\mathbf{v} = C_1 \begin{pmatrix} 5\cos x \\ -2\cos x - \sin x \end{pmatrix} + C_2 \begin{pmatrix} -5\sin x \\ -\cos x + 2\sin x \end{pmatrix} e^{-x}.$$

In terms of components we have

$$y = (5C_1 \cos x - 5C_2 \sin x)e^{-x}, \quad z = (-(2C_1+C_2)\cos x + (2C_2-C_1)\sin x)e^{-x}.$$

To find the solution of the Cauchy problem, we substitute the initial conditions into the general solution and obtain: $y = 5C_1 = -2$ and $z = -(2C_1 + C_2) = 1$. Therefore, $C_1 = -\frac{2}{5}$ and $C_2 = -\frac{1}{5}$. Thus, the solution of the Cauchy problem is

$$y = (-2\cos x + \sin x)e^{-x}, \quad z = \cos x e^{-x}.$$

We note that the initial conditions could also be used in the complex form of the general solution. In fact, substituting these conditions into the general solution

$$\mathbf{y}_{ghc} = A_1\mathbf{y}_1 + A_2\mathbf{y}_2 = A_1 \begin{pmatrix} 5 \\ -2-i \end{pmatrix} e^{(-1-i)x} + A_2 \begin{pmatrix} 5 \\ -2+i \end{pmatrix} e^{(-1+i)x},$$

where A_1 and A_2 are complex coefficients, we arrive at the following system:

$$\begin{cases} 5A_1 + 5A_2 = -2 \\ (-2-i)A_1 + (-2+i)A_2 = 1 \end{cases}.$$

From the first equation we get $A_1 + A_2 = -\frac{2}{5}$ and, using this relationship, we can rewrite the second equation as $-i(A_1 - A_2) = 1+2(A_1 + A_2) = \frac{1}{5}$ or $A_1 - A_2 = \frac{1}{5}i$. Now adding the two equations together, we find $2A_1 = -\frac{2}{5}+\frac{1}{5}i$ or $A_1 = \frac{-2+i}{10}$. Then, $A_2 = -A_1 - \frac{2}{5} = \frac{-2-i}{10}$. Taking these coefficients into the general solution, we get the solution to the Cauchy problem:

$$\mathbf{y} = \frac{-2+i}{10} \begin{pmatrix} 5 \\ -2-i \end{pmatrix} e^{(-1-i)x} - \frac{2+i}{10} \begin{pmatrix} 5 \\ -2+i \end{pmatrix} e^{(-1+i)x}$$

$$= \frac{1}{2} \begin{pmatrix} -2+i \\ 1 \end{pmatrix} (\cos x - i\sin x)e^{-x} - \frac{1}{2} \begin{pmatrix} 2+i \\ -1 \end{pmatrix} (\cos x + i\sin x)e^{-x}$$

$$= \begin{pmatrix} -2\cos x + \sin x \\ \cos x \end{pmatrix} e^{-x}.$$

Naturally, we have found the same solution of the problem.

We recommend to the reader to solve this example using the method of reduction and compare the solutions.

3. $\begin{cases} u' = v + w \\ v' = 3u + w \\ w' = 3u + v \end{cases}.$

We rewrite the system in the vector form $\mathbf{y}' = A\mathbf{y}$, where

$$\mathbf{y} = \begin{pmatrix} u \\ v \\ w \end{pmatrix} \quad \text{and} \quad A = \begin{pmatrix} 0 & 1 & 1 \\ 3 & 0 & 1 \\ 3 & 1 & 0 \end{pmatrix}.$$

To find the eigenvalues of the matrix A, we solve the characteristic equation

$$det(A-\lambda I) = det \begin{pmatrix} -\lambda & 1 & 1 \\ 3 & -\lambda & 1 \\ 3 & 1 & -\lambda \end{pmatrix} = -\lambda^3 + 7\lambda + 6 = -(\lambda+1)(\lambda^2-\lambda-6) = 0.$$

The roots of this equation (eigenvalues) are $\lambda_1 = -1$, $\lambda_2 = -2$, $\lambda_3 = 3$. Recall from Linear Algebra that at least one eigenvector corresponds to each distinct eigenvalue and the eigenvectors of different eigenvalues are linearly independent. So, in this case, we have the complete system of eigenvectors. Let us find them by substituting λ_j into $(A - \lambda_j I)\mathbf{s}_j = 0$, $j = 1, 2, 3$. For $\lambda_1 = -1$ and $\mathbf{s}_1 = \begin{pmatrix} a_1 \\ b_1 \\ c_1 \end{pmatrix}$ we have

$$\begin{cases} a_1 + b_1 + c_1 = 0 \\ 3a_1 + b_1 + c_1 = 0 \\ 3a_1 + b_1 + c_1 = 0 \end{cases}.$$

Since the second and third equations coincide, we have to consider the first and second. Subtracting one from the other, we get $a_1 = 0$. Consequently, $b_1 = -c_1$ and we find the eigenvector $\mathbf{s}_1 = \begin{pmatrix} 0 \\ -1 \\ 1 \end{pmatrix}$. For $\lambda_2 = -2$ and $\mathbf{s}_2 = \begin{pmatrix} a_2 \\ b_2 \\ c_2 \end{pmatrix}$ we get the system

$$\begin{cases} 2a_2 + b_2 + c_2 = 0 \\ 3a_2 + 2b_2 + c_2 = 0 \\ 3a_2 + b_2 + 2c_2 = 0 \end{cases}.$$

Since the equations are linearly dependent, we can choose any two linearly independent ones, for example, the first and second. Subtracting one from the other we get $b_2 = -a_2$ and, consequently, $c_2 = -a_2$. Using $a_2 = -1$ we find the eigenvector $\mathbf{s}_2 = \begin{pmatrix} -1 \\ 1 \\ 1 \end{pmatrix}$. For $\lambda_3 = 3$ and $\mathbf{s}_3 = \begin{pmatrix} a_3 \\ b_3 \\ c_3 \end{pmatrix}$ we obtain the system

$$\begin{cases} -3a_3 + b_3 + c_3 = 0 \\ 3a_3 - 3b_3 + c_3 = 0 \\ 3a_3 + b_3 - 3c_3 = 0 \end{cases}.$$

The first two equations are linearly independent, so we use them. Adding them, we get $-2b_3 + 2c_3 = 0$ and, consequently, $3a_3 = 2c_3$. Choosing $c_3 = 3$, we get the eigenvector $\mathbf{s}_3 = \begin{pmatrix} 2 \\ 3 \\ 3 \end{pmatrix}$. Therefore, the fundamental system of solutions has the form

$$\mathbf{y}_1 = \begin{pmatrix} 0 \\ -1 \\ 1 \end{pmatrix} e^{-x}, \ \mathbf{y}_2 = \begin{pmatrix} -1 \\ 1 \\ 1 \end{pmatrix} e^{-2x}, \ \mathbf{y}_3 = \begin{pmatrix} 2 \\ 3 \\ 3 \end{pmatrix} e^{3x}$$

and the general solution is

$$\mathbf{y}_{gh} = \begin{pmatrix} u \\ v \\ w \end{pmatrix} = C_1\mathbf{y}_1 + C_2\mathbf{y}_2 + C_3\mathbf{y}_3 = C_1 \begin{pmatrix} 0 \\ -1 \\ 1 \end{pmatrix} e^{-x} + C_2 \begin{pmatrix} -1 \\ 1 \\ 1 \end{pmatrix} e^{-2x} + C_3 \begin{pmatrix} 2 \\ 3 \\ 3 \end{pmatrix} e^{3x}.$$

The same system was solved using the method of reduction (Example 3 in section 1.1), where the general solution was found in the form

$$\begin{pmatrix} u \\ v \\ w \end{pmatrix} = A_1 \begin{pmatrix} 1 \\ -1 \\ -1 \end{pmatrix} e^{-2x} + A_2 \begin{pmatrix} 1 \\ 3/2 \\ 3/2 \end{pmatrix} e^{3x} + A_3 \begin{pmatrix} 0 \\ 1 \\ -1 \end{pmatrix} e^{-x}$$

(the arbitrary constants are denoted here by A_j, $j = 1, 2, 3$). It can be seen that the simple relations $A_1 = -C_2$, $A_2 = 2C_3$, $A_3 = -C_1$ transform the last solution into the one obtained by the Euler method and vice versa. Thus, the two solutions are equivalent, as they should be.

4. $\begin{cases} u' = v + w \\ v' = u + w \\ w' = u + v \end{cases}.$

We rewrite the system in the vector form $\mathbf{y}' = A\mathbf{y}$, where

$$\mathbf{y} = \begin{pmatrix} u \\ v \\ w \end{pmatrix} \text{ and } A = \begin{pmatrix} 0 & 1 & 1 \\ 1 & 0 & 1 \\ 1 & 1 & 0 \end{pmatrix}.$$

To find the eigenvalues of the matrix A, we solve the characteristic equation

$$det(A - \lambda I) = det \begin{pmatrix} -\lambda & 1 & 1 \\ 1 & -\lambda & 1 \\ 1 & 1 & -\lambda \end{pmatrix} = (\lambda + 1)(\lambda^2 - \lambda - 2) = 0,$$

which has the double root $\lambda_{1,2} = -1$ and the single root $\lambda_3 = 2$. The eigenvectors associated with $\lambda_{1,2} = -1$ we find from the system

$$\begin{cases} a + b + c = 0 \\ a + b + c = 0 \\ a + b + c = 0 \end{cases}.$$

Since all the equations are the same, it is sufficient to solve just one of them. This implies that we have two arbitrary parameters, for example b and c, and, consequently, two linearly independent eigenvectors. Taking $b = 0$, $c = 1$ we have $a = -1$ and eigenvector $\mathbf{s}_1 = \begin{pmatrix} -1 \\ 0 \\ 1 \end{pmatrix}$. Choosing $b = 1$, $c = 0$

we get $a = -1$ and the second eigenvector $\mathbf{s}_2 = \begin{pmatrix} -1 \\ 1 \\ 0 \end{pmatrix}$ linearly independent

of the first. For $\lambda_3 = 2$ and $\mathbf{s}_3 = \begin{pmatrix} a_3 \\ b_3 \\ c_3 \end{pmatrix}$ we get the system

$$\begin{cases} -2a_3 + b_3 + c_3 = 0 \\ a_3 - 2b_3 + c_3 = 0 \\ a_3 + b_3 - 2c_3 = 0 \end{cases}.$$

We can use the first two equations because they are linearly independent. Subtracting one from the other, we get $a_3 = b_3$ and therefore $c_3 = a_3$. By

choosing $a_3 = 1$, we get the eigenvector $\mathbf{s}_3 = \begin{pmatrix} 1 \\ 1 \\ 1 \end{pmatrix}$. Thus, we find the

fundamental set of solutions

$$\mathbf{y}_1 = \begin{pmatrix} -1 \\ 0 \\ 1 \end{pmatrix} e^{-x}, \ \mathbf{y}_2 = \begin{pmatrix} -1 \\ 1 \\ 0 \end{pmatrix} e^{-x}, \ \mathbf{y}_3 = \begin{pmatrix} 1 \\ 1 \\ 1 \end{pmatrix} e^{2x}$$

and the general solution

$$\mathbf{y}_{gh} = \begin{pmatrix} u \\ v \\ w \end{pmatrix} = C_1\mathbf{y}_1 + C_2\mathbf{y}_2 + C_3\mathbf{y}_3 = C_1 \begin{pmatrix} -1 \\ 0 \\ 1 \end{pmatrix} e^{-x} + C_2 \begin{pmatrix} -1 \\ 1 \\ 0 \end{pmatrix} e^{-x} + C_3 \begin{pmatrix} 1 \\ 1 \\ 1 \end{pmatrix} e^{2x}.$$

The solution of the same system was also found using the method of reduction (Example 4 in section 1.1) (here we use the arbitrary constants A_1, A_2, A_3):

$$\begin{pmatrix} u \\ v \\ w \end{pmatrix} = A_1 \begin{pmatrix} -1 \\ 0 \\ 1 \end{pmatrix} e^{-x} + A_2 \begin{pmatrix} 1 \\ 1 \\ 1 \end{pmatrix} e^{2x} + A_3 \begin{pmatrix} -1 \\ 1 \\ 0 \end{pmatrix} e^{-x}.$$

Obviously, the two solutions are equivalent and one is converted to the other by choosing $A_1 = C_1$, $A_2 = C_3$, $A_3 = C_2$.

5. $\begin{cases} u' = -u + v + w \\ v' = u - v + w \\ w' = u + v + w \end{cases}$.

We rewrite the system in the vector form $\mathbf{y}' = A\mathbf{y}$, where

$$\mathbf{y} = \begin{pmatrix} u \\ v \\ w \end{pmatrix} \text{ and } A = \begin{pmatrix} -1 & 1 & 1 \\ 1 & -1 & 1 \\ 1 & 1 & 1 \end{pmatrix}.$$

To find the eigenvalues of the matrix A, we solve the characteristic equation

$$det(A - \lambda I) = det \begin{pmatrix} -1 - \lambda & 1 & 1 \\ 1 & -1 - \lambda & 1 \\ 1 & 1 & 1 - \lambda \end{pmatrix} = -(\lambda + 1)(\lambda^2 - 4) = 0,$$

which has the three simple roots $\lambda_1 = -1$, $\lambda_{2,3} = \pm 2$. We find the eigenvector of $\lambda_1 = -1$ by solving the system

$$\begin{cases} b + c = 0 \\ a + c = 0 \\ a + b + 2c = 0 \end{cases}.$$

Since $b = -c$ and $a = -c$, it follows from the third equation that c is an arbitrary parameter. So, taking $c = 1$, we get $b = -1$ and $a = -1$, that is, the eigenvector $\mathbf{s}_1 = \begin{pmatrix} -1 \\ -1 \\ 1 \end{pmatrix}$. For the eigenvector $\lambda_2 = 2$ we have the system

$$\begin{cases} -3a + b + c = 0 \\ a - 3b + c = 0 \\ a + b - c = 0 \end{cases}.$$

Adding the third and second equations, we find $a = b$. Substituting this relationship into the third equation, we have $c = 2a$ (the first equation is automatically satisfied with these relationships). Therefore, choosing $a = 1$, we get $b = 1$ and $c = 2$, that is, the second eigenvector is $s_2 = \begin{pmatrix} 1 \\ 1 \\ 2 \end{pmatrix}$. Finally, for the eigenvector with $\lambda_3 = -2$ we have the system

$$\begin{cases} a + b + c = 0 \\ a + b + c = 0 \\ a + b + 3c = 0 \end{cases}.$$

Subtracting the second equation from the third, we get $c = 0$ and from the remaining relations it follows that $a = -b$. Therefore, by choosing $b = 1$, we get $a = -1$ and the eigenvector $s_3 = \begin{pmatrix} -1 \\ 1 \\ 0 \end{pmatrix}$. Thus, we find the fundamental system of solutions

$$\mathbf{y}_1 = \begin{pmatrix} -1 \\ -1 \\ 1 \end{pmatrix} e^{-x}, \ \mathbf{y}_2 = \begin{pmatrix} 1 \\ 1 \\ 2 \end{pmatrix} e^{2x}, \ \mathbf{y}_3 = \begin{pmatrix} -1 \\ 1 \\ 0 \end{pmatrix} e^{-2x}$$

and the general solution

$$\mathbf{y}_{gh} = \begin{pmatrix} u \\ v \\ w \end{pmatrix} = C_1\mathbf{y}_1 + C_2\mathbf{y}_2 + C_3\mathbf{y}_3 = C_1 \begin{pmatrix} -1 \\ -1 \\ 1 \end{pmatrix} e^{-x} + C_2 \begin{pmatrix} 1 \\ 1 \\ 2 \end{pmatrix} e^{2x} + C_3 \begin{pmatrix} -1 \\ 1 \\ 0 \end{pmatrix} e^{-2x}.$$

The solution of the same system was also found by the method of reduction (Example 6 in section 1.1, here we use the arbitrary constants A_1, A_2, A_3):

$$\begin{pmatrix} u \\ v \\ w \end{pmatrix} = A_1 \begin{pmatrix} -1 \\ -1 \\ 1 \end{pmatrix} e^{-x} + A_2 \begin{pmatrix} 1/2 \\ 1/2 \\ 1 \end{pmatrix} e^{2x} + A_3 \begin{pmatrix} -1 \\ 1 \\ 0 \end{pmatrix} e^{-2x}.$$

Obviously, the two solutions are equivalent and one is transformed into the other by choosing $A_1 = C_1$, $A_2 = 2C_2$, $A_3 = C_3$.

Case 2: There is no complete system of eigenvectors.
Recall from Linear Algebra that at least one eigenvector corresponds to each eigenvalue. The number of eigenvectors associated with a specific

eigenvalue can vary from 1 to the multiplicity of that eigenvalue. If the latter situation occurs with all eigenvalues, then there is a complete system of eigenvectors. Therefore, the lack of eigenvectors is caused by multiple eigenvalues that does not have a sufficient number of eigenvectors, that is, the number of the associated eigenvectors is smaller than the multiplicity of the eigenvalue. In this case, it is necessary to look for solutions in a different form and each of such eigenvalues can be treated separately. So, let us consider one of these eigenvalues and denote it λ. We assume that λ has the multiplicity $1 < k \leq n$ (n is the order of the system) and that there are $m < k$ linearly independent eigenvectors $\mathbf{s}_1, \ldots, \mathbf{s}_m$ associated with λ. In this case, we already have m linearly independent solutions $\mathbf{y}_1 = \mathbf{s}_1 e^{\lambda x}, \ldots, \mathbf{y}_m = \mathbf{s}_m e^{\lambda x}$ and it remains to complete this set with $k - m$ linearly independent solutions. The latter can be found in the form $\mathbf{P}_{k-m}(x)e^{\lambda x}$, where $\mathbf{P}_{k-m}(x)$ is a vector polynomial of degree $k - m$ and order n, that is,

$$\mathbf{P}_{k-m}(x) = \begin{pmatrix} a_{1k-m}x^{k-m} + \ldots + a_{11}x + a_{10} \\ \ldots \\ a_{nk-m}x^{k-m} + \ldots + a_{n1}x + a_{n0} \end{pmatrix}.$$

The coefficients of the vector polynomial a_{ij}, $i = 1, \ldots, n$, $j = 0, \ldots, k - m$ must be determined by substituting this proposed form of the solution into the original system in such a way that $k - m$ missing linearly independent solutions are obtained to complete the set associated with λ. It is usually recommended to construct the first complementary solution using a vector polynomial of degree 1, the second of degree 2, etc., the last of degree $k - m$. In this case, the linear independence of these solutions will be guaranteed automatically. Below we solve different examples to see how this works in practice.

Examples.

1. $\begin{cases} y' = 4y - z \\ z' = 2z + y \end{cases}$.

We rewrite this system in the vector form $\mathbf{y}' = A\mathbf{y}$, where $\mathbf{y} = \begin{pmatrix} y \\ z \end{pmatrix}$ and $A = \begin{pmatrix} 4 & -1 \\ 1 & 2 \end{pmatrix}$. Solving the characteristic equation

$$\det(A - \lambda I) = \det \begin{pmatrix} 4 - \lambda & -1 \\ 1 & 2 - \lambda \end{pmatrix} = \lambda^2 - 6\lambda + 9 = 0,$$

we find the double eigenvalue $\lambda_{1,2} = 3$ which we will denote λ for brevity. Substitution of this λ into the eigenvector equation $(A - \lambda I)\mathbf{s} = 0$ leads to the following system $\begin{cases} a - b = 0 \\ a - b = 0 \end{cases}$, where $\mathbf{s} = \begin{pmatrix} a \\ b \end{pmatrix}$. Since two equations coincide, we will only use one of them, which allows us to choose one parameter arbitrarily, indicating that there is only one eigenvector corresponding to $\lambda = 3$. Choosing $a = 1$, we get $b = 1$ and $\mathbf{s} = \begin{pmatrix} 1 \\ 1 \end{pmatrix}$. The corresponding solution comes in the form $\mathbf{y}_1 = \mathbf{s}e^{\lambda x} = \begin{pmatrix} 1 \\ 1 \end{pmatrix} e^{3x}$. One more linearly independent solution is missing. We can find it in the form $\mathbf{y}_2 = \mathbf{P}_1(x)e^{3x}$, $\mathbf{P}_1(x) = \begin{pmatrix} ax + b \\ cx + d \end{pmatrix}$ (the coefficients a and b of the vector $\mathbf{s}$ have already been found and will not appear again, so there is no confusion of notation). Substituting $\mathbf{y}_2$ and its derivative $\mathbf{y}_2' = \begin{pmatrix} 3ax + 3b + a \\ 3cx + 3d + c \end{pmatrix} e^{3x}$ into the original system, after dividing by the common factor e^{3x}, we get

$$\begin{cases} 3ax + 3b + a = 4(ax + b) - (cx + d) \\ 3cx + 3d + c = (ax + b) + 2(cx + d) \end{cases} ,$$

or simplifying

$$\begin{cases} ax + b - a = cx + d \\ ax + b = cx + d + c \end{cases} .$$

From the equality between the coefficients together with x, it follows the only relation $a = c$, and from the equality between the free coefficients, we have the two relations $b - a = d$ and $b = d + c$, but with the formula $a = c$ the former equals the latter. So we can choose two parameters a and b arbitrarily and determine the remaining two as a function of them. For example, if $a = 1$ we have $c = 1$ and if $b = 0$ we have $d = -1$. Thus, the second solution has the form $\mathbf{y}_2 = \begin{pmatrix} x \\ x - 1 \end{pmatrix} e^{3x}$ and the general solution is

$$\mathbf{y}_{gh} = \begin{pmatrix} y \\ z \end{pmatrix} = C_1 \mathbf{y}_1 + C_2 \mathbf{y}_2 = C_1 \begin{pmatrix} 1 \\ 1 \end{pmatrix} e^{3x} + C_2 \begin{pmatrix} x \\ x - 1 \end{pmatrix} e^{3x}.$$

The same system was solved using the method of reduction (Example 2 in section 1.1), where the following general solution was found

$$\begin{pmatrix} y \\ z \end{pmatrix} = A_1 \begin{pmatrix} 1 \\ 1 \end{pmatrix} e^{3x} + A_2 \begin{pmatrix} x \\ x - 1 \end{pmatrix} e^{3x}$$

(here we denote the arbitrary coefficients A_1 and A_2). As one can see, the two solutions coincide, with the relationship between the coefficients $A_1 = C_1$, $A_2 = C_2$.

2. $\begin{cases} u' = 2u + v + w \\ v' = -2u - w \\ w' = 2u + v + 2w \end{cases}$.

We rewrite this system in the vector form $\mathbf{y}' = A\mathbf{y}$, where

$$\mathbf{y} = \begin{pmatrix} u \\ v \\ w \end{pmatrix} \quad \text{and} \quad A = \begin{pmatrix} 2 & 1 & 1 \\ -2 & 0 & -1 \\ 2 & 1 & 2 \end{pmatrix}.$$

The eigenvalues of the matrix A satisfies the characteristic equation

$$det(A - \lambda I) = det \begin{pmatrix} 2 - \lambda & 1 & 1 \\ -2 & -\lambda & -1 \\ 2 & 1 & 2 - \lambda \end{pmatrix}$$

$$= -(\lambda^3 - 4\lambda^2 + 5\lambda - 2) = -(\lambda - 1)(\lambda^2 - 3\lambda + 2) = 0.$$

This equation has the single root $\lambda_1 = 2$ and the double root $\lambda_{2,3} = 1$. For $\lambda_1 = 2$ and $\mathbf{s}_1 = \begin{pmatrix} a_1 \\ b_1 \\ c_1 \end{pmatrix}$ we get the system

$$\begin{cases} b_1 + c_1 = 0 \\ -2a_1 - 2b_1 - c_1 = 0 \\ 2a_1 + b_1 = 0 \end{cases} .$$

We can use the first two equations, because they are linearly independent. Adding these equations, we have $b_1 = -2a_1$. By choosing $b_1 = -2$, we find the eigenvector $\mathbf{s}_1 = \begin{pmatrix} 1 \\ -2 \\ 2 \end{pmatrix}$ and the corresponding particular solution $\mathbf{y}_1 = \begin{pmatrix} 1 \\ -2 \\ 2 \end{pmatrix} e^{2x}$. For the eigenvectors $\mathbf{s} = \begin{pmatrix} a \\ b \\ c \end{pmatrix}$ associated with the double eigenvalue $\lambda = 1$ (we omit the indices for simplicity), we have the system

$$\begin{cases} a + b + c = 0 \\ -2a - b - c = 0 \\ 2a + b + c = 0 \end{cases} .$$

The last two equations coincide, but the first two are linearly independent, which means that we can choose only one parameter and, consequently, there exists only one eigenvector which corresponds to $\lambda = 1$. Adding the first and second equations, we determine that $a = 0$ and choosing $b = 1$, we get $c = -1$ and the eigenvector $\mathbf{s} = \begin{pmatrix} 0 \\ 1 \\ -1 \end{pmatrix}$ with the corresponding particular solution

$\mathbf{y}_2 = \begin{pmatrix} 0 \\ 1 \\ -1 \end{pmatrix} e^x$. Then we should find one more particular solution for $\lambda = 1$, which can be sought in the form

$$\mathbf{y}_3 = \mathbf{P}_1(x)e^x, \quad \mathbf{P}_1(x) = \begin{pmatrix} ax + b \\ cx + d \\ px + r \end{pmatrix}$$

(the components a, b and c of the vector $\mathbf{s}$ have already been found and will not appear anymore, so there is no confusion). Substituting $\mathbf{y}_3$ and its derivative $\mathbf{y}_3' = \begin{pmatrix} ax + b + a \\ cx + d + c \\ px + r + p \end{pmatrix} e^x$ into the original system, after eliminating the common factor e^x, we obtain

$$\begin{cases} ax + b + a = 2(ax + b) + (cx + d) + (px + r) \\ cx + d + c = -2(ax + b) - (px + r) \\ px + r + p = 2(ax + b) + (cx + d) + 2(px + r) \end{cases},$$

or simplifying

$$\begin{cases} -ax - b + a = (cx + d) + (px + r) \\ cx + d + c = -2(ax + b) - (px + r) \\ -px - r + p = 2(ax + b) + (cx + d) \end{cases}.$$

From the equality between the coefficients together with x we get

$$\begin{cases} -a = c + p \\ c = -2a - p \\ -p = 2a + c \end{cases}.$$

The last two relations coincide, while the first two are independent:

$$\begin{cases} a + c + p = 0 \\ 2a + c + p = 0 \end{cases}.$$

From these two relations it follows that $a = 0$ and $c = -p$. The equality between the free coefficients results in the system

$$\begin{cases} b + d + r = a \\ 2b + d + r = -c \\ 2b + d + r = p \end{cases}.$$

Since $c = -p$, the last two equations coincide and the first two remain

$$\begin{cases} b + d + r = 0 \\ 2b + d + r = -c \end{cases},$$

from which we find $b = -c$ and $r = c - d$. Therefore, by choosing $c = 1$ and $d = 0$ we obtain a complementary solution for $\lambda = 1$ in the form $\mathbf{y}_3 = \begin{pmatrix} -1 \\ x \\ -x+1 \end{pmatrix} e^x$. Thus, the general solution is

$$\mathbf{y}_{gh} = \begin{pmatrix} u \\ v \\ w \end{pmatrix} = C_1 \mathbf{y}_1 + C_2 \mathbf{y}_2 + C_3 \mathbf{y}_3 = C_1 \begin{pmatrix} 1 \\ -2 \\ 2 \end{pmatrix} e^{2x} + C_2 \begin{pmatrix} 0 \\ 1 \\ -1 \end{pmatrix} e^x + C_3 \begin{pmatrix} -1 \\ x \\ -x+1 \end{pmatrix} e^x.$$

We recommend to the reader to solve this system using the method of reduction and compare the solutions.

3. $\begin{cases} u' = u - w \\ v' = v + w \\ w' = -u - v - w \end{cases}.$

We rewrite the system in the vector form $\mathbf{y}' = A\mathbf{y}$, where

$$\mathbf{y} = \begin{pmatrix} u \\ v \\ w \end{pmatrix} \quad \text{and} \quad A = \begin{pmatrix} 1 & 0 & -1 \\ 0 & 1 & 1 \\ -1 & -1 & -1 \end{pmatrix}.$$

To find the eigenvalues of the matrix A, we solve the characteristic equation

$$\det(A - \lambda I) = \det \begin{pmatrix} 1 - \lambda & 0 & -1 \\ 0 & 1 - \lambda & 1 \\ -1 & -1 & -1 - \lambda \end{pmatrix} = -(\lambda - 1)(\lambda^2 - 1) = 0,$$

which has the single root $\lambda_1 = -1$ and the double root $\lambda_{2,3} = 1$. For $\lambda_1 = -1$

and $\mathbf{s}_1 = \begin{pmatrix} a_1 \\ b_1 \\ c_1 \end{pmatrix}$ we get the system

$$\begin{cases} 2a_1 - c_1 = 0 \\ 2b_1 + c_1 = 0 \\ -a_1 - b_1 = 0 \end{cases}.$$

From the first equation it follows that $c_1 = 2a_1$ and from the third $b_1 = -a_1$ (with these relations the second equation is automatically satisfied). By choosing $a_1 = 1$, we have the eigenvector $\mathbf{s}_1 = \begin{pmatrix} 1 \\ -1 \\ 2 \end{pmatrix}$ and the correspond-

ing particular solution $\mathbf{y}_1 = \begin{pmatrix} 1 \\ -1 \\ 2 \end{pmatrix} e^{-x}$. For the eigenvectors $\mathbf{s} = \begin{pmatrix} a \\ b \\ c \end{pmatrix}$

associated with the double eigenvalue $\lambda = 1$ (we omit the indices for simplicity), we find the system

$$\begin{cases} -c = 0 \\ c = 0 \\ -a - b - 2c = 0 \end{cases}.$$

So we have $c = 0$ and $b = -a$, that is, only one parameter can be chosen arbitrarily, and, consequently, there is only one eigenvector associated with $\lambda = 1$: choosing $a = 1$, we find $\mathbf{s} = \begin{pmatrix} 1 \\ -1 \\ 0 \end{pmatrix}$ with the corresponding particular

solution $\mathbf{y}_2 = \begin{pmatrix} 1 \\ -1 \\ 0 \end{pmatrix} e^x$. The second particular solution related to $\lambda = 1$

(linearly independent with the first one) can be found in the form

$$\mathbf{y}_3 = \mathbf{P}_1(x)e^x, \quad \mathbf{P}_1(x) = \begin{pmatrix} ax + b \\ cx + d \\ px + r \end{pmatrix}$$

(the components a, b and c of the vector $\mathbf{s}$ have already been found and will not be used anymore, so there is no confusion). Substituting $\mathbf{y}_3$ and its

derivative $\mathbf{y}_3' = \begin{pmatrix} ax + b + a \\ cx + d + c \\ px + r + p \end{pmatrix} e^x$ into the original system, after eliminating the common factor e^x, we obtain

$$\begin{cases} ax + b + a = (ax + b) - (px + r) \\ cx + d + c = (cx + d) + (px + r) \\ px + r + p = -(ax + b) - (cx + d) - (px + r) \end{cases}.$$

From the equality between the coefficients with x, we get

$$\begin{cases} p = 0 \\ p = 0 \\ 2p + a + c = 0 \end{cases},$$

whence $p = 0$ and $c = -a$. The equality between the constant terms results in the system

$$\begin{cases} r = -a \\ r = c \\ b + d + 2r = -p \end{cases}.$$

Since $c = -a$, the first two equations coincide and imply that $r = -a$. Taking this relationship and the condition $p = 0$ into the third equation, we get $b + d = 2a$. Therefore, by choosing $a = 1$ and $b = 0$ we get a complementary solution for $\lambda = 1$ in the form $\mathbf{y}_3 = \begin{pmatrix} x \\ -x + 2 \\ -1 \end{pmatrix} e^x$. Thus, the general solution is

$$\mathbf{y}_{gh} = \begin{pmatrix} u \\ v \\ w \end{pmatrix} = C_1 \mathbf{y}_1 + C_2 \mathbf{y}_2 + C_3 \mathbf{y}_3 = C_1 \begin{pmatrix} 1 \\ -1 \\ 2 \end{pmatrix} e^{-x} + C_2 \begin{pmatrix} 1 \\ -1 \\ 0 \end{pmatrix} e^x + C_3 \begin{pmatrix} x \\ -x+2 \\ -1 \end{pmatrix} e^x.$$

We recommend to the reader to solve the same system using the method of reduction and compare the solutions.

4. $\begin{cases} u' = 4u - v \\ v' = 3u + v - w \\ w' = u + w \end{cases}.$

We rewrite the system in the vector form $\mathbf{y}' = A\mathbf{y}$, where

$$\mathbf{y} = \begin{pmatrix} u \\ v \\ w \end{pmatrix} \quad \text{and} \quad A = \begin{pmatrix} 4 & -1 & 0 \\ 3 & 1 & -1 \\ 1 & 0 & 1 \end{pmatrix}.$$

To find the eigenvalues of the matrix A, we solve the characteristic equation

$$\det(A-\lambda I)=\det\begin{pmatrix} 4-\lambda & -1 & 0 \\ 3 & 1-\lambda & -1 \\ 1 & 0 & 1-\lambda \end{pmatrix}=-\lambda^3+6\lambda^2-12\lambda+8=-(\lambda-2)^3=0.$$

This equation has the single triple root $\lambda_{1,2,3}=2$. The system for eigenvectors is written in the form

$$\begin{cases} 2a-b=0 \\ 3a-b-c=0 \\ a-c=0 \end{cases}.$$

From the first equation it follows that $b=2a$, from the third $c=a$ and, with these relations, the second equation is automatically satisfied. Therefore, we have the only arbitrary parameter for determining the eigenvectors, which means that there is the only eigenvector. For example, by choosing $a=1$, we get the eigenvector $\mathbf{s}=\begin{pmatrix}1\\2\\1\end{pmatrix}$ and the corresponding particular solution

$$\mathbf{y}_1=\begin{pmatrix}1\\2\\1\end{pmatrix}e^{2x}.$$

Therefore, there are two more particular (linearly independent) solutions to be constructed, which can be sought in the form

$$\mathbf{y}=\mathbf{P}_2(x)e^{2x},\quad \mathbf{P}_2(x)=\begin{pmatrix} ax^2+bx+c \\ kx^2+mx+n \\ px^2+qx+r \end{pmatrix},$$

where a,b,c,k,m,n,p,q,r are constants to be determined. Substituting $\mathbf{y}$ into the original system, after eliminating the common factor e^{2x}, we get

$$\begin{cases} 2(ax^2+bx+c)+2ax+b=4(ax^2+bx+c)-(kx^2+mx+n) \\ 2(kx^2+mx+n)+2kx+m=3(ax^2+bx+c)+(kx^2+mx+n)-(px^2+qx+r), \\ 2(px^2+qx+r)+2px+q=(ax^2+bx+c)+(px^2+qx+r) \end{cases}$$

or simplifying,

$$\begin{cases} 2(ax^2+bx+c)-2ax-b-(kx^2+mx+n)=0 \\ 3(ax^2+bx+c)-(kx^2+mx+n)-2kx-m-(px^2+qx+r)=0 \\ (ax^2+bx+c)-(px^2+qx+r)-2px-q=0 \end{cases}.$$

From the equality between the coefficients together with x^2, we get the decoupled subsystem

$$\begin{cases} 2a - k = 0 \\ 3a - k - p = 0 \\ a - p = 0 \end{cases},$$

which coincides with the system for the eigenvector **s** and, consequently, we already know that there is only one arbitrary parameter, for example a, and the remaining two are defined through a in the form $k = 2a$ and $p = a$. From the equality between the coefficients together with x we get the subsystem

$$\begin{cases} 2b - m = 2a \\ 3b - m - q = 2k \\ b - q = 2p \end{cases}.$$

Substituting into this subsystem the expressions for k and p in terms of a, we get

$$\begin{cases} 2b - m = 2a \\ 3b - m - q = 4a \\ b - q = 2a \end{cases}.$$

The sum of the first and third equations results in the second, which is a linear consequence of these two. So, we have only the two equations to satisfy. From the first it follows that $m = 2b - 2a$ and from the third $q = b - 2a$. In this way, the parameters m and q are defined in terms of b and a. Finally, by equating the constant terms, we get the subsystem

$$\begin{cases} 2c - n = b \\ 3c - n - r = m \\ c - r = q \end{cases}.$$

Since $m = 2b - 2a$ and $q = b - 2a$, we have

$$\begin{cases} 2c - n = b \\ 3c - n - r = 2b - 2a \\ c - r = b - 2a \end{cases}.$$

Again, the sum of the first and third equations results in the second, so we only have the two equations to satisfy. From the first it follows that $n = 2c - b$ and from the third $r = c - b + 2a$. Thus, n and r are defined in terms of c, b and a. Thus, the solution $\mathbf{y} = \mathbf{P}_2(x)e^{2x}$, is defined in the form

$$\mathbf{y} = \left(\frac{ax^2 + bx + c}{2ax^2 + (2b - 2a)x + (2c - b)\ ax^2 + (b - 2a)x + (c - b + 2a)} \right) e^{2x}$$

with the three arbitrary parameters a, b, c. We now see that using the values $a = 0$, $b = 0$ and $c = 1$, we get the already found solution $\mathbf{y}_1$. To obtain the solutions linearly independent with $\mathbf{y}_1$, we have to choose a and/or b that are not zero. For example, for $a = 0$, $b = 1$ and $c = 0$ we get $\mathbf{y}_2 = \begin{pmatrix} x \\ 2x - 1 \\ x - 1 \end{pmatrix} e^{2x}$,

while for $a = 1$, $b = 0$ and $c = 0$ we obtain $\mathbf{y}_3 = \begin{pmatrix} x^2 \\ 2x^2 - 2x \\ x^2 - 2x + 2 \end{pmatrix} e^{2x}$,

which are two more linearly independent solutions. Thus, $\mathbf{y}_1$, $\mathbf{y}_2$, $\mathbf{y}_3$ form a system of linearly independent particular solutions and, therefore, the general solution is in the form

$$\mathbf{y}_{gh} = \begin{pmatrix} u \\ v \\ w \end{pmatrix} = C_1\mathbf{y}_1 + C_2\mathbf{y}_2 + C_3\mathbf{y}_3$$

$$= C_1 \begin{pmatrix} 1 \\ 2 \\ 1 \end{pmatrix} e^{2x} + C_2 \begin{pmatrix} x \\ 2x - 1 \\ x - 1 \end{pmatrix} e^{2x} + C_3 \begin{pmatrix} x^2 \\ 2x^2 - 2x \\ x^2 - 2x + 2 \end{pmatrix} e^{2x}.$$

The same system was solved using the method of reduction (Example 7 in section 1.1), where the general solution was found in the same form.

We recommend to the reader to solve Example 5 of section 1.1 using the Euler method and compare the solutions.

Exercises for section 1

Solve using the method of reduction and Euler method and compare the solutions:

1. $\begin{cases} u' = u - 5v \\ v' = 2u - v \end{cases}$.

2. $\begin{cases} u' = 2u + v \\ v' = 4v - u \end{cases}$.

3. $\begin{cases} u' = -9v \\ v' = u \end{cases}$.

4. $\begin{cases} u' = 8v - u \\ v' = v + u \end{cases}$.

5. $\begin{cases} u' = u - v \\ v' = v - u \end{cases}$.

6. $\begin{cases} u' = 2u + v \\ v' = 3u + 4v \end{cases}$.

7. $\begin{cases} u' = u - v \\ v' = v - 4u \end{cases}$.

8. $\begin{cases} u' = u + v \\ v' = 3v - 2u \end{cases}$.

9. $\begin{cases} u' = u - 3v \\ v' = 3u + v \end{cases}$.

10. $\begin{cases} u' = 2u + v \\ v' = 4v - u \end{cases}$.

11. $\begin{cases} u' = 2v - 3u \\ v' = v - 2u \end{cases}$.

12. $\begin{cases} u' = 3u - v + w \\ v' = -u + 5v - w \\ w' = u - v + 3w \end{cases}$.

13. $\begin{cases} u' = -v + w \\ v' = w \\ w' = -u + w \end{cases}$.

14. $\begin{cases} u' = 2u - v + w \\ v' = u + 2v - w \\ w' = -u - v + 2w \end{cases}$.

15. $\begin{cases} u' = u + w - v \\ v' = u + v - w \\ w' = 2u - v \end{cases}$.

16. $\begin{cases} u' = u - 2v - w \\ v' = v - u + w \\ w' = u - w \end{cases}$.

17. $\begin{cases} u' = u - v - w \\ v' = u + v \\ w' = 3u + w \end{cases}$.

18. $\begin{cases} u' = 2u + v \\ v' = u + 3v - w \\ w' = 2v + 3w - u \end{cases}$.

19. $\begin{cases} u' = 4u - v - w \\ v' = u + 2v - w \\ w' = u - v + 2w \end{cases}$.

20. $\begin{cases} u' = 2u - v - w \\ v' = 3u - 2v - 3w \\ w' = 2w - u + v \end{cases}$.

21. $\begin{cases} u' = 3u - 2v - w \\ v' = 3u - 4v - 3w \\ w' = 2u - 4v \end{cases}$.

22. $\begin{cases} u' = u - v + w \\ v' = u + v - w \\ w' = 2w - v \end{cases}$.

23. $\begin{cases} u' = v - 2w - u \\ v' = 4u + v \\ w' = 2u + v - w \end{cases}$.

24. $\begin{cases} u' = 2u - v - w \\ v' = 2u - v - 2w \\ w' = 2w - u + v \end{cases}$.

25. $\begin{cases} u' = 7u + v + 2w \\ v' = 2u + 3v + w \\ w' = -8u - 2v - w \end{cases}$.

26. $\begin{cases} u' = -2u - v - w \\ v' = -3v + w \\ w' = -v - w \end{cases}$.

2   Nonhomogeneous systems with constant co-efficients

Definition. A nonhomogeneous linear system with constant coefficients has the form

$$\mathbf{y}' = A\mathbf{y} + \mathbf{f},$$

where A is a constant matrix and $\mathbf{f}$ is a vector function of the independent variable.

2.1   Method of reduction

This method is a direct extension of the respective method for homogeneous systems and its application requires no further explanation. As before, two versions can be used, with the letter D and without it, each with its advantages depending on the order of the system and the structure of the equations.

Examples.

1. $\begin{cases} y' = 2z - 2x \\ z' = 3 - 2y \end{cases}$.

In this system, either of the two unknown functions can be easily eliminated. Let us eliminate y, using the second equation of the system, which we rewrite in the form $y = \frac{1}{2}(3 - z')$. Substituting this expression into the first equation, we find the second-order equation for a single unknown function z: $\frac{1}{2}(0 - z'') = 2z - 2x$ or, after simplification, $z'' + 4z = 4x$. The characteristic equation $\lambda^2 + 4 = 0$ has complex conjugate roots $\lambda_{1,2} = \pm 2i$ and, consequently, the two linearly independent solutions are $z_1 = \cos 2x$ and $z_2 = \sin 2x$, which form the general solution of the homogeneous equation $z_h = C_1 \cos 2x + C_2 \sin 2x$. The particular solution of the nonhomogeneous equation can be sought in the form $z_p = ax + b$. Substituting this function into the equation, we get $4(ax + b) = 4x$, whence $a = 1, b = 0$, that is, $z_p = x$. Therefore, the general solution of the equation for z comes in the form

$$z = z_h + z_p = C_1 \cos 2x + C_2 \sin 2x + x.$$

Returning to y we find

$$y = \frac{1}{2}(3 - z') = \frac{1}{2}(3 + 2C_1 \sin 2x - 2C_2 \cos 2x - 1) = C_1 \sin 2x - C_2 \cos 2x + 1.$$

Thus, the general solution of the system is

$$\begin{cases} y = C_1 \sin 2x - C_2 \cos 2x + 1 \\ z = C_1 \cos 2x + C_2 \sin 2x + x \end{cases}.$$

2. $\begin{cases} y' = z - 3y + e^{2x} \\ z' = -y - 5z + e^x \end{cases}.$

Let us eliminate y using the second equation of the system, which we rewrite in the form $y = -z' - 5z + e^x$. Substituting this expression into the first equation, we find the second order equation for z:

$$-z'' - 5z' + e^x = z - 3(-z' - 5z + e^x) + e^{2x}$$

or, after simplification, $z'' + 8z' + 16z = 4e^x - e^{2x}$. The characteristic equation $\lambda^2 + 8\lambda + 16 = 0$ has a double real root $\lambda = -4$ and, consequently, the two independent solutions of the homogeneous equation are $z_1 = e^{-4x}$ and $z_2 = xe^{-4x}$. The nonhomogeneous equation is separated into the two parts. First, we look for a particular solution to the equation $z'' + 8z' + 16z = 4e^x$ in the form $z_{p1} = ae^x$. Substituting this function into the equation, we get $a + 8a + 16a = 4$, from which it follows that $a = \frac{4}{25}$. Second, we seek a particular solution to the equation $z'' + 8z' + 16z = e^{2x}$ in the form $z_{p2} = be^{2x}$. Substitution of this function into the equation leads to the relation $4b + 16b + 16b = 1$, from which we get $b = \frac{1}{36}$. Therefore, the general solution of the equation for z has the form

$$z = C_1 z_1 + C_2 z_2 + z_{p1} + z_{p2} = (C_1 + C_2 x)e^{-4x} + \frac{4}{25}e^x - \frac{1}{36}e^{2x}.$$

Now we return to y:

$$y = -z' - 5z + e^x = (-C_1 - C_2 - C_2 x)e^{-4x} + \frac{1}{25}e^x + \frac{7}{36}e^{2x}.$$

Thus, the general solution of the system is

$$\begin{cases} y = (-C_1 - C_2 - C_2 x)e^{-4x} + \frac{1}{25}e^x + \frac{7}{36}e^{2x} \\ z = (C_1 + C_2 x)e^{-4x} + \frac{4}{25}e^x - \frac{1}{36}e^{2x} \end{cases}.$$

3. $\begin{cases} y' = y + 2z \\ z' = y - 5 \sin x \end{cases}.$

Using the expression $y = z' + 5 \sin x$ from the second equation, we eliminate

y from the first and obtain: $z'' - z' - 2z = 5(\sin x - \cos x)$. The characteristic equation $\lambda^2 - \lambda - 2 = 0$ has simple real roots $\lambda_1 = -1$ and $\lambda_2 = 2$, and, consequently, the two independent solutions of the homogeneous equation are $z_1 = e^{-x}$ and $z_2 = e^{2x}$. To find the particular solution of the nonhomogeneous equation, we consider the auxiliary equation $z'' - z' - 2z = 5e^{ix}$, whose particular solution can be found in the form $z_{pa} = ae^{ix}$. Substituting this function into the auxiliary equation, we get $(-3 - i)a = 5$, from which it follows that $a = \frac{-3+i}{2}$, that is, $z_{pa} = \frac{-3+i}{2}e^{ix}$. Therefore, the solution to the equation $z'' - z' - 2z = 5\sin x$ is given by the formula

$$z_{p1} = Im(z_{pa}) = Im(\frac{-3+i}{2}(\cos x + i\sin x)) = \frac{1}{2}\cos x - \frac{3}{2}\sin x$$

and the solution of $z'' - z' - 2z = 5\cos x$ by the formula

$$z_{p2} = Re(z_{pa}) = Re(\frac{-3+i}{2}(\cos x + i\sin x)) = -\frac{3}{2}\cos x - \frac{1}{2}\sin x.$$

Finally, the general solution of the equation for z has the form

$$z = C_1 z_1 + C_2 z_2 + z_{p1} - z_{p2} = C_1 e^{-x} + C_2 e^{2x} + 2\cos x - \sin x.$$

Now we go back to y:

$$y = z' + 5\sin x = -C_1 e^{-x} + 2C_2 e^{2x} - \cos x + 3\sin x.$$

Thus, the general solution of the system is

$$\begin{cases} y = -C_1 e^{-x} + 2C_2 e^{2x} - \cos x + 3\sin x \\ z = C_1 e^{-x} + C_2 e^{2x} + 2\cos x - \sin x \end{cases}.$$

2.2   Euler method

According to the theory of linear systems, the general solution of a nonhomogeneous system can be found as the sum of the general solution of the corresponding homogeneous system and any particular solution of the nonhomogeneous system. The search for the general solution of the homogeneous system by the Euler method follows the same scheme as before. The additional problem is to find a particular solution of the original system. Similarly to solution of higher-order linear equations, the Euler method for systems is applied to some special forms (although quite generic and frequent) of the right-hand sides $\mathbf{f}$ of the system $\mathbf{y'} = A\mathbf{y} + \mathbf{f}$. Next, we consider the two families of functions.

Euler method, Case 1: right-hand side $\mathbf{f} = \mathbf{P}_m(x)e^{\gamma x}$.

Recall that $\mathbf{P}_m(x)$ is the notation of the vector polynomial function of degree m, that is, each component of $\mathbf{P}_m(x)$ is a polynomial of degree less than or equal to m and at least one of them has degree m. The exponent γ can be either real or complex, although in the examples of this section we focus on real γ. For the right-hand side $\mathbf{f} = \mathbf{P}_m(x)e^{\gamma x}$, the Euler method suggests to find the particular solution of the system in the form $\mathbf{y}_p = \mathbf{Q}_{m+s}(x)e^{\gamma x}$, where s is the multiplicity of γ as a root (eigenvalue) of the characteristic equation ($s = 0$ if γ is not an eigenvalue, $s = 1$ if γ is a simple eigenvalue, etc.) and $\mathbf{Q}_{m+s}(x)$ is the vector polynomial of degree $m + s$ whose coefficients must be determined by substituting $\mathbf{y}_p$ into the original system.

Examples.

1. $\begin{cases} y' = 2z - 2x \\ z' = 3 - 2y \end{cases}$.

We rewrite this system in the vector form $\mathbf{y}' = A\mathbf{y} + \mathbf{f}$, where $\mathbf{y} = \begin{pmatrix} y \\ z \end{pmatrix}$,

$A = \begin{pmatrix} 0 & 2 \\ -2 & 0 \end{pmatrix}$ and $\mathbf{f} = \begin{pmatrix} -2x \\ 3 \end{pmatrix}$. We start with the homogeneous system $\mathbf{y}' = A\mathbf{y}$. The eigenvalues of A are found from the characteristic equation

$$det(A - \lambda I) = det \begin{pmatrix} -\lambda & 2 \\ -2 & -\lambda \end{pmatrix} = \lambda^2 + 4 = 0,$$

whence $\lambda_{1,2} = \pm 2i$. As we know, it is sufficient to consider only one of the pair of complex conjugate eigenvalues, for example, $\lambda_1 = 2i$. Substituting $\lambda_1 = 2i$ into the equation for eigenvector $\mathbf{s} = \begin{pmatrix} a \\ b \end{pmatrix}$, we get $\begin{cases} -2ia + 2b = 0 \\ -2a - 2ib = 0 \end{cases}$.

Using the first equation (the second is its linear consequence), we have $b = ia$ and then $\mathbf{s} = \begin{pmatrix} 1 \\ i \end{pmatrix}$ with the corresponding solution $\mathbf{y} = se^{\lambda_1 x} = \begin{pmatrix} 1 \\ i \end{pmatrix} e^{2ix}$.

To use the real form, we extract the real and imaginary parts of $\mathbf{y}$:

$$\mathbf{u} = Re(\mathbf{y}) = Re(\begin{pmatrix} 1 \\ i \end{pmatrix}(\cos 2x + i \sin 2x)) = \begin{pmatrix} \cos 2x \\ -\sin 2x \end{pmatrix},$$

$$\mathbf{v} = Im(\mathbf{y}) = Im(\begin{pmatrix} 1 \\ i \end{pmatrix}(\cos 2x + i \sin 2x))) = \begin{pmatrix} \sin 2x \\ \cos 2x \end{pmatrix}.$$

Therefore, the general solution of the homogeneous system takes the form

$$\mathbf{y}_{gh} = C_1\mathbf{u} + C_2\mathbf{v} = C_1 \begin{pmatrix} \cos 2x \\ -\sin 2x \end{pmatrix} + C_2 \begin{pmatrix} \sin 2x \\ \cos 2x \end{pmatrix}.$$

We proceed to finding the particular solution of the nonhomogeneous system in the form $\mathbf{y}_p = \mathbf{Q}_{1+0}(x)e^{0 \cdot x} = \begin{pmatrix} ax + b \\ cx + d \end{pmatrix}$ ($\gamma = 0$ is not an eigenvalue of the homogeneous system and the degree of the vector polynomial $\mathbf{f}$ is 1). Substituting $\mathbf{y}_p$ and its derivative $\mathbf{y}_p' = \begin{pmatrix} a \\ c \end{pmatrix}$ into the original system, we get $\begin{cases} a = 2(cx + d) - 2x \\ c = 3 - 2(ax + b) \end{cases}$. The relations of the coefficients with x specify $c = 1$, $a = 0$. The free coefficients form two equations $a = 2d$, $c = 3 - 2b$. Since $a = 0$, it follows that $d = 0$, and using $c = 1$ we get $b = 1$. Thus, the particular solution is found in the form $\mathbf{y}_p = \begin{pmatrix} 1 \\ x \end{pmatrix}$.

Finally, the general solution of the original system is

$$\mathbf{y}_{gn} = \mathbf{y}_{gh} + \mathbf{y}_p = C_1 \begin{pmatrix} \cos 2x \\ -\sin 2x \end{pmatrix} + C_2 \begin{pmatrix} \sin 2x \\ \cos 2x \end{pmatrix} + \begin{pmatrix} 1 \\ x \end{pmatrix},$$

or in the component form

$$\begin{cases} y = C_1 \cos 2x + C_2 \sin 2x + 1 \\ z = -C_1 \sin 2x + C_2 \cos 2x + x \end{cases}.$$

Recall that solving this system by the method of reduction (Example 1 in section 2.1) we obtained

$$\begin{cases} y = A_1 \sin 2x - A_2 \cos 2x + 1 \\ z = A_1 \cos 2x + A_2 \sin 2x + x \end{cases}.$$

We can see that the two solutions are equal (use $A_1 = C_2$ and $A_2 = -C_1$).

2. $\begin{cases} y' = z - 3y + e^{2x} \\ z' = -y - 5z + e^x \end{cases}$.

We rewrite this system in the vector form $\mathbf{y}' = A\mathbf{y} + \mathbf{f}$, where

$$\mathbf{y} = \begin{pmatrix} y \\ z \end{pmatrix}, \quad A = \begin{pmatrix} -3 & 1 \\ -1 & -5 \end{pmatrix}, \quad \mathbf{f} = \mathbf{f}_1 + \mathbf{f}_2, \quad \mathbf{f}_1 = \begin{pmatrix} 1 \\ 0 \end{pmatrix} e^{2x}, \quad \mathbf{f}_2 = \begin{pmatrix} 0 \\ 1 \end{pmatrix} e^x.$$

Since the entire right-hand side $\mathbf{f}$ does not fit directly into the type of functions dealt with in the Euler method, we need to divide $\mathbf{f}$ into two parts, each of which has a form that can be solved. We start with the homogeneous system $\mathbf{y}' = A\mathbf{y}$. The eigenvalues of A are found from the characteristic equation

$$det(A - \lambda I) = det\begin{pmatrix} -3-\lambda & 1 \\ -1 & -5-\lambda \end{pmatrix} = \lambda^2 + 8\lambda + 16 = 0,$$

from which we get the double eigenvalue $\lambda = -4$. For the eigenvector $\mathbf{s} = \begin{pmatrix} a \\ b \end{pmatrix}$ we have the system $\begin{cases} a+b = 0 \\ -a-b = 0 \end{cases}$, whence $\mathbf{s} = \begin{pmatrix} 1 \\ -1 \end{pmatrix}$ with the corresponding solution $\mathbf{y}_1 = \mathbf{s}e^{\lambda x} = \begin{pmatrix} 1 \\ -1 \end{pmatrix}e^{-4x}$. To complete the fundamental set of solutions, the second solution of the homogeneous system, linearly independent with $\mathbf{y}_1$, is found in the form $\mathbf{y}_2 = \begin{pmatrix} ax+b \\ cx+d \end{pmatrix}e^{-4x}$. Substituting it into the homogeneous system, we obtain the polynomial equations

$$\begin{cases} -4ax - 4b + a = -3(ax+b) + (cx+d) \\ -4cx - 4d + c = -(ax+b) - 5(cx+d) \end{cases},$$

or simplifying,

$$\begin{cases} -ax - b + a = cx + d \\ cx + d + c = -ax - b \end{cases}.$$

The coefficients with x have the only relation $c = -a$ and the free coefficients are related by the formulas $a - b - d = 0$ and $b + c + d = 0$, which, under the condition $c = -a$, represent the same constraint $a - b - d = 0$. By choosing $a = 1$ and $b = 0$, we get $c = -1$ and $d = 1$. Thus, $\mathbf{y}_2 = \begin{pmatrix} x \\ -x+1 \end{pmatrix}e^{-4x}$.

Therefore, the general solution of the homogeneous system takes the form

$$\mathbf{y}_{gh} = C_1\mathbf{y}_1 + C_2\mathbf{y}_2 = C_1\begin{pmatrix} 1 \\ -1 \end{pmatrix}e^{-4x} + C_2\begin{pmatrix} x \\ -x+1 \end{pmatrix}e^{-4x}.$$

Now we seek a particular solution $\mathbf{y}_{p1}$ of the system $\mathbf{y}' = A\mathbf{y} + \mathbf{f}_1$, $\mathbf{f}_1 = \begin{pmatrix} 1 \\ 0 \end{pmatrix}e^{2x}$. It can be found in the form $\mathbf{y}_{p1} = \mathbf{Q}_{0+0}(x)e^{2x} = \begin{pmatrix} a \\ b \end{pmatrix}e^{2x}$ ($\gamma = 4$ is not an eigenvalue of the homogeneous system and the degree of the vector polynomial is 0). Substituting $\mathbf{y}_{p1}$ and its derivative into the non-homogeneous system, we get $\begin{cases} 2a = -3a + b + 1 \\ 2b = -a - 5b \end{cases}$ or $\begin{cases} 5a - b = 1 \\ a = -7b \end{cases}$. So $b = -\frac{1}{36}$ and $a = \frac{7}{36}$. Therefore, $\mathbf{y}_{p1} = \begin{pmatrix} 7/36 \\ -1/36 \end{pmatrix}e^{2x}$.

Similarly, we find a particular solution $\mathbf{y}_{p2}$ of the system $\mathbf{y}' = A\mathbf{y} + \mathbf{f}_2$,

$\mathbf{f}_2 = \begin{pmatrix} 0 \\ 1 \end{pmatrix} e^x$ using the representation $\mathbf{y}_{p2} = \begin{pmatrix} a \\ b \end{pmatrix} e^x$. Substituting $\mathbf{y}_{p2}$ and its

derivative into the nonhomogeneous system, we obtain $\begin{cases} a = -3a + b \\ b = -a - 5b + 1 \end{cases}$

or $\begin{cases} b = 4a \\ a + 6b = 1 \end{cases}$. So, $a = \frac{1}{25}$ and $b = \frac{4}{25}$. Therefore, $\mathbf{y}_{p2} = \begin{pmatrix} 1/25 \\ 4/25 \end{pmatrix} e^x$.

Finally, we can compose the general solution of the original system:

$$\mathbf{y}_{gn} = \mathbf{y}_{gh} + \mathbf{y}_{p1} + \mathbf{y}_{p2}$$

$$= C_1 \begin{pmatrix} 1 \\ -1 \end{pmatrix} e^{-4x} + C_2 \left(x - x + 1 \right) e^{-4x} + \begin{pmatrix} 7/36 \\ -1/36 \end{pmatrix} e^{2x} + \begin{pmatrix} 1/25 \\ 4/25 \end{pmatrix} e^x,$$

or in the form of components

$$\begin{cases} y = (C_1 + C_2 x)e^{-4x} + \frac{7}{36}e^{2x} + \frac{1}{25}e^x \\ z = (C_2 - C_1 - C_2 x)e^{-4x} - \frac{1}{36}e^{2x} + \frac{4}{25}e^x \end{cases}.$$

Solution of this system by the method of reduction gives the following result (Example 2 in section 2.1):

$$\begin{cases} y = (-A_1 - A_2 - A_2 x)e^{-4x} + \frac{1}{25}e^x + \frac{7}{36}e^{2x} \\ z = (A_1 + A_2 x)e^{-4x} + \frac{4}{25}e^x - \frac{1}{36}e^{2x} \end{cases}.$$

The two solutions are equivalent, which can be verified by establishing the relationship $A_1 + A_2 = -C_1$, $A_2 = -C_2$.

3. $\begin{cases} u' = 2u + v - 2w + 2 - x \\ v' = 1 - u \\ w' = u + v - w + 1 - x \end{cases}$.

We rewrite this system in the vector form $\mathbf{y}' = A\mathbf{y} + \mathbf{f}$, where

$$\mathbf{y} = \begin{pmatrix} u \\ v \\ w \end{pmatrix}, \quad A = \begin{pmatrix} 2 & 1 & -2 \\ -1 & 0 & 0 \\ 1 & 1 & -1 \end{pmatrix}, \quad \mathbf{f} = \begin{pmatrix} -x + 2 \\ 1 \\ -x + 1 \end{pmatrix}.$$

As usual, we start with the homogeneous system. Solving the characteristic equation

$$\det(A - \lambda I) = \det \begin{pmatrix} 2 - \lambda & 1 & -2 \\ -1 & -\lambda & 0 \\ 1 & 1 & -1 - \lambda \end{pmatrix} = (1 - \lambda)(\lambda^2 + 1) = 0,$$

we find the real eigenvalue $\lambda_1 = 1$ and the two complex conjugate eigenvalues $\lambda_{2,3} = \pm i$. The eigenvector $\mathbf{s}_1 = \begin{pmatrix} a_1 \\ b_1 \\ c_1 \end{pmatrix}$ is determined from the system

$$\begin{cases} a_1 + b_1 + c_1 = 0 \\ -a_1 - b_1 = 0 \\ a_1 + b_1 - 2c_1 = 0 \end{cases}.$$

From the first and second equations (the third is their linear consequence) it follows that $c_1 = 0$ and $a_1 = -b_1$. So we have $\mathbf{s}_1 = \begin{pmatrix} 1 \\ -1 \\ 0 \end{pmatrix}$ and $\mathbf{y}_1 = \begin{pmatrix} 1 \\ -1 \\ 0 \end{pmatrix} e^x$. From the two complex conjugate eigenvalues we choose $\lambda_2 = i$ and find the corresponding complex eigenvector $\mathbf{s} = \begin{pmatrix} a \\ b \\ c \end{pmatrix}$ by solving the system

$$\begin{cases} (2 - i)a + b + c = 0 \\ -a - ib = 0 \\ a + b - (1 + i)c = 0 \end{cases}.$$

From the first and second equations (the third is their linear consequence) we get the relations $a = -ib$ and $c = -ib$, and consequently $\mathbf{s} = \begin{pmatrix} 1 \\ i \\ 1 \end{pmatrix}$. So, the two solutions in real form are obtained by taking the real and imaginary parts of the complex solution:

$$\mathbf{y}_2 = Re(\mathbf{s}e^{ix}) = Re\left(\begin{pmatrix} 1 \\ i \\ 1 \end{pmatrix} (\cos x + i \sin x) \right) = \begin{pmatrix} \cos x \\ -\sin x \\ \cos x \end{pmatrix},$$

$$\mathbf{y}_3 = Im(\mathbf{s}e^{ix}) = Im\left(\begin{pmatrix} 1 \\ i \\ 1 \end{pmatrix} (\cos x + i \sin x) \right) = \begin{pmatrix} \sin x \\ \cos x \\ \sin x \end{pmatrix}.$$

Thus, the general solution of the homogeneous system takes the form

$$\mathbf{y}_{gh} = C_1\mathbf{y}_1 + C_2\mathbf{y}_2 + C_3\mathbf{y}_3 = C_1 \begin{pmatrix} 1 \\ -1 \\ 0 \end{pmatrix} e^x + C_2 \begin{pmatrix} \cos x \\ -\sin x \\ \cos x \end{pmatrix} + C_3 \begin{pmatrix} \sin x \\ \cos x \\ \sin x \end{pmatrix}.$$

We now search for the particular solution of the nonhomogeneous system in the form

$$\mathbf{y}_p = \mathbf{Q}_{1+0}(x)e^{0 \cdot x} = \begin{pmatrix} ax + b \\ cx + d \\ px + r \end{pmatrix}$$

($\gamma = 0$ is not an eigenvalue of the homogeneous system and the degree of the vector polynomial is 1). Substituting $\mathbf{y}_p$ into the original system, we get

$$\begin{cases} a = 2(ax + b) + (cx + d) - 2(px + r) - x + 2 \\ c = -(ax + b) + 1 \\ p = (ax + b) + (cx + d) - (px + r) - x + 1 \end{cases}.$$

From the relationship between the coefficients with x it follows that

$$\begin{cases} 2a + c - 2p = 1 \\ a = 0 \\ a + c - p = 1 \end{cases}$$

and then $a = 0$, $c = 1$, $p = 0$. Consequently, for the free coefficients we have

$$\begin{cases} 2b + d - 2r = a - 2 = -2 \\ -b = c - 1 = 0 \\ b + d - r = p - 1 = -1 \end{cases},$$

whence $b = 0, d = 0, r = 1$. Therefore, $\mathbf{y}_p = \begin{pmatrix} 0 \\ x \\ 1 \end{pmatrix}$.

Collecting all the results, we obtain the general solution of the original system:

$$\mathbf{y}_{gn} = \mathbf{y}_{gh} + \mathbf{y}_p = C_1 \begin{pmatrix} 1 \\ -1 \\ 0 \end{pmatrix} e^x + C_2 \begin{pmatrix} \cos x \\ -\sin x \\ \cos x \end{pmatrix} + C_3 \begin{pmatrix} \sin x \\ \cos x \\ \sin x \end{pmatrix} + \begin{pmatrix} 0 \\ x \\ 1 \end{pmatrix}.$$

We recommend to the reader to solve this system using the method of reduction and compare the solutions.

4. $\begin{cases} u' = u - 2v - w - 2e^x \\ v' = -u + v + w + 2e^x \\ w' = u - w - e^x \end{cases}.$

We rewrite the system in the vector form $\mathbf{y}' = A\mathbf{y} + \mathbf{f}$, where

$$\mathbf{y} = \begin{pmatrix} u \\ v \\ w \end{pmatrix}, \quad A = \begin{pmatrix} 1 & -2 & -1 \\ -1 & 1 & 1 \\ 1 & 0 & -1 \end{pmatrix}, \quad \mathbf{f} = \begin{pmatrix} -2 \\ 2 \\ -1 \end{pmatrix} e^x.$$

As usual, we start with the homogeneous system. Solving the characteristic equation

$$\det(A - \lambda I) = \det \begin{pmatrix} 1-\lambda & -2 & -1 \\ -1 & 1-\lambda & 1 \\ 1 & 0 & -1-\lambda \end{pmatrix} = -(\lambda+1)\lambda(\lambda-2) = 0,$$

we find the three distinct real eigenvalues $\lambda_1 = -1$, $\lambda_2 = 0$, $\lambda_3 = 2$. The eigenvector $\mathbf{s}_1 = \begin{pmatrix} a_1 \\ b_1 \\ c_1 \end{pmatrix}$, corresponding to $\lambda_1 = -1$, is found from the system

$$\begin{cases} 2a_1 - 2b_1 - c_1 = 0 \\ -a_1 + 2b_1 + c_1 = 0 \\ a_1 = 0 \end{cases}.$$

From the third equation we get $a_1 = 0$ and the other two give the relationship $c_1 = -2b_1$. So we have $\mathbf{s}_1 = \begin{pmatrix} 0 \\ 1 \\ -2 \end{pmatrix}$ and $\mathbf{y}_1 = \begin{pmatrix} 0 \\ -1 \\ -2 \end{pmatrix} e^{-x}$. For the second eigenvector $\mathbf{s}_2 = \begin{pmatrix} a_2 \\ b_2 \\ c_2 \end{pmatrix}$ we have the system

$$\begin{cases} a_2 - 2b_2 - c_2 = 0 \\ -a_2 + b_2 + c_2 = 0 \\ a_2 - c_2 = 0 \end{cases}.$$

From the third equation we get $a_2 = c_2$ and then the first (or second) shows that $b_2 = 0$. Then, we have $\mathbf{s}_2 = \begin{pmatrix} 1 \\ 0 \\ 1 \end{pmatrix}$ and $\mathbf{y}_2 = \begin{pmatrix} 1 \\ 0 \\ 1 \end{pmatrix}$. Finally, the third eigenvector $\mathbf{s}_3 = \begin{pmatrix} a_3 \\ b_3 \\ c_3 \end{pmatrix}$ satisfies the system

$$\begin{cases} -a_3 - 2b_3 - c_3 = 0 \\ -a_3 - b_3 + c_3 = 0 \\ a_3 - 3c_3 = 0 \end{cases}.$$

From the third equation it follows that $a_3 = 3c_3$ and then the first (or second) gives $b_3 = -2c_3$. So, we have $\mathbf{s}_3 = \begin{pmatrix} 3 \\ -2 \\ 1 \end{pmatrix}$ and $\mathbf{y}_3 = \begin{pmatrix} 3 \\ -2 \\ 1 \end{pmatrix} e^{2x}$. Therefore, the general solution of the homogeneous system takes the form

$$\mathbf{y}_{gh} = C_1\mathbf{y}_1 + C_2\mathbf{y}_2 + C_3\mathbf{y}_3 = C_1 \begin{pmatrix} 0 \\ 1 \\ -2 \end{pmatrix} e^{-x} + C_2 \begin{pmatrix} 1 \\ 0 \\ 1 \end{pmatrix} + C_3 \begin{pmatrix} 3 \\ -2 \\ 1 \end{pmatrix} e^{2x}.$$

We now proceed to the nonhomogeneous part, whose particular solution we seek in the form

$$\mathbf{y}_p = \mathbf{Q}_{0+0}(x)e^x = \begin{pmatrix} a \\ b \\ c \end{pmatrix} e^x$$

($\gamma = 1$ is not an eigenvalue of the homogeneous system and the degree of the vector polynomial is 0). Substituting $\mathbf{y}_p$ into the original system, we get

$$\begin{cases} a = a - 2b - c - 2 \\ b = -a + b + c + 2 \\ c = a - c - 1 \end{cases}.$$

Simplifying the system, we have

$$\begin{cases} 2b + c = -2 \\ a - c = 2 \\ a - 2c = 1 \end{cases}.$$

From the last two equations we find $a = 3$, $c = 1$ and from the first equation $b = -\frac{3}{2}$. So, $\mathbf{y}_p = \begin{pmatrix} 3 \\ -3/2 \\ 1 \end{pmatrix} e^x.$

Collecting the results, we obtain the general solution of the original system:

$$\mathbf{y}_{gn} = \mathbf{y}_{gh} + \mathbf{y}_p = C_1 \begin{pmatrix} 0 \\ 1 \\ -2 \end{pmatrix} e^{-x} + C_2 \begin{pmatrix} 1 \\ 0 \\ 1 \end{pmatrix} + C_3 \begin{pmatrix} 3 \\ -2 \\ 1 \end{pmatrix} e^{2x} + \begin{pmatrix} 3 \\ -3/2 \\ 1 \end{pmatrix} e^x.$$

We recommend to the reader to solve this system using the method of reduction and compare the solutions.

5. $\begin{cases} u' = -5u + v - 2w + e^{-x} + e^{-2x} \\ v' = -u - v + 3e^{-x} + 2e^{-2x} \\ w' = 6u - 2v + 2w - 2e^{-x} - 3e^{-2x} \end{cases}.$

We rewrite the system in the vector form $\mathbf{y}' = A\mathbf{y} + \mathbf{f}_1 + \mathbf{f}_2$, where

$$\mathbf{y} = \begin{pmatrix} u \\ v \\ w \end{pmatrix}, \quad A = \begin{pmatrix} -5 & 1 & -2 \\ -1 & -1 & 0 \\ 6 & -2 & 2 \end{pmatrix}, \quad \mathbf{f}_1 = \begin{pmatrix} 1 \\ 3 \\ -2 \end{pmatrix} e^{-x}, \quad \mathbf{f}_2 = \begin{pmatrix} 1 \\ 2 \\ -3 \end{pmatrix} e^{-2x}.$$

As usual, we start with the homogeneous system. Solving the characteristic equation

$$det(A - \lambda I) = det \begin{pmatrix} -5 - \lambda & 1 & -2 \\ -1 & -1 - \lambda & 0 \\ 6 & -2 & 2 - \lambda \end{pmatrix} = -(\lambda + 2)((\lambda + 1)^2 + 1) = 0,$$

we find the real eigenvalue $\lambda_1 = -2$ and the two complex conjugate eigenvalues $\lambda_{2,3} = -1 \pm i$. The eigenvector $\mathbf{s}_1 = \begin{pmatrix} a_1 \\ b_1 \\ c_1 \end{pmatrix}$ is defined from the system

$$\begin{cases} -3a_1 + b_1 - 2c_1 = 0 \\ -a_1 + b_1 = 0 \\ 6a_1 - 2b_1 + 4c_1 = 0 \end{cases}.$$

From the second equation it follows that $b_1 = a_1$ and then from the first (or third) that $c_1 = -a_1$. Therefore, $\mathbf{s}_1 = \begin{pmatrix} 1 \\ 1 \\ -1 \end{pmatrix}$ and $\mathbf{y}_1 = \begin{pmatrix} 1 \\ 1 \\ -1 \end{pmatrix} e^{-2x}$. From the two complex conjugate eigenvalues we choose $\lambda_2 = -1 + i$ and find the corresponding complex eigenvector $\mathbf{s} = \begin{pmatrix} a \\ b \\ c \end{pmatrix}$ by solving the system

$$\begin{cases} (-4 - i)a + b - 2c = 0 \\ -a - ib = 0 \\ 6a - 2b + (3 - i)c = 0 \end{cases}.$$

From the second equation we have $b = ia$ and then from the first (or third) $c = -2a$. Therefore, $\mathbf{s} = \begin{pmatrix} 1 \\ i \\ -2 \end{pmatrix}$ and the corresponding complex solution is $\mathbf{y}_c = \mathbf{s}e^{(-1+i)x}$. Consequently, the two solutions in real form are obtained by taking the real and imaginary parts:

$$\mathbf{y}_2 = Re(\mathbf{y}_c) = Re\left(\begin{pmatrix} 1 \\ i \\ -2 \end{pmatrix} (\cos x + i \sin x)e^{-x} \right) = \begin{pmatrix} \cos x \\ -\sin x \\ -2\cos x \end{pmatrix} e^{-x},$$

$$\mathbf{y}_3 = Im(\mathbf{y}_c) = Im\left(\begin{pmatrix} 1 \\ i \\ -2 \end{pmatrix} (\cos x + i \sin x)e^{-x} \right) = \begin{pmatrix} \sin x \\ \cos x \\ -2\sin x \end{pmatrix}.$$

Thus, the general solution of the homogeneous system takes the form

$$\mathbf{y}_{gh}=C_1\mathbf{y}_1+C_2\mathbf{y}_2+C_3\mathbf{y}_3=C_1\begin{pmatrix}1\\1\\-1\end{pmatrix}e^{-2x}+C_2\begin{pmatrix}\cos x\\-\sin x\\-2\cos x\end{pmatrix}e^{-x}+C_3\begin{pmatrix}\sin x\\\cos x\\-2\sin x\end{pmatrix}e^{-x}.$$

We now proceed to the nonhomogeneous system. For the first right-hand part $\mathbf{f}_1$, we look for a particular solution in the form

$$\mathbf{y}_{p1} = \mathbf{Q}_{0+0}(x)e^{-x} = \begin{pmatrix}a\\b\\c\end{pmatrix}e^{-x}$$

($\gamma = -1$ is not an eigenvalue of the homogeneous system and the degree of the vector polynomial is 0). Substituting $\mathbf{y}_{p1}$ into the system with $\mathbf{f}_1$, we get

$$\begin{cases}-a = -5a + b - 2c + 1\\-b = -a - b + 3\\-c = 6a - 2b + 2c - 2\end{cases}.$$

Simplifying the system, we have

$$\begin{cases}4a - b + 2c = 1\\a = 3\\6a - 2b + 3c = 2\end{cases}.$$

Since $a = 3$, the first and third equations take the form $\begin{cases}b - 2c = 11\\2b - 3c = 16\end{cases}$,

whence $b = -1$ and $c = -6$. So, $\mathbf{y}_{p1} = \begin{pmatrix}3\\-1\\-6\end{pmatrix}e^{-x}$.

For the second right-hand side $\mathbf{f}_2$, we seek a particular solution in the form

$$\mathbf{y}_{p2} = \mathbf{Q}_{0+1}(x)e^{-2x} = \begin{pmatrix}ax + \alpha\\bx + \beta\\cx + \sigma\end{pmatrix}e^{(}-2x)$$

($\gamma = -2$ is the simple eigenvalue of the homogeneous system and the degree of the vector polynomial is 0). (The parameters a, b, c and the polynomial Q of the solution $\mathbf{y}_{p1}$ will no longer be used, so there will be no confusion of notations). Substituting $\mathbf{y}_{p2}$ into the system with $\mathbf{f}_2$, we get

$$\begin{cases}-2ax - 2\alpha + a = -5(ax + \alpha) + (bx + \beta) - 2(cx + \sigma) + 1\\-2bx - 2\beta + b = -(ax + \alpha) - (bx + \beta) + 2\\-2cx - 2\sigma + c = 6(ax + \alpha) - 2(bx + \beta) + 2(cx + \sigma) - 3\end{cases}.$$

For the coefficients with x, we have the separate subsystem

$$\begin{cases} 3a - b + 2c = 0 \\ a - b = 0 \\ 6a - 2b + 4c = 0 \end{cases}.$$

From the second equation it follows that $b = a$ and then from the first (or third) that $c = -a$. The constant coefficients generate the second subsystem

$$\begin{cases} 3\alpha - \beta + 2\sigma = 1 - a \\ \alpha - \beta = 2 - b \\ 6\alpha - 2\beta + 4\sigma = 3 + c \end{cases}.$$

Taking into account the relations $b = a$ and $c = -a$, we get

$$\begin{cases} 3\alpha - \beta + 2\sigma = 1 - a \\ \alpha - \beta = 2 - a \\ 6\alpha - 2\beta + 4\sigma = 3 - a \end{cases}.$$

Substituting the expression $\beta = \alpha + a - 2$ from the second equation into the first and third, we get

$$\begin{cases} 3\alpha - (\alpha + a - 2) + 2\sigma = 1 - a \\ 6\alpha - 2(\alpha + a - 2) + 4\sigma = 3 - a \end{cases},$$

or simplifying $\begin{cases} 2\alpha + 2\sigma = -1 \\ 4\alpha + 4\sigma = a - 1 \end{cases}$. For these two equations to be compatible, it must be $a = -1$. So the other main coefficients of the polynomial $\mathbf{Q}_{0+1}(x)$ are $b = a = -1$ and $c = -a = 1$. The relations for the constant terms become $2\sigma = -1 - 2\alpha$ and $\beta = \alpha - 3$. Choosing, for example, $\alpha = 0$, we get $\beta = -3$ and $\sigma = -\frac{1}{2}$. Therefore, the solution we search for has the

form $\mathbf{y}_{p2} = \begin{pmatrix} -x \\ -x - 3 \\ x - 1/2 \end{pmatrix} e^{-2x}$.

Collecting all the results, we obtain the general solution of the original system:

$$\mathbf{y}_{gn} = \mathbf{y}_{gh} + \mathbf{y}_{p1} + \mathbf{y}_{p1} = C_1 \begin{pmatrix} 1 \\ 1 \\ -1 \end{pmatrix} e^{-2x} + C_2 \begin{pmatrix} \cos x \\ -\sin x \\ -2\cos x \end{pmatrix} e^{-x} + C_3 \begin{pmatrix} \sin x \\ \cos x \\ -2\sin x \end{pmatrix} e^{-x}$$

$$+ \begin{pmatrix} 3 \\ -1 \\ -6 \end{pmatrix} e^{-x} + \begin{pmatrix} -x \\ -x-3 \\ x-1/2 \end{pmatrix} e^{-2x}.$$

Euler method, Case 2: right-hand side $\mathbf{f} = \mathbf{P}_m(x)e^{\alpha x}\{\cos\beta x, \sin\beta x\}$.

Recall that $\mathbf{P}_m(x)$ is the vector polynomial of degree m and the key brackets mean that we consider in parallel either of the two cases on the right side: $\mathbf{f}_c = \mathbf{P}_m(x)e^{\alpha x}\cos\beta x$ and $\mathbf{f}_s = \mathbf{P}_m(x)e^{\alpha x}\sin\beta x$. Here we only consider the real exponent $\alpha \in \mathbb{R}$. As in the case of a linear equation, we reduce this right-hand side to the previous Case 1 with the exponent $\gamma = \alpha + i\beta \in \mathbb{C}$. To do so, we introduce the auxiliary right-hand side $\mathbf{g} = \mathbf{P}_m(x)e^{\gamma x}$, $\gamma = \alpha+i\beta$ with the property that $\mathbf{f}_c = Re(\mathbf{g})$ and $\mathbf{f}_s = Im(\mathbf{g})$. Following the procedure in the Case 1, we look for the auxiliary solution in the form $\mathbf{z}_{pn} = \mathbf{Q}_{m+s}(x)e^{\gamma x}$, where the coefficients of the vector polynomial $\mathbf{Q}_{m+s}(x)$ are found by substituting $\mathbf{z}_{pn}$ into the auxiliary equation $\mathbf{z}' = A\mathbf{z} + \mathbf{g}$. After finding $\mathbf{z}_{pn}$, we can determine the solutions $\mathbf{y}_c$ of the equation $\mathbf{y}' = A\mathbf{y} + \mathbf{f}_c$ and $\mathbf{y}_s$ of $\mathbf{y}' = A\mathbf{y} + \mathbf{f}_s$, by applying Theorem 3: $\mathbf{y}_c = Re(\mathbf{z}_{pn})$ and $\mathbf{y}_s = Im(\mathbf{z}_{pn})$. This completes the search for the particular solution of the original equation.

Examples.

1. $\begin{cases} y' = y + 2z \\ z' = y - 5\sin x \end{cases}$.

We rewrite the system in the vector form $\mathbf{y}' = A\mathbf{y} + \mathbf{f}$, where

$$\mathbf{y} = \begin{pmatrix} y \\ z \end{pmatrix}, \quad A = \begin{pmatrix} 1 & 2 \\ 1 & 0 \end{pmatrix}, \quad \mathbf{f} = \begin{pmatrix} 0 \\ -5 \end{pmatrix} \sin x.$$

We start with the homogeneous system $\mathbf{y}' = A\mathbf{y}$. The eigenvalues of A are found by solving the characteristic equation

$$det(A - \lambda I) = det\begin{pmatrix} 1-\lambda & 2 \\ 1 & -\lambda \end{pmatrix} = \lambda^2 - \lambda - 2 = 0,$$

from which we have the two real eigenvalues $\lambda_1 = -1$, $\lambda_2 = 2$. For the first eigenvector $\mathbf{s}_1 = \begin{pmatrix} a_1 \\ b_1 \end{pmatrix}$ we have the system $\begin{cases} 2a_1 + 2b_1 = 0 \\ a_1 + b_1 = 0 \end{cases}$, whence $\mathbf{s}_1 = \begin{pmatrix} 1 \\ -1 \end{pmatrix}$ with the corresponding solution $\mathbf{y}_1 = \mathbf{s}_1 e^{\lambda_1 x} = \begin{pmatrix} 1 \\ -1 \end{pmatrix} e^{-x}$. For the second eigenvector $\mathbf{s}_2 = \begin{pmatrix} a_2 \\ b_2 \end{pmatrix}$ we have the system $\begin{cases} -a_2 + 2b_2 = 0 \\ a_2 - 2b_2 = 0 \end{cases}$,

whence $\mathbf{s}_2 = \begin{pmatrix} 2 \\ 1 \end{pmatrix}$ and $\mathbf{y}_2 = \mathbf{s}_2 e^{\lambda_2 x} = \begin{pmatrix} 2 \\ 1 \end{pmatrix} e^{2x}$. Therefore, the general solution of the homogeneous system has the form

$$\mathbf{y}_{gh} = C_1 \mathbf{y}_1 + C_2 \mathbf{y}_2 = C_1 \begin{pmatrix} 1 \\ -1 \end{pmatrix} e^{-x} + C_2 \begin{pmatrix} 2 \\ 1 \end{pmatrix} e^{2x}.$$

We now find a particular solution $\mathbf{y}_p$ of the nonhomogeneous system $\mathbf{y}' = A\mathbf{y} + \mathbf{f}$. To do this, we introduce the auxiliary function $\mathbf{g} = \begin{pmatrix} 0 \\ -5 \end{pmatrix} e^{ix}$ and consider the auxiliary problem $\mathbf{z}' = A\mathbf{z} + \mathbf{g}$. The solution to the latter can be found in the form

$$\mathbf{z}_p = \mathbf{Q}_{0+0}(x) e^{ix} = \begin{pmatrix} a \\ b \end{pmatrix} e^{ix}$$

($\gamma = i$ is not an eigenvalue of the homogeneous system and the degree of the vector polynomial is 0). Substituting $\mathbf{z}_p$ into the auxiliary system, we get $\begin{cases} ia = a + 2b \\ ib = a - 5 \end{cases}$. The solution of this system is $a = 3 - i$, $b = -1 + 2i$.

So $\mathbf{z}_p = \begin{pmatrix} 3 - i \\ -1 + 2i \end{pmatrix} e^{ix}$. Extracting the imaginary part of $\mathbf{z}_p$, we find the solution $\mathbf{y}_p$:

$$\mathbf{y}_p = Im(\{bfz_p\}) = Im(\begin{pmatrix} 3 - i \\ -1 + 2i \end{pmatrix}(\cos x + i \sin x)) = \begin{pmatrix} -\cos x + 3\sin x \\ 2\cos x - \sin x \end{pmatrix}.$$

Finally, we compose the general solution of the original system:

$$\mathbf{y}_{gn} = \mathbf{y}_{gh} + \mathbf{y}_p = C_1 \begin{pmatrix} 1 \\ -1 \end{pmatrix} e^{-x} + C_2 \begin{pmatrix} 2 \\ 1 \end{pmatrix} e^{2x} + \begin{pmatrix} -\cos x + 3\sin x \\ 2\cos x - \sin x \end{pmatrix},$$

or in the component form

$$\begin{cases} y = C_1 e^{-x} + 2C_2 e^{2x} - \cos x + 3\sin x \\ z = -C_1 e^{-x} + C_2 e^{2x} + 2\cos x - \sin x \end{cases}.$$

Solving this system by the method of reduction, we found the following general solution (Example 3 in section 2.1):

$$\begin{cases} y = -A_1 e^{-x} + 2A_2 e^{2x} - \cos x + 3\sin x \\ z = A_1 e^{-x} + A_2 e^{2x} + 2\cos x - \sin x \end{cases}.$$

Obviously, the two solutions coincide.

2. $\begin{cases} y' = 4y - 3z + \sin x \\ z' = 2y - z - 2\cos x \end{cases}$.

We rewrite the system in the vector form $\mathbf{y}' = A\mathbf{y} + \mathbf{f}$, where

$$\mathbf{y} = \begin{pmatrix} y \\ z \end{pmatrix}, \quad A = \begin{pmatrix} 4 & -3 \\ 2 & -1 \end{pmatrix}, \quad \mathbf{f} = \mathbf{f}_1 + \mathbf{f}_2, \mathbf{f}_1 = \begin{pmatrix} 1 \\ 0 \end{pmatrix} \sin x, \quad \mathbf{f}_2 = \begin{pmatrix} 0 \\ -2 \end{pmatrix} \cos x.$$

The right side has been split into two, because the Euler method does not allow to treat $\sin x$ and $\cos x$ at the same time. We start with the homogeneous system $\mathbf{y}' = A\mathbf{y}$. Solving the characteristic equation

$$det(A - \lambda I) = det \begin{pmatrix} 4 - \lambda & -3 \\ 2 & -1 - \lambda \end{pmatrix} = \lambda^2 - 3\lambda + 2 = 0,$$

we find the eigenvalues $\lambda_1 = 1$, $\lambda_2 = 2$. The first eigenvector $\mathbf{s}_1 = \begin{pmatrix} a_1 \\ b_1 \end{pmatrix}$

satisfies the system $\begin{cases} 3a_1 - 3b_1 = 0 \\ 2a_1 - 2b_1 = 0 \end{cases}$, whence $\mathbf{s}_1 = \begin{pmatrix} 1 \\ 1 \end{pmatrix}$ with the corre-

sponding solution $\mathbf{y}_1 = \mathbf{s}_1 e^{\lambda_1 x} = \begin{pmatrix} 1 \\ 1 \end{pmatrix} e^x$. To find the second eigenvec-

tor $\mathbf{s}_2 = \begin{pmatrix} a_2 \\ b_2 \end{pmatrix}$ we solve the system $\begin{cases} 2a_2 - 3b_2 = 0 \\ 2a_2 - 3b_2 = 0 \end{cases}$, whose solution is

$\mathbf{s}_2 = \begin{pmatrix} 3 \\ 2 \end{pmatrix}$. So, the second solution has the form $\mathbf{y}_2 = \mathbf{s}_2 e^{\lambda_2 x} = \begin{pmatrix} 3 \\ 2 \end{pmatrix} e^{2x}$.

Therefore, we have the following general solution of the homogeneous system:

$$\mathbf{y}_{gh} = C_1\mathbf{y}_1 + C_2\mathbf{y}_2 = C_1 \begin{pmatrix} 1 \\ 1 \end{pmatrix} e^x + C_2 \begin{pmatrix} 3 \\ 2 \end{pmatrix} e^{2x}.$$

Next, we search for a particular solution $\mathbf{y}_p$ of the nonhomogeneous system $\mathbf{y}' = A\mathbf{y} + \mathbf{f}$, $\mathbf{f} = \mathbf{f}_1 + \mathbf{f}_2$. To solve the system with the right-hand side $\mathbf{f}_1 = \begin{pmatrix} 1 \\ 0 \end{pmatrix} \sin x$, we consider the auxiliary function $\mathbf{g}_1 = \begin{pmatrix} 1 \\ 0 \end{pmatrix} e^{ix}$ and the corresponding auxiliary problem $\mathbf{z}' = A\mathbf{z} + \mathbf{g}_1$. The particular solution of the last can be found in the form $\mathbf{z}_{p1} = \begin{pmatrix} a \\ b \end{pmatrix} e^{ix}$ ($\gamma = i$ is not an eigenvalue of the homogeneous system and the degree of the vector polynomial is 0). To specify a and b, we substitute $\mathbf{z}_{p1}$ into the auxiliary system and obtain $\begin{cases} ia = 4a - 3b + 1 \\ ib = 2a - b \end{cases}$. The solution of this system is $a = \frac{1}{5}(-1 + 2i)$,

$b = \frac{1}{5}(1 + 3i)$. So, $\mathbf{z}_{p1} = \frac{1}{5}\begin{pmatrix} -1 + 2i \\ 1 + 3i \end{pmatrix} e^{ix}$. Extracting the imaginary part of $\mathbf{z}_{p1}$, we find the solution $\mathbf{y}_{p1}$:

$$\mathbf{y}_{p1} = Im(\{\mathbf{z}_{p1}\}) = Im\left(\frac{1}{5}\begin{pmatrix} -1 + 2i \\ 1 + 3i \end{pmatrix}(\cos x + i\sin x)\right) = \frac{1}{5}\begin{pmatrix} 2\cos x - \sin x \\ 3\cos x + \sin x \end{pmatrix}.$$

Similarly, to solve the system with the right-hand side $\mathbf{f}_2 = \begin{pmatrix} 0 \\ -2 \end{pmatrix}\cos x$, we consider the auxiliary function $\mathbf{g}_2 = \begin{pmatrix} 0 \\ -2 \end{pmatrix} e^{ix}$ and the corresponding auxiliary problem $\mathbf{z}' = A\mathbf{z} + \mathbf{g}_2$, whose particular solution we find in the form $\mathbf{z}_{p2} = \begin{pmatrix} c \\ d \end{pmatrix} e^{ix}$. Substituting $\mathbf{z}_{p2}$ in the second auxiliary system, we get $\begin{cases} ia = 4a - 3b \\ ib = 2a - b - 2 \end{cases}$ and find $a = \frac{1}{5}(3 + 9i)$, $b = \frac{1}{5}(7 + 11i)$. So, $\mathbf{z}_{p2} = \frac{1}{5}\begin{pmatrix} 3 + 9i \\ 7 + 11i \end{pmatrix} e^{ix}$. Extracting the real part of $\mathbf{z}_{p2}$, we obtain the solution $\mathbf{y}_{p2}$:

$$\mathbf{y}_{p2} = Re(\mathbf{z}_{p2}) = Im\left(\frac{1}{5}\begin{pmatrix} 3 + 9i \\ 7 + 11i \end{pmatrix}(\cos x + i\sin x)\right) = \frac{1}{5}\begin{pmatrix} 3\cos x - 9\sin x \\ 7\cos x - 11\sin x \end{pmatrix}.$$

Therefore, the particular solution of the original system has the form

$$\mathbf{y}_p = \mathbf{y}_{p1} + \mathbf{y}_{p2} = \frac{1}{5}\begin{pmatrix} 2\cos x - \sin x \\ 3\cos x + \sin x \end{pmatrix} + \frac{1}{5}\begin{pmatrix} 3\cos x - 9\sin x \\ 7\cos x - 11\sin x \end{pmatrix} = \begin{pmatrix} \cos x - 2\sin x \\ 2\cos x - 2\sin x \end{pmatrix}.$$

Collecting all the results, we obtain the general solution of the original system:

$$\mathbf{y}_{gn} = \mathbf{y}_{gh} + \mathbf{y}_p = C_1 \begin{pmatrix} 1 \\ 1 \end{pmatrix} e^x + C_2 \begin{pmatrix} 3 \\ 2 \end{pmatrix} e^{2x} + \begin{pmatrix} \cos x - 2\sin x \\ 2\cos x - 2\sin x \end{pmatrix}.$$

We recommend to the reader to solve this system using the method of reduction and compare the solutions.

3. $\begin{cases} u' = -3u - 4v + 4w + \sin x + \cos x \\ v' = 3u + 4v - 5w - \sin x - \cos x \\ w' = u + v - 2w \end{cases}$.

We rewrite the system in the vector form $\mathbf{y}' = A\mathbf{y} + \mathbf{f}_1 + \mathbf{f}_2$, where

$$\mathbf{y} = \begin{pmatrix} u \\ v \\ w \end{pmatrix}, \quad A = \begin{pmatrix} -3 & -4 & 4 \\ 3 & 4 & -5 \\ 1 & 1 & -2 \end{pmatrix}, \quad \mathbf{f}_1 = \begin{pmatrix} 1 \\ -1 \\ 0 \end{pmatrix}\sin x, \quad \mathbf{f}_2 = \begin{pmatrix} 1 \\ -1 \\ 0 \end{pmatrix}\cos x.$$

As usual, we start with the homogeneous system. Solving the characteristic equation

$$det(A - \lambda I) = det \begin{pmatrix} -3 - \lambda & -4 & 4 \\ 3 & 4 - \lambda & -5 \\ 1 & 1 & -2 - \lambda \end{pmatrix}$$

$$= -(\lambda + 3)[(\lambda - 4)(\lambda + 2) + 5] + 4[-3(\lambda + 2) + 5] + 4[3 - (4 - \lambda)]$$

$$= -(\lambda + 3)(\lambda + 1)(\lambda - 3) - 8(\lambda + 1) = -(\lambda + 1)(\lambda^2 - 1) = 0,$$

we find the single eigenvalue $\lambda_1 = 1$ and the double eigenvalue $\lambda_{2,3} = -1$.

The first eigenvector $\mathbf{s}_1 = \begin{pmatrix} a_1 \\ b_1 \\ c_1 \end{pmatrix}$ satisfies the system

$$\begin{cases} -4a_1 - 4b_1 + 4c_1 = 0 \\ 3a_1 + 3b_1 - 5c_1 = 0 \\ a_1 + b_1 - 3c_1 = 0 \end{cases} .$$

It follows that $c_1 = 0$ and $b_1 = -a_1$. Therefore, $\mathbf{s}_1 = \begin{pmatrix} 1 \\ -1 \\ 0 \end{pmatrix}$ and $\mathbf{y}_1 = \begin{pmatrix} 1 \\ -1 \\ 0 \end{pmatrix} e^x$. For the second eigenvector $\mathbf{s}_2 = \begin{pmatrix} a_2 \\ b_2 \\ c_2 \end{pmatrix}$ we have the system

$$\begin{cases} -2a_2 - 4b_2 + 4c_2 = 0 \\ 3a_2 + 5b_2 - 5c_2 = 0 \\ a_2 + b_2 - c_2 = 0 \end{cases} ,$$

whence $a_2 = 0$ and $b_2 = c_2$. Therefore, $\mathbf{s}_2 = \begin{pmatrix} 0 \\ 1 \\ 1 \end{pmatrix}$ and $\mathbf{y}_2 = \begin{pmatrix} 0 \\ 1 \\ 1 \end{pmatrix} e^{-x}$.

Since there is no other eigenvector for the eigenvalue $\lambda_{2,3} = -1$, linearly independent with $\mathbf{s}_2$, the second solution related to the exponent e^{-x} must be sought in the form $\mathbf{y}_3 = \begin{pmatrix} ax + \alpha \\ bx + \beta \\ cx + \gamma \end{pmatrix} e^{-x}$, where the coefficients are found by substituting this function into the homogeneous system. This leads to the following algebraic system:

$$\begin{cases} -ax - \alpha + a = -3(ax + \alpha) - 4(bx + \beta) + 4(cx + \gamma) \\ -bx - \beta + b = 3(ax + \alpha) + 4(bx + \beta) - 5(cx + \gamma) \\ -cx - \gamma + c = (ax + \alpha) + (bx + \beta) - 2(cx + \gamma) \end{cases} .$$

For the coefficients with x we get the separate subsystem

$$\begin{cases} -2a - 4b + 4c = 0 \\ 3a + 5b - 5c = 0 \\ a + b - c = 0 \end{cases} ,$$

whence $a = 0$ and $b = c$. The second subsystem, for the constant terms, has the form

$$\begin{cases} -2\alpha - 4\beta + 4\gamma = a = 0 \\ 3\alpha + 5\beta - 5\gamma = b = c \\ \alpha + \beta - \gamma = c \end{cases} .$$

Multiplying the third equation by 5 and subtracting the second from the result, we get $2\alpha = 4c$ or $\alpha = 2c$. The first equation then takes the form $-4\beta + 4\gamma = 4c$ or $\beta = \gamma - c$. Choosing, for example, $c = 1$ and $\gamma = 0$, we get $b = c = 1$, $\alpha = 2c = 2$, $\beta = \gamma - c = -1$, and we specify the third solution in the form $\mathbf{y}_3 = \begin{pmatrix} 2 \\ x-1 \\ x \end{pmatrix} e^{-x}$. Thus, the general solution of the homogeneous system takes the form

$$\mathbf{y}_{gh} = C_1 \mathbf{y}_1 + C_2 \mathbf{y}_2 + C_3 \mathbf{y}_3 = C_1 \begin{pmatrix} 1 - 1 \\ 0 \end{pmatrix} e^x + C_2 \begin{pmatrix} 0 \\ 1 \end{pmatrix} e^{-x} + C_3 \begin{pmatrix} 2 \\ x-1 \\ x \end{pmatrix} e^{-x}.$$

We now proceed to the nonhomogeneous equation. Both functions, $\mathbf{f}_1$ and $\mathbf{f}_2$, can be replaced by the same auxiliary right-hand side $\mathbf{g} = \begin{pmatrix} 1 \\ -1 \\ 0 \end{pmatrix} e^{ix}$. Since i is not a root of the characteristic equation and the vector polynomial together with the exponential is of degree 0, we look for a particular auxiliary solution in the form $\mathbf{y}_{pa} = \mathbf{Q}_{0+0}(x)e^{ix} = \begin{pmatrix} a \\ b \\ c \end{pmatrix} e^{ix}$. Substituting $\mathbf{y}_{pa}$ into the system with the right-hand side $\mathbf{g}$, we get

$$\begin{cases} ia = -3a - 4b + 4c + 1 \\ ib = 3a + 4b - 5c - 1 \\ ic = a + b - 2c \end{cases} .$$

Simplifying the system, we have

$$\begin{cases} (3+i)a + 4b - 4c = 1 \\ 3a + (4-i)b - 5c = 1 \\ a + b - (2+i)c = 0 \end{cases}.$$

From the third equation we express a by the formula $a = (2+i)c - b$ and substitute it into the first two equations:

$$\begin{cases} (3+i)((2+i)c - b) + 4b - 4c = 1 \\ 3((2+i)c - b) + (4-i)b - 5c = 1 \end{cases},$$

or after simplification

$$\begin{cases} (1-i)b + (1+5i)c = 1 \\ (1-i)b + (1+3i)c = 1 \end{cases}.$$

Subtracting the second equation from the first, we find $(1+5i)c-(1+3i)c = 0$, hence $c = 0$. So $b = \frac{1}{1-i} = \frac{1+i}{2}$ and $a = -b = -\frac{1+i}{2}$. Consequently, the auxiliary particular solution has the form $\mathbf{y}_{pa} = \frac{1}{2} \begin{pmatrix} -1-i \\ 1+i \\ 0 \end{pmatrix} e^{ix}$. The two particular solutions of the original system are obtained by taking the real and imaginary parts of $\mathbf{y}_{pa}$:

$$\mathbf{y}_{p1} = Re(\mathbf{y}_{pa}) = \frac{1}{2}Re\left(\begin{pmatrix} -1-i \\ 1+i \\ 0 \end{pmatrix} (\cos x + i\sin x) \right) = \frac{1}{2} \begin{pmatrix} -\cos x + \sin x \\ \cos x - \sin x \\ 0 \end{pmatrix}$$

$$\mathbf{y}_{p2} = Im(\mathbf{y}_{pa}) = \frac{1}{2}Im\left(\begin{pmatrix} -1-i \\ 1+i \\ 0 \end{pmatrix} (\cos x + i\sin x) \right) = \frac{1}{2} \begin{pmatrix} -\cos x - \sin x \\ \cos x + \sin x \\ 0 \end{pmatrix}.$$

Adding two particular solutions, we get

$$\mathbf{y}_p = \mathbf{y}_{p1} + \mathbf{y}_{p2} = \begin{pmatrix} -\cos x \\ \cos x \\ 0 \end{pmatrix}.$$

Finally, we find the general solution of the original system in the form

$$\mathbf{y}_{gn} = \mathbf{y}_{gh} + \mathbf{y}_p = C_1 \begin{pmatrix} 1 \\ -1 \\ 0 \end{pmatrix} e^x + C_2 \begin{pmatrix} 0 \\ 1 \\ 1 \end{pmatrix} e^{-x} + C_3 \begin{pmatrix} 2 \\ x-1 \\ x \end{pmatrix} e^{-x} + \begin{pmatrix} -\cos x \\ \cos x \\ 0 \end{pmatrix}.$$

Exercises for sections 2.1 and 2.2

Solve using the method of reduction and Euler method and compare the solutions:

1. $\begin{cases} u' = v + 1 \\ v' = u + 1 \end{cases}$.

2. $\begin{cases} u' = v + x \\ v' = u - t \end{cases}$.

3. $\begin{cases} u' = v + 2e^x \\ v' = u + x^2 \end{cases}$.

4. $\begin{cases} u' = v - 5\cos x \\ v' = 2u + v \end{cases}$.

5. $\begin{cases} u' = 2u - 4v + 4e^{-2x} \\ v' = 2u - 2v \end{cases}$.

6. $\begin{cases} u' = 4u + v - e^{2x} \\ v' = v - 2u \end{cases}$.

7. $\begin{cases} u' = 2v - u + 1 \\ v' = 3v - 2u \end{cases}$.

8. $\begin{cases} u' = 2u + v + e^x \\ v' = -2u + 2x \end{cases}$.

9. $\begin{cases} u' = 2u - v \\ v' = v - 2u + 18x \end{cases}$.

10. $\begin{cases} u' = u - v + 2\sin x \\ v' = 2u - v \end{cases}$.

11. $\begin{cases} u' = 2u - v \\ v' = u + 2e^x \end{cases}$.

12. $\begin{cases} u' = 2u + v + 2e^x \\ v' = u + 2v - 3e^{4x} \end{cases}$.

13. $\begin{cases} u' = 2u - v \\ v' = 2v - u - 5e^x \sin x \end{cases}$.

14.
$$
\begin{cases}
u' = -2u + 3v + 4w - 3x \\
v' = -6u + 7v + 6w + 1 - 7x \\
w' = u - v + w + x
\end{cases}
\;.
$$

15.
$$
\begin{cases}
u' = 4u + 3v - 3w \\
v' = -3u - 2v + 3w \\
w' = 3u + 3v - 2w + 2e^{-x}
\end{cases}
\;.
$$

16.
$$
\begin{cases}
u' = u - 2v - w - 2e^{x} \\
v' = -u + v + w + 2e^{x} \\
w' = u - w - e^{x}
\end{cases}
\;.
$$

2.3 Method of variation of parameters (Lagrange method)

This method is an extension of the method of variation of parameters already applied to solution of nonhomogeneous linear equations of the first order and higher order. Following the idea of the method, we first find the general solution of the corresponding homogeneous system, and then employ the form of this solution with unknown functions instead of arbitrary constants to find the solution of the nonhomogeneous system. Theoretically, this method allows you to find the particular solution for any right-hand side, but it usually requires more technical work than the method of reduction or Euler method, in the cases when all the methods are applicable.

Let us start with the general description of the algorithm. Given a non-homogeneous linear system with constant coefficients $\mathbf{y}' = A\mathbf{y} + \mathbf{f}$, we first find the general solution of the corresponding homogeneous system $\mathbf{y}' = A\mathbf{y}$, using one of the methods we have already studied. Recall that this general solution has the form $\mathbf{y}_{gh} = C_1\mathbf{y}_1 + \dots C_n\mathbf{y}_n$, where $\mathbf{y}_1, \dots, \mathbf{y}_n$ are linearly independent particular solutions of the homogeneous system and $C_1, \dots, C_n$ are arbitrary constants. In the second step, we look for the general solution of the original system in the form $\mathbf{y}_{gn} = C_1(x)\mathbf{y}_1 + \dots C_n(x)\mathbf{y}_n$, where $C_1, \dots, C_n$ are functions to be determined. To find these functions, we substitute the proposed form of $\mathbf{y}_{gn}$ into the original system and obtain

$$
\mathbf{y_{gn}}' = [C_1'\mathbf{y}_1 + \dots C_n'\mathbf{y}_n] + [C_1\mathbf{y}_1' + \dots C_n\mathbf{y}_n']
$$

$$
= [C_1'\mathbf{y}_1 + \dots C_n'bf y_n] + \mathbf{y_{gh}}' = A[C_1\mathbf{y}_1 + \dots C_n\mathbf{y}_n] + \mathbf{f} = A\mathbf{y}_{gh} + \mathbf{f}.
$$

Since $\mathbf{y}_{gh}' = A\mathbf{y}_{gh}$, the system simplifies to the form $C_1'\mathbf{y}_1 + \dots C_n'\mathbf{y}_n = \mathbf{f}$. The last system does not have the normal form with respect to the unknown functions $C_1, \dots, C_n$ (recall that in the normal form the k-th equation contains only the derivative C_k' in the explicit form). Therefore, this system is first solved with respect to the derivatives $C_1', \ldots, C_n'$, using, for example, the Gauss elimination method (since, for any fixed x, this system is an algebraic linear system for unknowns $C_1', \ldots, C_n'$). The result is a split set of n equations, each of which contains the unknown function C_k' in isolated form, expressed in terms of known functions $\mathbf{y}_1, \ldots, \mathbf{y}_n$ and $\mathbf{f}$, that is, $C_k' = F_k(x)$, $k = 1, \ldots, n$, where $F_k(x)$ are given functions. Then, we solve each of these equations separately by integrating the function on the right-hand side. We substitute the solutions $C_k(x)$ into the proposed form of the general solution $\mathbf{y}_{gn}$ and, thus, the nonhomogeneous linear system is solved. Note that the general solution $\mathbf{y}_{gn}$ contains all the particular solutions.

Examples.

1. $\begin{cases} u' = u - v + \frac{1}{\cos x} \\ v' = 2u - v \end{cases}$.

First, we solve the corresponding homogeneous system $\begin{cases} u' = u - v \\ v' = 2u - v \end{cases}$ using any known method. For example, applying the Euler method, we find the eigenvalues of the characteristic equation

$$det(A - \lambda I) = det \begin{pmatrix} 1 - \lambda & -1 \\ 2 & -1 - \lambda \end{pmatrix} = \lambda^2 + 1 = 0,$$

whence $\lambda_{1,2} = \pm i$. It is sufficient to consider only one of the pair of complex conjugate eigenvalues. Choosing, for instance, $\lambda_1 = -i$, we get the following system for components of the eigenvector $\begin{cases} (1 + i)a - b = 0 \\ 2a - (1 - i)b = 0 \end{cases}$. Using the first equation (the second is a linear consequence of this), we have $b = (1 + i)a$ and then $\mathbf{s} = \begin{pmatrix} 1 \\ 1 + i \end{pmatrix}$ with the corresponding solution $\mathbf{y} = \mathbf{s}e^{\lambda_1 x} = \begin{pmatrix} 1 \\ 1 + i \end{pmatrix} e^{-ix}$. To obtain the real form of solutions, we extract the real and imaginary parts of $\mathbf{y}$:

$$\mathbf{y}_1 = Re(\mathbf{y}) = Re\left(\begin{pmatrix} 1 \\ 1 + i \end{pmatrix} (\cos x - i \sin x) \right) = \begin{pmatrix} \cos x \\ \cos x + \sin x \end{pmatrix},$$

$$\mathbf{y}_2 = Im(\mathbf{y}) = Im\left(\begin{pmatrix} 1 \\ 1 + i \end{pmatrix} (\cos x - i \sin x) \right) = \begin{pmatrix} -\sin x \\ \cos x - \sin x \end{pmatrix}.$$

Therefore, the general solution of the homogeneous system takes the form

$$\mathbf{y}_{gh} = C_1 \mathbf{y}_1 + C_2 \mathbf{y}_2 = C_1 \begin{pmatrix} \cos x \\ \cos x + \sin x \end{pmatrix} + C_2 \begin{pmatrix} -\sin x \\ \cos x - \sin x \end{pmatrix}.$$

In the second step of the algorithm, we look for the general solution of the nonhomogeneous system in the form

$$u = C_1(x)\cos x - C_2(x)\sin x, \quad v = (C_1(x)+C_2(x))\cos x + (C_1(x)-C_2(x))\sin x.$$

Substituting these functions into the original system, we get

$$
\begin{cases}
C_1' \cos x - C_2' \sin x - C_1 \sin x - C_2 \cos x \\
\quad = [C_1 \cos x - C_2 \sin x] - [(C_1+C_2)\cos x + (C_1-C_2)\sin x] + \frac{1}{\cos x} \\
(C_1' + C_2')\cos x + (C_1' - C_2')\sin x - (C_1 + C_2)\sin x + (C_1 - C_2)\cos x \\
\quad = 2[C_1 \cos x - C_2 \sin x] - [(C_1 + C_2)\cos x + (C_1 - C_2)\sin x]
\end{cases}.
$$

Simplifying, we have

$$
\begin{cases}
C_1' \cos x - C_2' \sin x = \frac{1}{\cos x} \\
C_1'(\cos x + \sin x) + C_2'(\cos x - \sin x) = 0
\end{cases}.
$$

Solving the algebraic linear system with respect to C_1', C_2', we find the two decoupled equations: $C_1' = 1 - \tan x$, $C_2' = 1 + \tan x$. Integration of the right-hand sides gives the solutions for C_1 and C_2:

$$C_1 = x + \ln|\tan x| + A_1, \quad C_2 = x - \ln|\tan x| + A_2$$

(here A_1 and A_2 are arbitrary constants). To finish the algorithm of solution, all that remains is to substitute these expressions into the formula of the general solution:

$$
\begin{cases}
u = (x + \ln|\tan x| + A_1)\cos x - (x - \ln|\tan x| + A_2)\sin x \\
v = (A_1 + A_2 + 2x)\cos x + (A_1 - A_2 + 2\ln|\tan x|)\sin x
\end{cases}.
$$

2. $\begin{cases} u' = -2u - 4v + 1 + 4x \\ v' = -u + v + \frac{3}{2}x^2 \end{cases}.$

First, we solve the corresponding homogeneous system $\begin{cases} u' = -2u - 4u \\ v' = -u + v \end{cases}$ using any known method. For example, applying the Euler method, we solve the characteristic equation

$$\det(A - \lambda I) = \det\begin{pmatrix} -2 - \lambda & -4 \\ -1 & 1 - \lambda \end{pmatrix} = \lambda^2 + \lambda - 6 = 0$$

and find the eigenvalues $\lambda_{1,2} = -3, 2$. For $\lambda_1 = 2$ we have the system for the components of the eigenvector $\begin{cases} -4a - 4b = 0 \\ -a - b = 0 \end{cases}$, whence $a = -b$. Therefore,

$s_1 = \begin{pmatrix} -1 \\ 1 \end{pmatrix}$ and the corresponding solution is $y_1 = s_1 e^{\lambda_1 x} = \begin{pmatrix} -1 \\ 1 \end{pmatrix} e^{2x}$.

For $\lambda_2 = -3$ we have the system for the components of the eigenvector $\begin{cases} a - 4b = 0 \\ -a + 4b = 0 \end{cases}$, from which it follows the relation $a = 4b$. Therefore, $s_2 = \begin{pmatrix} 1 \\ 4 \end{pmatrix}$ and the corresponding solution is $y_2 = s_2 e^{\lambda_2 x} = \begin{pmatrix} 4 \\ 1 \end{pmatrix} e^{-3x}$. So, the general solution of the homogeneous system takes the form

$$y_{gh} = C_1 y_1 + C_2 y_2 = C_1 \begin{pmatrix} -1 \\ 1 \end{pmatrix} e^{2x} + C_2 \begin{pmatrix} 4 \\ 1 \end{pmatrix} e^{-3x}.$$

In the second step of the algorithm, we look for the general solution of the nonhomogeneous system in the form

$$u = -C_1(x) e^{2x} + 4C_2(x) e^{-3x}, \quad v = C_1(x) e^{2x} + C_2(x) e^{-3x}.$$

Substituting these functions into the original system, we get

$$\begin{cases} -C_1' e^{2x} + 4C_2' e^{-3x} - 2C_1 e^{2x} - 12C_2 e^{-3x} \\ \quad = -2[-C_1 e^{2x} + 4C_2 e^{-3x}] - 4[C_1 e^{2x} + C_2 e^{-3x}] + 1 + 4x \\ C_1' e^{2x} + C_2' e^{-3x} + 2C_1 e^{2x} - 3C_2 e^{-3x} \\ \quad = -[-C_1 e^{2x} + 4C_2 e^{-3x}] + [C_1 e^{2x} + C_2 e^{-3x}] + \frac{3}{2} x^2 \end{cases},$$

or simplifying,

$$\begin{cases} -C_1' e^{2x} + 4C_2' e^{-3x} = 1 + 4x \\ C_1' e^{2x} + C_2' e^{-3x} = \frac{3}{2} x^2 \end{cases}.$$

Solving the algebraic linear system with respect to C_1', C_2', we find the two decoupled equations:

$$C_1' = \frac{1}{5}(6x^2 - 4x - 1) e^{-2x}, \quad C_2' = \frac{1}{10}(3x^2 + 8x + 2) e^{3x}.$$

Integration of the right-hand sides gives the solutions for C_1 and C_2:

$$C_1 = -\frac{1}{5}(3x^2 + x) e^{-2x} + A_1, \quad C_2 = \frac{1}{10}(x^2 + 2x) e^{3x} + A_2$$

(A_1 and A_2 are arbitrary constants). All that remains is to substitute these expressions into the formula of the general solution:

$$\begin{cases} u = -C_1 e^{2x} + 4C_2 e^{-3x} + x^2 + x \\ v = C_1 e^{2x} + C_2 e^{-3x} - \frac{x^2}{2} \end{cases}.$$

We recommend to solve this system using the reduction method and Euler method and compare the solutions.

$$3. \quad \begin{cases} u' = v + \tan^2 x - 1 \\ v' = -u + \tan x \end{cases}.$$

First, we solve the corresponding homogeneous system $\begin{cases} u' = v \\ v' = -u \end{cases}$ using any known method. For example, applying the Euler method, we solve the characteristic equation

$$det(A - \lambda I) = det \begin{pmatrix} -\lambda & 1 \\ -1 & -\lambda \end{pmatrix} = \lambda^2 + 1 = 0$$

and find the eigenvalues $\lambda_{1,2} = \pm i$. Using $\lambda_1 = -i$, we obtain the system for the components of the eigenvector $\begin{cases} ia + b = 0 \\ -a + ib = 0 \end{cases}$, from which it follows the relation $a = ib$. Consequently, $\mathbf{s} = \begin{pmatrix} i \\ 1 \end{pmatrix}$ and the corresponding solution is $\mathbf{y} = \mathbf{s}e^{\lambda_1 x} = \begin{pmatrix} i \\ 1 \end{pmatrix} e^{-ix}$. We obtain the real form of solutions, extracting the real and imaginary parts of $\mathbf{y}$:

$$\mathbf{y}_1 = Re(\mathbf{y}) = Re\left(\begin{pmatrix} i \\ 1 \end{pmatrix} (\cos x - i \sin x) \right) = \begin{pmatrix} \sin x \\ \cos x \end{pmatrix},$$

$$\mathbf{y}_2 = Im(\mathbf{y}) = Im\left(\begin{pmatrix} i \\ 1 \end{pmatrix} (\cos x - i \sin x) \right) = \begin{pmatrix} \cos x \\ -\sin x \end{pmatrix}.$$

Therefore, the general solution of the homogeneous system takes the form

$$\mathbf{y}_{gh} = C_1 \mathbf{y}_1 + C_2 \mathbf{y}_2 = C_1 \begin{pmatrix} \sin x \\ \cos x \end{pmatrix} + C_2 \begin{pmatrix} \cos x \\ -\sin x \end{pmatrix}.$$

In the second step of the algorithm, we look for the general solution of the nonhomogeneous system in the form

$$u = C_1(x) \sin x + C_2(x) \cos x, \quad v = C_1(x) \cos x - C_2(x) \sin x.$$

Substituting these functions into the original system, we get

$$\begin{cases} C_1' \sin x + C_2' \cos x + C_1 \cos x - C_2 \sin x = [C_1 \cos x - C_2 \sin x] + \tan^2 x - 1 \\ C_1' \cos x - C_2' \sin x - C_1 \sin x - C_2 \cos x = -[C_1 \sin x + C_2 \cos x] + \tan x \end{cases},$$

or simplifying,

$$\begin{cases} C_1' \sin x + C_2' \cos x = \tan^2 x - 1 \\ C_1' \cos x - C_2' \sin x = \tan x \end{cases}.$$

Solving the algebraic linear system with respect to C_1', C_2', we find the two decoupled equations: $C_1' = \tan^2 x \sin x$, $C_2' = -\cos x$. Integration of the right-hand sides gives the solutions for C_1 and C_2:

$$C_1 = \frac{1}{\cos x} + \cos x + A_1, \ C_2 = -\sin x + A_2$$

(A_1 and A_2 are arbitrary constants). All that remains is to substitute these expressions into the formula for the general solution, which, after simplification, results in:

$$\begin{cases} u = A_1 \sin x + A_2 \cos x + \tan x \\ v = A_1 \cos x - A_2 \sin x + 2 \end{cases}.$$

4. $\begin{cases} u' = 2v - u \\ v' = 4v - 3u + \frac{e^{3x}}{e^{2x}+1} \end{cases}.$

First, we solve the corresponding homogeneous system $\begin{cases} u' = 2v - u \\ v' = 4v - 3u \end{cases}$ us-

ing any known method. For example, applying the Euler method, we solve the characteristic equation

$$det(A - \lambda I) = det \begin{pmatrix} -1 - \lambda & 2 \\ -3 & 4 - \lambda \end{pmatrix} = \lambda^2 - 3\lambda + 2 = 0$$

and find the eigenvalues $\lambda_{1,2} = 1, 2$. For $\lambda_1 = 1$ we have the system for the components of the eigenvector $\begin{cases} -2a + 2b = 0 \\ -3a + 3b = 0 \end{cases}$, from which it follows

the relation $a = b$. Therefore, $\mathbf{s}_1 = \begin{pmatrix} 1 \\ 1 \end{pmatrix}$ and the corresponding solution is $\mathbf{y}_1 = \mathbf{s}_1 e^{\lambda_1 x} = \begin{pmatrix} 1 & 1 \end{pmatrix} e^x$. For $\lambda_2 = 2$ we have the system for the eigenvector $\begin{cases} -3a + 2b = 0 \\ -3a + 2b = 0 \end{cases}$, from which it follows the relation $3a = 2b$. Therefore, $\mathbf{s}_2 = \begin{pmatrix} 2 \\ 3 \end{pmatrix}$ and the corresponding solution is $\mathbf{y}_2 = \mathbf{s}_2 e^{\lambda_2 x} = \begin{pmatrix} 2 \\ 3 \end{pmatrix} e^{2x}$. So, the general solution of the homogeneous system takes the form

$$\mathbf{y}_{gh} = C_1 \mathbf{y}_1 + C_2 \mathbf{y}_2 = C_1 \begin{pmatrix} 1 \\ 1 \end{pmatrix} e^x + C_2 \begin{pmatrix} 2 \\ 3 \end{pmatrix} e^{2x}.$$

In the second step of the algorithm, we look for the general solution of the nonhomogeneous system in the form

$$u = C_1(x)e^x + 2C_2(x)e^{2x}, \quad v = C_1(x)e^x + 3C_2(x)e^{2x}.$$

Substituting these functions into the original system, we get

$$\begin{cases} C_1'e^x + 2C_2'e^{2x} + C_1e^x + 4C_2e^{2x} \\ \quad = 2[C_1e^x + 3C_2e^{2x}] - [C_1(x)e^x + 2C_2(x)e^{2x}] \\ C_1'e^x + 3C_2'e^{2x} + C_1e^x + 6C_2e^{2x} \\ \quad = 4[C_1(x)e^x + 3C_2(x)e^{2x}] - 3[C_1(x)e^x + 2C_2(x)e^{2x}] + \frac{e^{3x}}{e^{2x}+1} \end{cases},$$

or simplifying,

$$\begin{cases} C_1'e^x + 2C_2'e^{2x} = 0 \\ C_1'e^x + 3C_2'e^{2x} = \frac{e^{3x}}{e^{2x}+1} \end{cases}.$$

Solving the algebraic linear system with respect to C_1', C_2', we find the two decoupled equations: $C_1' = -2\frac{e^{2x}}{e^{2x}+1}$, $C_2' = \frac{e^x}{e^{2x}+1}$. Integration of the right-hand sides gives the solutions for C_1 and C_2:

$$C_1 = -\ln(e^{2x} + 1) + A_1, \quad C_2 = \arctan e^x + A_2$$

(A_1 and A_2 are arbitrary constants). All that remains is to substitute these expressions into the formula for the general solution:

$$\begin{cases} u = (-\ln(e^{2x} + 1) + A_1)e^x + 2(\arctan e^x + A_2)e^{2x} \\ v = (-\ln(e^{2x} + 1) + A_1)e^x + 3(\arctan e^x + A_2)e^{2x} \end{cases}.$$

5. $\begin{cases} u' = v + \frac{1}{\sin x} \\ v' = -u \end{cases}.$

First, we solve the corresponding homogeneous system $\begin{cases} u' = v \\ v' = -u \end{cases}$ using any known method. For example, applying the Euler method, we solve the characteristic equation

$$\det(A - \lambda I) = \det\begin{pmatrix} -\lambda & 1 \\ -1 & -\lambda \end{pmatrix} = \lambda^2 + 1 = 0$$

and find the eigenvalues $\lambda_{1,2} = \pm i$. Using $\lambda_1 = -i$, we obtain the system for the components of the eigenvector $\begin{cases} ia + b = 0 \\ -a + ib = 0 \end{cases}$, from which it follows the relation $a = ib$. Consequently, $\mathbf{s} = \begin{pmatrix} 1 \\ -i \end{pmatrix}$ and the corresponding solution is $\mathbf{y} = \mathbf{s}e^{\lambda_1 x} = \begin{pmatrix} 1 \\ -i \end{pmatrix} e^{-ix}$. We obtain the real form of solutions, extracting the real and imaginary parts of $\mathbf{y}$:

$$\mathbf{y}_1 = Re(\mathbf{y}) = Re\left(\begin{pmatrix} 1 \\ -i \end{pmatrix} (\cos x - i \sin x) \right) = \begin{pmatrix} \cos x \\ \sin x \end{pmatrix},$$

$$\mathbf{y}_2 = Im(\mathbf{y}) = Im\left(\begin{pmatrix} 1 \\ -i \end{pmatrix} (\cos x - i \sin x) \right) = \begin{pmatrix} -\sin x \\ -\cos x \end{pmatrix}.$$

Therefore, the general solution of the homogeneous system takes the form

$$\mathbf{y}_{gh} = C_1\mathbf{y}_1 + C_2\mathbf{y}_2 = C_1 \begin{pmatrix} \cos x \\ -\sin x \end{pmatrix} + C_2 \begin{pmatrix} \sin x \\ \cos x \end{pmatrix}.$$

In the second step of the algorithm, we look for the general solution of the nonhomogeneous system in the form

$$u = C_1(x) \cos x + C_2(x) \sin x, \quad v = -C_1(x) \sin x + C_2(x) \cos x.$$

Substituting these functions into the original system, we get

$$\begin{cases} C_1' \cos x + C_2' \sin x - C_1 \sin x + C_2 \cos x = [-C_1(x) \sin x + C_2(x) \cos x] + \frac{1}{\sin x} \\ -C_1' \sin x + C_2' \cos x - C_1 \cos x - C_2 \sin x = -[C_1(x) \cos x + C_2(x) \sin x] \end{cases},$$

or simplifying,

$$\begin{cases} C_1' \cos x + C_2' \sin x = \frac{1}{\sin x} \\ -C_1' \sin x + C_2' \cos x = 0 \end{cases}.$$

Solving the algebraic linear system with respect to C_1', C_2', we find the two decoupled equations: $C_1' = \cot x \sin x$, $C_2' = 1$. Integration of the right-hand sides gives the solutions for C_1 and C_2:

$$C_1 = \ln|\sin x| + A_1, \quad C_2 = x + A_2$$

(A_1 and A_2 are arbitrary constants). To finish the algorithm of solution, all that remains is to substitute these expressions into the formula of the general solution, which, after simplification, results in:

$$\begin{cases} u = (\ln|\sin x| + A_1) \cos x + (x + A_2) \sin x \\ v = -(\ln|\sin x| + A_1) \sin x + (x + A_2) \cos x \end{cases}.$$

6. $\begin{cases} u' = v - \frac{\sin x}{\cos^2 x} + \cos x \\ v' = -u + 2v + 2w - \frac{1}{\cos x} - \sin x \\ w' = u + \frac{1}{\cos x} + \sin x \end{cases}$.

First, we solve the corresponding homogeneous system

$$\begin{cases} u' = v \\ v' = -u + 2v + 2w \\ w' = u \end{cases}$$

using any known method. Applying the Euler method, we solve the characteristic equation

$$det(A - \lambda I) = det \begin{pmatrix} -\lambda & 1 & 0 \\ -1 & 2-\lambda & 2 \\ 1 & 0 & -\lambda \end{pmatrix} = -\lambda \cdot \lambda(\lambda - 2) - (\lambda - 2) = -(\lambda - 2)(\lambda^2 + 1) = 0$$

and we find the eigenvalues $\lambda_1 = 2$, $\lambda_{2,3} = \pm i$. Then, for the components of the eigenvector $\mathbf{s}_1$ we have the system

$$\begin{cases} -2a + b = 0 \\ -a + 2c = 0 \\ a - 2c = 0 \end{cases},$$

from which the relations $b = 2a$ and $a = 2c$ follow. Therefore, $\mathbf{s}_1 = \begin{pmatrix} 2 \\ 4 \\ 1 \end{pmatrix}$

and the corresponding solution is $\mathbf{y}_1 = \mathbf{s}_1 e^{\lambda_1 x} = \begin{pmatrix} 2 \\ 4 \\ 1 \end{pmatrix} e^{2x}$. For the complex

eigenvector $\mathbf{s}$ (corresponding to $\lambda_2 = i$) we have the system

$$\begin{cases} -ia + b = 0 \\ -a + (2 - i)b + 2c = 0 \\ a - ic = 0 \end{cases},$$

whence $b = ia$ and $a = ic$. Consequently, $\mathbf{s} = \begin{pmatrix} i \\ -1 \\ 1 \end{pmatrix}$ and the corresponding

solution in the complex form is $\mathbf{y} = \mathbf{s} e^{\lambda_2 x} = \begin{pmatrix} i \\ -1 \\ 1 \end{pmatrix} e^{ix}$. We obtain the real form of solutions, extracting the real and imaginary parts of $\mathbf{y}$:

$$\mathbf{y}_2 = Re(\mathbf{y}) = Re(\begin{pmatrix} i \\ -1 \\ 1 \end{pmatrix}(\cos x + i\sin x)) = \begin{pmatrix} -\sin x \\ -\cos x \\ \cos x \end{pmatrix},$$

$$\mathbf{y}_3 = Im(\mathbf{y}) = Im(\begin{pmatrix} i \\ -1 \\ 1 \end{pmatrix}(\cos x + i\sin x)) = \begin{pmatrix} \cos x \\ -\sin x \\ \sin x \end{pmatrix}.$$

Therefore, the general solution of the homogeneous system takes the form

$$\mathbf{y}_{gh} = C_1\mathbf{y}_1 + C_2\mathbf{y}_2 + C_3\mathbf{y}_3 = C_1\begin{pmatrix} 2 \\ 4 \\ 1 \end{pmatrix}e^{2x} + C_2\begin{pmatrix} -\sin x \\ -\cos x \\ \cos x \end{pmatrix} + C_3\begin{pmatrix} \cos x \\ -\sin x \\ \sin x \end{pmatrix}.$$

In the second step of the algorithm, we look for the general solution of the nonhomogeneous system in the form

$$u = 2C_1(x)e^{2x} - C_2(x)\sin x + C_3(x)\cos x, \quad v = 4C_1(x)e^{2x} - C_2(x)\cos x - C_3(x)\sin x,$$

$$w = C_1(x)e^{2x} + C_2(x)\cos x + C_3(x)\sin x.$$

Substituting these functions into the original system and simplifying, we get

$$\begin{cases} 2C_1'e^{2x} - C_2'\sin x + C_3'\cos x = -\frac{\sin x}{\cos^2 x} + \cos x \\ 4C_1'e^{2x} - C_2'\cos x - C_3'\sin x = -\frac{1}{\cos x} - \sin x \\ C_1'e^{2x} + C_2'\cos x + C_3'\sin x = \frac{1}{\cos x} + \sin x \end{cases}.$$

Now we solve the algebraic linear system with respect to C_1', C_2', C_3'. Adding the third and second equations, we find $5C_1'e^{2x} = 0$, whence $C_1' = 0$ and $C_1 = A_1$ (A_1 is a constant). With $C_1' = 0$, the second and third equations coincide and it remains to solve the system of the first and second equations:

$$\begin{cases} -C_2'\sin x + C_3'\cos x = -\frac{\sin x}{\cos^2 x} + \cos x \\ -C_2'\cos x - C_3'\sin x = -\frac{1}{\cos x} - \sin x \end{cases}.$$

Multiplying the first by $\sin x$ and adding it to the second multiplied by $\cos x$, we get $-C_2' = -\frac{\sin^2 x}{\cos^2 x} - 1$. So, $C_2' = \frac{1}{\cos^2 x}$ and $C_2 = \tan x + A_2$ (A_2 is a constant). Similarly, multiplying the first by $\cos x$ and subtracting the second multiplied by $\sin x$, we get $C_3' = 1$, whence $C_3 = x + A_3$ (A_3 is a constant).

Finally, replacing the found functions C_1, C_2, C_3 in the proposed formula for the general solution, we find the general solution of the original system:

$$\mathbf{y}_{gn} = A_1 \begin{pmatrix} 2 \\ 4 \\ 1 \end{pmatrix} e^{2x} + (A_2 + \tan x) \begin{pmatrix} -\sin x \\ -\cos x \\ \cos x \end{pmatrix} + (A_3 + x) \begin{pmatrix} \cos x \\ -\sin x \\ \sin x \end{pmatrix}.$$

7. $\begin{cases} u' = -u + 2v + w + \frac{2}{\cos^2 x} \\ v' = u - 2v + 3w + \frac{1}{\cos^2 x} \\ w' = 4u - 8v + 6w \end{cases}$.

First, we solve the corresponding homogeneous system

$\begin{cases} u' = -u + 2v + w \\ v' = u - 2v + 3w \\ w' = 4u - 8v + 6w \end{cases}$ using any known method. Applying the Euler method,

we solve the characteristic equation

$$\det(A - \lambda I) = \det \begin{pmatrix} -1 - \lambda & 2 & 1 \\ 1 & -2 - \lambda & 3 \\ 4 & -8 & 6 - \lambda \end{pmatrix}$$

$$= (1 - \lambda)((\lambda + 2)(\lambda - 6) + 24) - 2(6 - \lambda - 12) + (-8 + 4(2 + \lambda))$$
$$= -(1 + \lambda)(\lambda^2 - 4\lambda + 12) + (6\lambda + 12) = -(\lambda^3 - 3\lambda^2 + 2\lambda) = -\lambda(\lambda - 1)(\lambda - 2) = 0,$$

whose roots are $\lambda_1 = 0$, $\lambda_2 = 1$, $\lambda_3 = 2$. For the components of the eigenvector $\mathbf{s}_1$ (which corresponds to $\lambda_1 = 0$) we have the system

$$\begin{cases} -a + 2b + c = 0 \\ a - 2b + 3c = 0 \\ 4a - 8b + 6c = 0 \end{cases}.$$

From the first two equations it follows that $c = 0$, and consequently, $a = 2b$. Therefore, $\mathbf{s}_1 = \begin{pmatrix} 2 \\ 1 \\ 0 \end{pmatrix}$ and the corresponding solution is $\mathbf{y}_1 = \mathbf{s}_1 e^{\lambda_1 x} = \begin{pmatrix} 2 \\ 1 \\ 0 \end{pmatrix}$.

For the components of the eigenvector $\mathbf{s}_2$ (which corresponds to $\lambda_2 = 1$) we have the system

$$\begin{cases} -2a + 2b + c = 0 \\ a - 3b + 3c = 0 \\ 4a - 8b + 5c = 0 \end{cases}.$$

Eliminating a from the first two equations, we get $-4b + 7c = 0$, and substituting this relationship into the third equation, we get $4a - 9c = 0$. Therefore,

$$\mathbf{s}_2 = \begin{pmatrix} 9 \\ 7 \\ 4 \end{pmatrix}$$ and the corresponding solution is $\mathbf{y}_2 = \mathbf{s}_2 e^{\lambda_2 x} = \begin{pmatrix} 9 \\ 7 \\ 4 \end{pmatrix} e^x$. Finally,

for the eigenvector $\mathbf{s}_3$ (which corresponds to $\lambda_3 = 2$) we have the system

$$\begin{cases} -3a + 2b + c = 0 \\ a - 4b + 3c = 0 \\ 4a - 8b + 4c = 0 \end{cases}.$$

Eliminating a from the first two equations, we get $-10b + 10c = 0$, that is, $b = c$, and substituting this relationship into the third equation, we get $4a - 4c = 0$, that is, $a = c$. Therefore, $\mathbf{s}_3 = \begin{pmatrix} 1 \\ 1 \\ 1 \end{pmatrix}$ and the corresponding solution

is $\mathbf{y}_3 = \mathbf{s}_3 e^{\lambda_3 x} = \begin{pmatrix} 1 \\ 1 \\ 1 \end{pmatrix} e^{2x}$. Thus, the general solution of the homogeneous

system takes the form

$$\mathbf{y}_{gh} = C_1 \mathbf{y}_1 + C_2 \mathbf{y}_2 + C_3 \mathbf{y}_3 = C_1 \begin{pmatrix} 2 \\ 1 \\ 0 \end{pmatrix} + C_2 \begin{pmatrix} 9 \\ 7 \\ 4 \end{pmatrix} e^x + C_3 \begin{pmatrix} 1 \\ 1 \\ 1 \end{pmatrix} e^{2x}.$$

In the second step of the algorithm, we look for the general solution of the nonhomogeneous system in the form

$$u = 2C_1(x) + 9C_2(x)e^x + C_3(x)e^{2x}, \quad v = C_1(x) + 7C_2(x)e^x + C_3(x)e^{2x},$$

$$w = 4C_2(x)e^x + C_3(x)e^{2x}.$$

Substituting these functions into the original system and simplifying, we get

$$\begin{cases} 2C_1' + 9C_2'e^x + C_3'e^{2x} = \frac{2}{\cos^2 x} \\ C_1' + 7C_2'e^x + C_3'e^{2x} = \frac{1}{\cos^2 x} \\ 4C_2'e^x + C_3'e^{2x} = 0 \end{cases}.$$

Now we solve the algebraic linear system with respect to C_1', C_2', C_3'. Subtracting the first equation from the second multiplied by 2, we eliminate C_1': $5C_2'e^x + C_3'e^{2x} = 0$. Subtracting the third equation from the last one, we get $C_2' = 0$ and $C_2 = A_2$ (A_2 is a constant). Therefore, from the third equation it follows that $C_3' = 0$ and $C_3 = A_3$ (A_3 is a constant). Finally, from the second equation we get $C_1' = \frac{1}{\cos^2 x}$ and, therefore, $C_1 = \tan x + A_1$ (A_1 is a constant).

Substituting the functions C_1, C_2, C_3 into the proposed formula for the general solution, we find the general solution of the original system:

$$\mathbf{y}_{gn} = (A_1 + \tan x) \begin{pmatrix} 2 \\ 1 \\ 0 \end{pmatrix} + A_2 \begin{pmatrix} 9 \\ 7 \\ 4 \end{pmatrix} e^x + A_3 \begin{pmatrix} 1 \\ 1 \\ 1 \end{pmatrix} e^{2x}.$$

8. $\begin{cases} u' = -3u + 3v + 2w + \frac{1}{e^x(1+x^2)} \\ v' = v + w - \frac{2}{e^x(1+x^2)} \\ w' = -8u + 6v + 4w + \frac{4}{e^x(1+x^2)} \end{cases}$.

First, we solve the corresponding homogeneous system

$$\begin{cases} u' = -3u + 3v + 2w \\ v' = v + w \\ w' = -8u + 6v + 4w \end{cases}$$

using any known method. Applying the Euler method, we solve the characteristic equation

$$det(A - \lambda I) = det \begin{pmatrix} -3-\lambda & 3 & 2 \\ 0 & 1-\lambda & 1 \\ -8 & 6 & 4-\lambda \end{pmatrix}$$

$$= (1 - \lambda)((\lambda + 3)(\lambda - 4) + 16) - (-6(\lambda + 3) + 24)$$

$$= (1-\lambda)(\lambda^2 - \lambda + 4) - 6(1-\lambda) = (1-\lambda)(\lambda^2 - \lambda - 2) = (1-\lambda)(\lambda+1)(\lambda-2) = 0,$$

whose roots are $\lambda_1 = -1$, $\lambda_2 = 1$, $\lambda_3 = 2$. For the components of the eigenvector $\mathbf{s}_1$ (which corresponds to $\lambda_1 = -1$) we have the system

$$\begin{cases} -2a + 3b + 2c = 0 \\ 2b + c = 0 \\ -8a + 6b + 5c = 0 \end{cases} .$$

From the second equation we get $c = -2b$, and taking this relationship into the first, we get $b = -2a$. Therefore, $\mathbf{s}_1 = \begin{pmatrix} 1 \\ -2 \\ 4 \end{pmatrix}$ and the corresponding

solution is $\mathbf{y}_1 = \mathbf{s}_1 e^{\lambda_1 x} = \begin{pmatrix} 1 \\ -2 \\ 4 \end{pmatrix} e^{-x}$. For the components of the eigenvector

s_2 (which corresponds to $\lambda_2 = 1$) we have the system

$$\begin{cases} -4a + 3b + 2c = 0 \\ c = 0 \\ -8a + 6b + 3c = 0 \end{cases}.$$

Since $c = 0$, the two remaining equations simplify to $3b = 4a$. Therefore,

$s_2 = \begin{pmatrix} 3 \\ 4 \\ 0 \end{pmatrix}$ and the corresponding solution is $y_2 = s_2 e^{\lambda_2 x} = \begin{pmatrix} 3 \\ 4 \\ 0 \end{pmatrix} e^x$. Finally,

for the eigenvector s_3 (which corresponds to $\lambda_3 = 2$) we have the system

$$\begin{cases} -5a + 3b + 2c = 0 \\ -b + c = 0 \\ -8a + 6b + 2c = 0 \end{cases}.$$

Due to the relationship $b = c$ in the second equation, the remaining two

equations simplify to $a = b$. Therefore, $s_3 = \begin{pmatrix} 1 \\ 1 \\ 1 \end{pmatrix}$ and the correspond-

ing solution is $y_3 = s_3 e^{\lambda_3 x} = \begin{pmatrix} 1 \\ 1 \\ 1 \end{pmatrix} e^{2x}$. Thus, the general solution of the

homogeneous system takes the form

$$y_{gh} = C_1 y_1 + C_2 y_2 + C_3 y_3 = C_1 \begin{pmatrix} 1 \\ -2 \\ 4 \end{pmatrix} e^{-x} + C_2 \begin{pmatrix} 3 \\ 4 \\ 0 \end{pmatrix} e^x + C_3 \begin{pmatrix} 1 \\ 1 \\ 1 \end{pmatrix} e^{2x}.$$

In the second step of the algorithm, we look for the general solution of the nonhomogeneous system in the form

$$u = C_1(x)e^{-x} + 3C_2(x)e^x + C_3(x)e^{2x}, \; v = -2C_1(x)e^{-x} + 4C_2(x)e^x + C_3(x)e^{2x},$$

$$w = 4C_1(x)e^{-x} + C_3(x)e^{2x}.$$

Substituting these functions into the original system and simplifying, we get

$$\begin{cases} C_1' e^{-x} + 3C_2' e^x + C_3' e^{2x} = \frac{1}{e^x(1+x^2)} \\ -2C_1' e^{-x} + 4C_2' e^x + C_3' e^{2x} = -\frac{2}{e^x(1+x^2)} \\ 4C_1' e^{-x} + C_3' e^{2x} = \frac{4}{e^x(1+x^2)} \end{cases}.$$

Now, we solve the algebraic linear system with respect to C_1', C_2', C_3'. Adding the second equation to the first equation multiplied by 2, we eliminate C_1':

$10C_2'e^x + 3C_3'e^{2x} = 0$. Subtracting the third equation from the first multiplied by 4, we get another equation without C_1': $12C_2'e^x + 3C_3'e^{2x} = 0$. From the last two equations it follows that $C_2' = 0$ and $C_3' = 0$, that is, $C_2 = A_2$ and $C_3 = A_3$ (A_2 and A_3 are arbitrary constants). Then, $C_1'e^{-x} = \frac{1}{e^x(1+x^2)}$ or $C_1' = \frac{1}{1+x^2}$, whence $C_1 = \arctan x + A_1$ (A_1 is a constant).

Substituting the functions C_1, C_2, C_3 into the proposed formula of the general solution, we find the general solution of the original system:

$$\mathbf{y}_{gn} = (A_1 + \arctan x) \begin{pmatrix} 1 \\ -2 \\ 4 \end{pmatrix} e^{-x} + A_2 \begin{pmatrix} 3 \\ 4 \\ 0 \end{pmatrix} e^x + A_3 \begin{pmatrix} 1 \\ 1 \end{pmatrix} e^{2x}.$$

Exercises for section 2.3

Solve applying the Lagrange method:

1. $\begin{cases} u' = -4u - 2v + \frac{2}{e^x - 1} \\ v' = 6u + 3v - \frac{3}{e^x - 1} \end{cases}$.

2. $\begin{cases} u' = 3u - 2v \\ v' = 2u - v + 15e^x \sqrt{x} \end{cases}$.

3. $\begin{cases} u' = u - 2v \\ v' = u - v + \frac{1}{2\sin x} \end{cases}$.

4. $\begin{cases} u' = 3u - 4v + \frac{e^x}{\sin 2x} \\ v' = 2u - v \end{cases}$.

5. $\begin{cases} u' = 3u + v \\ v' = -4u - v + \frac{e^x}{2\sqrt{x}} \end{cases}$.

6. $\begin{cases} u' = 3u - 2v + \frac{e^{3x}}{e^x + 1} \\ v' = u - \frac{e^{3x}}{e^x + 1} \end{cases}$.

7. $\begin{cases} u' = -u - 2v + 2e^{-x} \\ v' = 3u + 4v + e^{-x} \end{cases}$.

8. $\begin{cases} u' = -u - v + 4\cos 2x \\ v' = 3u - 2v + 8\cos 2x + 5\sin 2x \end{cases}$.

9. $\begin{cases} u' = v \\ v' = 4v - 5u + \frac{e^{2x}}{\cos x} \end{cases}$.

10. $\begin{cases} u' = u + 2v - 9x \\ v' = 2u + v + 4e^x \end{cases}$.

11. $\begin{cases} u' = -5u + v - 2w + e^{-x} + e^{-2x} \\ v' = -u - v + 3e^{-x} + 2e^{-2x} \\ w' = 6u - 2v + 2w - 2e^{-x} - 3e^{-2x} \end{cases}$.

12. $\begin{cases} u' = v \\ v' = w - \frac{\sin^3 x}{\cos^6 x} \\ w' = -u - v - w \end{cases}$.

13. $\begin{cases} u' = 2u + 4v + w + e^{2x} \ln x \\ v' = 2v + w \\ w' = 4v - w \end{cases}$.

14. $\begin{cases} u' = 2u + v - 3w + \tan x + \tan^2 x \\ v' = 3u - 2v - 3w + 1 \\ w' = u + v - 2w + \tan x + \tan^2 x \end{cases}$.

Exercises for section 2

Solve the following systems; if possible, apply different methods and compare the obtained solutions:

1. $\begin{cases} u' = v + \cos 2x - 2\sin 2x \\ v' = -u + 2v + 2\sin 2x + 3\cos 2x \end{cases}$.

2. $\begin{cases} u' = u - 2v - 2xe^x \\ v' = 5u - v - (2x + 6)e^x \end{cases}$.

3. $\begin{cases} u' = u - v + \cos xe^x \\ v' = u + v + \sin xe^x \end{cases}$.

4. $\begin{cases} u' = 4u - 2v + \frac{1}{x^3} \\ v' = 8u - 4v - \frac{1}{x^2} \end{cases}$.

5. $\begin{cases} u' = -2u + 4v + \frac{1}{1+e^x} \\ v' = -2u + 4v - \frac{1}{1+e^x} \end{cases}$.

6. $\begin{cases} u' = 4u - 8v + \tan 4x \\ v' = 4u - 4v \end{cases}$.

7. $\begin{cases} u' = 3u - 6v + \frac{1}{\cos^3 3x} \\ v' = 3u - 3v \end{cases}$.

8. $\begin{cases} u' = -3u + v \\ v' = -4u + v + \frac{1}{xe^x} \end{cases}$.

9. $\begin{cases} u' = 2u + v - \ln x \\ v' = -4u - 2v + \ln x \end{cases}$.

10. $\begin{cases} u' = -3u - 2v + \frac{e^{2x}}{1+e^x} \\ v' = 10u + 6v \end{cases}$.

11. $\begin{cases} u' = 5u - 6v + \frac{3e^{2x}}{\cos^3 3x} \\ v' = 3u - v \end{cases}$.

12. $\begin{cases} u' = -2u + v + x \ln x \\ v' = -4u + 2v + 2x \ln x \end{cases}$.

13. $\begin{cases} u' = 4u - 2v \\ v' = 8u - 4v + \sqrt{x} \end{cases}$.

14. $\begin{cases} u' = 3u + 2v - \frac{1}{1+e^{-x}} \\ v' = -3u - 2v - \frac{1}{1+e^{-x}} \end{cases}$.

15. $\begin{cases} u' = 2u - v + w + \cos x \\ v' = 5u - 4v + 3w + \sin x \\ w' = 4u - 4v + 3w + 2\sin x - 2\cos x \end{cases}$.

16. $\begin{cases} u' = 2u + v - 3w + 2e^{2x} \\ v' = 3u - 2v - 3w - 2e^{2x} \\ w' = u + v - 2w \end{cases}$.

17. $\begin{cases} u' = 2u + v - 2w - x + 2 \\ v' = -u + 1 \\ w' = u + v - w - x + 1 \end{cases}$.

18. $\begin{cases} u' = -u + v + w + e^x \\ v' = u - v + w + e^{3x} \\ w' = u + v + w + 4 \end{cases}$.

19. $\begin{cases} u' = 2u - v + w - 2e^{-x} \\ v' = u + 2v - w - e^{-x} \\ w' = u - v + 2w - 3e^{-x} \end{cases}$.

$$20. \begin{cases} u' = 2u - 3v + x \\ v' = u - 2w - 3x^2 \\ w' = -v + 2w + 3x - 2 \end{cases}.$$

3   Solution of the Cauchy problem (initial value problem)

Solution of the Cauchy problem for a normal system of order n

$$\begin{cases} \mathbf{y}' = \mathbf{f}(x, \mathbf{y}) \\ \mathbf{y}(x_0) = \mathbf{b} \end{cases}$$

is usually carried out in the same way as for individual equations: first the general solution is found, containing n arbitrary coefficients $C_1, \ldots, C_n$, and after that, these coefficients are specified by substituting the general solution into the initial conditions. In general, this requires to solve a linear algebraic system of n equations for n unknowns $C_1, \ldots, C_n$, although sometimes this system can be split into subsystems of lower order.

Recall that, according to Theorem 4, the Cauchy problem for a linear system

$$\begin{cases} \mathbf{y}' = A\mathbf{y} + \mathbf{f} \\ \mathbf{y}(x_0) = \mathbf{b} \end{cases}$$

has a unique solution if the matrix of coefficients A and the right-hand side $\mathbf{f}$ are continuous. Translating this result to the system of coefficients $C_1, \ldots, C_n$, we can say that the latter always has a unique solution.

We illustrate the application of initial conditions to some systems, whose general solution has already been found.

Examples.

1. $\begin{cases} y' = 4y - z \\ z' = 2z + y \end{cases}$, $\begin{cases} y(1) = e^3 \\ z(1) = 2e^3 \end{cases}.$

The general solution of the system was found in the form $\begin{cases} y = (C_1 + C_2 x)e^{3x} \\ z = (C_1 - C_2 + C_2 x)e^{3x} \end{cases}$ (see Example 2 in section 1.1). Substituting this solution into the initial conditions, we get the linear algebraic system for the coefficients C_1, C_2: $\begin{cases} (C_1 + C_2)e^3 = e^3 \\ C_1 e^3 = 2e^3 \end{cases}$ or $\begin{cases} C_1 + C_2 = 1 \\ C_1 = 2 \end{cases}$. In this case, the second equation is decoupled and determines $C_1 = 2$. Then from the first equation it follows that $C_2 = -1$. Therefore, the solution of the Cauchy problem is

$$\begin{cases} y = (2 - x)e^{3x} \\ z = (3 - x)e^{3x} \end{cases}.$$

2. $\begin{cases} u' = 8v \\ v' = -2w \\ w' = 2u + 8v - 2w \end{cases}$, $\begin{cases} u(0) = -1 \\ v(0) = 2 \\ w(0) = 4 \end{cases}$.

The following general solution of the system was found in Example 5 of section 1.1:

$$\begin{cases} u = -4C_1 e^{-2x} + 2C_2 \sin 4x - 2C_3 \cos 4x \\ v = C_1 e^{-2x} + C_2 \cos 4x + C_3 \sin 4x \\ w = C_1 e^{-2x} + 2C_2 \sin 4x - 2C_3 \cos 4x \end{cases}.$$

Applying the initial conditions, we obtain the linear algebraic system for C_1, C_2, C_3:

$$\begin{cases} -4C_1 - 2C_3 = -1 \\ C_1 + C_2 = 2 \\ C_1 - 2C_3 = 4 \end{cases}.$$

The first and third equations form a separate subsystem, whose solution is $C_1 = 1$, $C_3 = -\frac{3}{2}$. Substituting $C_1 = 1$ into the second equation gives $C_2 = 1$. Then, the solution of the Cauchy problem has the form

$$\begin{cases} u = -4e^{-2x} + 2\sin 4x + 3\cos 4x \\ v = e^{-2x} + \cos 4x - \frac{3}{2}\sin 4x \\ w = e^{-2x} + 2\sin 4x + 3\cos 4x \end{cases}.$$

3. $\begin{cases} u' = u - w \\ v' = v + w \\ w' = -u - v - w \end{cases}$, $\begin{cases} u(0) = 1 \\ v(0) = 1 \\ w(0) = -1 \end{cases}$.

In Example 3 (Case 2 in section 1.2) the following general solution of the system was found:

$$\begin{pmatrix} u \\ v \\ w \end{pmatrix} = C_1 \begin{pmatrix} 1 \\ -1 \\ 2 \end{pmatrix} e^{-x} + C_2 \begin{pmatrix} 1 \\ -1 \\ 0 \end{pmatrix} e^{x} + C_3 \begin{pmatrix} x \\ -x + 2 \\ -1 \end{pmatrix} e^{x}.$$

Applying the initial conditions, we obtain the linear algebraic system for C_1, C_2, C_3:

$$\begin{cases} C_1 + C_2 = 1 \\ -C_1 - C_2 + 2C_3 = 1 \\ 2C_1 - C_3 = -1 \end{cases},$$

whose solution is $C_1 = 0$, $C_2 = 1$, $C_3 = 1$. Therefore, the solution of the Cauchy problem has the form

$$\begin{pmatrix} u \\ v \\ w \end{pmatrix} = \begin{pmatrix} 1 \\ -1 \\ 0 \end{pmatrix} e^x + \begin{pmatrix} x \\ -x+2 \\ -1 \end{pmatrix} e^x = \begin{pmatrix} x+1 \\ -x+1 \\ -1 \end{pmatrix} e^x.$$

4. $\begin{cases} u' = 2u + v - 2w + 2 - x \\ v' = 1 - u \\ w' = u + v - w + 1 - x \end{cases}$, $\begin{cases} u(0) = 2 \\ v(0) = -2 \\ w(0) = -3 \end{cases}$.

The following general solution of the system was found in Example 3 of section 2.2:

$$\begin{cases} u = C_1 e^x + C_2 \cos x + C_3 \sin x \\ v = -C_1 e^x - C_2 \sin x + C_3 \cos x + x \\ w = C_2 \cos x + C_3 \sin x + 1 \end{cases}.$$

Applying the initial conditions, we have the the linear algebraic system

$$\begin{cases} C_1 + C_2 = 2 \\ -C_1 + C_3 = -2 \\ C_2 + 1 = -3 \end{cases},$$

whose solution is $C_1 = 6, C_2 = -4, C_3 = 4$. Therefore, the solution of the Cauchy problem has the form

$$\begin{cases} u = 6e^x - 4\cos x + 4\sin x \\ v = -6e^x + 4\sin x + 4\cos x + x \\ w = -4\cos x + 4\sin x + 1 \end{cases}.$$

Exercises for section 3

Solve the following Cauchy problems:

1. $\begin{cases} u' = 3u + 8v \\ v' = -u - 3v \end{cases}$, $\begin{cases} u(0) = 6 \\ v(0) = -2 \end{cases}$.

2. $\begin{cases} u' + 3u + 4v = 0 \\ v' + 2u + 5v = 0 \end{cases}$, $\begin{cases} u(0) = 1 \\ v(0) = 4 \end{cases}$.

3. $\begin{cases} u' = u + v \\ v' = 4v - 2u \end{cases}$, $\begin{cases} u(0) = 0 \\ v(0) = -1 \end{cases}$.

4. $\begin{cases} u' = 4u - 5v \\ v' = u \end{cases}$, $\begin{cases} u(0) = 0 \\ v(0) = 1 \end{cases}$.

5. $\begin{cases} u' = 2u - v + w \\ v' = u + w \\ w' = v - 2w - 3u \end{cases}$, $\begin{cases} u(0) = 0 \\ v(0) = 0 \\ w(0) = 1 \end{cases}$.

6. $\begin{cases} u' = v - w \\ v' = -v + w \\ w' = u - w \end{cases}$, $\begin{cases} u(0) = 0 \\ v(0) = 0 \\ w(0) = 1 \end{cases}$.

7. $\begin{cases} u' = u - w \\ v' = v + w \\ w' = -u - v - w \end{cases}$, $\begin{cases} u(0) = 1 \\ v(0) = 1 \\ w(0) = -1 \end{cases}$.

8. $\begin{cases} u' = 2u - v + w + 1 + e^{-x} \\ v' = 2u - v - 2w + 1 \\ w' = -u + v + 2w - 1 + e^{-x} \end{cases}$, $\begin{cases} u(0) = 0 \\ v(0) = 0 \\ w(0) = 0 \end{cases}$.

9. $\begin{cases} u' = u + 2v - 9x \\ v' = 2u + v + 4e^{x} \end{cases}$, $\begin{cases} u(0) = 1 \\ v(0) = 2 \end{cases}$.

10. $\begin{cases} u' = v + \tan^2 x - 1 \\ v' = -u + \tan x \end{cases}$, $\begin{cases} u(0) = 1 \\ v(0) = 3 \end{cases}$.

11. $\begin{cases} u' = 4u - 5v + 4x + 1 \\ v' = u - 2v + x \end{cases}$, $\begin{cases} u(0) = 1 \\ v(0) = 2 \end{cases}$.

12. $\begin{cases} u' = 3u - v + w + e^{x} \\ v' = u + v + w - x \\ w' = 4u - v + 4w \end{cases}$, $\begin{cases} u(0) = \frac{41}{100} \\ v(0) = \frac{166}{100} \\ w(0) = -\frac{2}{100} \end{cases}$.

Chapter 14

Systems of equations: applications

In this section we will consider some application problems, namely, coupled spring-mass models, predator-prey problem, arms race models and vibrating systems. Solution of these problems requires construction of systems of differential equations, which reflect the essential properties of the considered phenomena. The focus of the study is on derivation of differential models and analysis of important properties of solutions of these models, which help to explain the behavior of physical/biological/social complexes.

1 Coupled spring-mass models

Consider two masses m_1 and m_2 attached to two flexible springs A and B, whose spring constants are k_1 and k_2, respectively. The first spring A, with weight m_1, is attached at the top to a rigid support, while the second spring B, with weight m_2, is attached to the first spring, as shown in Fig.14.1. Let us call this model the first spring-mass model. We denote $x_1(t)$ and $x_2(t)$ the vertical displacements of the first and second mass from their equilibrium positions L_1 and L_2, where L_1 is the distance from the fixed upper support to m_1 and L_2 is the distance from m_1 to m_2. We use the vertical axis x with a positive downward direction and the origin at the balance point of the first mass to measure these displacements.

The forces acting on the second mass m_2 are the gravitational force $F_{2g} = m_2 g$ (directed downwards) and the force of the resistance of the second

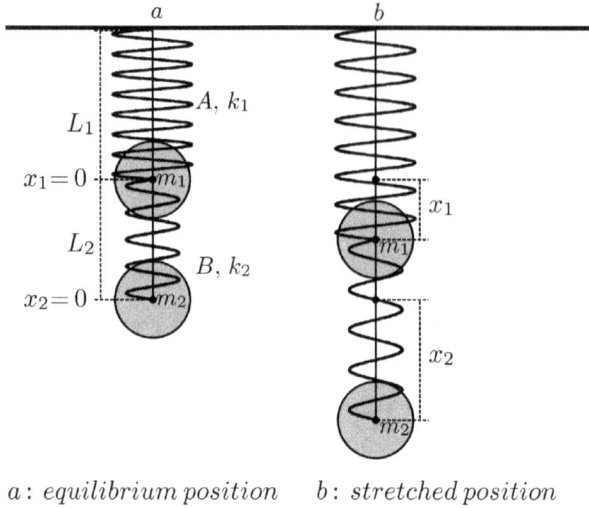

a : equilibrium position   b : stretched position

Figure 14.1: Coupled mass-spring model: two springs attached at the top

spring F_{2r} which, according to Hooke's law, is proportional to the elongation of the spring l_2 from the original position: $F_{2r} = -k_2 l_2$. Since the first spring is also moving and its current position is $x_1(t)$, then the elongation of the second spring (the distance between m_2 and m_1) is $l_2 = L_2 + x_2(t) - x_1(t)$ (see Fig.14.1). In the equilibrium position $l_2 = L_2$, the actions of the two forces compensate each other, that is, $F_{2g} + F_{2r} = m_2 g - k_2 L_2 = 0$. So, according to Newton's second law, the vertical displacement of the second spring $x_2(t)$ satisfies the relation $m_2 x_{2tt} = F_{2g} + F_{2r}$ or

$$m_2 x_{2tt} = m_2 g - k_2(L_2 + x_2 - x_1).$$

Taking into account that $m_2 g - k_2 L_2 = 0$, we get

$$m_2 x_{2tt} = -k_2(x_2 - x_1).$$

The negative sign, together with the positive coefficient of the spring k_2, indicates that the force of the spring resistance acts in the opposite direction to the direction of displacement from equilibrium.

In relation to the first mass m_1, the following forces are present: the gravitational force $F_{1g} = m_1 g$, the force of resistance of the first spring F_{1r} which, according to Hooke's law, is proportional to the elongation $l_1 = L_1 + x_1$ from the original position: $F_{1r} = -k_1 l_1 = -k_1(L_1 + x_1)$, and also the resultant force of the second spring, acting in the opposite direction $F_2 = k_2(x_2 - x_1)$. In the equilibrium position $l_1 = L_1$, the joint action of the forces F_{1g} and F_{1r} cancel each other, that is, $F_{1g} + F_{1r} = m_1 g - k_1 L_1 = 0$. Therefore, according to Newton's second law, the vertical displacement of the first spring $x_1(t)$ satisfies the relation $m_1 x_{1tt} = F_{1g} + F_{1r} + F_2$ or

$$m_1 x_{1tt} = m_1 g - k_1(x_1 + L_1) + k_2(x_2 - x_1).$$

Taking into account that $m_1 g - k_1 L_1 = 0$, we obtain

$$m_1 x_{1tt} = -k_1 x_1 + k_2(x_2 - x_1).$$

The negative sign, together with the positive coefficient of the spring k_1, indicates that the force of the spring resistance acts in the opposite direction to the direction of displacement from equilibrium.

Thus, we arrive at the following system of two linear equations of the second order:

$$\begin{cases} m_1 x_1'' = -k_1 x_1 + k_2(x_2 - x_1) \\ m_2 x_2'' = -k_2(x_2 - x_1) \end{cases}.$$

Dividing the first equation by m_1 and the second by m_2, denoting $a = \frac{k_1}{m_1}$, $b = \frac{k_2}{m_1}$ and $c = \frac{k_2}{m_2}$, and introducing the functions $y_1 = x_1, y_2 = x_1', y_3 = x_2, y_4 = x_2'$, we represent the last system in the normal form of a system of linear equations of the first order:

$$\begin{cases} y_1' = y_2 \\ y_2' = -(a + b)y_1 + by_3 \\ y_3' = y_4 \\ y_4' = cy_1 - cy_3 \end{cases}$$

or in the vector form

$$y' = Ay, \ y = \begin{pmatrix} y_1 \\ y_2 \\ y_3 \\ y_4 \end{pmatrix}, \ A = \begin{pmatrix} 0 & 1 & 0 & 0 \\ -(a+b) & 0 & b & 0 \\ 0 & 0 & 0 & 1 \\ c & 0 & -c & 0 \end{pmatrix}.$$

To find the eigenvalues of the matrix A, we solve the characteristic equation

$$\det(A - \lambda I) = \det \begin{pmatrix} -\lambda & 1 & 0 & 0 \\ -(a+b) & -\lambda & b & 0 \\ 0 & 0 & -\lambda & 1 \\ c & 0 & -c & -\lambda \end{pmatrix} = \lambda^2(\lambda^2 + c) + (a+b)(\lambda^2 + c) - bc = 0.$$

We have a bi-quadratic equation that can be solved using the substitution $\lambda^2 = \mu$. The quadratic equation for μ has the canonical form

$$\mu^2 + \mu(a + b + c) + ac = 0.$$

The two roots are

$$\mu_{1,2} = \frac{1}{2}[-(a + b + c) \pm \sqrt{(a + b + c)^2 - 4ac}].$$

The discriminant is positive:

$$\Delta = (a + b + c)^2 - 4ac = (a - c)^2 + b^2 + 2ab + 2cb > 0$$

and is smaller than $(a+b+c)^2$, which guarantees that the roots are real and negative, satisfying the evaluation

$$-(a + b + c) < \mu_1 < -\frac{1}{2}(a + b + c) < \mu_2 < 0.$$

Therefore, all the eigenvalues are imaginary numbers, found by the formulas

$$\lambda_{1,2} = \pm i\sqrt{|\mu_1|}, \ \lambda_{3,4} = \pm i\sqrt{|\mu_2|}$$

with the evaluation

$$0 < |\lambda_{3,4}| < |\lambda_{1,2}| < \sqrt{a + b + c}.$$

In what follows we will denote $|\lambda_{1,2}| = \sqrt{|\mu_1|} = \alpha$ and $|\lambda_{3,4}| = \sqrt{|\mu_2|} = \beta$. Since the eigenvalues of the real matrix A are complex, the corresponding eigenvectors must be complex conjugates: if $\mathbf{s} = \mathbf{p} + i\mathbf{q}$ is an eigenvector of $\lambda_1 = -i\alpha$, then $\bar{\mathbf{s}} = \mathbf{p} - i\mathbf{q}$ is an eigenvector of $\lambda_2 = -i\alpha$, and similarly, if $\mathbf{w} = \mathbf{u} + i\mathbf{v}$ is an eigenvector of $\lambda_3 = -i\beta$, then $\bar{\mathbf{w}} = \mathbf{u} - i\mathbf{v}$ is an eigenvector of $\lambda_4 = -i\beta$. Therefore, the general solution of the system can be represented in the form

$$\mathbf{y} = C_1\mathbf{s}e^{i\alpha t} + C_2\bar{\mathbf{s}}e^{-i\alpha t} + C_3\mathbf{w}e^{i\beta t} + C_4\bar{\mathbf{w}}e^{-i\beta t}.$$

Besides, in view of the physical significance of the problem, the sought solution must have the real form, which implies that the complex constants C_1, C_2, C_3, C_4 must be complex conjugates in pairs, that is, $C_2 = \bar{C}_1$ and $C_4 = \bar{C}_3$. Then, using Euler's formula $e^{i\nu} = \cos\nu + i\sin\nu$ and performing elementary algebraic transformations, we can write the general solution in real form:

$$\mathbf{y} = A_1(\mathbf{p}\cos\alpha t - \mathbf{q}\sin\alpha t) + B_1(\mathbf{q}\cos\alpha t + \mathbf{p}\sin\alpha t)$$

$$+A_2(\mathbf{u}\cos\beta t - \mathbf{v}\sin\beta t) + B_2(\mathbf{v}\cos\beta t + \mathbf{u}\sin\beta t),$$

or equivalently,

$$\mathbf{y} = (A_1\cos\alpha t + B_1\sin\alpha t))\mathbf{p} + (B_1\cos\alpha t - A_1\sin\alpha t))\mathbf{q}$$

$$+(A_2\cos\beta t + B_2\sin\beta t)\mathbf{u} + (B_2\cos\beta t - A_2\sin\beta t)\mathbf{v},$$

where

$$A_1 = C_1 + C_2, \ \ B_1 = i(C_1 - C_2), \ \ A_2 = C_3 + C_4, \ \ B_2 = i(C_3 - C_4)$$

are real parameters (due to the relations $C_2 = \overline{C}_1$ and $C_4 = \overline{C}_3$).

To find eigenvector $\mathbf{s} = \begin{pmatrix} s_1 \\ s_2 \\ s_3 \\ s_4 \end{pmatrix}$, we solve the system $(A - i\alpha I)\mathbf{s} = \mathbf{0}$, that

is,

$$\begin{cases} -i\alpha s_1 + s_2 = 0 \\ -(a+b)s_1 - i\alpha s_2 + bs_3 = 0 \\ -i\alpha s_3 + s_4 = 0 \\ cs_1 - cs_3 - i\alpha s_2 = 0 \end{cases}.$$

Due to the specific structure of the system, the solution is easy to find: from the third equation follows $s_4 = i\alpha s_3$; substituting this relation into the fourth equation, we get $cs_1 = (c-\alpha^2)s_3$; and then from the first we get $s_2 = i\alpha s_1 = i\alpha\frac{c-\alpha^2}{c}s_3$. We note that with these relations the second equation is also satisfied. Since an eigenvector is determined up to a multiplicative (non-zero) constant, we can choose $s_3 = c$ and then

$$\mathbf{s} = \begin{pmatrix} c-\alpha^2 \\ i\alpha(c-\alpha^2) \\ c \\ i\alpha c \end{pmatrix} = \begin{pmatrix} c-\alpha^2 \\ 0 \\ c \\ 0 \end{pmatrix} + i\begin{pmatrix} 0 \\ \alpha(c-\alpha^2) \\ 0 \\ \alpha c \end{pmatrix} = \mathbf{p} + i\mathbf{q}.$$

Similarly, the eigenvector $\mathbf{w}$ can be found in the following form:

$$\mathbf{w} = \begin{pmatrix} c-\beta^2 \\ i\beta(c-\beta^2) \\ c \\ i\beta c \end{pmatrix} = \begin{pmatrix} c-\beta^2 \\ 0 \\ c \\ 0 \end{pmatrix} + i\begin{pmatrix} 0 \\ \beta(c-\beta^2) \\ 0 \\ \beta c \end{pmatrix} = \mathbf{u} + i\mathbf{v}.$$

In this way, all the elements of the general solution are specified, leaving only arbitrary constants A_1, B_1, A_2, B_2, which are determined by the initial conditions.

Let us consider the Cauchy problem with initial conditions in the form $y_1(0) = g_1, y_2(0) = 0, y_3(0) = g_3, y_4(0) = 0$ (that is, the positions of the masses m_1 and m_2 are g_1 and g_3 and the initial velocities are zero). Substituting these conditions into the general solution, we get the following algebraic linear system for unknowns A_1, B_1, A_2, B_2:

$$\begin{cases} (c - \alpha^2)A_1 + (c - \beta^2)A_2 = g_1 \\ \alpha(c - \alpha^2)B_1 + \beta(c - \beta^2)B_2 = 0 \\ cA_1 + cA_2 = g_3 \\ \alpha cB_1 + \alpha cB_2 = 0 \end{cases}.$$

This system can be decoupled into two subsystems of order 2. The first is

$$\begin{cases} (c-\alpha^2)A_1+(c-\beta^2)A_2 = g_1 \\ cA_1 + cA_2 = g_3 \end{cases},$$

whose solution is

$$A_1 = \frac{1}{c(\beta - \alpha^2)}[cg_1 - (c - \beta^2)g_3], \quad A_2 = \frac{1}{c(\beta - \alpha^2)}[-cg_1 + (c - \alpha^2)g_3].$$

The second is

$$\begin{cases} \alpha(c - \alpha^2)B_1 + \beta(c - \beta^2)B_2 = 0 \\ \alpha cB_1 + \beta cB_2 = 0 \end{cases},$$

whence $B_1 = B_2 = 0$. Thus, the solution to this Cauchy problem has the following form:

$$\mathbf{y} = \begin{pmatrix} y_1 \\ y_2 \\ y_3 \\ y_4 \end{pmatrix} = \begin{pmatrix} A_1(c - \alpha^2) \cos \alpha t + A_2(c - \beta^2) \cos \beta t \\ -A_1\alpha(c - \alpha^2) \sin \alpha t - A_2\beta(c - \beta^2) \sin \beta t \\ A_1 c \cos \alpha t + A_2 c \cos \beta t \\ -A_1 \alpha c \sin \alpha t - A_2 \beta c \sin \beta t \end{pmatrix}$$

with the parameters $\alpha, \beta, A_1, B_1, A_2, B_2$ specified above.

Let us use the considered general model to solve the specific problem of oscillations of two coupled springs with masses $m_1 = 3g$, $m_2 = 5g$ and coefficients $k_1 = 18g/s^2$, $k_2 = 3g/s^2$. In this case, the parameters are $a = \frac{k_1}{m_1} = 6$, $b = \frac{k_2}{m_1} = 1$ and $c = \frac{k_2}{m_2} = \frac{3}{5}$ (the units of measurement are omitted for brevity) and the system matrix is

$$A = \begin{pmatrix} 0 & 1 & 0 & 0 \\ -7 & 0 & 1 & 0 \\ 0 & 0 & 0 & 1 \\ 0.6 & 0 & -0.6 & 0 \end{pmatrix}.$$

The eigenvalues of the matrix A are

$$\lambda_{1,2} = \pm i\alpha = \pm 2.6632i, \quad \lambda_{3,4} = \pm i\beta = \pm 0.7124i.$$

The corresponding eigenvectors are

$$\mathbf{s} = \mathbf{p} + i\mathbf{q} = \begin{pmatrix} -6.4924 \\ 0 \\ 0.6000 \\ 0 \end{pmatrix} + i \begin{pmatrix} 0 - 17.2903 \\ 0 \\ 1.5979 \end{pmatrix} \quad \text{and} \quad \bar{\mathbf{s}} = \mathbf{p} - i\mathbf{q}$$

for $\lambda_{1,2}$, and

$$\mathbf{w} = \mathbf{u} + i\mathbf{v} = \begin{pmatrix} 0.0924 \\ 0 \\ 0.6000 \\ 0 \end{pmatrix} + i \begin{pmatrix} 0 \\ 0.0658 \\ 0 \\ 0.4275 \end{pmatrix} \quad \text{and} \quad \bar{\mathbf{w}} = \mathbf{u} - i\mathbf{v}$$

for $\lambda_{3,4}$. (All these values are approximate.) Adding the initial conditions $y_1(0) = 1cm$, $y_2(0) = 0cm/s$, $y_3(0) = -2cm$, $y_4(0) = 0cm/s$, we find the constants $A_1 = -0.1986$, $A_2 = -3.1347$, $B_1 = 0$, $B_2 = 0$. Thus, the solution to the problem has the form

$$\mathbf{y} = \begin{pmatrix} y_1 \\ y_2 \\ y_3 \\ y_4 \end{pmatrix} = \begin{pmatrix} 1.2897 \cos \alpha t - 0.2897 \cos \beta t \\ -3.4347 \sin \alpha t + 0.2064 \sin \beta t \\ -0.1192 \cos \alpha t - 1.8808 \cos \beta t \\ 0.3174 \sin \alpha t + 1.3400 \sin \beta t \end{pmatrix},$$

where $\alpha = 2.6632$, $\beta = 0.7124$. The graphs of components y_1 and y_3 (the vibrations of masses m_1 and m_2) are shown in Fig.14.2.

Another coupled spring-mass system, with three springs fixed at both ends, is shown in Fig.14.3. Let us call this the second model.

We leave it to the reader to show that the equations of this system have the form

$$\begin{cases} m_1 x_1'' = -k_1 x_1 + k_2(x_2 - x_1) \\ m_2 x_2'' = -k_2(x_2 - x_1) - k_3 x_2 \end{cases}$$

and find their solutions.

2   Predator-prey models

Let us investigate the model of two species when one of them (the predator) feeds only on the other (the prey), while the prey feeds on another type

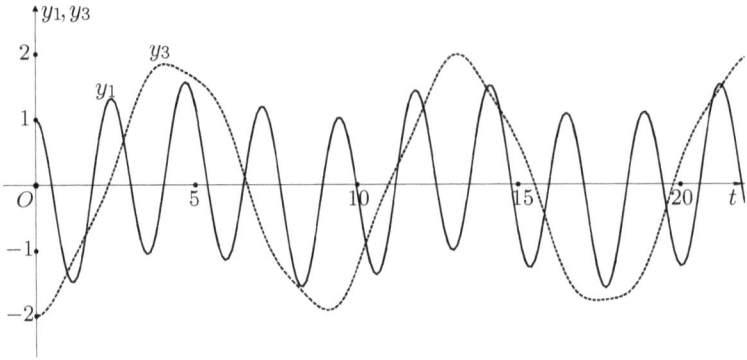

Figure 14.2: Coupled spring-mass model: oscillations of masses.

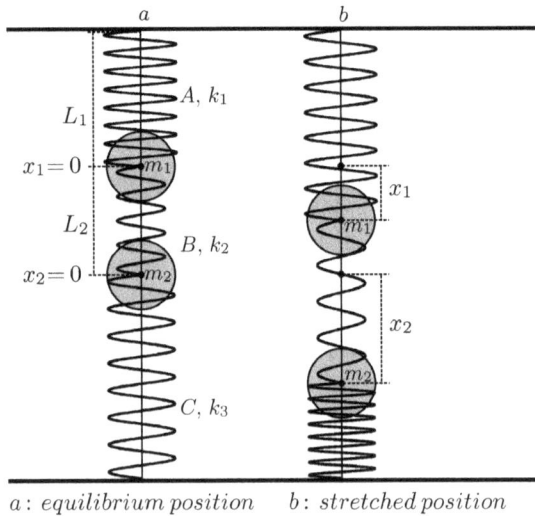

a : equilibrium position   b : stretched position

Figure 14.3: Coupled spring-mass model: three springs attached at both ends.

of food that there is in abundance. The first mathematical model of this scenario was introduced and studied by the biologist and statistician Lotka and the mathematician Volterra.

During the period from 1910 to 1923, biologist D'Ancona analyzed the growth of predatory fish population (such as sharks, skates, rays) and the decrease in the population of other species of fish in the Adriatic Sea. He noticed that the percentage of predatory fish in the total fish population was higher during and immediately after the First World War (1914-1918). He decided that the cause of this was a reduction in fishing during the war. He reasoned that the lower level of fishing during the war led to an increase in the prey fish population, which in turn led to a greater increase in the number of predator fish, thus resulting in an increase in the percentage of predators. However, D'Ancona was unable to give any biological or ecological reasons why the increase in fishing should benefit more prey than predators. So he turned to his father-in-law, Volterra, a famous Italian mathematician, and suggested to create a mathematical model to explain the situation. Within a few months, Volterra had formulated a model of the dynamics of two interacting populations and performed an analysis of the obtained system, which offered a solution to the problem. His results were published in 1926.

Around the same time, the American biologist and statistician Lotka formulated and studied different models of interaction between two species, the results of which were presented in his book in 1925.

Below we will consider the basic Lotka-Volterra model, involving two species, the prey and the predator. Let us consider that the unknown functions in relation to time t are the population (number of individuals) of the prey $x(t)$ and the population of the predator $y(t)$. We set up the equations that express the following assumptions:

1. In the absence of the predator, the prey population increases according to the exponential law, that is, $x' = ax$, $a > 0$ when $y = 0$.

2. In the absence of prey, the predator population decreases according to the exponential law, that is, $y' = -by$, $b > 0$ when $x = 0$.

3. The interaction between two species takes place by means of encounters between their individuals. The number of encounters between predator and prey is proportional to the product of the two populations. Each of these encounters results in an increase in the predator population and a decrease in the prey population. This means that the rate of growth of the predator population is proportional to dxy, while the rate of decrease of the prey population is proportional to $-cxy$, where c and d are positive coefficients.

4. There are no other sources that influence the dynamics (increase or decrease) of the two populations.

Using the above assumptions, we arrive at the following differential equations:

$$\begin{cases} x' = ax - cxy \\ y' = -by + dxy \end{cases}$$

where a, b, c, d are positive constants. In addition, according to the meaning of the unknown functions, we have to put the non-negativity constraints

$x \geq 0$, $y \geq 0$.

To find the critical points (where the derivatives are zero), we solve the system

$$\begin{cases} ax - cxy = x(a - cy) = 0 \\ -by + dxy = y(-b + dx) = 0 \end{cases}.$$

The solutions of the first equation are $x = 0$ and $y = \frac{a}{c}$ and of the second $y = 0$ and $x = \frac{b}{d}$. If $x = 0$, then $x \neq \frac{b}{d}$ and therefore the second equation is only satisfied when $y = 0$. If $y = \frac{a}{c}$, then $y \neq 0$ and so the second equation requires that $x = \frac{b}{d}$. Therefore, the system has only two critical points $(0,0)$ and $(\frac{b}{d}, \frac{a}{c})$.

Let us determine the location of the trajectories of the system – the curves (x, y) in the xy-plane, where $x(t)$ and $y(t)$ are solutions of the differential problem (these pictures are called phase portraits). First, we note that the functions $f(x, y) = ax - cxy$ and $g(x, y) = -by + dxy$ of the right-hand sides are continuous in $\mathbb{R}^2$ and their partial derivatives $f_x = a - cy$, $f_y = -cx$, $g_x = dy$, $g_y = -b + dx$ are also continuous functions in $\mathbb{R}^2$. Furthermore, f and g do not depend directly on t and therefore their continuity and that of their derivatives extends to $\mathbb{R}^3$ (the space of points (t, x, y)). Therefore, by the Cauchy Theorem, the solution of the differential system exists and is unique for any initial conditions.

Let us consider specific initial conditions. If at a given moment t_0 it occurs that $x_0 = x(t_0) = 0$, then the first equation of the system shows that $x'(t_0) = 0$ and, by differentiating this equation, we can see that $x^{(k)}(t_0) = 0$, $\forall k \in \mathbb{N}$ (we assume that $x(t)$ is an analytic function). So, $x(t) = 0$ for all $t \geq t_0$. By replacing t with $-t$, we keep practically the same first equation, only changing the sign of the derivative on the left-hand side. Therefore, if $x(t_0) = 0$, this implies that $x(t) = 0$ at all times. In this case, the second equation simplifies to $y' = -by$, whose solution is $y(t) = y_0 e^{-bt}$, $y_0 = y(t_0)$. This reflects the second assumption of the model. Along with the existence and uniqueness of the solution, this property shows that any trajectory that has at least one point on the y-axis always remains on that axis and no other trajectory can intersect the y-axis. The same considerations apply to the condition $y_0 = y(t_0) = 0$: in this case, due to the second equation, $y(t) = 0$ at all times, the first equation has the solution $x(t) = x_0 e^{at}$, $x_0 = x(t_0)$ (which reflects the first assumption of the model), any trajectory that has at least one point on the x-axis remains on that axis and no other trajectory can intercept the x-axis. From this it follows, in particular, that the critical point $(0,0)$ is the equilibrium point: once the trajectory is at this point, it will remain there. Therefore, given the non-negativity condition $x_0 \geq 0$, $y_0 \geq 0$, we can conclude the following:

1) any trajectory that has at least one point on the non-negative part of the y-axis belongs to the non-negative part of this axis;

2) any trajectory that has at least one point on the non-negative part of the x-axis belongs to the non-negative part of this axis;

3) any other trajectory (which has no point on the x- or y-axis) remains within the first quadrant.

Using the following approximations of the right-hand sides in a neighborhood of an arbitrary point $(\bar{x}, \bar{y})$:

$$f(x,y) = ax - cxy \approx f(\bar{x}, \bar{y}) + df(\bar{x}, \bar{y}), \quad g(x,y) = -by + dxy \approx g(\bar{x}, \bar{y}) + dg(\bar{x}, \bar{y}),$$

where

$$df = f_x \cdot (x - \bar{x}) + f_y \cdot (y - \bar{y}) = (a - c\bar{y}) \cdot (x - \bar{x}) - c\bar{x} \cdot (y - \bar{y})$$

$$dg = g_x \cdot (x - \bar{x}) + g_y \cdot (y - \bar{y}) = d\bar{y} \cdot (x - \bar{x}) + (-b + d\bar{x}) \cdot (y - \bar{y}),$$

we can represent the equations of the system in a neighborhood of $(\bar{x}, \bar{y})$, with a certain degree of precision, in the form

$$\begin{cases} x' = a\bar{x} - c\bar{x}\bar{y} + (a - c\bar{y}) \cdot (x - \bar{x}) - c\bar{x} \cdot (y - \bar{y}) \\ y' = -b\bar{y} + d\bar{x}\bar{y} + d\bar{y} \cdot (x - \bar{x}) + (-b + d\bar{x}) \cdot (y - \bar{y}) \end{cases}.$$

In particular, in a neighborhood of the point $(\bar{x}, \bar{y}) = (\frac{b}{d}, \frac{a}{c})$ we have

$$\begin{cases} x' = a\frac{b}{d} - c\frac{b}{d}\frac{a}{c} + (a - c\frac{a}{c}) \cdot (x - \frac{b}{d}) - c\frac{b}{d} \cdot (y - \frac{a}{c}) \\ y' = -b\frac{a}{c} + d\frac{b}{d}\frac{a}{c} + d\frac{a}{c} \cdot (x - \frac{b}{d}) + (-b + d\frac{b}{d}) \cdot (y - \frac{a}{c}) \end{cases},$$

or simplifying

$$\begin{cases} x' = -\frac{cb}{d}y + \frac{ab}{d} \\ y' = \frac{ad}{c}x - \frac{ab}{c} \end{cases}.$$

The eigenvalues of the matrix of this system are $\lambda_{1,2} = \pm i\sqrt{ab}$, which indicates that the critical point $(\frac{b}{d}, \frac{a}{c})$ has neutral stability: the trajectories neither approach nor flee from this point. Therefore, in order to find out what happens to the trajectories when they return to this point, it is necessary to carry out additional studies.

The results that shed light on the behavior of the trajectories of the differential system were obtained by Volterra and can be summarized in the two principles. The first states that the trajectories of the differential system are closed and periodic, containing the critical point $(x_c, y_c) = (\frac{b}{d}, \frac{a}{c})$ inside. The period T depends on the coefficients a, b, c, d and the initial conditions

$x_0 > 0, y_0 > 0$. To demonstrate this statement, let us simplify a bit the original system by scaling the unknown functions and independent variable. First we introduce the new functions by the formulas $u = \frac{d}{b}x$, $v = \frac{c}{a}y$ and obtain the system

$$\begin{cases} u_t = au(1 - v), \\ v_t = -bv(1 - u) \end{cases}.$$

Next, we introduce the new independent variable $\tau = at$ and obtain a slightly simplified system

$$\begin{cases} u_\tau = u(1 - v), \\ v_\tau = -\alpha v(1 - u) \end{cases},$$

where $u \geq 0$, $v \geq 0$ and $\alpha > 0$. Since the sign of each derivative changes at the point 1, let us divide the first quadrant of the coordinate plane (u, v) in the four parts as it is shown in Fig.14.4: the first $(u > 1, v > 1)$ (labeled A), the second $(u < 1, v > 1)$ (labeled B), the third $(u < 1, v < 1)$ (labeled C) and the fourth $(u > 1, v < 1)$ (labeled D). Consider a point $P_0 = (u_0, v_0)$ on the boundary of the first and fourth parts (that is, $u_0 > 1$, $v_0 = 1$) and suppose that this is the initial position of the trajectory of a particle traversed in the uv-plane following the equations of the system. At P_0 we have $u_\tau(P_0) = 0$, $v_\tau(P_0) > 0$, which means that the particle will move upward and enter in the first part, where $u_\tau(P) < 0$, $v_\tau(P) > 0$ and will continue going upward and to the left until arrive at a point $P_1 = (u_1, v_1)$ on the boundary of the first and second parts with the coordinates $u_1 = 1$, $v_1 > 1$. Then, the particle will enter the second part where $u_\tau(P) < 0$, $v_\tau(P) < 0$ and will move downward and to the left in this part until arrive at the point $P_2 = (u_2 < 1, v_2 = 1)$ on the boundary of the second and third parts. Passing through this point, the particle will enter the third part where $u_\tau(P) > 0$, $v_\tau(P) < 0$, and consequently, it will go downward and to the right in this part until arrive at the point $P_3 = (u_3 = 1, v_2 < 1)$ on the boundary of the third and fourth parts. Next, the particle will pass through P_3 and enter the fourth part where $u_\tau(P) > 0$, $v_\tau(P) > 0$, and consequently, it will go upward and to the right until arrive at some point $P_4 = (u_4 > 1, v_4 = 1)$ on the boundary of the third and fourth parts. The principal question now is if this point will coincide with the initial point P_0 or not.

To answer this question, let us derive additional information about the trajectories of the system. Recalling that $v_u = \frac{v_\tau}{u_\tau}$, we have the differential equation $v_u = \frac{-\alpha v(1-u)}{u(1-v)}$, where τ is eliminated. Solving this separable equation, we get $\int \frac{1-v}{v} dv = -\alpha \int \frac{1-u}{u} du$ or, after integration,

$$\ln v - v = -\alpha(\ln u - u) - A,$$

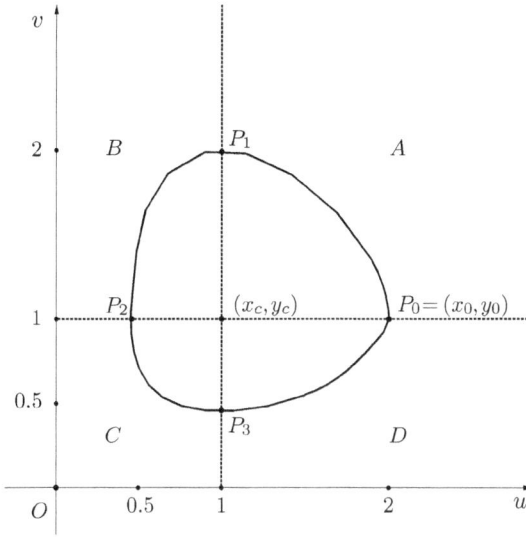

Figure 14.4: Lotka-Volterra model for the simplified system: phase portrait.

where A is arbitrary constant. The same expression can be written in the form $\alpha u + v - \ln(u^\alpha v) = A$. Eliminating logarithm, we can represent this solution in the exponential form $u^\alpha v e^{-\alpha u - v} = C$ or

$$\left(\frac{u}{e^u}\right)^\alpha \cdot \frac{v}{e^v} = C$$

(the solutions $v \equiv 0$ and $u \equiv 0$ are recovered in this form under the choice $C = 0$, but these are out of our interest). We cannot solve analytically this transcendental equation, but we can establish some of its properties. Let us consider the function

$$F(s) = \frac{s}{e^s}, \quad s \in [0, +\infty).$$

First of all, notice that $F(s) \geq 0$ on $[0, +\infty)$. From the equation of the critical point

$$F_s = e^{-s} - se^{-s} = (1 - s)e^{-s} = 0$$

it follows that the unique critical point of $F(s)$ is $s_c = 1$. Since $F_s > 0$ on $[0, 1)$ and $F_s < 0$ on $(1, +\infty)$, the point $s_c = 1$ is a strict global maximum of $F(s)$ on $[0, +\infty)$. Besides,

$$\lim_{s \to +\infty} F(s) = \lim_{s \to +\infty} \frac{s}{e^s} = 0$$

(by L'Hospital's rule). Therefore, the range of $F(s)$ is $[0, F(1)] = [0, e^{-1}]$ and, on the interval $(0, +\infty)$, the function $F(s)$ takes each value of its range exactly twice at the points s_1 and s_2 such that $s_1 < 1 < s_2$, except for the global maximum $F(1) = e^{-1}$, which occurs only at the point $s_c = 1$. It follows immediately from the range of F that possible values of the constant C belong to the interval $[0, e^{-1-\alpha}]$. Besides, if $C \neq 0$, both u and v are positive. In terms of trajectories this means that if the initial point is located inside the first quadrant, the trajectory will never touch the u- and v-axis.

Taking these results into account, we can see that for $v = 1$ the function $F(v)$ takes the maximum value $F(v = 1) = e^{-1}$ and the corresponding value of u is defined by the equation $(\frac{u}{e^u})^\alpha = C \cdot e$ or $F(u) = (C \cdot e)^{1/\alpha}$ (of course, the value of the constant C should be chosen among the admissible values, that is, $(C \cdot e)^{1/\alpha} \in (0, e^{-1})$). According to the properties of $F(u)$, this equation has exactly two solutions $u_1 < 1$ and $u_2 > 1$. This implies that the considered trajectory of a particle, after making a turn around the point $(1, 1)$ should arrive at the point $P_4 = P_0$. Otherwise, we would have a third point $(u_4, 1)$ with the same coordinate $v = 1$, that is, $F(u)$ would have three different points at which it takes the same value, that contradicts its properties.

Thus, we have shown that any trajectory of the simplified differential system, which passes through a point in the first quadrant, is a closed curve around the point $(1, 1)$ (see Fig.14.4). Consequently, the solutions of this system are periodic functions (see Fig.14.5).

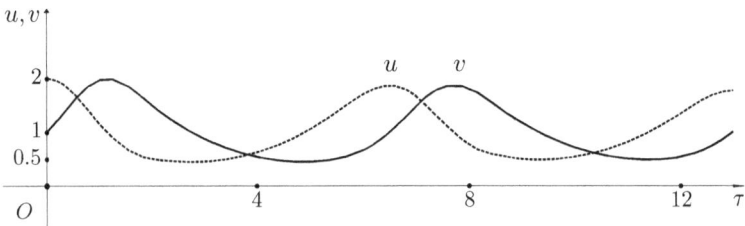

Figure 14.5: Lotka-Volterra model for the simplified system: graphs of $u(\tau)$ and $v(\tau)$.

Finally, since u, v and τ are just the scaled values of x, y and t, the same conclusion is true for the original quantities: for any initial point $P_0 = (x_0, y_0)$ lying inside the first quadrant of the xy-plane, the trajectory of the original differential system represents a closed curve (contained in the first quadrant) around the critical point $(x_c, y_c) = (\frac{b}{d}, \frac{a}{c})$. This means that the solutions of the Lotka-Voltera model are periodic.

The second principle of Volterra deals with the average value of prey and predator populations during the time interval equal to the period T: $m_x = \frac{1}{T} \int_0^T x(t)dt$ and $m_y = \frac{1}{T} \int_0^T y(t)dt$. This principle states that $m_x = \frac{b}{d}$ and $m_y = \frac{a}{c}$, that is, the average values are equal to the critical point values. This result can be demonstrated as follows. Dividing the first equation of the differential system by x and integrating over the interval $[0, T]$, we get

$$\int_0^T \frac{x'}{x} dt = \ln x |_0^T = \ln x(T) - \ln x(0) = \int_0^T (a - cy) dt.$$

Since, according to the first principle, $x(T) = x(0)$, then $\int_0^T (a - cy) dt = 0$ and, consequently,

$$aT - c \int_0^T y dt = 0 \quad \text{or} \quad \frac{1}{T} \int_0^T y(t) dt = \frac{a}{c}.$$

The second formula is obtained in the same way.

One more result obtained by Volterra (the fishing principle) clarifies the effect of fishing on the interaction of prey and predator populations. Let us assume that an indiscriminate fishery is carried out, in which the fishermen keep any type of fish that has been caught, and that the number of fish extracted in the fishery is constant in time and is proportional to the population of each of the two species. In this case, the predator-prey model with the inclusion of fishing effects is formulated as follows:

$$\begin{cases} x' = ax - cxy - \alpha x \\ y' = -by + dxy - \alpha y \end{cases},$$

where the terms αx and αy, $\alpha > 0$ represent the losses of populations as a result of fishing. This system can be rewritten in the form

$$\begin{cases} x' = (a - \alpha)x - cxy \\ y' = -(b + \alpha)y + dxy \end{cases}$$

which for $a - \alpha > 0$ coincides with the original Lotka-Volterra system with a replaced by $a - \alpha$ and b by $b + \alpha$. So the critical point of the latter system is $(\frac{b+\alpha}{d}, \frac{a-\alpha}{c})$ (besides the origin point). Since $\alpha > 0$, we see that $\frac{b+\alpha}{d} > \frac{b}{d}$ and $\frac{a-\alpha}{c} < \frac{a}{c}$. Thus, we arrive at the formulation of the third principle of Volterra, which states that indiscriminate fishing, constant in time, results in an increase in the average value of the prey and a decrease in the average value of the predator. This principle allows us to explain, within the model used, the empirical data obtained by D'Ancona that the predator population increases when fishing is reduced and its population decreases when fishing intensifies. The consequence of fishing on the prey population is the opposite.

Due to the non-linearity, a general analytical solution of the Lotka-Volterra model is unknown. Approximate solutions are usually found using numerical methods. The numerical solution in a particular case when $a = 1, b = 2, c = 0.5, d = 0.25$ is shown in Figs.14.6 and 14.7. The first shows the graphs of $x(t)$ and $y(t)$, and the second shows the trajectory of the model (phase portrait) relative to the initial conditions $x_0 = 10$, $y_0 = 5$. In this case, the critical point within the trajectory is $(x_c, y_c) = (8, 2)$ and the period of the solution is approximately 10.

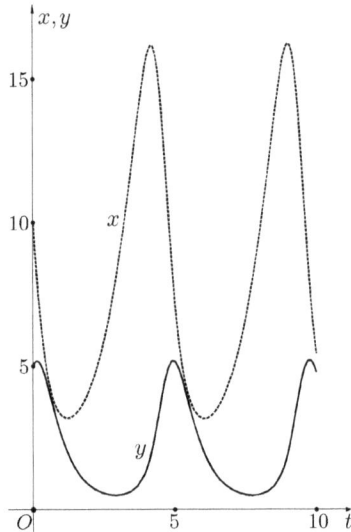

Figure 14.6: Lotka-Volterra model with $a = 1, b = 2, c = 0.5, d = 0.25$: graphs of $x(t)$ and $y(t)$.

3   Arms race models

The arms race model considered in this section was developed by Lewis Fry Richardson in the 1930s, initially to describe the competition between two nations, and later the model was generalized to the case of n nations and applied to analyze the situation on the eve of World War II. In an attempt

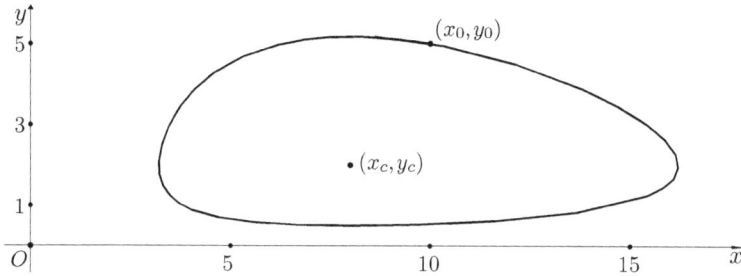

Figure 14.7: Lotka-Volterra model with $a = 1, b = 2, c = 0.5, d = 0.25$: phase portrait.

to avoid this imminent war, Richardson sent his study to a US journal with a request for immediate publication, but the work was rejected.

The main idea behind the two-nation arms race model is the concept of mutual fear. One of the nations begins to increase arms production, claiming that the reason for this is to strengthen its defense (due to suspicions of invasion by another nation). The second nation, fearing the growing strength of the first and its offensive intentions, also begins to produce more weapons, stating that its objective is self-defense. Due to mutual fear, the armaments of both nations continue to increase over a period of time. As neither nation has infinite resources, this situation cannot last forever. When the costs of the production of weapons start to become too great, requiring a significant part of the budget of one of the nations, the population of that country forces the government to limit the resources spent on weapons and, in this way, slow down the growth of weapons or even limit them.

Let us consider that the unknown functions are expenses of each nation for arms production, for example, the amount of money spent per year, and we set up the equations that express the following assumptions:
1. Expenses increase due to mutual fear.
2. Civilian societies in each country resist the increase in weapons production.
3. There are other (external) sources, besides the volume of arms produced, which influence the dynamics (increase or decrease) of arms expenses.

Denoting by $x(t)$ and $y(t)$ the amount of money spent per year on arms production by the first and second nations, respectively (in a universal monetary unit), we obtain the following differential equations:

$$\begin{cases} x' = -cx + ay + r \\ y' = bx - dy + s \end{cases},$$

where a, b, c, d are non-negative constants and r, s are arbitrary constants (representing the external forces). The terms ay and bx represent the growth in arms production by the first and second nations, respectively, due to their fear of the strengthening of another nation. The terms $-cx$ and $-dy$ reflect the decrease in arms production by the first and second nations, respectively, due to the resistance of their social communities. The terms r and s can represent the demands (requirements, debts) of the first and second nations, respectively, in relation to each other. For example, if the first nation has complaints (in the form of debts or old wars) against the second, then r is positive; if the first nation has good relations with the second and has no old disputes, then r is negative. Thus, Richardson's model establishes that, in each nation, the amount spent on arms production increases proportionally to the current spending of another nation, also decreases proportionally to the current spending of the nation itself, and increases or decreases according to feelings towards another nation. In addition, according to the meaning of the unknown functions, we have to place the restrictions of non-negativity $x \geq 0$, $y \geq 0$.

The system obtained is nonhomogeneous linear with constant coefficients, and consequently, it can be solved using the standard procedure. First, we find the general solution of the homogeneous system using the Euler method. The characteristic equation

$$det(A - \lambda I) = det \begin{pmatrix} -c - \lambda & a \\ b & -d - \lambda \end{pmatrix} = \lambda^2 + \lambda(c + d) + cd - ab = 0$$

has the roots $\lambda_{1,2} = \frac{-(c+d)\pm\sqrt{\Delta}}{2}$, where the discriminant

$$\Delta = (c + d)^2 - 4(cd - ab) = (c - d)^2 + 4ab$$

is positive, and consequently, both roots are real and simple. Therefore, the solution of the homogeneous part is

$$\mathbf{u}_h = C_1 \mathbf{s}_1 e^{\lambda_1 t} + C_2 \mathbf{s}_2 e^{\lambda_2 t},$$

where $\mathbf{s}_{1,2}$ are eigenvectors corresponding to the eigenvalues $\lambda_{1,2}$ and $C_{1,2}$ are arbitrary constants. To find the eigenvector $\mathbf{s}_1 = \begin{pmatrix} \alpha_1 \\ \beta_1 \end{pmatrix}$, we need to solve the algebraic linear system $A\mathbf{s}_1 = \lambda_1\mathbf{s}_1$. Since the determinant of this homogeneous system is zero, it is sufficient to solve only one of the two equations. For example, the first equation $(-c-\lambda_1)\alpha_1 + a\beta_1 = 0$ establishes the following relationship: $\beta_1 = \frac{c+\lambda_1}{a}\alpha_1$. Choosing $\alpha_1 = 2a$, we find

$$\mathbf{s}_1 = \begin{pmatrix} 2a \\ 2c + 2\lambda_1 \end{pmatrix} = \begin{pmatrix} 2a \\ c - d - \sqrt{\Delta} \end{pmatrix}.$$

Similarly, we determine that

$$\mathbf{s}_2 = \begin{pmatrix} 2a \\ 2c + 2\lambda_2 \end{pmatrix} = \begin{pmatrix} 2a & c - d + \sqrt{\Delta} \end{pmatrix}.$$

Second, we find a particular solution of the original nonhomogeneous system. Eliminating from consideration the singular situation when $cd = ab$, we can find the particular solution in the form of constants $x = \bar{x}$, $x = \bar{y}$. Substituting into the original system, we find the algebraic linear system
$$\begin{cases} -cx + ay + r = 0 \\ bx - dy + s = 0 \end{cases}$$ with the determinant $cd - ab$ different from 0. So, this system has a single solution

$$\bar{x} = \frac{rd + sa}{cd - ab}, \quad \bar{y} = \frac{rb + sc}{cd - ab},$$

and therefore,

$$\bar{\mathbf{u}} = \begin{pmatrix} \bar{x} \\ \bar{y} \end{pmatrix} = \begin{pmatrix} \frac{rd+sa}{cd-ab} \\ \frac{rb+sc}{cd-ab} \end{pmatrix}$$

is the particular solution of the differential system. The solution $\bar{\mathbf{u}}$ is the equilibrium point (the only one) of the differential system, that is, if the differential system is in this position at the initial moment, it will remain in this position forever. Note that, in the xy-plane, the solution $\bar{x}, \bar{y}$ represents the point of intersection of the lines $R_1 : -cx + ay + r = 0$ and $R_2 : bx - dy + s = 0$. Finally, the general solution of the original system has the form
$$\mathbf{u} = \mathbf{u}_h + \bar{\mathbf{u}} = C_1\mathbf{s}_1 e^{\lambda_1 t} + C_2\mathbf{s}_2 e^{\lambda_2 t} + \bar{\mathbf{u}}.$$

Applying the initial conditions $\mathbf{u}_0 = \begin{pmatrix} x_0 \\ y_0 \end{pmatrix}$, we obtain the following system for specifying the constants C_1 and C_2: $C_1\mathbf{s}_1 + C_2\mathbf{s}_2 + \bar{\mathbf{u}} = \mathbf{u}_0$ or in component form

$$\begin{cases} C_1 \cdot 2a + C_2 \cdot 2a = x_0 - \bar{x} \\ C_1 \cdot (c - d - \sqrt{\Delta}) + C_2 \cdot (c - d + \sqrt{\Delta}) = y_0 - \bar{y} \end{cases}.$$

The determinant of this system is not zero: $4a\sqrt{\Delta} > 0$, and so the system has the only solution that can be expressed in the form

$$C_1 = \frac{1}{2\sqrt{\Delta}} \left[\frac{x_0 - \bar{x}}{2a}(c - d + \sqrt{\Delta}) - (y_0 - \bar{y}) \right],$$

$$C_2 = \frac{1}{2\sqrt{\Delta}} \left[-\frac{x_0 - \bar{x}}{2a}(c - d - \sqrt{\Delta}) + (y_0 - \bar{y}) \right].$$

To simplify further analysis, let us assume that $c = d$. In this case,

$$\Delta = 4ab, \lambda_1 = -c - \sqrt{ab}, \lambda_2 = -c + \sqrt{ab}, \mathbf{s}_1 = \begin{pmatrix} 2a \\ -2\sqrt{ab} \end{pmatrix}, \mathbf{s}_2 = \begin{pmatrix} 2a \\ 2\sqrt{ab} \end{pmatrix}$$

and the constants C_1, C_2 are calculated using the formulas

$$C_1 = \frac{1}{4\sqrt{a}} \left[\frac{x_0 - \bar{x}}{\sqrt{a}} - \frac{y_0 - \bar{y}}{\sqrt{b}} \right], C_2 = \frac{1}{4\sqrt{a}} \left[\frac{x_0 - \bar{x}}{\sqrt{a}} + \frac{y_0 - \bar{y}}{\sqrt{b}} \right].$$

Substituting the values of $C_1, C_2, \lambda_1, \lambda_2$ and $\mathbf{s}_1, \mathbf{s}_2$ into the general solution and writing it in terms of components, we get the following solution to the Cauchy problem:

$$\begin{cases} x = \frac{\sqrt{a}}{2} \left[\frac{x_0 - \bar{x}}{\sqrt{a}} - \frac{y_0 - \bar{y}}{\sqrt{b}} \right] e^{(-c - \sqrt{ab})t} + \frac{\sqrt{a}}{2} \left[\frac{x_0 - \bar{x}}{\sqrt{a}} + \frac{y_0 - \bar{y}}{\sqrt{b}} \right] e^{(-c + \sqrt{ab})t} + \bar{x} \\ y = -\frac{\sqrt{b}}{2} \left[\frac{x_0 - \bar{x}}{\sqrt{a}} - \frac{y_0 - \bar{y}}{\sqrt{b}} \right] e^{(-c - \sqrt{ab})t} + \frac{\sqrt{b}}{2} \left[\frac{x_0 - \bar{x}}{\sqrt{a}} + \frac{y_0 - \bar{y}}{\sqrt{b}} \right] e^{(-c + \sqrt{ab})t} + \bar{y} \end{cases}.$$

The solution of the problem must still be subject to the constraints $x \geq 0$, $y \geq 0$. This can generate yet another equilibrium point, unrelated to the differential system. Indeed, if at instant t_1 the values $x = 0, y = 0$ have been obtained and r, s are negative, then, according to the differential system, the values of x, y at the next instants must be negative (since their derivatives are negative). However, due to the non-negativity constraints, x, y will remain zero starting from t_1.

Let us analyze where the arms race can lead over a long period of time, that is, how the solutions of the considered model behave when $t \to \infty$. The real situations that can happen are:
1) an explosive race, when arms increase without restriction, which in practice leads to war, because states cannot bear the weight of the unlimited increase of the arms expenses;
2) a total disarmament, when weapons are extinguished;
3) a balance, when the quantity of weapons of each nation does not change over time.

In the model under analysis, each of these situations can occur, depending on the values of the coefficients of the differential system and the initial conditions. We note that the smallest root, $\lambda_1 = \frac{-(c+d)-\sqrt{\Delta}}{2}$, is always negative, but the sign of the largest, $\lambda_2 = \frac{-(c+d)+\sqrt{\Delta}}{2}$, depends on the relationship between the system parameters. Let us consider the main cases that can occur.

If $cd > ab$ (the joint action of the two nations against armaments is stronger than the causes of armaments), then $\sqrt{\Delta} < \sqrt{(c+d)^2} = c+d$ and therefore $\lambda_2 < 0$. In this case, whatever the initial conditions (the initial reserves of weapons) and the values of the right-hand sides r, s (the effect of external forces) happen to be, the differential solution will approach to the equilibrium point $(\bar{x}, \bar{y})$ because $e^{\lambda_1 t} \underset{t \to +\infty}{\to} 0$ and $e^{\lambda_2 t} \underset{t \to +\infty}{\to} 0$. This is a stable situation. We can see that the non-negativity constraint $x \geq 0$, $y \geq 0$ does not affect this trend. In fact, for the derivative of the differential solution we have

$$
\begin{cases}
x' = \frac{\sqrt{a}}{2}\left[\frac{x_0-\bar{x}}{\sqrt{a}} - \frac{y_0-\bar{y}}{\sqrt{b}}\right](-c-\sqrt{ab})e^{(-c-\sqrt{ab})t} \\
\quad + \frac{\sqrt{a}}{2}\left[\frac{x_0-\bar{x}}{\sqrt{a}} + \frac{y_0-\bar{y}}{\sqrt{b}}\right](-c+\sqrt{ab})e^{(-c+\sqrt{ab})t} \\
y' = -\frac{\sqrt{b}}{2}\left[\frac{x_0-\bar{x}}{\sqrt{a}} - \frac{y_0-\bar{y}}{\sqrt{b}}\right](-c-\sqrt{ab})e^{(-c-\sqrt{ab})t} \\
\quad + \frac{\sqrt{b}}{2}\left[\frac{x_0-\bar{x}}{\sqrt{a}} + \frac{y_0-\bar{y}}{\sqrt{b}}\right](-c+\sqrt{ab})e^{(-c+\sqrt{ab})t}
\end{cases},
$$

or, regrouping the terms,

$$
\begin{cases}
x' = \frac{\sqrt{a}}{2}e^{-ct}\left[\frac{x_0-\bar{x}}{\sqrt{a}}\left(-(c+\sqrt{ab})e^{-\sqrt{ab}t} - (c-\sqrt{ab})e^{\sqrt{ab)t}}\right)\right. \\
\quad \left. + \frac{y_0-\bar{y}}{\sqrt{b}}\left((c+\sqrt{ab})e^{-\sqrt{ab}t} - (c-\sqrt{ab})e^{\sqrt{ab}t}\right)\right] \\
y' = \frac{\sqrt{b}}{2}e^{-ct}\left[\frac{x_0-\bar{x}}{\sqrt{a}}\left((c+\sqrt{ab})e^{-\sqrt{ab}t} - (c-\sqrt{ab})e^{\sqrt{ab)t}}\right)\right. \\
\quad \left. + \frac{y_0-\bar{y}}{\sqrt{b}}\left(-(c+\sqrt{ab})e^{-\sqrt{ab}t} - (c-\sqrt{ab})e^{\sqrt{ab}t}\right)\right]
\end{cases}.
$$

Both coefficients $-(c+\sqrt{ab})$ and $-(c-\sqrt{ab})$ together with $x_0-\bar{x}$ are negative, while the coefficients $c+\sqrt{ab}$ and $-(c-\sqrt{ab})$ together with $y_0-\bar{y}$ have opposite signs, but the second one, with the negative sign, is the main one because it has an exponential multiplier increasing with time. Therefore, the sign of the derivative x' is determined by the sign of $x_0-\bar{x}$ and $y_0-\bar{y}$, at least for fairly large values of t. The same is true for the sign of y'. This can be seen even more clearly if we use the approximation for large values of t:

$$
\begin{cases}
x' \approx -\frac{\sqrt{a}}{2}e^{-ct} \cdot e^{\sqrt{ab}t} \cdot (c-\sqrt{ab}) \cdot \left[\frac{x_0-\bar{x}}{\sqrt{a}} + \frac{y_0-\bar{y}}{\sqrt{b}}\right] \\
y' \approx -\frac{\sqrt{b}}{2}e^{-ct} \cdot e^{\sqrt{ab}t} \cdot (c-\sqrt{ab}) \cdot \left[\frac{x_0-\bar{x}}{\sqrt{a}} + \frac{y_0-\bar{y}}{\sqrt{b}}\right]
\end{cases}.
$$

So, considering the values x, y as the current initial conditions $x_0 = x$, $y_0 = y$, we can conclude that if $\bar{x} > 0, \bar{y} > 0$ (this occurs, for example, when $r > 0, s > 0$), then the points $x_0 < \bar{x}, y_0 < \bar{y}$, including $x_0 = 0, y_0 = 0$, cannot be stable, because at these points the derivatives x', y' are positive. Therefore, the solution will not be close to these points and will tend to the equilibrium point $(\bar{x}, \bar{y})$. If $\bar{x} = 0, \bar{y} = 0$ (this occurs when $r = s = 0$), then the equilibrium point is the origin itself and the solution of the problem converges to $(0,0)$. Finally, if $\bar{x} < 0, \bar{y} < 0$ (which occurs, for example, when $r < 0, s < 0$), then the equilibrium point is not unattainable among solutions of the problem. In this case, for any current point $(x, y) = (x_0, y_0)$ we have $x_0 \geq 0 > \bar{x}$, $y_0 \geq 0 > \bar{y}$, and therefore, the derivatives x', y' are negative, which brings the solution to the stationary point $(x, y) = (0,0)$, due to the non-negativity condition.

In the case $cd < ab$ (the joint action of the two nations against armaments is weaker than the causes of armaments), $\sqrt{\Delta} > c + d$ and therefore $\lambda_2 > 0$. This means that the absolute values of the components of the differential solution will increase without restriction, because the first exponential term $e^{\lambda_1 t}$ will converge to zero, while the second, $e^{\lambda_2 t}$ will tend to infinity. This is an unstable situation. This will not happen only if $C_2 = 0$, which occurs only in the very special case when the initial condition belongs to the line passing through the equilibrium point in the direction of the eigenvector $\mathbf{s}_1$: $\begin{pmatrix} x_0 \\ y_0 \end{pmatrix} = \begin{pmatrix} \bar{x} \\ \bar{y} \end{pmatrix} + \tau \mathbf{s}_1$. Any other initial condition leads to an infinite increase of the absolute value of the differential solution. However, this does not always result in explosive arming, because the non-negativity conditions restrict the admissible values of x, y, which can lead to a disarming situation even when $\lambda_2 > 0$. Let us analyze what can happen in this case in more detail.

The part of the solution with factor C_1 has a negative exponent and approaches 0 as t increases. So, this term does not causes an increase of weapons. The problem of explosive armament can arise because of the second term, with factor C_2, which has a positive exponent. If $C_2 > 0$, then x and y will increase tending to $+\infty$ (for positive values) which will result in unrestricted armament. If $C_2 = 0$ (that is, the initial conditions belong to the line $\begin{pmatrix} \bar{x} \\ \bar{y} \end{pmatrix} + \tau \mathbf{s}_1$), then there will be only the first part of the solution, with a negative exponent, and the solution will converge to the equilibrium point $\begin{pmatrix} \bar{x} \\ \bar{y} \end{pmatrix}$. Finally, if $C_2 < 0$, the x and y components of the differential solution will decrease infinitely at $-\infty$. But the complementary condition $x \geq 0, y \geq 0$ will force the components to tend to the origin point and the total disarmament will occur. In fact, for any initial conditions $\begin{pmatrix} x_0 \\ y_0 \end{pmatrix} \neq \begin{pmatrix} \bar{x} \\ \bar{y} \end{pmatrix} + \tau \mathbf{s}_1$, the first term $C_1 \mathbf{s}_1 e^{\lambda_1 t}$ will tend to zero, while the second $C_2 \mathbf{s}_2 e^{\lambda_2 t}$ will increase in absolute value, keeping negative values. However, when one of the components of the solution $\mathbf{u}$ vanishes at instant t_1, it will not be able to decrease any further (due to the non-negativity constraint), keeping its zero value for all subsequent instants. In fact, let us assume that $x(t_1) = 0$ and $y(t_1) > 0$. The point $(x(t_1), y(t_1))$ corresponds to the path of the solution with the initial condition $(x(t_0), y(t_0))$ which generated the constant $C_2 < 0$. Therefore, considering $(x(t_1), y(t_1))$ as the new initial condition, we conclude that it will generate the same constant C_2 and, therefore, in the moment t_1 the solution will have the same tendency to reduce its components. However, x cannot decrease any further, so only the y component will decrease. At the next moments $t > t_1$ the constant C_2 will be different, but it will still be negative. This can be seen by solving the inequality $\frac{x_0 - \bar{x}}{\sqrt{a}} + \frac{y_0 - \bar{y}}{\sqrt{b}} < 0$ which guarantees the negativity of C_2. The solution region of this inequality is located below (and to the left of) the line $y_0 = \bar{y} - \sqrt{\frac{b}{a}}(x_0 - \bar{x})$ (see the grayish region in Fig.14.8) and all solutions associated with initial conditions in this region of the xy-plane will remain there, because of decrease of their components (due to the differential solution) or their equality to 0 (due to the non-negativity condition).

Finally, in the singular case $cd = ab$, the lines $R_1 : -cx + ay + r = 0$ and $R_2 : bx - dy + s = 0$ are parallel and the point $(\bar{x}, \bar{y})$ is no longer the solution of the original system. In this case, the first root of the characteristic equation is still negative $\lambda_1 = -c - d$, with the eigenvector $\mathbf{s}_1 = \begin{pmatrix} 2a \\ -2d \end{pmatrix}$, but the second is zero $\lambda_2 = 0$, with the eigenvector $\mathbf{s}_2 = \begin{pmatrix} 2a \\ 2c \end{pmatrix}$. Therefore, the particular solution can be found in the form $\bar{\mathbf{u}} = \begin{pmatrix} \bar{x} \\ \bar{y} \end{pmatrix} = \begin{pmatrix} kt + p \\ mt + q \end{pmatrix}$, where the constants k, m, p, q must be found by substituting this vector function into the original system. This leads to the following system of polynomial equations

$$\begin{cases} k = -c(kt + p) + a(mt + q) + r \\ m = b(kt + p) - d(mt + q) + s \end{cases}.$$

From the subsystem of the linear terms we have $\begin{cases} -ck + am = 0 \\ bk - dm = 0 \end{cases}$, from which it follows the unique relation $m = \frac{c}{a}k$. The subsystem for the free

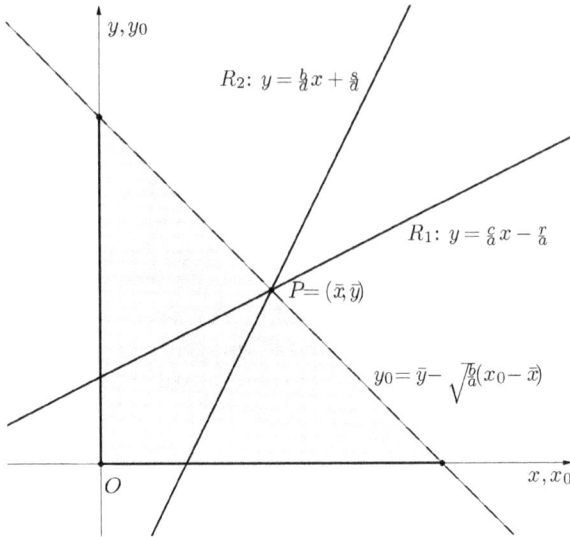

Figure 14.8: Arms race model of Richardson, instable case.

coefficients has the form $\begin{cases} -cp + aq = k - r \\ bp - dq = m - s \end{cases}$, Taking into account that $d = \frac{ab}{c}$ and $m = \frac{c}{a}k$, we rewrite the second equation in the form $bp - \frac{ab}{c}q = \frac{c}{a}k - s$ or, multiplying by $-\frac{c}{b}$, we get $-cp + aq = -\frac{c^2}{ab}k + \frac{c}{b}s$. Comparing with the first equation, we conclude that the last system is compatible only when $k - r = -\frac{c^2}{ab}k + \frac{c}{b}s$, from where we find the value of k: $k = a\frac{br+cs}{ab+c^2}$. So, $m = c\frac{br+cs}{ab+c^2}$ and the coefficients p, q satisfy the only relation $-cp+aq = k-r$. For example, by choosing $q = 0$ and using the value found for k, we get $p = \frac{cr-as}{ab+c^2}$. Therefore, the general solution of the differential system has the form

$$\begin{cases} x = C_1 \cdot 2ae^{(-c-d)t} + C_2 \cdot 2a + kt + p \\ y = C_1 \cdot (-2d)e^{(-c-d)t} + C_2 \cdot 2c + mt + q \end{cases} .$$

It can be seen that for large values of t the behavior of the general solution depends essentially on the particular solution. If $br + cs > 0$ (for instance, $r > 0, s > 0$), that is, the external causes stimulate armament growth, then $k > 0, m > 0$ and the two components of the solution go to $+\infty$. If $br+cs < 0$ (for instance, $r < 0, s < 0$), which occurs when external forces restrict arms race, then $k < 0, m < 0$ and the two components of the solution converges to 0 (taking into account the non-negativity constraint). If $br + cs = 0$, that is, the joint action of external forces is neutral, then the solution will approaches the point $(x, y) = (2aC_2 + p, 2cC_2 + q)$, where C_2 is determined from the initial conditions.

4 Vibrating systems

4.1 Vibrating beads

Consider n small beads, each having mass m, spaced at equal intervals of length l on a tightly stretched string under a constant tension T. At equilibrium, all of the beads are positioned on a horizontal string. Each bead is initially displaced from its equilibrium position by a small vertical distance and then the beads are released so that they can vibrate freely. The problem is to find a position of each bead after the initial perturbation. In solution of this problem we neglect gravitation and restrict motion to small vertical vibrations. The displacements of the beads, relative to the equilibrium position, are denoted by $x_1, x_2, \ldots, x_n$ (see Fig.14.9).

Let θ_k be the angle formed by the segment of the string between x_k and x_{k+1} with the horizontal line (see Fig.14.9). Then, for any $k = 2, \ldots, n-1$ (for each inner bead) the upward force on the k-th bead is $T \sin \theta_k$, while the downward force is $T \sin \theta_{k-1}$. If we introduce additionally two fictional beads x_0 and x_{n+1} at the endpoints of the string with zero displacements ($x_0 = 0$ and $x_{n+1} = 0$), then the same forces can be calculated for the 1-st and n-th beads. Using the approximation of small vibrations, we can evaluate the total force on the k-th bead as follows:

$$F = T(\sin \theta_k - \sin \theta_{k-1}) \approx T(\tan \theta_k - \tan \theta_{k-1})$$

$$= T \left(\frac{x_{k+1} - x_k}{l} - \frac{x_k - x_{k-1}}{l} \right) = \frac{T}{l}(x_{k-1} - 2x_k + x_{k+1}).$$

Therefore, according to Newton's second law, the equation of motion of k-th bead is

$$m x_k'' = \frac{T}{l}(x_{k-1} - 2x_k + x_{k+1}), \quad k = 1, \ldots, n.$$

These equations can be written in matrix form as follows:

$$\mathbf{x}'' = -\omega^2 A \mathbf{x},$$

where $\omega^2 = \frac{T}{lm}$, $\mathbf{x} = (x_1, x_2, \ldots, x_n)^T$ (as usual, the superscript "T" means

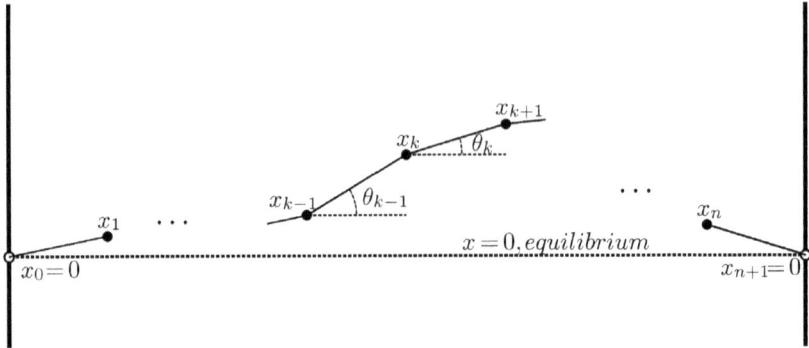

Figure 14.9: Oscillating beads on the horizontal string

transpose) and

$$A = \begin{pmatrix} 2 & -1 & 0 & 0 & \ldots & 0 & 0 & 0 \\ -1 & 2 & -1 & 0 & \ldots & 0 & 0 & 0 \\ 0 & -1 & 2 & -1 & \ldots & 0 & 0 & 0 \\ & & & \ldots & & & & \\ 0 & 0 & 0 & 0 & \ldots & -1 & 2 & -1 \\ 0 & 0 & 0 & 0 & \ldots & 0 & -1 & 2 \end{pmatrix}.$$

This system is not a normal one. Although we can reduce it to a normal form of order $2n$ (by introducing new functions $y_{2k-1} = x_k, y_{2k} = x'_k, k = 1, \ldots, n$), it is more convenient and simpler to use another approach, splitting the system in n separate equations of the second order. Let us describe how it can be accomplished, taking advantage of the properties of the matrix A.

The matrix A is symmetric and positive definite. The fist is obvious, since $A^T = A$, and the second can be shown by direct calculation with an arbitrary non-zero vector $v = (v_1, \ldots, v_n)^T$:

$$\mathbf{v}^T A \mathbf{v} = v_1(2v_1 - v_2) + v_2(-v_1 + 2v_2 - v_3) + v_3(-v_2 + 2v_3 - v_4) + \ldots$$

$$+ v_{n-1}(-v_{n-2} + 2v_{n-1} - v_n) + v_n(-v_{n-1} + 2v_n)$$

$$= v_1^2 + v_n^2 + (v_1^2 - 2v_1v_2 + v_2^2) + (v_2^2 - 2v_2v_3 + v_3^2) + \ldots + (v_{n-1}^2 - 2v_{n-1}v_n + v_n^2)$$

$$= v_1^2 + v_n^2 + (v_1 - v_2)^2 + (v_2 - v_3)^2 + \ldots (v_{n-1} - v_n)^2 > 0.$$

Therefore, the matrix A has n positive eigenvalues λ_k, $k = 1, \ldots, n$ and complete system of n real eigenvectors $\mathbf{v}_k$, $k = 1, \ldots, n$, orthogonal to each other (just recall the properties of symmetric positive definite matrices from Linear Algebra). Furthermore, for this specific matrix A we can find the set of eigenvalues λ_k and eigenvectors $\mathbf{v}_k$:

$$\lambda_k = 4\sin^2\frac{\alpha_k}{2}, \quad \mathbf{v}_k = (\sin\alpha_k j)_{j=1}^n, \quad \alpha_k = \frac{k\pi}{n+1}, \quad k = 1,\ldots,n,$$

where index j indicates the elements of the eigenvector $\mathbf{v}_k$. The derivation of this result is not very simple, but the verification of these formulas is straightforward and that is what we do:

1) for any $j = 2,\ldots,n-1$

$$(A\mathbf{v}_k)_j = -\sin\alpha_k(j-1) + 2\sin\alpha_k j - \sin\alpha_k(j+1) = -2\sin\alpha_k j\cos\alpha_k + 2\sin\alpha_k j$$

$$= 2(1-\cos\alpha_k)\sin\alpha_k j = 4\sin^2\frac{\alpha_k}{2}\cdot\sin\alpha_k j;$$

2) for $j = 1$

$$(A\mathbf{v}_k)_1 = 2\sin\alpha_k - \sin 2\alpha_k = 2\sin\alpha_k - 2\sin\alpha_k\cos\alpha_k$$

$$= 2(1-\cos\alpha_k)\sin\alpha_k = 4\sin^2\frac{\alpha_k}{2}\cdot\sin\alpha_k;$$

3) for $j = n$

$$(A\mathbf{v}_k)_n = -\sin\alpha_k(n-1) + 2\sin\alpha_k n = -\sin\alpha_k(n-1) + 2\sin\alpha_k n - \sin\alpha_k(n+1)$$

$$= -2\sin\alpha_k n\cos\alpha_k + 2\sin\alpha_k n = 2(1-\cos\alpha_k)\sin\alpha_k n = 4\sin^2\frac{\alpha_k}{2}\cdot\sin\alpha_k n.$$

Next, let us calculate the norm of each vector $\mathbf{v}_k$:

$$|\mathbf{v}_k|^2 = \mathbf{v}_k\cdot\mathbf{v}_k = \sum_{j=1}^n\sin^2\alpha_k j = \sum_{j=1}^n\frac{1-\cos 2\alpha_k j}{2} = \frac{n}{2} - \frac{1}{2}\sum_{j=1}^n\cos 2\alpha_k j.$$

The last sum can be evaluated as follows:

$$\sum_{j=1}^n\cos 2\alpha_k j = \frac{1}{\sin 2\alpha_k}\sum_{j=1}^n\cos 2\alpha_k j\cdot\sin 2\alpha_k$$

$$= \frac{1}{\sin 2\alpha_k}\sum_{j=1}^n\frac{1}{2}(\sin(2j+2)\alpha_k - \sin(2j-2)\alpha_k)$$

$$= \frac{1}{2\sin 2\alpha_k}(\sin(2n+2)\alpha_k + \sin 2n\alpha_k - \sin 2\alpha_k) = -\frac{1}{2\sin 2\alpha_k}2\sin 2\alpha_k = -1.$$

Therefore,

$$|\mathbf{v}_k|^2 = \frac{n}{2} + \frac{1}{2} = \frac{n+1}{2}, \text{ that is, } |\mathbf{v}_k| = \sqrt{\frac{n+1}{2}}.$$

Using this result we can find the orthonormal eigenvectors

$$\mathbf{u}_k = \frac{\mathbf{v}_k}{|\mathbf{v}_k|} = \sqrt{\frac{2}{n+1}} \left(\sin \alpha_k j\right)_{j=1}^{n},$$

which are frequently called normal modes in the problems of oscillations.

Now we can compose the matrix S whose columns are the eigenvectors $\mathbf{u}_k$: $S = (\mathbf{u}_1, \mathbf{u}_2, \ldots, \mathbf{u}_n)$. Since $\mathbf{u}_k$ are orthonormal, the matrix S is orthogonal, that is, $S^T = S^{-1}$ (as usual, S^{-1} denotes the matrix inverse to S). As it is known from Linear Algebra, the matrix S together with the diagonal matrix of eigenvalues $\Lambda = diag(\lambda_1, \lambda_2, \ldots, \lambda_n)$ allow the following decomposition of A: $A = S\Lambda S^T$. Employing this representation in the original system, we get

$$\mathbf{x}'' + \omega^2 S\Lambda S^T \mathbf{x} = \mathbf{0}.$$

Then, multiplying this system by S^T and introducing the new functions $\mathbf{y} = (y_1, \ldots, y_n)^T = S^T \mathbf{x}$, we obtain

$$\mathbf{y}'' + \omega^2 \Lambda \mathbf{y} = \mathbf{0}.$$

In this way, we reduce the original system to the following set of the separate equations of second order:

$$y_k'' + \omega^2 \lambda_k y_k = 0, \ k = 1, \ldots, n.$$

The inverse relationship between $\mathbf{y}$ and $\mathbf{x}$ is given by the formula $\mathbf{x} = S\mathbf{y} = y_1 \mathbf{u}_1 + \ldots + y_n \mathbf{u}_n$, which is called the normal mode representation (the set of displacements $\mathbf{x}$ is expanded in the linear combination of the normal modes). The corresponding coefficients (functions) of expansion y_k are called the amplitudes of normal modes. So, we derived n separate equations for amplitudes of normal modes. Equations of this type were already solved analytically and met in applications (see Chapters 9 and 11). Each equation describes simple harmonic oscillations with the circular frequency $\varpi_k = \omega\sqrt{\lambda_k}$, whose solution is $y_k = A_k \cos \varpi_k t + B_k \sin \varpi_k t$, where A_k and B_k are arbitrary constants determined by initial conditions. Solving the equations for amplitudes of each normal mode, the required displacements are found by the formula $\mathbf{x} = S\mathbf{y}$.

For example, if $n = 3$, then

$$A = \begin{pmatrix} 2 & -1 & 0 \\ -1 & 2 & -1 \\ 0 & -1 & 2 \end{pmatrix}.$$

The eigenvalues of A are

$$\lambda_k = 4\sin^2 \frac{\alpha_k}{2}, \quad \alpha_k = \frac{k\pi}{4}, \quad k = 1, 2, 3,$$

that is,

$$\lambda_1 = 2 - \sqrt{2}, \quad \lambda_2 = 2, \quad \lambda_3 = 2 + \sqrt{2}.$$

The corresponding normalized eigenvectors are

$$\mathbf{u}_k = \sqrt{\frac{1}{2}} \, (\sin \alpha_k j)_{j=1}^3, \quad k = 1, 2, 3,$$

that is,

$$\mathbf{u}_1 = \frac{1}{2}(1, \sqrt{2}, 1)^T, \quad \mathbf{u}_2 = \frac{1}{\sqrt{2}}(1, 0, -1)^T, \quad \mathbf{u}_3 = \frac{1}{2}(1, -\sqrt{2}, 1)^T.$$

The matrices of eigenvalues and eigenvectors are

$$\Lambda = diag(\lambda_1, \lambda_2, \lambda_3) = \begin{pmatrix} 2 - \sqrt{2} & 0 & 0 \\ 0 & 2 & 0 \\ 0 & 0 & 2 + \sqrt{2} \end{pmatrix},$$

$$S = (\mathbf{u}_1, \mathbf{u}_2, \mathbf{u}_3) = \frac{1}{2} \begin{pmatrix} 1 & \sqrt{2} & 1 \\ \sqrt{2} & 0 & -\sqrt{2} \\ 1 & -\sqrt{2} & 1 \end{pmatrix}.$$

The three equations for the amplitudes of the normal modes take the form

$$y_1'' + \varpi_1^2 y_1 = 0, \quad y_2'' + \varpi_2^2 y_2 = 0, \quad y_3'' + \varpi_3^2 y_3 = 0,$$

where

$$\varpi_1 = \sqrt{2 - \sqrt{2}}\, \omega, \quad \varpi_2 = \sqrt{2}\, \omega, \quad \varpi_3 = \sqrt{2 + \sqrt{2}}\, \omega$$

and their solutions are

$$y_1 = A_1 \cos \varpi_1 t + B_1 \sin \varpi_1 t, \quad y_2 = A_2 \cos \varpi_2 t + B_2 \sin \varpi_2 t,$$

$$y_3 = A_3 \cos \varpi_3 t + B_3 \sin \varpi_3 t,$$

where A_k, B_k, $k = 1, 2, 3$ are arbitrary constants. Finally, the displacements of beads x_k are found through the linear combination of $\mathbf{u_k}$:

$$\mathbf{x} = S\mathbf{y} = y_1\mathbf{u_1} + y_2\mathbf{u_2} + y_3\mathbf{u_3},$$

that is,

$$x_1 = \frac{1}{2}y_1 + \frac{1}{\sqrt{2}}y_2 + \frac{1}{2}y_3 \ , \ x_2 = \frac{1}{\sqrt{2}}y_1 - \frac{1}{\sqrt{2}}y_3 \ , \ x_3 = \frac{1}{2}y_1 - \frac{1}{\sqrt{2}}y_2 + \frac{1}{2}y_3 \ .$$

4.2   Earthquakes and tall buildings

Consider a building as a set of n floors, each of mass m, connected together by vertical walls. Let us analyze the response of such a building to horizontal seismic motion at the foundation generated by an earthquake of the force f. We disconsider gravitation and restrict motion to the horizontal direction, denoting displacements of the floors, relative to a fixed frame of reference, by $x_1, x_2, \ldots, x_n$ (see Fig.14.10). At equilibrium, all of the displacements and velocities are zero and the floors are in vertical alignment.

When adjacent floors are not in alignment, we assume that the walls exert a restoring force proportional to the difference in displacements between the floors with proportionality coefficient κ. Then, by Newton's second law, the equation of motion for the inner k-th floor is

$$mx_k'' = -\kappa(x_k - x_{k-1}) - \kappa(x_k - x_{k+1}), k = 2, \ldots, n-1$$

whereas the equations for the first and top floors are

$$mx_1'' = -\kappa(x_1 - f) - \kappa(x_1 - x_2)$$

and

$$mx_n'' = -\kappa(x_n - x_{n-1}),$$

respectively.

These equations can be written in matrix form as follows:

$$\mathbf{x}'' + \omega^2 A\mathbf{x} = \omega^2\mathbf{f},$$

where $\omega^2 = \frac{\kappa}{m}$, $\mathbf{x} = (x_1, x_2, \ldots, x_n)^T$, $\mathbf{f} = (f, 0, \ldots, 0)^T$ and

$$A = \begin{pmatrix} 2 & -1 & 0 & 0 & \ldots & 0 & 0 & 0 \\ -1 & 2 & -1 & 0 & \ldots & 0 & 0 & 0 \\ 0 & -1 & 2 & -1 & \ldots & 0 & 0 & 0 \\ & & & \ldots & & & & \\ 0 & 0 & 0 & 0 & \ldots & -1 & 2 & -1 \\ 0 & 0 & 0 & 0 & \ldots & 0 & -1 & 1 \end{pmatrix}.$$

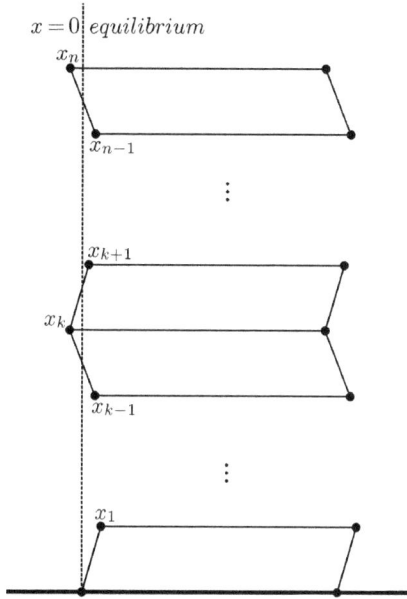

Figure 14.10: Tall building vibrations

Notice that the elements of the matrix A are constant, but the right-hand side depends on time t. One can see that the obtained system is very similar to that found in the problem of vibrating beads. Consequently, the forthcoming analysis follows closely the study of the previous problem.

First, notice that the system is not a normal one. Although we can reduce this system to a normal form of order $2n$ (by introducing new functions $y_{2k-1} = x_k, y_{2k} = x'_k, k = 1, \ldots, n$), it is more convenient and simpler to use another approach, splitting the system in n separate equations of the second order. Let us describe how it can be accomplished, taking advantage of the properties of the matrix A.

The matrix A is symmetric and positive definite. The fist is obvious, since $A^T = A$, and the second can be shown by direct calculation with an arbitrary non-zero vector $v = (v_1, \ldots, v_n)^T$:

$$\mathbf{v}^T A\mathbf{v} = v_1(2v_a - v_2) + v_2(-v_1 + 2v_2 - v_3) + v_3(-v_2 + 2v_3 - v_4) + \ldots$$

$$+ v_{n-1}(-v_{n-2} + 2v_{n-1} - v_n) + v_n(-v_{n-1} + v_n)$$

$$= v_1^2 + (v_1^2 - 2v_1v_2 + v_2^2) + (v_2^2 - 2v_2v_3 + v_3^2) + \ldots + (v_{n-1}^2 - 2v_{n-1}v_n + v_n^2)$$

$$= v_1^2 + (v_1 - v_2)^2 + (v_2 - v_3)^2 + \dots (v_{n-1} - v_n)^2 > 0.$$

Therefore, the matrix A has n positive eigenvalues λ_k, $k = 1, \dots, n$ and complete system of n real eigenvectors $\mathbf{v}_k$, $k = 1, \dots, n$, orthogonal to each other (just recall the properties of symmetric positive matrices from Linear Algebra). Furthermore, for this specific matrix A we can find the set of eigenvalues λ_k and eigenvectors $\mathbf{v}_k$:

$$\lambda_k = 4\sin^2\frac{\alpha_k}{2}, \quad \mathbf{v}_k = (\sin\alpha_k j)_{j=1}^n, \quad \alpha_k = \frac{\pi(2k-1)}{2n+1}, \quad k = 1, \dots, n,$$

where index j indicates the elements of the vector $\mathbf{v}_k$. The derivation of this result is rather complicated, but the verification of these formulas is quite simple and we show it below:

1) for any $j = 2, \dots, n-1$

$$(A\mathbf{v}_k)_j = -\sin\alpha_k(j-1) + 2\sin\alpha_k j - \sin\alpha_k(j+1) = -2\sin\alpha_k j \cos\alpha_k + 2\sin\alpha_k j$$

$$= 2(1 - \cos\alpha_k)\sin\alpha_k j = 4\sin^2\frac{\alpha_k}{2} \cdot \sin\alpha_k j;$$

2) for $j = 1$

$$(A\mathbf{v}_k)_1 = 2\sin\alpha_k - \sin\alpha_k \cdot 2 = 2\sin\alpha_k - 2\sin\alpha_k\cos\alpha_k$$

$$= 2(1 - \cos\alpha_k)\sin\alpha_k = 4\sin^2\frac{\alpha_k}{2} \cdot \sin\alpha_k;$$

3) for $j = n$

$$(A\mathbf{v}_k)_n = -\sin\alpha_k(n-1) + \sin\alpha_k n = \sin(\alpha_k(n-1) + \pi(2k-1)) + \sin\alpha_k n$$

$$= \sin\alpha_k 3n + \sin\alpha_k n = 2\sin\alpha_k 2n\cos\alpha_k n = 4\sin\alpha_k n\cos^2\alpha_k n$$

$$= 4\sin\alpha_k n\cos^2\left(\frac{\pi}{2}(2k-1) - \frac{\alpha_k}{2}\right) = 4\sin^2\frac{\alpha_k}{2}\sin\alpha_k n.$$

Next, let us calculate the norm of each vector $\mathbf{v}_k$:

$$|\mathbf{v}_k|^2 = \mathbf{v}_k \cdot \mathbf{v}_k = \sum_{j=1}^n \sin^2\alpha_k j = \sum_{j=1}^n \frac{1 - \cos 2\alpha_k j}{2} = \frac{n}{2} - \frac{1}{2}\sum_{j=1}^n \cos 2\alpha_k j.$$

The last sum can be evaluated as follows:

$$\sum_{j=1}^n \cos 2\alpha_k j = \frac{1}{\sin\alpha_k}\sum_{j=1}^n \cos 2\alpha_k j \cdot \sin\alpha_k$$

$$= \frac{1}{\sin \alpha_k} \sum_{j=1}^{n} \frac{1}{2} \left(\sin(2j+1)\alpha_k - \sin(2j-1)\alpha_k \right)$$

$$= \frac{1}{2 \sin \alpha_k} \left(\sin(2n+1)\alpha_k - \sin \alpha_k \right) = -\frac{1}{2}.$$

Therefore,

$$|\mathbf{v}_k|^2 = \frac{n}{2} + \frac{1}{4} = \frac{2n+1}{4}, \text{ that is, } |\mathbf{v}_k| = \frac{\sqrt{2n+1}}{2}.$$

Using this result we can find the orthonormal eigenvectors

$$\mathbf{u}_k = \frac{\mathbf{v}_k}{|\mathbf{v}_k|} = \frac{2}{\sqrt{2n+1}} \left(\sin \alpha_k j \right)_{j=1}^{n}$$

(also called normal modes).

Now we can compose the matrix S whose columns are the eigenvectors $\mathbf{u}_k$: $S = (\mathbf{u}_1, \mathbf{u}_2, \ldots, \mathbf{u}_n)$. Since $\mathbf{u}_k$ are orthonormal, the matrix S is orthogonal, that is, $S^T = S^{-1}$. As it is known from Linear Algebra, the matrix S together with the diagonal matrix of eigenvalues $\Lambda = diag(\lambda_1, \lambda_2, \ldots, \lambda_n)$ allow the following decomposition of A: $A = S \Lambda S^T$. Employing this representation in the original system, we get

$$\mathbf{x}'' + \omega^2 S \Lambda S^T \mathbf{x} = \omega^2 \mathbf{f}.$$

Then, multiplying this system by S^T and introducing the new functions $\mathbf{y} = (y_1, \ldots, y_n)^T = S^T \mathbf{x}$, we obtain

$$\mathbf{y}'' + \omega^2 \Lambda \mathbf{y} = \omega^2 S^T \mathbf{f}.$$

In this way, we reduce the original system to the following set of the separate equations of second order:

$$y_k'' + \omega^2 \lambda_k y_k = \omega^2 g_k, \quad k = 1, \ldots, n,$$

where $g_k = (S^T \mathbf{f})_k = \mathbf{u}_k \mathbf{f}$.

The inverse relationship between $\mathbf{y}$ and $\mathbf{x}$ is given by the formula $\mathbf{x} = S\mathbf{y} = y_1 \mathbf{u}_1 + \ldots + y_n \mathbf{u}_n$, which is called the normal mode representation (the set of displacements $\mathbf{x}$ is expanded in the linear combination of the normal modes). The corresponding coefficients (functions) of expansion y_k are called the amplitudes of normal modes. So, we derived n separate equations for amplitudes of normal modes. Equations of this type were already solved analytically and met in applications (see Chapters 9 and 11). For instance, if $\mathbf{f} = \mathbf{0}$, then each equation describes simple harmonic oscillations with the circular frequency $\omega\sqrt{\lambda_k}$. For non-zero $\mathbf{f}$, we have forced oscillations, whose solution is represented as superposition of harmonic oscillations and forced motions (see section 1 in Chapter 11). Solving the equations for amplitudes of each normal mode, the required displacements are found by the formula $\mathbf{x} = S\mathbf{y}$.

For example, if $n = 3$, then

$$A = \begin{pmatrix} 2 & -1 & 0 \\ -1 & 2 & -1 \\ 0 & -1 & 1 \end{pmatrix}.$$

The eigenvalues of A are

$$\lambda_k = 4\sin^2\frac{\alpha_k}{2}, \quad \alpha_k = \frac{(2k-1)\pi}{7}, \quad k = 1, 2, 3,$$

that is,

$$\lambda_1 \approx 0.198, \quad \lambda_2 \approx 1.555, \quad \lambda_3 \approx 3.247.$$

The corresponding normalized eigenvectors are

$$\mathbf{u}_k = \frac{2}{\sqrt{7}}\left(\sin\alpha_k j\right)_{j=1}^3, \quad k = 1, 2, 3,$$

that is,

$$\mathbf{u}_1 \approx (0.328, 0.591, 0.737)^T, \quad \mathbf{u}_2 \approx (0.737, 0.328, -0.591)^T,$$

$$\mathbf{u}_3 \approx (0.591, -0.737, 0.328)^T.$$

The matrices of eigenvalues and eigenvectors are

$$\Lambda = diag(\lambda_1, \lambda_2, \lambda_3) = \begin{pmatrix} 0.198 & 0 & 0 \\ 0 & 1.555 & 0 \\ 0 & 0 & 3.247 \end{pmatrix},$$

$$S = (\mathbf{u}_1, \mathbf{u}_2, \mathbf{u}_3) = \begin{pmatrix} 0.328 & 0.737 & 0.591 \\ 0.591 & 0.328 & -0.737 \\ 0.737 & -0.591 & 0.328 \end{pmatrix}.$$

The three equations for the amplitudes of the normal modes take the form

$$y_1'' + \varpi_1^2 y_1 = 0, \quad y_2'' + \varpi_2^2 y_2 = 0, \quad y_3'' + \varpi_3^2 y_3 = 0,$$

where $\varpi_k^2 = \lambda_k \omega^2$, $k = 1, 2, 3$, and their solutions are

$$y_1 = A_1 \cos\varpi_1 t + B_1 \sin\varpi_1 t, \quad y_2 = A_2 \cos\varpi_2 t + B_2 \sin\varpi_2 t,$$

$$y_3 = A_3 \cos \varpi_3 t + B_3 \sin \varpi_3 t,$$

where A_k, B_k, $k = 1, 2, 3$ are arbitrary constants. Finally, the displacements of beads x_k are found through the linear combination of $\mathbf{u_k}$:

$$\mathbf{x} = S\mathbf{y} = y_1 \mathbf{u_1} + y_2 \mathbf{u_2} + y_3 \mathbf{u_3},$$

that is,

$$x_1 = 0.328 y_1 + 0.737 y_2 + 0.591 y_3 \;, \quad x_2 = 0.591 y_1 + 0.328 y_2 - 0.737 y_3 \;,$$

$$x_3 = 0.737 y_1 - 0.591 y_2 + 0.328 y_3 \;.$$

In the presence of earthquake force, modeled as $f = (F \sin \gamma t, 0, 0)^T$, the normal mode forces are given by the formula $g_k = \mathbf{u_k} f$, $k = 1, 2, 3$, that is,

$$g_1 = 0.328 F \sin \gamma t, \;\; g_2 = 0.737 F \sin \gamma t, \;\; g_3 = 0.591 F \sin \gamma t.$$

Consequently, the three equations for the amplitudes take the form

$$y_1'' + \varpi_1^2 y_1 = 0.328 F \sin \gamma t, \;\; y_2'' + \varpi_2^2 y_2 = 0.737 F \sin \gamma t,$$

$$y_3'' + \varpi_3^2 y_3 = 0.591 F \sin \gamma t.$$

If $\gamma \neq \varpi_k$, $k = 1, 2, 3$, their solutions are

$$y_1 = A_1 \cos \varpi_1 t + B_1 \sin \varpi_1 t + \frac{0.328 F}{\varpi_1^2 - \gamma^2} \sin \gamma t,$$

$$y_2 = A_2 \cos \varpi_2 t + B_2 \sin \varpi_2 t + \frac{0.737 F}{\varpi_2^2 - \gamma^2} \sin \gamma t,$$

$$y_3 = A_3 \cos \varpi_3 t + B_3 \sin \varpi_3 t + \frac{0.591 F}{\varpi_3^2 - \gamma^2} \sin \gamma t.$$

Problems for Chapter 14

1. Find the general solution of the first spring-mass model (shown in Fig.14.1) in the case when the masses are $m_1 = 5g$, $m_2 = 10g$ and the coefficients $k_1 = 10g/s^2$, $k_2 = 10g/s^2$. Solve this problem when the initial displacement of the first mass is $3cm$, while the second mass is in equilibrium and the initial velocities of the two masses are zero.

2. Considering the presence of damping of the masses in the model shown in Fig.14.1, verify that the equations of the system take the form

$$\begin{cases} m_1 x_1'' = -k_1 x_1 + k_2(x_2 - x_1) - \kappa_1 x_1' \\ m_2 x_2'' = -k_2(x_2 - x_1) - \kappa_2 x_2' \end{cases}$$

where $\kappa_1 > 0$ and $\kappa_2 > 0$ are the damping coefficients of the first and second masses. Transform this system into an equivalent system of four first-order equations. Find the general solution of the system when the masses are $m_1 = 10g$, $m_2 = 10g$, spring coefficients $k_1 = 5g/s^2$, $k_2 = 10g/s^2$ and damping coefficients $\kappa_1 = 15g/s$, $\kappa_2 = 15g/s$.

3. Transform the system of the second spring-mass model (shown in Fig.14.3)

$$\begin{cases} m_1 x_1'' = -k_1 x_1 + k_2(x_2 - x_1) \\ m_2 x_2'' = -k_2(x_2 - x_1) - k_3 x_2 \end{cases}$$

in an equivalent system of four first-order equations.

Find the general solution of this model when the masses are $m_1 = 2g$, $m_2 = 2g$ and the coefficients $k_1 = 4g/s^2$, $k_2 = 6g/s^2$, $k_3 = 4g/s^2$.

4. Considering the presence damping of both masses in the model shown in Fig.14.3, verify that the equations of the system take the form

$$\begin{cases} m_1 x_1'' = -k_1 x_1 + k_2(x_2 - x_1) - \kappa_1 x_1' \\ m_2 x_2'' = -k_2(x_2 - x_1) - k_3 x_2 - \kappa_2 x_2' \end{cases}$$

where $\kappa_1 > 0$ and $\kappa_2 > 0$ are the damping coefficients of the first and second masses. Transform this system into an equivalent system of four first-order equations. Find the general solution of the system when the masses are $m_1 = 2g$, $m_2 = 2g$, spring coefficients $k_1 = 10g/s^2$, $k_2 = 3g/s^2$, $k_3 = 10g/s^2$, and damping coefficients $\kappa_1 = 12g/s$, $\kappa_2 = 12g/s$.

5. Find the critical point and the numerical solution of the Lotka-Volterra model with $a = 1, b = 2, c = 0.5, d = 0.25$ in the presence of fishing corresponding to the coefficient $\alpha = 0.1$. Use the initial conditions $x_0 = 10$, $y_0 = 5$. Compare the obtained solution with the one shown in Figs.14.6 and 14.7.

6. Find the critical point and the numerical solution of the Lotka-Volterra model with $a = 0.2, b = 0.1, c = 0.0025, d = 0.002$ without the presence of fishing. Carry out the same study in the presence of fishing with the coefficient $\alpha = 0.001$. Use the two pairs of initial conditions: $x_0 = 40$, $y_0 = 70$ and $x_0 = 60$, $y_0 = 90$. Compare the results of the models with fishing and without.

7. Find the equilibrium point and the general solution of the arms race model without external sources ($r = s = 0$) when $c = d = 1$, $a = b = 0.5$. Solve this problem with the initial condition $x_0 = 10$, $y_0 = 15$. Carry out a similar study for the model with external sources, setting that $r = s = 0.5$. Compare the obtained results with the theoretical analysis.

8. Find the equilibrium point and the general solution of the arms race model without external sources ($r = s = 0$) when $c = d = 0.5$, $a = b = 1$. Solve this problem with the initial condition $x_0 = 10$, $y_0 = 15$. Carry out a similar study for the model with external sources, setting that $r = s = -0.5$. Compare the obtained results with the theoretical analysis.

9. Find the general solution of the arms race model without external sources ($r = s = 0$) when $c = d = 1$, $a = b = 1$. Solve this problem with the initial condition $x_0 = 10$, $y_0 = 15$. Carry out a similar study for the model with external sources, setting that $r = s = -0.5$. Compare the obtained results with the theoretical analysis.

10. Find the solution for the system of vibrating beads in the case of $n = 3$, $\omega = 5$ if the initial displacements are $x_1 = 0.1$, $x_2 = 0.2$, $x_3 = 0.1$ and the initial velocities are zero.

11. Find a general solution for the earthquake problem in the case of $n = 3$ and $\gamma = \varpi_1$.

12. Find the solution for the earthquake problem in the case of $n = 3$, $\omega = 5$, $f = (\sin t, 0, 0)^T$ if the initial displacements are zero and the initial velocities are $(1, 0.5, 0.2)^T$.

Answers to selected exercises

Chapter 1 Initial concepts

Section 3 Differential equation for family of functions

Find ODE whose general solution is a given family of functions:
1. $y = (x - C)^3$. Solution: $y' = 3y^{2/3}$
2. $y = e^{Cx}$. Solution: $y = e^{y'x/y}$
3. $x^2 + Cy^2 = 2y$. Solution: $xy + (2y - x^2)y' = yy'$
4. $y = Ax^2 + Be^x$. Solution: $y'' - y = (y'' - y')\frac{x^2-2}{2(x-1)}$
5. $\ln y = Ax + By$. Solution: $\ln y = \frac{xy'}{y} + \frac{y''-y'^2}{y^2 y''}(y - xy')$
6. $y = Ax^3 + Bx^2 + Cx$. Solution: $y = xy' - \frac{3x^2}{2}y'' + \frac{x^3}{6}y'''$

Chapter 3 Explicit equations of the first order: methods of solution

Section 1.1 Separable equations

Solve the following separable equations:
1. $\sin x \cdot \tan y\, dx - \frac{dy}{\sin x} = 0$. Solution: $\ln|\sin y| = \frac{x}{2} - \frac{1}{4}\sin 2x + C$
2. $(xy^3 + x)dx + (x^2y^2 - y^2)dy = 0$. Solution: $\sqrt[3]{y^3 + 1} = \frac{C}{\sqrt{x^2-1}}$
3. $(1 + y^2)dx - (y + yx^2)dy = 0$. Solution: $\frac{1}{2}\ln(1 + y^2) = \arctan x + C$
4. $y' = 2xy + x$. Solution: $\ln|2y + 1| = x^2 + C$
5. $2xyy' = 1 - x^2$. Solution: $y^2 = \ln|x| - \frac{x^2}{2} + C$
6. $y' = e^{x^2}x(1 + y^2)$. Solution: $\arctan y = \frac{1}{2}e^{x^2} + C$
7. $y'\cot x + y = 2$. Solution: $y = C\cos x + 2$
8. $(1 + e^{3y})xdx = e^{3y}dy$. Solution: $\frac{x^2}{2} = \frac{1}{3}\ln(1 + e^{3y}) + C$

9. $y - xy' = 1 + x^2y'$. Solution: $y - 1 = \frac{Cx}{x+1}$

10. $2x^2yy' + y^2 = 2$. Solution: $\ln|2 - y^2| = \frac{1}{x} + C$

11. $y - xy' = 2(1 + x^2y')$. Solution: $y - 2 = \frac{Cx}{2x+1}$

12. $(1 + e^x)ydy - e^ydx = 0$. Solution: $-e^{-y}(y + 1) = x - \ln|1 + e^x| + C$

13. $(x + xy^2)dy = (y^2 - y)dx$. Solution: $y + \frac{(y-1)^2}{|y|} = \ln|x| + C$

14. $y' = \sin^2 y$. Solution: $\cot y = -x + C$, $y = k\pi, k \in \mathbb{Z}$

15. $y' = (y - 1)x$. Solution: $y - 1 = Ce^x$

16. $x^2y^2y' + 1 = y$. Solution: $\frac{y^2}{2} + y + \ln|y - 1| = -\frac{1}{x} + C$, $y = 1$

17. $xy' + y = y^2$. Solution: $\frac{y-1}{y} = Cx$, $y = 0$

18. $dy = e^{x+y}dx$. Solution: $-e^{-y} = e^x + C$

Section 1.2 Equations reducible to separable

Solve the following equations, transforming them into separable:

1. $y' = \cos(x - y - 1)$. Solution: $\cot\frac{x-y-1}{2} = C - x$, $y = x - 1 + 2k\pi, k \in \mathbb{Z}$

2. $y' + 1 = \frac{1}{(x+y)+(x+y)^2}$. Solution: $\frac{(x+y)^2}{2} + \frac{(x+y)^3}{3} = x + C$

3. $y' - y = 2x - 3$. Solution: $2x + y - 1 = Ce^x$

4. $y' = \sqrt{4x + 2y - 1}$.
   Solution: $\sqrt{4x + 2y - 1} - 2\ln(\sqrt{4x + 2y - 1} + 2) = x + C$

5. $y' = x + y + 1$. Solution: $\ln|x + y + 2| = x + C$, $x + y + 2 = 0$

6. $y' = \sqrt{y - x} + 1$. Solution: $y = x + \frac{(x+C)^2}{4}$, $y = x$

7. $(2x + y + 2)dx - (4x + 2y + 9)dy = 0$. Solution: $y + 2x + 4 = Ce^{x-2y}$

8. $(y - 3x + 2)dx + (3x - y - 1)dy = 0$. Solution: $2y - 6x + 1 = Ce^{2y-2x}$

Section 2.1 Homogeneous equations

Solve the following homogeneous equations:

1. $xy' = \sqrt{x^2 - y^2} + y$. Solution: $\arcsin\frac{y}{x} = \ln|Cx|$

2. $y = x(y' - \sqrt[x]{e^y})$. Solution: $-e^{-y/x} = \ln|Cx|$

3. $ydx + (2\sqrt{xy} - x)dy = 0$. Solution: $-\ln|\frac{y}{x}| - \sqrt{\frac{y}{x}} = \ln|Cx|$

4. $xy + y^2 = (2x^2 + xy)y'$. Solution: $2\ln|\frac{y}{x}| + \frac{y}{x} = \ln|\frac{C}{x}|$

5. $xy' + y(\ln\frac{y}{x} - 1) = 0$. Solution: $\ln\frac{y}{x} = \frac{C}{x}$

6. $(2x - y)dx + (x + y)dy = 0$. Solution: $\frac{1}{\sqrt{2}}\arctan\frac{y}{\sqrt{2}x} + \frac{1}{2}ln\frac{y^2+2x^2}{x^2} = \ln|\frac{C}{x}|$

7. $(4x^2 + 3xy + y^2)dx + (4y^2 + 3xy + x^2)dy = 0$.
   Solution: $\frac{1}{4}\ln|\frac{y+x}{x}| + \frac{3}{8}\ln\frac{y^2+x^2}{x^2} = \ln|\frac{C}{x}|$

8. $(2\sqrt{xy} - y)dx + xdy = 0$. Solution: $\frac{y}{x} = \ln^2|\frac{C}{x}|$

9. $y^2 + x^2y' = xyy'$. Solution: $\ln|Cy| = \frac{y}{x}$

10. $(y^2 - 2xy)dx + x^2dy = 0$. Solution: $\frac{y}{y-x} = Cx$

11. $2x^3y' = 2x^2y - y^3$. Solution: $\frac{x^2}{y^2} = \ln|x| + C$ e $y = 0$
12. $(9x^2 + y^2)y' = 2xy$. Solution: $\frac{xy}{y^2-x^2} = Cx$
13. $y' = \frac{y}{x} - e^{y/x}$. Solution: $y = -x\ln|\ln|Cx||$
14. $xy' - y = x\tan\frac{y}{x}$. Solution: $\sin\frac{y}{x} = Cx$

Section 2.2 Equations reducible to homogeneous

Solve the following equations, transforming them into homogeneous or separable:

1. $(x - y)dx + (2y - x + 1)dy = 0$. Solution: $\frac{x^2}{2} + y^2 - xy + y = C$
2. $(3y-7x+7)dx-(3x-7y-3)dy = 0$. Solution: $(x+y-1)^5(x-y-1)^2 = C$
3. $(y + 2)dx = (2x + y - 4)dy$. Solution: $(y + 2)^2 = C(x + y - 1)$, $y = 1 - x$
4. $(x - 2y - 1)dx + (3x - 6y + 2)dy = 0$. Solution: $x + 3y - \ln|x - 2y| = C$
5. $(x + y + 1)dx + (x - y + 3)dy = 0$. Solution: $x^2 + 2xy - y^2 + 2x + 6y = C$
6. $(2x - y - 2)dx + (x + y - 4)dy = 0$.
Solution: $y^2 + 2x^2 - 4y - 8x + 12 = Ce^{\sqrt{2}\arctan\frac{y-2}{\sqrt{2}(x-2)}}$
7. $(2x + y - 3)y' + y + 1 = 0$. Solution: $(y + 1)(x - 2)^3 = C(3x + y - 5)$
8. $(x + 4y)y' = 2x + 3y - 5$. Solution: $(y - x + 5)^5(x + 2y - 2) = C$

Section 3.1 Exact equations

Solve the following exact equations:
1. $(3x^2 - 2x - y)\,dx + (2y - x + 3y^2)dy = 0$. Solution: $x^3 - x^2 - xy + y^2 + y^3 = C$
2. $\left(\frac{y}{\sqrt{1-x^2y^2}} - 2x\right)dx + \frac{x}{\sqrt{1-x^2y^2}}dy = 0$. Solution: $\arcsin(xy) - x^2 = C$
3. $\left(\frac{x}{\sqrt{x^2-y^2}} - 1\right)dx - \frac{y}{\sqrt{x^2-y^2}}dy = 0$. Solution: $\sqrt{x^2 - y^2} - x = C$
4. $(1 + e^{x/y})\,dx + e^{x/y}\left(1 - \frac{x}{y}\right)dy = 0$. Solution: $x + ye^{x/y} = C$
5. $2xydx + (x^2 - y^2)dy = 0$. Solution: $3x^2y - y^3 = C$
6. $e^{-y}dx - (2y + xe^{-y})dy = 0$. Solution: $e^{-y}x - y^2 = C$
7. $\frac{3x^2+y^2}{y^2}dx - \frac{2x^3+5y}{y^3}dy = 0$. Solution: $\frac{x^3}{y^2} + x + \frac{5}{y} = C$
8. $(3x^2 + 6xy^2)dx + (6x^2y + 4y^3)dy = 0$. Solution: $x^3 + 3x^2y^2 + y^4 = C$
9. $\left(\frac{xy}{\sqrt{1+x^2}} + 2xy - \frac{y}{x}\right)dx + (\sqrt{1 + x^2} + x^2 - \ln x)dy = 0$.
Solution: $y\sqrt{1 + x^2} + x^2y - y\ln x = C$
10. $\left(\sin y + y\sin x + \frac{1}{x}\right)dx + \left(x\cos y - \cos x + \frac{1}{y}\right)dy = 0$.
Solution: $x\sin y - y\cos x + \ln|xy| = C$
11. $(3x^2 + y - 1)dx + (x + 3y^2 - 1)dy = 0$. Solution: $x^3 + xy + y^3 - x - y = C$

12. $(y + \sin x)dx + (x + \cos y)dy = 0$. Solution: $xy - \cos x + \sin y = C$

13. $(y^2 + \ln x)dx + (2xy - \ln y)dy = 0$. Solution: $xy^2 + x(\ln x - 1) - y(\ln y - 1) = C$

14. $(1 + 3x^2 \ln y)dx + (3y^2 + \frac{x^3}{y})dy = 0$. Solution: $x^3 \ln y + x + y^3 = C$

15. $\left(2x - \frac{\sin^2 y}{x^2}\right) dx + \left(2y + \frac{\sin 2y}{x}\right) dy = 0$. Solution: $x^2 + y^2 + \frac{\sin^2 y}{x} = C$

16. $\left(\frac{y}{x^2} + \frac{1}{y}\right) dx - \left(\frac{x}{y^2} + \frac{1}{x} + 2y\right) dy = 0$. Solution: $\frac{x}{y} - \frac{y}{x} - y^2 = C$

17. $\frac{y}{x}dx + (1 + \ln(xy)) dy = 0$. Solution: $y \ln(xy) = C$

18. $2x(1 + \sqrt{x^2 - y})dx - \sqrt{x^2 - y}dy = 0$. Solution: $x^2 + \frac{2}{3}(x^2 - y)^{3/2} = C$

Section 3.2 Equations reducible to exact equation. Integration factor

Find an integration factor, transform the given equation into exact one, and solve:

1. $\left(\frac{x}{y} + 1\right) dx + \left(\frac{x}{y} - 1\right) dy = 0$. Solution: $\mu = y$, $x^2 - y^2 + 2xy = C$

2. $\left(x^2 + y\right) dx - xdy = 0$. Solution: $\mu = \frac{1}{x^2}$, $x - \frac{y}{x} = C$

3. $(xy^2 + y)dx - xdy = 0$. Solution: $\mu = \frac{1}{y^2}$, $\frac{x^2}{2} + \frac{x}{y} = C$

4. $(x \cos y - y \sin y)dy + (x \sin y + y \cos y)dx = 0$.
Solution: $\mu = e^x$, $e^x(x \sin y - \sin y + y \cos y) = C$

5. $(x^2 + y^2 + x)dx + ydx = 0$. Solution: $\mu = e^{2x}$, $(y^2 + x^2)e^{2x} = C$

6. $ydy = (xdy + ydx)\sqrt{1 + y^2}$. Solution: $\mu = \frac{1}{1+y^2}$, $\sqrt{1 + y^2} - xy = C$

7. $y^2dx - (xy + x^3)dy = 0$. Solution: $\mu = \frac{1}{x^3}$, $y^2 = x^2(C - 2y)$

8. $y(x + y)dx + (xy + 1)dy = 0$. Solution: $\mu = \frac{1}{y}$, $\frac{x^2}{2} + xy + \ln|y| = C$

9. $(3x + 2y + y^2)dx + (x + 4xy + 5y^2)dy = 0$.
Solution: $\mu = x + y^2$, $(x + y)(x + y^2)^2 = C$

10. $2xy \ln ydx + (x^2 + y^2\sqrt{y^2 + 1})dy = 0$.
Solution: $\mu = \frac{1}{y}$, $x^2 \ln y + \frac{1}{3}(y^2 + 1)^{3/2} = C$

11. $(x + y^2)dx - 2xydy = 0$. Solution: $\mu = \frac{1}{x^2}$, $\ln|x| - \frac{y^2}{x} = C$

12. $\left(1 - \frac{x}{y}\right) dx + \left(2xy + \frac{x}{y} + \frac{x^2}{y^2}\right) dy = 0$.
Solution: $\mu = \frac{1}{x}$, $\ln|x| + \ln|y| + y^2 - \frac{x}{y} = C$

13. $\left(x^2 - \sin^2 y\right) dx + x \sin 2ydy = 0$. Solution: $\mu = \frac{1}{x^2}$, $x^2 + \sin^2 y = Cx$

14. $ydx - (x + x^2 + y^2)dy = 0$. Solution: $\mu = \frac{1}{x^2+y^2}$, $\arctan \frac{x}{y} - y = C$, $y = 0$

15. $(-xy \sin x + 2y \cos x)dx + 2x \cos xdy = 0$. Solution: $\mu = xy$, $x^2y^2 \cos x = C$

16. $(x^2 + 2xy - y^2)dx + (y^2 + 2xy - x^2)dy = 0$.
Solution: $\mu = \frac{1}{(x+y)^2}$, $x^2 + y^2 = C(x + y)$

17. $y(x + y + 1)dx + (x + 2y)dy = 0$. Solution: $\mu = e^x$, $(xy + y^2)e^x = C$

18. $xdx + (x^2y + 4y)dy = 0$, $y(1) = 0$.

Solution: $\mu = \frac{1}{4+x^2}$, $\ln(4+x^2) + y^2 = C$, $\ln(4+x^2) + y^2 = \ln 5$

Section 3.3 Formation of a differential

Solve the following equations by forming exact differentials:

1. $(y - 4xy^3)dx = (2x^2y^2 + x)dy$.

Solution: $\frac{ydx - xdy}{y^2} = 2(x^2dy + 2xydx)$, $d\left(\frac{x}{y}\right) = 2d(x^2y)$, $x = 2x^2y^2 + Cy$, $y = 0$

2. $(x + y^2)dx - 2xydy = 0$.

Solution: $\frac{dx}{x} - \frac{2xydy - y^2dx}{x^2} = 0$, $d\left(\ln x - \frac{y^2}{x}\right) = 0$, $x = Ce^{y^2/x}$

3. $(x^2 + y)dx - xdy = 0$.

Solution: $dx + \frac{ydx - xdy}{x^2} = 0$, $d\left(x - \frac{y}{x}\right) = 0$, $x - \frac{y}{x} = C$

4. $(x + y^2)dx - 2xydy = 0$. Solution: $\frac{1}{x}dx + \frac{y^2dx - 2xydy}{x^2} = 0$, $\ln|x| - \frac{y^2}{x} = C$

5. $(2x^2y + 2y + 5)dx + (2x^3 + 2x)dy = 0$.

Solution: $2y(x^2 + 1)dx + 5dx + 2x(x^2 + 1)dy = 0$, $\frac{5}{x^2 + 1}dx + 2(ydx + xdy) = 0$, $5\arctan x + 2xy = C$

6. $(2xy^2 - 3y^3)dx + (7 - 3xy^2)dy = 0$.

Solution: $(2x - 3y)dx + \frac{7}{y^2}dy - 3xdy = 0$, $2xdx + \frac{7}{y^2}dy - 3(ydx + xdy) = 0$, $x^2 - \frac{7}{y} - 3xy = C$, $y = 0$

7. $2xydx = (x^2 - 2y^3)dy$.

Solution: $\frac{2x}{y}dx - \frac{x^2}{y^2}dy + 2ydy = 0$, $\frac{2xydx - x^2dy}{y^2} + 2ydy = 0$, $\frac{x^2}{y} + y^2 = C$, $y = 0$

8. $(y - 3x^2y^3)dx - (x + x^3y^2)dy = 0$.

Solution: $\frac{ydx - xdy}{y^2} - (3x^2ydx + x^3dy) = 0$, $\frac{x}{y} - x^3y = C$, $y = 0$

9. $(2xy^2 + y)dx - (x^2y + 2x)dy = 0$.

Solution: $\frac{ydx - 2xdy}{y^3} + \frac{2xy^2dx - x^2ydy}{y^3} = 0$, $\frac{y^2dx - 2xydy}{y^4} + \frac{2xydx - x^2dy}{y^2} = 0$, $\frac{x}{y^2} + \frac{x^2}{y} = C$, $y = 0$

10. $(2xy^3 + y)dx - 2xdy = 0$.

Solution: $\frac{ydx - 2xdy}{y^3} + 2xdx = 0$, $\frac{y^2dx - 2xydy}{y^4} + 2xdx = 0$, $\frac{x}{y^2} + x^2 = C$, $y = 0$

11. $x^3dy + 2(y - x^2)ydx = 0$.

Solution: $\frac{x^3}{y^2}dy - 2\frac{x^2}{y}dx + 2dx = 0$, $\frac{x^2dy - 2xydx}{y^2} + \frac{2}{x}dx = 0$, $\frac{x^2}{y} - 2\ln|x| = C$, $y = 0$

12. $xdy = y(1 - ye^x)dx$. Solution: $\frac{xdy - ydx}{y^2} + e^xdx = 0$, $-\frac{x}{y} + e^x = C$, $y = 0$

Section 4.1 Linear equations

Solve the following linear equations (with respect to y or x):

1. $x^2y' = 2xy + 3$. Solution: $y = Cx^2 - \frac{1}{x}$

2. $ydx = (3x - y^2)dy$. Solution: $x = Cy^3 + y^2$, $y = 0$

3. $xy' + (x + 1)y = 3x^2e^{-x}$. Solution: $y = \frac{x^3 + C}{x}e^{-x}$

4. $xy' - 2y + x^2 = 0$. Solution: $y = (C - \ln|x|)x^2$

5. $y' - y = e^x$. Solution: $y = (x + C)e^x$

6. $y' + \frac{1}{x}y = 3x$. Solution: $y = \frac{1}{x}(C + x^3)$

7. $x\,dy + (x^2 - y)dx = 0$. Solution: $y = x(C - x)$

8. $2y\,dx + (y^2 - 2x)dy = 0$. Solution: $x = Cy - \frac{y^2}{2}$

9. $y' - y\sin x = \sin x\cos x$. Solution: $y = Ce^{-\cos x} - \cos x + 1$

10. $(1 + x^2)y' - 2xy = (1 + x^2)^2$. Solution: $y = (1 + x^2)(C + x)$

11. $(x - 2xy - y^2)y' + y^2 = 0$. Solution: $x = y^2(Ce^{1/y} + 1)$

12. $dx = (2x + e^y)dy$. Solution: $x = Ce^{2y} - e^y$

13. $y' + y\tan x = e^x\cos x$. Solution: $y = (C + e^x)\cos x$

14. $x(y - \sqrt{1 + x^2})dx + (1 + x^2)dy = 0$. Solution: $y = \frac{C + x^2}{2\sqrt{1 + x^2}}$

15. $(1 + y^2)dx + (xy - y^3)dy = 0$. Solution: $(3x - y^2 + 2)\sqrt{1 + y^2} = C$

16. $(\sin x - 1)y' + y\cos x = \sin x$. Solution: $y = \frac{C - \cos x}{\sin x - 1}$

Section 4.2 Bernoulli equation

Solve the following Bernoulli equations (with respect to y or x):

1. $y' - \frac{y}{x} = \frac{1}{2y}$. Solution: $y^2 = Cx^2 - x$

2. $(xy + x^2y^3)y' = 1$. Solution: $x = \frac{1}{Ce^{-y^2/2} - y^2 + 2}$

3. $y' + 2xy = 2x^3y^3$. Solution: $\frac{1}{y^2} = Ce^{2x^2} + x^2 + \frac{1}{2}$

4. $3y^2y' + y^3 + x = 0$. Solution: $y^3 = Ce^{-x} - x + 1$

5. $(1 + x^2)y' - 2xy = 4\sqrt{y(1 + x^2)}\arctan x$.
Solution: $\sqrt{y} = \sqrt{1 + x^2}(\arctan^2 x + C)$ e $y = 0$

6. $x^3\sin y\,y' + 2y = xy'$. Solution: $y = (C - \cos^2 y)x^2$, $y = 0$

7. $y' = y^4\cos x + y\tan x$. Solution: $y^{-3} = C\cos^3 x - 3\sin x\cos^2 x$, $y = 0$

8. $(x + 1)(y' + y^2) = -y$. Solution: $y(x + 1)(\ln|x + 1| + C) = 1$

9. $xy' - 2x^2\sqrt{y} = 4y$. Solution: $y = x^4\ln^2(Cx)$, $y = 0$

10. $2y' - \frac{x}{y} = \frac{xy}{x^2 - 1}$. Solution: $y^2 = x^2 - 1 + C\sqrt{|x^2 - 1|}$

11. $y'x^3\sin y = xy' - 2y$. Solution: $x^2(C - \cos y) = y$, $y = 0$

12. $(2x^2y\ln y - x)y' = y$. Solution: $xy(C - \ln^2 y) = 1$

13. $8y' + 3x^2y(y^2 - 4) = 0$. Solution: $y^2(e^{-x^3} + C) = 4C$

14. $(y^2 - 1)dx - y(x + (y^2 - 1)\sqrt{x})dy = 0$.
Solution: $3\sqrt{x} = y^2 - 1 + C\sqrt[4]{|y^2 - 1|}$, $y = 1$, $y = -1$

Exercises for Chapter 3

Solve the following explicit equations of the first order and the corresponding initial value problems; if possible, use different methods for the same equation and compare the obtained solutions:

1. $3e^x\sin y\,dx + (1 - e^x)\cos y\,dy = 0$. Solution: $\sin y = C(e^x - 1)^3$

2. $y' = \frac{y}{x} + e^{-y/x}$. Solution: $y = x \ln |\ln |Cx||$

3. $\frac{2x}{y^3} dx + \frac{y^2 - 3x^2}{y^4} dy = 0$, $y(1) = 1$. Solution: $\frac{x^2}{y^3} - \frac{1}{y} = C$; $y = x$

4. $(x + 2y)y' = 1$, $y(0) = -1$. Solution: $x + 2y + 2 = Ce^y$; $x + 2y + 2 = 0$

5. $xy' + y = \sin x$, $y(\frac{\pi}{2}) = \frac{2}{\pi}$. Solution: $xy = C - \cos x$; $xy = 1 - \cos x$

6. $(x^2 - \sin^2 y)dx + x \sin 2y dy = 0$. Solution: $\sin^2 y + x^2 = Cx$.

7. $(1 - x)(y' + y) = e^{-x}$, $y(0) = 0$.
Solution: $y = C - \ln|1 - x| \cdot e^{-x}$; $y = -\ln|1 - x| \cdot e^{-x}$

8. $y' = \frac{y+2}{x+1} + \tan \frac{y-2x}{x+1}$. Solution: $\sin \frac{y-2x}{x+1} = C(x+1)$

9. $xy' + y = y^2 \ln x$, $y(1) = 1$. Solution: $\frac{1}{y} = Cx + \ln x + 1$; $\frac{1}{y} = \ln x + 1$

10. $(x + y - 1)^2 y' = 2(y + 2)^2$. Solution: $y + 2 = Ce^{-2 \arctan \frac{y+2}{x-3}}$

11. $y' = 2x(x^2 + y)$, $y(0) = 0$. Solution: $y = Ce^{x^2} - x^2 - 1$; $y = e^{x^2} - x^2 - 1$

12. $xy^2 dx + (x^2 y - x)dy = 0$, $y(-1) = 1$.
Solution: $\mu = \frac{1}{xy}$, $xy + \ln|y| = C$; $xy + \ln|y| = -1$

13. $y' + 2y = e^x y^2$, $y(1) = 0$. Solution: $y(e^x + Ce^{2x}) = 1$, $y = 0$; $y = 0$

14. $(4x^2 - xy + y^2)dx + (x^2 - xy + 4y^2)dy = 0$. Solution: $(\frac{y}{x} + 1)(\frac{y^3}{x^3} + 1) = Cx^4$

15. $y(1 + xy)dx + (5y - x + y^2 \sin y)dy = 0$.
Solution: $\mu = \frac{1}{y^2}$, $\frac{x}{y} + \frac{x^2}{2} + 5 \ln|y| - \cos y = C$

16. $\cos y dx = (x + 2\cos y) \sin y dy$, $y(0) = \frac{\pi}{4}$.
Solution: $x = (C - \frac{1}{2} \cos 2y) \frac{1}{\cos y}$; $x = -\frac{\cos 2y}{2 \cos y}$

17. $(x+2)(1+y^2)dx + (x+1)y^2 dy = 0$. Solution: $x + y + \ln|x+1| - \arctan y = C$

18. $\left(\frac{1}{x} - \frac{y^2}{(x-y)^2}\right) dx + \left(\frac{x^2}{(x-y)^2} - \frac{1}{y}\right) dy = 0$. Solution: $\ln \frac{x}{y} + \frac{xy}{x-y} = C$

19. $(x^2 + y^2 - x)dx - ydy = 0$. Solution: $\mu = \frac{1}{x^2+y^2}$, $2x - \ln(x^2 + y^2) = C$

20. $ydx + (3x - xy + 2)dy = 0$. Solution: $xy^3 = 2y^2 + 4y + 4 + Ce^y$, $y = 0$

Chapter 4 Implicit equations of the first order: methods of solution

Section 1 Polynomial equations with respect to the derivative

Solve the following polynomial equations:

1. $y'^2 = y^3 - y^2$. Solution: $\sqrt{y - 1} = \pm \tan(\frac{x}{2} + C)$, $y = 0$, $y = 1$
2. $y'^2 - y^2 = 0$. Solution: $y = Ce^{\pm x}$
3. $(y' + 1)^3 = 27(x + y)^2$. Solution: $y + x = (x + C)^3$, $y = -x$
4. $y'^2 = 4y^3(1 - y)$. Solution: $y(1 + (x + C)^2) = 1$, $y = 0$, $y = 1$
5. $y'^3 + y^2 = yy'(y' + 1)$. Solution: $4y = (x + A)^2$, $y = Be^x$
6. $y'^2 + xy = y^2 + xy'$. Solution: $y = Ae^x$, $y = Be^{-x} + x - 1$
7. $y'^3 + (x + 2)e^y = 0$. Solution: $(x + 2)^{4/3} + C = 4e^{-y/3}$
8. $y'(2y - y') = y^2 \sin^2 x$. Solution: $\ln(Cy) = x \pm \sin x$

9. $y(xy' - y)^2 = y - 2xy'$. Solution: $(Cx + 1)^2 = 1 - y^2$, $y = \pm 1$

10. $yy'(yy' - 2x) = x^2 - 2y^2$. Solution: $2(x - C)^2 + 2y^2 = C^2$, $y = \pm x$

11. $y'^2 + 4xy' - y^2 - 2x^2y = x^4 - 4x^2$. Solution: $y = Ce^{\pm x} - x^2$

12. $xy'(xy' + y) = 2y^2$. Solution: $x^2y = A$, $y = Bx$

Section 2.1 Explicit equations in y in general form

Solve the following explicit with respect to y equations:

1. $y = y'^2 + 2y'^3$. Solution: $\begin{cases} x = 3p^2 + 2p + C \\ y = 2p^3 + p^2 \end{cases}$, $y = 0$

2. $y = \ln(1 + y'^2)$. Solution: $\begin{cases} x = 2\arctan p + C \\ y = \ln(1 + p^2) \end{cases}$, $y = 0$

3. $y = (y' - 1)e^{y'}$. Solution: $\begin{cases} x = e^p + C \\ y = (p - 1)e^p \end{cases}$, $y = -1$

4. $y'^4 - y'^2 = y^2$. Solution: $\begin{cases} x = \pm(2\sqrt{p^2 - 1} + \arcsin \frac{1}{|p|} + C \\ y = \pm p\sqrt{p^2 - 1} \end{cases}$, $y = 0$

5. $5y + y'^2 = x(x + y')$. Solution: $\begin{cases} x = -\frac{p}{2} + C \\ 5y = C^2 - \frac{5p^2}{4} \end{cases}$, $4y = x^2$

6. $y'^3 + y^2 = xyy'$. Solution: $\begin{cases} pxy = y^2 + p^3 \\ y^2(2p + C) = p^4 \end{cases}$, $y = 0$

7. $y' \sin y' + \cos y' - y = 0$. Solution: $\begin{cases} x = \sin p + C \\ y = p\sin p + \cos p \end{cases}$, $y = 1$

8. $x^4y'^2 - xy' - y = 0$. Solution: $y = C^2 - \frac{C}{x}$, $y = -\frac{1}{4x^2}$

9. $y = xy' - x^2y'^3$. Solution: $\begin{cases} xp^2 = C\sqrt{|p|} - 1 \\ y = xp - x^2p^3 \end{cases}$, $y = 0$

10. $y = 2xy' + y^2y'^3$. Solution: $\begin{cases} 2xp^2 = C - C^2p^2 \\ py = C \end{cases}$, $27y^4 = -32x^3$, $y = 0$

Section 2.2 Lagrange and Clairaut equations

Solve the following Lagrange and Clairaut equations:

1. $2yy' = x(y'^2 + 4)$. Solution: $y = Cx^2 + \frac{1}{C}$

2. $y = -xy' + y'^2$. Solution: $\begin{cases} x = \frac{C}{\sqrt{p}} + \frac{2p}{3} \\ y = -xp + p^2 \end{cases}$.

3. $y = -xy' - a\sqrt{1 + y'^2}$, $a \in \mathbb{R}$.
Solution: $y = Cx - a\sqrt{1 + C^2}$, $x^2 + y^2 = a^2$ $(ay < 0)$

4. $y'^3 + xy'^2 - y = 0$. Solution: $\begin{cases} x = -\frac{1}{2} - p + \frac{C}{(p-1)^2} \\ y = -\frac{p^2}{2} + \frac{Cp^2}{(p-1)^2} \end{cases}$; $y = 0$, $y = x + 1$

5. $y'^2 - 2xy' + y = 0$. Solution: $\begin{cases} x = \frac{2}{3}p + \frac{C}{p^2} \\ y = 2px - p^2 \end{cases}$, $y = 0$

6. $y + xy' = 4\sqrt{y'}$. Solution: $\begin{cases} x\sqrt{p} = \ln p + C \\ y = \sqrt{p}(4 - \ln p - C) \end{cases}$, $y = 0$

7. $xy' - y = \ln y'$. Solution: $y = Cx - \ln C$, $y = \ln x + 1$

8. $xy'(y' + 2) = y$. Solution: $y = \pm 2\sqrt{Cx} + C$, $y = -x$

9. $2y'^2(y - xy') = 1$. Solution: $2C^2(y - Cx) = 1$, $8y^3 = 27x^2$

10. $y = xy' - (2 + y')$. Solution: $y = Cx - C - 2$

Section 3 Explicit equations in x

Solve the following explicit with respect to x equations:

1. $3y'^3 - xy' + 1 = 0$. Solution: $\begin{cases} x = 3p^2 + p^{-1} \\ y = 2p^3 - \ln|p| + C \end{cases}$

2. $x = \frac{y}{2y'} + e^{yy'}$. Solution: $x = \frac{y^2}{2C} + e^C$, $\begin{cases} x = -\frac{1}{2p^2}\ln(2p^2) + \frac{1}{2p^2} \\ y = -\frac{1}{p}\ln(2p^2) \end{cases}$

3. $y'^2 - 4xyy' + 8y^2 = 0$. Solution: $y = C(x - C)^2$, $27y = 4x^3$, $y = 0$

4. $x = \frac{y}{y'}\ln y - \frac{y'^2}{y^2}$. Solution: $x = \frac{1}{C}\ln y - C^2$, $x = -3\left(\frac{\ln y}{2}\right)^{2/3}$

5. $e^{y'} + y' = x$. Solution: $\begin{cases} x = e^p + p \\ y = e^p(p - 1) + \frac{p^2}{2} + C \end{cases}$

6. $xy'^3 = 1 + y'$. Solution: $\begin{cases} x = p^3 + p^2 \\ y = \frac{3}{2}p^2 + 2p + C \end{cases}$

7. $y'^3 - y' = x + 1$. Solution: $\begin{cases} x = p^3 - p - 1 \\ y = \frac{3}{4}p^4 - \frac{p^2}{2} + C \end{cases}$

8. $x = y'\sqrt{y'^2 + 1}$. Solution: $\begin{cases} x = p\sqrt{p^2 + 1} \\ y = \frac{2}{3}(\sqrt{p^2 + 1})^3 - \sqrt{p^2 + 1} + C \end{cases}$

Exercises for Chapter 4

Solve the following implicit equations of the first order and the corresponding initial values problems; if possible, use different methods for the same equation and compare the obtained solutions:

1. $y = xy'^2 - 2y'^3$. Solution: $\begin{cases} x = 2p + 1 + \frac{C}{(p-1)^2} \\ y = p^2 + \frac{Cp^2}{(p-1)^2} \end{cases}$; $y = 0$, $y = x - 2$

2. $y'^2 - 2yy' = y^2(e^x - 1)$, $y(0) = 1$.
Solution: $\ln(Cy) = x \pm 2e^{x/2}$; $\ln y + 2 = x + 2e^{x/2}$, $\ln y - 2 = x - 2e^{x/2}$

3. $y'^3 - xy^4y' - y^5 = 0$. Solution: $x = C^2 - \frac{1}{Cy}$, $x = (\sqrt[3]{2} + \frac{1}{\sqrt[3]{4}})\frac{1}{y^{2/3}}$

4. $y'^2 - 2xy' = x^2 - 4y$, $y(1) = 0$. Solution: $4y = C^2 - 2(x - C)^2$, $2y = x^2$;
$4y = (2 + \sqrt{2})^2 - 2(x - 2 - \sqrt{2})^2$, $4y = (2 - \sqrt{2})^2 - 2(x - 2 + \sqrt{2})^2$

5. $y'^3 = 3(xy' - y)$. Solution: $C^3 = 3(Cx - y)$, $9y^2 = 4x^3$

6. $x = y' \sin y'$. Solution: $\begin{cases} x = p \sin p \\ y = (p^2 - 1) \sin p + p \cos p + C \end{cases}$

7. $y = (2 + y')\sqrt{1 - y'}$. Solution: $y = x + C - \frac{1}{27}(x + C)^3$

8. $y'^2 - (y + x^2)y' + x^2 y = 0$, $y(0) = 0$.
Solution: $y = Ae^x$, $y = \frac{x^3}{3} + B$; $y = 0$, $y = \frac{x^3}{3}$

9. $y'^4 = 2yy' + y^2$. Solution: $\begin{cases} x = \pm 2\sqrt{p^2 + 1} - \ln(\sqrt{p^2 + 1} \pm 1) + C \\ y = -p \pm p\sqrt{p^2 + 1} \end{cases}$, $y = 0$

10. $y = xy'^2 - 2y'^3$, $y(1) = 0$.
Solution: $\begin{cases} x = C(p - 1)^{-2} + 2p + 1 \\ y = Cp^2(p - 1)^{-2} + p^2 \end{cases}$, $y = x - 2$, $y = 0$;
$\begin{cases} x = -\frac{1}{4}(p - 1)^{-2} + 2p + 1 \\ y = -\frac{1}{4}p^2(p - 1)^{-2} + p^2 \end{cases}$, $y = \frac{(x-1)^2}{4}$, $y = 0$

11. $y'^2 + 2yy' \cot x - y^2 = 0$. Solution: $y = \frac{A}{\cos^2 \frac{x}{2}}$, $y = \frac{B}{\sin^2 \frac{x}{2}}$

12. $(3x + 5)y'^2 - (3y + x)y' + y = 0$. Solution: $(x - 3y)^2 = 20y$

Chapter 7 Higher order equations: methods of reduction of order

Section 1 Equation $y^{(n)} = f(x)$

Solve the following initial value problems:

1. $y'' = x + \sin x$, $y(0) = -3, y'(0) = 0$.
Solution: $y = \frac{x^3}{6} - \sin x + Ax + B$; $y = \frac{x^3}{6} - \sin x + x - 3$

2. $y'' = \frac{x}{e^{2x}}$, $y(0) = \frac{1}{4}, y'(0) = -\frac{1}{4}$.
Solution: $y = \frac{x+1}{4}e^{-2x} + Ax + B$, $y = \frac{x+1}{4}e^{-2x}$

3. $y''' = \frac{\ln x}{x^2}$, $y(1) = 0, y'(1) = 1, y''(1) = 2$.
Solution: $y = -\frac{x}{2}\ln^2 x + A\frac{x^2}{2} + Bx + C$, $y = -\frac{x}{2}\ln^2 x + \frac{3x^2}{2} - 2x + \frac{1}{2}$

Section 2 Equation $F(x, y^{(k)}, \ldots, y^{(n)}) = 0$, $k > 0$ (equation without y and lower derivatives)

Solve the following equations:

1. $y'' + 2xy'^2 = 0$. Solution: $y = \frac{1}{C}\arctan\frac{x}{C} + B$

2. $x^2 y'' + xy' = 1$. Solution: $y = \frac{1}{2}\ln^2 |x| + C\ln |x| + B$

3. $x^2 y'' = y'^2$. Solution: $Ax - A^2 y = \ln |Ax + 1| + B$, $2y = x^2 + C$, $y = C$

4. $y''(e^x + 1) + y' = 0$. Solution: $y = A(x - e^{-x}) + B$

5. $y''' = 2(y'' - 1)\cot x$. Solution: $2y = A\cos 2x + (1 + 2A)x^2 + Bx + C$

6. $y''' = \sqrt{1 + y''^2}$. Solution: $y = \frac{e^{x+A} - e^{-(x+A)}}{2} + Bx + C$

7. $yy'' = y'^2 - y'^3$. Solution: $y + A \ln |y| = x + B$, $y = C$
8. $xy''' = y'' - xy''$. Solution: $y = A(x+2)e^{-x} + Bx + C$

Section 3 Equation $F(y, y', \ldots, y^{(n)}) = 0$ (equation without x)

Solve the following equations and initial value problems:

1. $2y'^2 = (y-1)y''$, $y(0) = 2, y'(0) = 2$.
Solution: $-\frac{1}{y_g - 1} = Cx + B$, $-\frac{1}{y-1} = 2x - 1$
2. $y'' + y'^2 = 2e^{-y}$. Solution: $e^y + A = (x+B)^2$
3. $y'' = 2y^3$, $y(0) = 1, y'(0) = 1$.
Solution: $x + A = \int (y^4 + B)^{-1/2} dy$, $y = \frac{1}{1-x}$
4. $2yy'' = y^2 + y'^2$. Solution: $y = A(1 \pm \frac{e^{x+B} + e^{-(x+B)}}{2})$, $y = Ce^{\pm x}$
5. $y^4 - y^3 y'' = 1$. Solution: $\ln |y^2 + A \pm \sqrt{y^4 + 2Ay^2 + 1}| = 2x + B$, $y = \pm 1$
6. $yy'' = 1 + y'^2$. Solution: $y = A\frac{1}{2}(e^{(x+B)/A} + e^{-(x+B)/A})$
7. $y'' = e^{2y}$, $y(0) = 0, y'(0) = 1$.
Solution: $x = -\frac{1}{\sqrt{A}} \ln \left| \frac{2 \pm \sqrt{1 + Ae^{-2y}}}{\sqrt{1 + Ae^{-2y}}} \right| + B$, $y = -\ln |C \pm x|$; $y = -\ln |x - 1|$
8. $2yy'' - 3y'^2 = 4y^2$. Solution: $y \cos^2(x + A) = B$

Chapter 9 Linear equations with constant coefficients: methods of solution

Section 1 Homogeneous equations

Solve the following homogeneous equations:

1. $y'' + y' - 2y = 0$. Solution: $\lambda = 1, -2$, $y = C_1 e^x + C_2 e^{-2x}$
2. $y'' + 4y' + 3y = 0$. Solution: $\lambda = -1, -3$, $y = C_1 e^{-x} + C_2 e^{-3x}$
3. $y'' - 2y' = 0$. Solution: $\lambda = 0, 2$, $y = C_1 + C_2 e^{2x}$
4. $2y'' - 5y' + 2y = 0$. Solution: $\lambda = 2, 1/2$, $y = C_1 e^{2x} + C_2 e^{x/2}$
5. $y'' - 3y' + 10y = 0$. Solution: $\lambda_{1,2} = -2; 5$, $y = C_1 e^{-2x} + C_2 e^{5x}$
6. $y'' - 2y' + y = 0$. Solution: $\lambda = 1, 1$, $y = e^x(C_1 + C_2 x)$
7. $4y'' + 4y' + y = 0$. Solution: $\lambda = -1/2, -1/2$, $y = e^{-x/2}(C_1 + C_2 x)$
8. $y^{(5)} - 6y^{(4)} + 9y''' = 0$.
Solution: $\lambda = 0, 0, 0, 3, 3$, $y = C_1 + C_2 x + C_3 x^2 + e^{3x}(C_4 + C_5 x)$
9. $y''' - 3y'' + 3y' - y = 0$. Solution: $\lambda = 1, 1, 1$, $y = e^x(C_1 + C_2 x + C_3 x^2)$
10. $y''' - 3y' + 2y = 0$. Solution: $\lambda = -2, 1, 1$, $y = C_1 e^{-2x} + e^x(C_2 + C_3 x)$
11. $y'' - 4y' + 5y = 0$. Solution: $\lambda_{1,2} = 2 \pm i$, $y = e^{2x}(C_1 \cos x + C_2 \sin x)$
12. $y'' + 2y' + 10y = 0$.
Solution: $\lambda_{1,2} = -1 \pm 3i$, $y = e^{-x}(C_1 \cos 3x + C_2 \sin 3x)$
13. $y'' + 4y = 0$. Solution: $\lambda_{1,2} = \pm 2i$, $y = C_1 \cos 2x + C_2 \sin 2x$
14. $y''' - 8y = 0$.

Solution: $\lambda = 2, \lambda_{1,2} = -1 \pm \sqrt{3}i$, $y = C_1 e^{2x} + e^{-x}(C_2 \cos \sqrt{3}x + C_3 \sin \sqrt{3}x)$

15. $y^{(4)} - y = 0$.

Solution: $\lambda = -1, 1, \lambda_{1,2} = \pm i$, $y = C_1 e^{-x} + C_2 e^x + C_3 \cos x + C_4 \sin x$

16. $y^{(4)} + 4y = 0$.

Solution: $\lambda_{1,2} = 1 \pm i, \lambda_{3,4} = -1 \pm i$,

$y = e^x(C_1 \cos x + C_2 \sin x) + e^{-x}(C_3 \cos x + C_4 \sin x)$

17. $y'' + 4y' + 20y = 0$.

Solution: $\lambda_{1,2} = -2 \pm 4i$, $y = e^{-2x}(C_1 \cos 4x + C_2 \sin 4x)$

18. $y''' - y'' + 4y' - 4y = 0$.

Solution: $\lambda_1 = 1, \lambda_{2,3} = \pm 2i$, $y = C_1 e^x + C_2 \cos 2x + C_3 \sin 2x$

19. $y^{(4)} + 2y'' + y = 0$.

Solution: $\lambda = -i, -i, +i, +i$, $y = (C_1 + C_2 x) \cos x + (C_3 + C_4 x) \sin x$

20. $y^{(5)} + 8y''' + 16y' = 0$.

Solution: $\lambda = 0, -2i, -2i, +2i, +2i$,

$y = C_1 + (C_2 + C_3 x) \cos 2x + (C_4 + C_5 x) \sin 2x$

Section 2 Nonhomogeneous equations: method of undetermined coefficients

Solve the following nonhomogeneous equations by the method of undetermined coefficients:

1. $y'' - 5y' - 6y = 3 \cos x + 19 \sin x$.
Solution: $y = C_1 e^{-x} + C_2 e^{6x} + \cos x - 2 \sin x$

2. $y'' + 2y' - 3y = (12x^2 + 6x - 4)e^x$. Solution: $y = C_1 e^{-3x} + C_2 e^x + (x^3 - x)e^x$

3. $y'' - 9y' + 20y = 126e^{-2x}$. Solution: $y = C_1 e^{4x} + C_2 e^{5x} + 3e^{-2x}$

4. $y'' + 10y' + 25y = 40 + 52x - 240x^2 - 200x^3$.
Solution: $y = (C_1 + C_2 x)e^{-5x} - 8x^3 + 4x$

5. $y'' + 36y = 36 + 66x - 36x^3$. Solution: $y = C_1 \cos 6x + C_2 \sin 6x - x^3 + 2x + 1$

6. $y'' + 4y' + 20y = -4 \cos 4x - 52 \sin 4x$.
Solution: $y = (C_1 \cos 4x + C_2 \sin 4x)e^{-2x} + 3 \cos 4x - \sin 4x$

7. $y'' + 5y' = 39 \cos 3x - 105 \sin 3x$.
Solution: $y = C_1 + C_2 e^{-5x} + 4 \cos 3x + 5 \sin 3x$

8. $y'' + 6y' + 9y = 72e^{3x}$. Solution: $y = (C_1 + C_2 x)e^{-3x} + 2e^{3x}$

9. $y'' + y' - 2y = 9 \cos x - 7 \sin x$. Solution: $y = C_1 e^{-2x} + C_2 e^x - 2 \cos x + 3 \sin x$

10. $y'' + 2y' + y = 6e^{-x}$. Solution: $y = (C_1 + C_2 x)e^{-x} + 3x^2 e^{-x}$

11. $4y'' - 4y' + y = -25 \cos x$. Solution: $y = (C_1 + C_2 x)e^{x/2} + 3 \cos x + 4 \sin x$

12. $y'' - 2y' - 8y = 12 \sin 2x - 36 \cos 2x$.
Solution: $y = C_1 e^{-2x} + C_2 e^{4x} + 3 \cos 2x$

13. $y'' - 7y' + 12y = 3e^{4x}$. Solution: $y = C_1 e^{3x} + C_2 e^{4x} + 3xe^{4x}$

14. $y'' - 6y' + 10y = 51e^{-x}$. Solution: $y = (C_1 \cos x + C_2 \sin x)e^{3x} + 3e^{-x}$

15. $y'' - 2y' = (4x + 4)e^{2x}$. Solution: $y = C_1 + C_2 e^{2x} + (x^2 + x)e^{2x}$

16. $y'' - 3y' + 2y = (34 - 12x)e^{-x}$. Solution: $y = C_1 e^x + C_2 e^{2x} + (4 - 2x)e^{-x}$

17. $y'' - 6y' + 34y = 18\cos 5x + 60\sin 5x$.
Solution: $y = (C_1\cos 5x + C_2\sin 5x)e^{3x} + 2\cos 5x$
18. $y'' + 6y' + 10y = 74e^{3x}$. Solution: $y = (C_1\cos x + C_2\sin x)e^{-3x} + 2e^{3x}$
19. $y'' - 4y' = 8 - 16x$. Solution: $y = C_1 + C_2e^{4x} + 2x^2 - x$
20. $y'' - 2y' + 3y = x^3 + \sin x$.
Solution: $y = (C_1\cos\sqrt{2}x + C_2\sin\sqrt{2}x)e^x + \frac{1}{27}(9x^3 + 18x^2 + 6x - 8) + \frac{1}{4}(\sin x + \cos x)$
21. $y''' + 2y'' - y' - 2y = e^x + x^2$.
Solution: $y = C_1e^x + C_2e^{-x} + C_3e^{-2x} - \frac{x^2}{2} + \frac{x}{2} - \frac{5}{4} + \frac{1}{6}xe^x$
22. $y'' - 4y' + 4y = (x^3 + x)e^{2x}$. Solution: $y = C_1e^{2x} + C_2xe^x + (\frac{x^5}{20} + \frac{x^3}{6})e^{2x}$
23. $y'' + 4y = x^2\sin 2x$.
Solution: $y = C_1\cos 2x + C_2\sin 2x - \frac{x^3}{12}\cos 2x + \frac{x}{32}\cos 2x + \frac{x^2}{16}\sin 2x$
24. $y'' + 2y' + 2y = x^2 + \sin x$.
Solution: $y = (C_1\cos x + C_2\sin x)e^{-x} + \frac{1}{2}(x-1)^2 + \frac{1}{5}(\sin x - 2\cos x)$
25. $y'' - 9y = x + e^{2x} - \sin 2x$.
Solution: $y = C_1e^{3x} + C_2e^{-3x} - \frac{x}{9} - \frac{1}{5}e^{2x} + \frac{1}{13}\sin 2x$
26. $y''' + 3y'' + 2y' = x^2 + 4x + 8$.
Solution: $y = C_1 + C_2e^{-x} + C_3e^{-2x} + \frac{x^3}{6} + \frac{x^2}{4} + \frac{11x}{4}$
27. $y'' + y = -2\sin x + 4x\cos x$.
Solution: $y = C_1\cos x + C_2\sin x + 2x\cos x + x^2\sin x$
28. $y''' - y'' - 4y' + 4y = 2x^2 - 4x - 1 + (2x^2 + 5x + 1)e^{2x}$.
Solution: $y = C_1e^x + C_2e^{2x} + C_3e^{-2x} + \frac{x^2}{2} + \frac{x^3}{6}e^{2x}$

Section 3 Nonhomogeneous equations: method of variation of parameters (Lagrange method))

Solve the following nonhomogeneous equations by the Lagrange method:
1. $y'' - 2y' + y = \frac{e^x}{x}$. Solution: $y = (-x + C_1)e^x + (\ln|x| + C_2)xe^x$
2. $y'' + 4y = \tan 2x$.
Solution: $y = C_1\cos 2x + C_2\sin 2x - \frac{1}{4}\cos 2x \cdot \ln\left|\tan\left(x + \frac{\pi}{4}\right)\right|$
3. $y'' + 9y = \frac{1}{\sin 3x}$.
Solution: $y = C_1\cos 3x + C_2\sin 3x - \frac{1}{3}x\cos 3x + \frac{1}{9}\sin 3x\ln|\sin 3x|$
4. $y'' - 2y' + 2y = \frac{e^x}{\sin^2 x}$.
Solution: $y = \ln\left|\cot\frac{x}{2}\right| \cdot e^x\cos x - e^x + (C_1\cos x + C_2\sin x)e^x$
5. $y'' + 3y' + 2y = \frac{e^{-x}}{e^x + 2}$.
Solution: $y = C_1e^{-x} + C_2e^{-2x} + \frac{1}{2}xe^{-x} - \frac{1}{2}e^{-x}\ln(e^x + 2) - e^{-2x}\ln(e^x + 2)$
6. $y'' + y' = e^{2x}\cos e^x$. Solution: $y = C_1 + C_2e^{-x} + 2e^{-x}\sin e^x - \cos e^x$
7. $y'' + 4y' = \frac{1}{\sin^2 x}$.
Solution: $y = C_1\cos 2x + C_2\sin 2x - \cos 2x\ln\sin x + \sin 2x(\frac{1}{2}\cot x + x)$
8. $y'' - 2y' + 5y = 3e^x + e^x\tan 2x$.

Solution: $y = e^x(C_1 \cos 2x + C_2 \sin 2x) + \frac{1}{4}e^x \left(3 + \ln \left|\tan\left(\frac{\pi}{4} - x\right)\right| \cdot \cos 2x\right)$

9. $y'' + 4y' + 4y = e^{-2x}\ln x$.
Solution: $y = (C_1 + C_2 x)e^{-2x} + \left(\frac{1}{2}x^2 \ln x - \frac{3}{4}x^2\right)e^{-2x}$

10. $y'' - 2y' + y = \frac{e^x}{x^2+1}$.
Solution: $y = (C_1 + C_2 x)e^x + (x \arctan x - \ln\sqrt{x^2+1})e^x$

11. $y''' + y' = \frac{\sin x}{\cos^2 x}$.
Solution: $y = C_1 + C_2 \cos x + C_3 \sin x + \frac{1}{\cos x} + \cos x \ln |\cos x| + \sin x(x - \tan x)$

12. $y'' + 2y' + 2y = \frac{e^x}{\cos x}$.
Solution: $y = (C_1 \cos x + C_2 \sin x)e^{-x} + e^{-x}\cos x \ln |\cos x| + xe^{-x}\sin x$

13. $y'' - 2y' + y = \frac{e^x}{x^2}$. Solution: $y = (C_1 + C_2 x)e^x - \ln x \cdot e^x - e^x$

14. $y'' + 4y' + 4y = \frac{e^{-2x}}{x^3}$. Solution: $y = (C_1 + C_2 x)e^{-2x} + \frac{1}{2x}e^{-2x}$

15. $y'' + y = \frac{1}{\sin^2 x}$. Solution: $y = C_1 \cos x + C_2 \sin x + 1 + \frac{1}{2}\ln \left|\frac{1+\cos x}{1-\cos x}\right| - 1$

16. $y'' - 3y' + 2y = \frac{1}{1+e^x}$.
Solution: $y = C_1 e^x + C_2 e^{2x} + (e^x + e^{2x})(x - \ln(1 + e^x)) + e^x + \frac{1}{2}$

17. $y'' - y = \frac{e^x - e^{-x}}{e^x + e^{-x}}$. Solution: $y = C_1 e^{-x} + C_2 e^x + 2\arctan e^x \cdot \cosh x - 1$

18. $y'' - 2y' = 5(3 - 4x)\sqrt{x}$. Solution: $y = C_1 + C_2 e^{2x} + 4x^2\sqrt{x}$

19. $y'' - 2y' + 10y = \frac{9e^x}{\cos 3x}$.
Solution: $y = e^x(C_1 \cos 3x + C_2 \sin 3x) + e^x(\ln |\cos 3x| \cdot \cos 3x + 3x \sin 3x)$

20. $y'' - 4y' + 8y = 4(7 - 21x + 18x^2)\sqrt[3]{x}$.
Solution: $y = e^{2x}(C_1 \cos 2x + C_2 \sin 2x) + 9x^2\sqrt[3]{x}$

21. $y'' + y = -\cot^2 x$. Solution: $y = C_1 \cos x + C_2 \sin x - \frac{1}{2}\ln \left|\frac{1+\cos x}{1-\cos x}\right| \cdot \cos x + 2$

22. $y'' + 3y' = \frac{3x-1}{x^2}$. Solution: $y = C_1 e^{-3x} + C_2 + \ln |x|$

23. $y'' - 4y' + 4y = \frac{2e^{2x}}{1+x^2}$.
Solution: $y = e^{2x}(C_1 + C_2 x) + e^{2x}(2x \arctan x - \ln(1 + x^2))$

24. $y'' + y' = 7(4 + 3x)\sqrt[3]{x}$. Solution: $y = C_1 e^{-x} + C_2 + 9x^2\sqrt[3]{x}$

25. $y'' + 2y' + 2y = \frac{e^{-x}}{\sin x}$.
Solution: $y = e^{-x}(C_1 \cos x + C_2 \sin x) + e^{-x}(\ln |\sin x| \cdot \sin x - x \cos x)$

26. $y'' + 2y = 2 - 4x^2 \sin x^2$. Solution: $y = C_1 \cos(\sqrt{2}x) + C_2 \sin(\sqrt{2}x) + \sin x^2$

27. $y'' + 2y' + y = (x + 2)(\ln x + \frac{1}{x})$. Solution: $y = e^{-x}(C_1 + C_2 x) + x \ln x$

28. $y'' - y' = \frac{x+1}{x^2}$. Solution: $y = C_1 + C_2 e^x + \ln |x|$

Exercises for sections 2 and 3

Solve the following nonhomogeneous equations and the corresponding initial value problems; if possible, use both the method of undetermined coefficients and variation of parameters and compare the obtained solutions:

1. $y'' + 2y' + y = \frac{e^{-x}}{x}$. Solution: $y = (-x + C_1)e^{-x} + (\ln |x| + C_2)xe^{-x}$

2. $y'' + 4y = \frac{1}{\sin 2x}$. Solution: $y = C_1 \cos 2x + C_2 \sin 2x - \frac{x}{2}\cos 2x + \frac{1}{4}\ln |\sin 2x|$

3. $y'' + 10y' + 34y = -9e^{-5x}$, $y(0) = 0, y'(0) = 6$.

Solution: $y = (C_1\cos 3x + C_2\sin 3x)e^{-5x} - e^{-5x}$; $y = (\cos 3x + 2\sin 3x)e^{-5x} - e^{-5x}$

4. $y'' + 2y' + 5y = -8e^{-x}\sin 2x$, $y(0) = 2, y'(0) = 6$.
Solution: $y = (C_1\cos 2x + C_2\sin 2x)e^{-x} + 2xe^{-x}\cos 2x$; $y = (2\cos 2x + 3\sin 2x)e^{-x} + 2xe^{-x}\cos 2x$

5. $y'' - y' = e^{2x}\sin e^x$. Solution: $y = C_1 + C_2 e^x - e^x\sin e^x$

6. $y'' - y = \frac{e^x}{e^x+1}$.
Solution: $y = C_1 e^{-x} + C_2 e^x - \frac{1}{2} + \frac{1}{2}e^{-x}\ln(e^x+1) + \frac{1}{2}e^x\ln\frac{e^x}{e^x+1}$

7. $y'' - 10y' + 25y = e^{5x}$, $y(0) = 1, y'(0) = 0$.
Solution: $y = (C_1 + C_2 x)e^{5x} + \frac{x^2}{2}e^{5x}$; $y = (1 - 5x)e^{5x} + \frac{x^2}{2}e^{5x}$

8. $y'' - 2y' + 37y = 36e^x\cos 6x$, $y(0) = 0, y'(0) = 6$.
Solution: $y = (C_1\cos 6x + C_2\sin 6x)e^x + 3x\sin 6xe^x$; $y = \sin 6xe^x + 3x\sin 6xe^x$

9. $y'' - 2y' + 2y = \frac{e^x}{\sin^2 x}$.
Solution: $y = (C_1\cos x + C_2\sin x)e^x + e^x\cos x\ln\cot\frac{x}{2} + e^x$

10. $y'' + 3y' = (40x + 58)e^{2x}$, $y(0) = 0, y'(0) = -2$.
Solution: $y = C_1 + C_2 e^{-3x} + (4x + 3)e^{2x}$; $y = -7 + 4e^{-3x} + (4x + 3)e^{2x}$

11. $y'' + 2y' = 6x^2 + 2x + 1$, $y(0) = 2, y'(0) = 2$.
Solution: $y = C_1 + C_2 e^{-2x} + x^3 - x^2 + \frac{3}{2}x$; $y = \frac{9}{4} - \frac{1}{4}e^{-2x} + x^3 - x^2 + \frac{3}{2}x$

12. $y'' - 3y' + 2y = \frac{e^x}{1+e^x}$.
Solution: $y = C_1 e^x + C_2 e^{2x} + (e^x - e^{2x})(x - \ln(1 + e^x)) - e^x$

13. $y'' + 2y' + 2y = 2x^2 + 8x + 6$, $y(0) = 1, y'(0) = 4$.
Solution: $y = (C_1\cos x + C_2\sin x)e^{-x} + x^2 + 2x$; $y = (\cos x + 3\sin x)e^{-x} + x^2 + 2x$

14. $y'' - 4y = (15 - 16x^2)\sqrt{x}$. Solution: $y = C_1 e^{-2x} + C_2 e^{2x} + 4x^2\sqrt{x}$

15. $y'' + 4y' + 4y = \frac{e^{-2x}}{x+1}$.
Solution: $y = e^{-2x}(C_1 + C_2 x) + e^{-2x}((x + 1)\ln|x + 1| - x)$

16. $y'' + 16y = (\cos 4x - 8\sin 4x)e^x$, $y(0) = 0, y'(0) = 5$.
Solution: $y = C_1\cos 4x + C_2\sin 4x + e^x\cos 4x$; $y = -\cos 4x + \sin 4x + e^x\cos 4x$

17. $y'' - 14y' + 53y = 53x^3 - 42x^2 + 59x - 14$, $y(0) = 0, y'(0) = 7$.
Solution: $y = (C_1\cos 2x + C_2\sin 2x)e^{7x} + x^3 + x$; $y = 3\sin 2xe^{7x} + x^3 + x$

18. $y'' + 2y' + 5y = \frac{2e^{-x}}{\cos 2x}$.
Solution: $y = e^{-x}(C_1\cos 2x + C_2\sin 2x) + e^{-x}(x\sin 2x + \frac{1}{2}\ln|\cos 2x| \cdot \cos 2x)$

19. $y'' - 12y' + 36y = 32\cos 2x + 24\sin 2x$, $y(0) = 2, y'(0) = 4$.
Solution: $y_{gn} = C_1 e^{6x} + C_2 xe^{6x} + \cos 2x$; $y = e^{6x} - 2xe^{6x} + \cos 2x$

20. $y'' - 2y = -2 - 4x^2\cos x^2$. Solution: $y = C_1 e^{-\sqrt{2}x} + C_2 e^{\sqrt{2}x} + \cos x^2$

21. $y'' - 2y' = \frac{1}{x} - 2\ln(ex)$. Solution: $y = C_1 + C_2 e^{2x} + x\ln|x|$

22. $y'' + 2y' + 2y = xe^{-x}$, $y(0) = 0, y'(0) = 0$.
Solution: $y = e^{-x}(C_1\cos x + C_2\sin x) + xe^{-x}$; $y = e^{-x}(x - \sin x)$

23. $y''' - 3y' - 2y = 9e^{2x}$, $y(0) = 0, y'(0) = -3, y''(0) = 3$.
Solution: $y = (C_1 + C_2 x)e^{-x} + C_3 e^{2x} + xe^{2x}$; $y = (x - 1)(e^{2x} - e^{-x})$

24. $y'' + y = \frac{1}{\cos^2 x}$. Solution: $y = C_1 \cos x + C_2 \sin x + \frac{1}{2} \ln \left| \frac{1+\sin x}{1-\sin x} \right| \cdot \sin x - 1$

Section 4 Cauchy problem (initial value problem)

Solve the following initial value problems using the traditional method and the Laplace transform; compare the found solutions:

1. $y'' - 3y' + 2y = e^{-x}$, $y(0) = 0, y'(0) = 1$.
Solution: $y = C_1 e^x + C_2 e^{2x} + \frac{1}{6} e^{-x}$; $y = -\frac{3}{2} e^x + \frac{4}{3} e^{2x} + \frac{1}{6} e^{-x}$

2. $y'' - y' - 2y = 3xe^x$, $y(0) = 0, y'(0) = 0$.
Solution: $y = C_1 e^{-x} + C_2 e^{2x} - \frac{3}{4}(1 + 2x)e^x$; $y = -\frac{1}{4} e^{-x} + e^{2x} - \frac{3}{4}(1 + 2x)e^x$

3. $y'' - 5y' + 4y = (10x + 1)e^{-x}$, $y(0) = 0, y'(0) = 0$.
Solution: $y = C_1 e^x + C_2 e^{4x} + (\frac{4}{5} + x)e^{-x}$; $y = -e^x + \frac{1}{5} e^{4x} + (\frac{4}{5} + x)e^{-x}$

4. $y'' + 5y' + 6y = e^{-2x}$, $y(0) = -1, y'(0) = 0$.
Solution: $y = C_1 e^{-3x} + C_2 e^{-2x} + xe^{-2x}$; $y = 3e^{-3x} + (x - 4)e^{-2x}$

5. $y'' - 2y' + y = 2e^x$, $y(0) = 1, y'(0) = 1$.
Solution: $y = (C_1 + C_2 x)e^x + x^2 e^x$; $y = (x^2 + 1)e^x$

6. $y'' + 2y' + y = (x + 2)e^{-x}$, $y(0) = 1, y'(0) = -1$.
Solution: $y = (C_1 + C_2 x)e^{-x} + (\frac{1}{6} x^3 + x^2)e^{-x}$; $y = (1 + x^2 + \frac{1}{6} x^3)e^{-x}$

7. $y'' - 2y' - 3y = 4e^{3x} - 4e^{-x}$, $y(0) = 2, y'(0) = 0$.
Solution: $y = C_1 e^{-x} + C_2 e^{3x} + xe^{3x} + xe^{-x}$; $y = (2 + x)e^{-x} + xe^{3x}$

8. $y'' + y = 4\cos x$, $y(0) = 1, y'(0) = -1$.
Solution: $y = C_1 \cos x + C_2 \sin x + 2x \sin x$; $y = \cos x + (2x - 1)\sin x$

9. $y'' + y = 5xe^{2x}$, $y(0) = 0, y'(0) = 1$.
Solution: $y = C_1 \cos x + C_2 \sin x + (x - \frac{4}{5})e^{2x}$; $y = \frac{4}{5}(\cos x + 2\sin x) + (x - \frac{4}{5})e^{2x}$

10. $y'' + 9y = 6\cos 3x + 9\sin 3x$, $y(0) = 1, y'(0) = 0$.
Solution: $y = C_1 \cos 3x + C_2 \sin 3x - \frac{3}{2} x \cos 3x + x \sin 3x$; $y = (1 - \frac{3}{2} x) \cos 3x + (x + \frac{1}{2}) \sin 3x$

11. $y'' + 4y = 4(\cos 2x + \sin 2x)$, $y(0) = 0, y'(0) = 1$.
Solution: $y = C_1 \cos 2x + C_2 \sin 2x + x(\sin 2x - \cos 2x)$; $y = \sin 2x + x(\sin 2x - \cos 2x)$

12. $y'' + y = 2(\cos x - \sin x)$, $y(0) = 1, y'(0) = 2$.
Solution: $y = C_1 \cos x + C_2 \sin x + x(\cos x + \sin x)$; $y = (1 + x)(\cos x + \sin x)$

Section 5 Boundary value problem for 2nd order equations

Solve the following boundary value problems using the traditional method and the Green method; compare the found solutions:

1. $y'' + 4y = 2x$, $y(0) = 0, y(\frac{\pi}{8}) = 0$.
Solution: $G(x, s) = \begin{cases} (\sin 2s - \cos 2s) \sin 2x, 0 \le x \le s \\ \sin 2s(\sin 2x - \cos 2x), s \le x \le \frac{\pi}{8} \end{cases}$;
$y = C_1 \cos 2x + C_2 \sin 2x + \frac{1}{2} x$; $y = -\frac{\pi}{8\sqrt{2}} \sin 2x + \frac{1}{2} x$

2. $y'' = e^{3x}$, $y(0) = 0, 3y(1) + y'(1) = 0$.

Solution: $G(x,s) = \begin{cases} (\frac{3s}{4} - 1)x, 0 \leq x \leq s \\ \frac{s}{4}(3x - 4), s \leq x \leq 1 \end{cases}$;

$y = C_1 x + C_2 + \frac{1}{9}e^{3x}$; $y = (\frac{1}{12} - \frac{1}{6}e^3)x - \frac{1}{9} + \frac{1}{9}e^{3x}$

3. $y'' - y = \sin x$, $y'(0) = 0$, $y'(2) + y(2) = 0$.

Solution: $G(x,s) = \begin{cases} -e^{-s}\cosh x, 0 \leq x \leq s \\ -e^{-x}\cosh s, s \leq x \leq 2 \end{cases}$;

$y = C_1 e^{-x} + C_2 e^x - \frac{1}{2}\sin x$; $y = (\frac{1}{4e^2}(\cos 2 + \sin 2) - \frac{1}{2})e^{-x} + \frac{1}{4e^2}(\cos 2 + \sin 2)e^x - \frac{1}{2}\sin x$

4. $y'' - y = \cos x$, $y'(0) = 0$, $y'(2) + y(2) = 0$.

Solution: $G(x,s) = \begin{cases} -e^{-s}\cosh x, 0 \leq x \leq s \\ -e^{-x}\cosh s, s \leq x \leq 2 \end{cases}$;

$y = C_1 e^{-x} + C_2 e^x - \frac{1}{2}\cos x$; $y = \frac{1}{4e^2}(\cos 2 - \sin 2)(e^{-x} + e^x) - \frac{1}{2}\cos x$

5. $y'' + y = x^2 + 2x$, $y(0) = 0$, $y'(1) = 0$.

Solution: $G(x,s) = -\frac{1}{\cos 1} \begin{cases} \sin x \cos(1 - s), 0 \leq x \leq s \\ \cos(1 - x)\sin s, s \leq x \leq 1 \end{cases}$;

$y = C_1 \cos x + C_2 \sin x + x^2 + 2x - 2$; $y = 2\cos x + \frac{2\sin 1 - 4}{\cos 1}\sin x + x^2 + 2x - 2$

6. $y'' + 4y = 2^{-x}$, $y'(0) = 0$, $y(1) = 0$.

Solution: $G(x,s) = \frac{1}{2\cos 2} \begin{cases} \cos 2x \sin(2s - 2), 0 \leq x \leq s \\ \sin(2x - 2)\cos 2s, s \leq x \leq 1 \end{cases}$;

$y = C_1 \cos 2x + C_2 \sin 2x + \frac{1}{2}e^{-2x}$; $y = -\frac{\sin 2 + e^{-2}}{2\cos 2}\cos 2x + \frac{1}{2}\sin 2x + \frac{1}{2}e^{-2x}$

7. $y'' - 4y = xe^x$, $y'(0) = 0$, $2y(1) - y'(1) = 0$.

Solution: $G(x,s) = \begin{cases} e^{2s}\cosh 2x, 0 \leq x \leq s \\ e^{2x}\cosh 2s, s \leq x \leq 1 \end{cases}$;

$y = C_1 e^{-2x} + C_2 e^{2x} - (\frac{1}{3}x + \frac{2}{9})e^x$; $y = \frac{1}{18}e^{3-2x} + \frac{1}{18}(e^3 + 5)e^{2x} - (\frac{1}{3}x + \frac{2}{9})e^x$

8. $y'' - y' = e^x$, $y(0) = 0$, $y(1) - y'(1) = 0$.

Solution: $G(x,s) = \begin{cases} 1 - e^x, 0 \leq x \leq s \\ e^x(e^{-s} - 1), s \leq x \leq 1 \end{cases}$;

$y = C_1 + C_2 e^x + xe^x$; $y = e - e^{1+x} + xe^x$

9. $y'' - y = 2x$, $y(0) = 0$, $y(1) = 0$.

Solution: $G(x,s) = -\frac{1}{\sinh 1} \begin{cases} \sinh x \sinh(1 - s), 0 \leq x \leq s \\ \sinh s \sinh(1 - x), s \leq x \leq 1 \end{cases}$;

$y = C_1 e^{-x} + C_2 e^x - 2x$; $y = \frac{2}{e - e^{-1}}(e^x - e^{-x}) - 2x$

10. $x^2 y'' + 3xy' - 3y = 12x^3 - 21x^4$, $y(1) = 0$, $y(2) - 2y'(2) = 0$.

Solution: $G(x,s) = \begin{cases} \frac{1}{4}s^2(\frac{1}{x^3} - x), 1 \leq x \leq s \\ \frac{1}{4}x(\frac{1}{s^2} - s^2), s \leq x \leq 2 \end{cases}$;

$y = \frac{C_1}{x^3} + C_2 x + x^3 - x^4$; $y = -\frac{64}{x^3} + 64x + x^3 - x^4$

Section 6 Sturm-Liouville problem for 2nd order equations

Solve the following Sturm-Liouville problems:

1. $y'' = \lambda y$, $x \in [0, d]$, $y(0) = y'(d) = 0$.

Solution: $\lambda_k = -\left(\frac{(2k-1)\pi}{2d}\right)^2$, $y_k = \sin\frac{(2k-1)\pi}{2d}x$, $k \in \mathbb{N}$

2. $y'' = \lambda y$, $x \in [0, d]$, $y'(0) = y(d) = 0$.

Solution: $\lambda_k = -\left(\frac{(2k-1)\pi}{2d}\right)^2$, $y_k = \cos\frac{(2k-1)\pi}{2d}x$, $k \in \mathbb{N}$

3. $y'' = \lambda y$, $x \in [0, d]$, $y(0) = y'(d) + y(d) = 0$.

Solution: $\lambda_k = -\frac{\theta_k^2}{d^2}$, $\tan\theta_k = -\frac{\theta_k}{d}$, $\theta_k \in ((k-1/2)\pi, k\pi)$, $y_k = \sin\frac{\theta_k}{d}x$, $k \in \mathbb{N}$

4. $y'' = \lambda y$, $x \in [0, d]$, $y'(0) - y(0) = y(1) = 0$.

Solution: $\lambda_k = -\theta_k^2$, $\tan\theta_k = -\frac{1}{\theta_k}$, $\theta_k \in ((k-1/2)\pi, k\pi)$, $y_k = \theta_k\cos\theta_k x + \sin\theta_k x$, $k \in \mathbb{N}$

5. $y'' + 2y' + (1 - \lambda)y = 0$, $x \in [0, 1]$, $y(0) = y(1) = 0$.

Solution: $\lambda_k = -(k\pi)^2$, $y_k = e^{-x}\sin k\pi x$, $k \in \mathbb{N}$

6. $x^2 y'' - xy' + y = \lambda y$, $x \in [1, 2]$, $y(1) = y(2) = 0$.

Solution: $\lambda_k = -\left(\frac{k\pi}{\ln 2}\right)^2$, $y_k = x\sin\frac{k\pi\ln x}{\ln 2}$, $k \in \mathbb{N}$

Chapter 10 Linear equations with variable coefficients: power series method

Section 3 Solution about ordinary points

Verify whether the indicated center point is ordinary and solve in the corresponding power series; find the interval of validity of the obtained solution:

1. Solve $xy' - y - x - 1 = 0$ about the center point 1.

Solution: $y = Cx + 2(x - 1) + \sum_{n=2}^{\infty}(-1)^n\frac{1}{n(n-1)}(x - 1)^n$

2. Solve $y'' + xy' + y = 0$ about the center point 0.

Solution: $y_1 = \sum_{n=0}^{\infty}(-1)^n\frac{x^{2n}}{(2n)!!}$, $y_2 = \sum_{n=0}^{\infty}(-1)^n\frac{x^{2n+1}}{(2n+1)!!}$

3. Solve $y'' - xy' - y = 0$ about the center point 0.

Solution: $y_1 = \sum_{n=0}^{\infty}\frac{x^{2n}}{(2n)!!}$, $y_2 = \sum_{n=0}^{\infty}\frac{x^{2n+1}}{(2n+1)!!}$

4. Solve $y'' + xy' + 2y = 0$, $y(0) = 3$, $y'(0) = -2$.

Solution: $y = 3\sum_{n=0}^{\infty}(-1)^n\frac{x^{2n}}{(2n-1)!!} - 2\sum_{n=0}^{\infty}(-1)^n\frac{x^{2n+1}}{(2n)!!}$

5. Solve $y'' + x^2 y' + xy = 0$ about the center point 0.

Solution: $y_1 = 1 + \sum_{n=1}^{\infty}(-1)^n\frac{1^2\cdot4^2\cdot\ldots\cdot(3n-2)^2}{(3n)!}x^{3n}$,

$y_2 = x + \sum_{n=1}^{\infty}(-1)^n\frac{2^2\cdot5^2\cdot\ldots\cdot(3n-1)^2}{(3n+1)!}x^{3n+1}$

6. Solve $(x - 1)y'' + y' = 0$ about the center point 0.

Solution: $y_1 = 1$, $y_2 = \sum_{n=1}^{\infty}\frac{1}{n}x^n$

7. Solve $(x^2 - 1)y'' + 4xy' + 2y = 0$ about the center point 0.

Solution: $y_1 = \sum_{n=0}^{\infty}x^{2n} = \frac{1}{1-x^2}$, $y_2 = \sum_{n=0}^{\infty}x^{2n+1} = \frac{x}{1-x^2}$

8. Solve $(x - 1)y'' - xy' + y = 0$, $y(0) = -2$, $y'(0) = 6$.

Solution: $y = -2 + 6x - 2\sum_{n=2}^{\infty}\frac{x^n}{n!} = 8x - 2e^x$

9. Solve $y'' - 2xy' + 8y = 0$, $y(0) = 3$, $y'(0) = 0$. Solution: $y = 3 - 12x^2 + 4x^4$

10. Solve $y'' - xy = 1$ about the center point 0.

Solution: $y = C_1 \left[1 + \sum_{n=1}^{\infty} \frac{1 \cdot 4 \cdot \ldots \cdot (3n-2)}{(3n)!} x^{3n} \right]$

$+ C_2 \left[x + \sum_{n=1}^{\infty} \frac{2 \cdot 5 \cdot \ldots \cdot (3n-1)}{(3n+1)!} x^{3n+1} \right] + \sum_{n=1}^{\infty} \frac{1 \cdot 3 \cdot \ldots \cdot (3n-3)}{(3n-1)!} x^{3n-1}$

11. Solve $y'' - xy' + 2y = 0$ about the center point 0.

Solution: $y_1 = 1 - x^2$, $y_2 = x - \sum_{n=1}^{\infty} \frac{(2n-3)!!}{(2n+1)!} x^{2n+1}$

12. Solve $y'' + 2xy' + 2y = 0$ about the center point 0.

Solution: $y_1 = \sum_{n=0}^{\infty} (-1)^n \frac{2^n}{(2n)!!} x^{2n}$, $y_2 = \sum_{n=0}^{\infty} (-1)^n \frac{2^n}{(2n+1)!!} x^{2n+1}$

13. Solve $(x^2 + 1)y'' - 6y = 0$ about the center point 0.

Solution: $y_1 = 1 + 3x^2 + x^4 + \sum_{n=3}^{\infty} (-1)^n \frac{3}{(2n-1)(2n-3)} x^{2n}$, $y_2 = x + x^3$

14. Solve $(x^2 + 1)y'' + 2xy' = 0$, $y(0) = 0$, $y'(0) = 1$.

Solution: $y = \sum_{n=0}^{\infty} (-1)^n \frac{x^{2n+1}}{2n+1}$

Section 5 Euler equation

Solve the following Euler equations and the corresponding initial value problems:

1. $x^2 y'' - 3xy' - 5y = 0$. Solution: $y = C_1 x^5 + C_2 x^{-1}$

2. $x^2 y'' + 5xy' + 3y = 0$. Solution: $y = C_1 x^{-3} + C_2 x^{-1}$

3. $x^2 y'' - xy' + y = 0$. Solution: $y = C_1 x + C_2 x \ln x$

4. $x^2 y'' + 9xy' + 16y = 0$. Solution: $y = C_1 x^{-4} + C_2 x^{-4} \ln x$

5. $x^2 y'' + 2xy' - 6y = 0$, $y(1) = 3$, $y'(1) = 1$.

Solution: $y = C_1 x^{-3} + C_2 x^2$; $y = x^{-3} + 2x^2$

6. $(x - 3)^2 y'' + 5(x - 3)y' + 4y = 0$, $y(4) = 1$, $y'(4) = 1$.

Solution: $y = C_1 (x - 3)^{-2} + C_2 (x - 3)^{-2} \ln(x - 3)$; $y = (x - 3)^{-2} + 3(x - 3)^{-2} \ln(x - 3)$

Section 6 Solution about singular points: Frobenius method

Classify the indicated center point and solve using the corresponding series; find the set of validity of the obtained solution:

1. Solve $2x^2 y'' + (3x - x^2)y' - (x + 1)y = 0$ about the center point 0.

Solution: $y_1 = |x|^{1/2} \left[1 + \sum_{n=1}^{\infty} \frac{3 \cdot (2x)^n}{(2n+3)!!} \right]$, $y_2 = \frac{1}{x} \sum_{n=0}^{\infty} \frac{x^n}{n!} = \frac{e^x}{x}$

2. Solve $4xy'' + 2y' + y = 0$ about the center point 0.

Solution: $y_1 = \sum_{n=0}^{\infty} (-1)^n \frac{x^n}{(2n)!} = \cos \sqrt{x}$,

$y_2 = \sqrt{x} \sum_{n=0}^{\infty} (-1)^n \frac{x^n}{(n+1)!} = \sin \sqrt{x}$

3. Solve $(1 + x)y' - py = 0$ about the center point 0.

Solution: $y = C \sum_{n=0}^{\infty} \frac{p(p-1) \cdot \ldots \cdot (p-n+1)}{n!} x^n$

4. Solve $9x(1 - x)y'' - 12y' + 4y = 0$ about the center point 0.

Solution: $y_1 = 1 + \sum_{n=1}^{\infty} \frac{1 \cdot 4 \cdot 7 \cdot \ldots \cdot (3n-2)}{3 \cdot 6 \cdot 9 \cdot \ldots \cdot 3n} x^n$,

$y_2 = |x|^{7/3} \left[1 + \sum_{n=1}^{\infty} \frac{8 \cdot 11 \cdot 14 \cdot \ldots \cdot (3n+5)}{10 \cdot 13 \cdot 16 \cdot \ldots \cdot (3n+7)} x^n \right]$

5. Solve $x^2 y'' + xy' + (4x^2 - \frac{1}{9})y = 0$ about the center point 0.
Solution: $y_1 = J_{1/3}(2x)$, $y_2 = J_{-1/3}(2x)$

6. Solve $x^2 y'' + xy' + (x^2 - \frac{1}{9})y = 0$ about the center point 0.
Solution: $y_1 = J_{1/3}(x)$, $y_2 = J_{-1/3}(x)$

7. Solve $x^2 y'' + xy' + (4x^2 - \frac{1}{9})y = 0$ about the center point 0.
Solution: $y_1 = J_{1/3}(2x)$, $y_2 = J_{-1/3}(2x)$

8. Solve $2xy'' + (1+x)y' + y = 0$ about the center point 0.
Solution: $y_1 = |x|^{1/2} \sum_{n=0}^{\infty} (-1)^n \frac{1}{2^n n!} x^n$, $y_2 = \sum_{n=0}^{\infty} (-1)^n \frac{1}{(2n-1)!!} x^n$

9. Solve $xy'' + (x-6)y' - 3y = 0$ about the center point 0.
Solution: $y_1 = 1 - \frac{1}{2}x + \frac{1}{10}x^2 - \frac{1}{120}x^3$, $y_2 = x^7 + \sum_{n=1}^{\infty} (-1)^n \frac{4 \cdot 5 \cdot 6 \cdot \ldots \cdot (n+3)}{n! 8 \cdot 9 \cdot 10 \cdot \ldots \cdot (n+7)} x^{n+7}$

10. Solve $2xy'' - y' + 2y = 0$ about the center point 0.
Solution: $y_1 = |x|^{3/2} \left[1 + \sum_{n=1}^{\infty} (-1)^n \frac{3 \cdot 2^n}{n!(2n+3)!!} x^n \right]$,

$y_2 = 1 + 2x + \sum_{n=2}^{\infty} (-1)^{n+1} \frac{2^n}{n! \cdot (2n-3)!!} x^n$

11. Solve $2xy'' - y' + 2y = 0$ about the center point 0.
Solution: $y_1 = |x|^{1/3} \sum_{n=0}^{\infty} \frac{1}{n! \cdot 3^n} x^n$, $y_2 = 1 + \sum_{n=1}^{\infty} \frac{1}{2 \cdot 5 \cdot 8 \cdot \ldots \cdot (3n-1)} x^n$

12. Solve $2x^2 y'' - x(x-1)y' - y = 0$ about the center point 0.
Solution: $y_1 = x \left[1 + \sum_{n=1}^{\infty} \frac{1}{5 \cdot 7 \cdot 9 \cdot \ldots \cdot (2n+3)} x^n \right]$,

$y_2 = |x|^{-1/2} \left[1 + \sum_{n=1}^{\infty} \frac{1}{2 \cdot 4 \cdot 4 \cdot \ldots \cdot 2n} x^n \right]$

13. Solve $xy'' + 2y' - xy = 0$ about the center point 0.
Solution: $y_1 = x^{-1} \sum_{n=0}^{\infty} \frac{1}{(2n)!} x^{2n} = \frac{\cosh x}{x}$,

$y_2 = x^{-1} \sum_{n=0}^{\infty} \frac{1}{(2n+1)!} x^{2n+1} = \frac{\sinh x}{x}$

14. Solve $x(x-1)y'' + 3y' - 2y = 0$ about the center point 0.
Solution: $y_1 = 1 + \frac{2}{3}x + \frac{1}{3}x^2$, $y_2 = \sum_{n=0}^{\infty} (n+1)x^{n+4}$

15. Solve $xy'' + (1-x)y' - y = 0$ about the center point 0.
Solution: $y_1 = \sum_{n=0}^{\infty} \frac{1}{n!} x^n = e^x$,
$y_2 = y_1(x) \ln x - \left[x + \sum_{n=2}^{\infty} \frac{1}{n!} \cdot \left(\sum_{k=1}^{n} \frac{1}{k}\right) x^n \right]$

16. Solve $2x^2 y'' + xy' - (x+1)y = 0$ about the center point 0.
Solution: $y_1 = x \left[1 + \sum_{n=1}^{\infty} \frac{3}{n!(2n+3)!!} x^n \right]$,

$y_2 = |x|^{-1/2} \left[1 - x - \sum_{n=2}^{\infty} \frac{1}{n!(2n-3)!!} x^n \right]$

17. Solve $xy'' - (2x-1)y' + (x-1)y = 0$ about the center point 0.
Solution: $y_1 = e^x$, $y_2 = y_1(x) \ln x$

18. Solve $x^2 y'' + x(2+3x)y' - 2y = 0$ about the center point 0.
Solution: $y_1 = x^{-2} - 3x^{-1} + \frac{9}{2}$, $y_2 = x + 6 \sum_{n=4}^{\infty} \frac{(-3)^{n-3}}{n!} x^{n-2}$

19. Solve $x^2 y'' + xy' - xy = 0$ about the center point 0.

Solution: $y_1 = \sum_{n=0}^{\infty} \frac{x^n}{(n!)^2}$, $y_2 = y_1 \ln x - 2 \sum_{n=1}^{\infty} \frac{H(n)}{n!} x^n$, $H(n) = \sum_{k=1}^{n} \frac{1}{k}$

20. Solve $4x^2 y'' + (1 - 2x)y = 0$ about the center point 0.

Solution: $y_1 = x^{1/2} \sum_{n=0}^{\infty} \frac{x^n}{2^n (n!)^2}$, $y_2 = y_1 \ln x - x^{1/2} \sum_{n=1}^{\infty} \frac{H(n)}{2^{n-1}(n!)^2} x^n$, $H(n) = \sum_{k=1}^{n} \frac{1}{k}$

Exercises for sections 3 and 6

Classify the indicated center point and solve using the corresponding series; find the set of validity of the obtained solution:

1. Solve $(1 - x^2)y'' - 6xy' - 4y = 0$ about the center point 0.

Solution: $y_1 = \sum_{n=0}^{\infty}(n+1)x^{2n} = \frac{1}{(1-x^2)^2}$, $y_2 = \sum_{n=0}^{\infty} \frac{2n+3}{3} x^{2n+1} = \frac{3x - x^3}{3(1-x^2)^2}$

2. Solve $2xy'' + (1 + x)y' - 2y = 0$ about the center point 0.

Solution: $y_1 = x^{1/2} \sum_{n=0}^{\infty}(-1)^n \frac{3x^n}{2^n n!(2n-3)(2n-1)(2n+1)}$, $y_2 = 1 + 2x + \frac{1}{3}x^2$

3. Solve $xy'' + y = 0$ about the center point 0.

Solution: $y_1 = \sum_{n=1}^{\infty}(-1)^n \frac{x^n}{n!(n-1)!}$,

$y_2 = y_1 \ln x + 1 + x - \sum_{n=2}^{\infty}(-1)^n \frac{(H_n + H_{n-1})x^n}{n!(n-1)!}$, $H_n = \sum_{k=1}^{n} \frac{1}{k}$

4. Solve $y'' + (x - 1)^2 y' - 4(x - 1)y = 0$ about the center point 1.

Solution: $y_1 = \sum_{n=0}^{\infty}(-1)^n \frac{4(x-1)^{3n}}{3^n(3n-1)(3n-4)n!}$, $y_2 = (x - 1) + \frac{1}{4}(x - 1)^4$

5. Solve $y'' + 3xy' + 3y = 0$ about the center point 0.

Solution: $y_1 = \sum_{n=0}^{\infty}(-1)^n \frac{3^n x^{2n}}{(2n)!!}$, $y_2 = \sum_{n=0}^{\infty}(-1)^n \frac{3^n x^{2n+1}}{(2n+1)!!}$

6. Solve $x^2 y'' + 3xy' + (1 - 2x)y = 0$ about the center point 0.

Solution: $y_1 = x^{-1} \sum_{n=0}^{\infty} \frac{2^n x^n}{(n!)^2}$, $y_2 = y_1 \ln x - x^{-1} \sum_{n=1}^{\infty} \frac{2^{n+1} H_n x^n}{(n!)^2}$, $H_n = \sum_{k=1}^{n} \frac{1}{k}$

7. Solve $2x(x + 1)y'' + 3(x + 1)y' - y = 0$ about the center point 0.

Solution: $y_1 = \sum_{n=0}^{\infty}(-1)^{n+1} \frac{x^n}{4n^2 - 1}$, $y_2 = x^{-1/2} + x^{1/2}$

8. Solve $(x^2 + 4)y'' + 2xy' - 12y = 0$ about the center point 0.

Solution: $y_1 = \sum_{n=0}^{\infty}(-1)^n \frac{3(n+1)x^{2n}}{2^{2n}(2n-1)(2n-3)}$, $y_2 = x + \frac{5}{12}x^3$

9. Solve $(1 + 4x^2)y'' - 8y = 0$ about the center point 0.

Solution: $y_1 = 1 + 4x^2$, $y_2 = \sum_{n=0}^{\infty}(-1)^{n+1} \frac{4^n x^{2n+1}}{4n^2 - 1}$

10. Solve $x(x - 2)y'' + 2(x - 1)y' - 2y = 0$ about the center point 2.

Solution: $y_1 = 1 + (x - 2)$, $y_2 = y_1 \ln(x - 2) - \frac{5}{2}(x - 2) + \sum_{n=2}^{\infty}(-1)^n \frac{(n+1)(x-2)^n}{2^n n(n-1)}$

11. Solve $4xy'' + 3y' + 3y = 0$ about the center point 0.

Solution: $y_1 = x^{1/4} \sum_{n=0}^{\infty}(-1)^n \frac{3^n x^n}{n! \cdot 5 \cdot 9 \cdot 13 \cdots (4n+1)}$,

$y_2 = 1 + \sum_{n=1}^{\infty}(-1)^n \frac{3^n x^n}{n! \cdot 3 \cdot 7 \cdot 11 \cdots (4n-1)}$

12. Solve $y'' - 2(x + 3)y' - 3y = 0$ about the center point -3.

Solution: $y_1 = 1 + \sum_{n=1}^{\infty} \frac{3 \cdot 7 \cdot 11 \cdots (4n-1)(x+3)^{2n}}{(2n)!}$,

$y_2 = \sum_{n=0}^{\infty} \frac{1 \cdot 5 \cdot 9 \cdots (4n+1)(x+3)^{2n+1}}{(2n+1)!}$

13. Solve $x(1+x)y'' + (1+5x)y' + 3y = 0$ about the center point 0.
Solution: $y_1 = \sum_{n=0}^{\infty}(-1)^n(n+1)(n+2)x^n$,
$y_2 = y_1 \ln x - \frac{3}{2}(y_1 - 1) + \frac{1}{2}\sum_{n=1}^{\infty}(-1)^n(2n+3)x^n$
14. Solve $(1+x^2)y'' + 10xy' + 20y = 0$ about the center point 0.
Solution: $y_1 = \sum_{n=0}^{\infty}(-1)^n(n+1)(2n+1)(2n+3)x^{2n}$,
$y_2 = \sum_{n=0}^{\infty}(-1)^n(n+1)(n+2)(2n+3)x^{2n+1}$
15. Solve $2x^2(1-x)y'' - x(1+7x)y' + y = 0$ about the center point 0.
Solution: $y_1 = \sum_{n=0}^{\infty}(2n+1)(2n+5)x^{n+1}$, $y_2 = x^{1/2}\sum_{n=0}^{\infty}(n+1)(n+2)x^n$
16. Solve $x(1+x)y'' + (x+5)y' - 4y = 0$ about the center point -1.
Solution: $y_1 = 1+(x+1)+\frac{1}{2}(x+1)^2$, $y_2 = \sum_{n=5}^{\infty}(n-4)(n-3)(n+1)(x+1)^n$
17. Solve $y'' + (x-2)y = 0$ about the center point 2.
Solution: $y_1 = 1 + \sum_{n=1}^{\infty}(-1)^n\frac{(x-2)^{3n}}{3^n n!\cdot 2\cdot 5\cdot 8\cdots\cdot(3n-1)}$,
$y_2 = \sum_{n=0}^{\infty}(-1)^n\frac{(x-2)^{3n+1}}{3^n n!\cdot 1\cdot 4\cdot 7\cdots\cdot(3n+1)}$
18. Solve $x^2(1+2x)y'' + 2x(1+6x)y' - 2y = 0$ about the center point 0.
Solution: $y_1 = x^{-2} - 6x^{-1} + 24$, $y_2 = \sum_{n=1}^{\infty}(-2)^{n-1}(n+3)(n+4)x^n$
19. Solve $x^2y'' + x(1-x)y' - (1+3x)y = 0$ about the center point 0.
Solution: $y_1 = \sum_{n=0}^{\infty}\frac{(n+3)x^{n+1}}{n!}$,
$y_2 = y_1 \ln x + x^{-1} - 2 - \sum_{n=0}^{\infty}\frac{(1-(n+3)H_n)x^{n+1}}{n!}$, $H_n = \sum_{k=1}^{n}\frac{1}{k}$, $H_0 = 0$
20. Solve $x^2y'' + x(x-1)y' + (1-x)y = 0$ about the center point 0.
Solution: $y_1 = x$, $y_2 = y_1 \ln x + \sum_{n=1}^{\infty}(-1)^n\frac{x^{n+1}}{n!\cdot n}$

Chapter 13 Linear systems: methods of solution

Section 1 Homogeneous systems with constant coefficients

Solve using the method of reduction and Euler method and compare the solutions:

1. $\begin{cases} u' = u - 5v \\ v' = 2u - v \end{cases}$.

Solution: $\lambda = \pm 3i$, $\begin{cases} u = 5C_1\cos 3x + 5C_2\sin 3x \\ v = C_1(\cos 3x + 3\sin 3x) + C_2(\sin 3x - 3\cos 3x) \end{cases}$

2. $\begin{cases} u' = 2u + v \\ v' = 4v - u \end{cases}$. Solution: $\lambda = 3, 3$, $\begin{cases} u = (C_1 + C_2x)e^{3x} \\ v = (C_1 + C_2 + C_2x)e^{3x} \end{cases}$

3. $\begin{cases} u' = -9v \\ v' = u \end{cases}$. Solution: $\lambda = -3i, 3i$, $\begin{cases} u = 3C_1\cos 3x - 3C_2\sin 3x \\ v = C_2\cos 3x + C_1\sin 3x \end{cases}$

4. $\begin{cases} u' = 8v - u \\ v' = v + u \end{cases}$. Solution: $\lambda = -3, 3$, $\begin{cases} u = 2C_1e^{3x} - 4C_2e^{-3x} \\ v = C_1e^{3x} + C_2e^{-3x} \end{cases}$

5. $\begin{cases} u' = u - v \\ v' = v - u \end{cases}$. Solution: $\lambda = 0, 2,$ $\begin{cases} u = C_1 + C_2 e^{2x} \\ v = C_1 - C_2 e^{2x} \end{cases}$

6. $\begin{cases} u' = 2u + v \\ v' = 3u + 4v \end{cases}$. Solution: $\lambda = 1, 5,$ $\begin{cases} u = C_1 e^x + C_2 e^{5x} \\ v = -C_1 e^x + 3C_2 e^{5x} \end{cases}$

7. $\begin{cases} u' = u - v \\ v' = v - 4u \end{cases}$. Solution: $\lambda = -1, 3,$ $\begin{cases} u = C_1 e^{-x} + C_2 e^{3x} \\ v = 2C_1 e^{-x} - 2C_2 e^{3x} \end{cases}$

8. $\begin{cases} u' = u + v \\ v' = 3v - 2u \end{cases}$.

Solution: $\lambda = 2 \pm i,$ $\begin{cases} u = (C_1 \cos x + C_2 \sin x)e^{2x} \\ v = ((C_1 + C_2) \cos x + (C_2 - C_1) \sin x)e^{2x} \end{cases}$

9. $\begin{cases} u' = u - 3v \\ v' = 3u + v \end{cases}$. Solution: $\lambda = 1 \pm 3i,$ $\begin{cases} u = (C_1 \cos 3x + C_2 \sin 3x)e^x \\ v = (C_1 \sin 3x - C_2 \cos 3x)e^x \end{cases}$

10. $\begin{cases} u' = 2u + v \\ v' = 4v - u \end{cases}$. Solution: $\lambda = 3, 3,$ $\begin{cases} u = (C_1 + C_2 x)e^{3x} \\ v = (C_1 + C_2 + C_2 x)e^{3x} \end{cases}$

11. $\begin{cases} u' = 2v - 3u \\ v' = v - 2u \end{cases}$. Solution: $\lambda = -1, -1,$ $\begin{cases} u = (C_1 + 2C_2 x)e^{-x} \\ v = (C_1 + C_2 + 2C_2 x)e^{-x} \end{cases}$

12. $\begin{cases} u' = 3u - v + w \\ v' = -u + 5v - w \\ w' = u - v + 3w \end{cases}$.

Solution: $\lambda = 2, 3, 6,$ $\begin{cases} u = C_1 e^{2x} + C_2 e^{3x} + C_3 e^{6x} \\ v = C_2 e^{3x} - 2C_3 e^{6x} \\ w = -C_1 e^{2x} + C_2 e^{3x} + C_3 e^{6x} \end{cases}$

13. $\begin{cases} u' = -v + w \\ v' = w \\ w' = -u + w \end{cases}$.

Solution: $\lambda = 1, \pm i,$ $\begin{cases} u = (C_1 - C_2) \cos x + (C_1 + C_2) \cos x \\ v = C_1 \sin x - C_2 \cos x + C_3 e^x \\ w = C_1 \cos x + C_2 \sin x + C_3 e^x \end{cases}$

14. $\begin{cases} u' = 2u - v + w \\ v' = u + 2v - w \\ w' = -u - v + 2w \end{cases}$.

Solution: $\lambda = 1, 2, 3$, $\begin{cases} u = C_1 e^{2x} - C_2 e^{3x} \\ v = C_1 e^{2x} - C_3 e^{x} \\ w = C_1 e^{2x} - C_2 e^{3x} - C_3 e^{x} \end{cases}$

15. $\begin{cases} u' = u + w - v \\ v' = u + v - w \\ w' = 2u - v \end{cases}$.

Solution: $\lambda = -1, 1, 2$, $\begin{cases} u = C_1 e^{x} + C_2 e^{2x} + C_3 e^{-x} \\ v = C_1 e^{x} - 3C_3 e^{-x} \\ w = C_1 e^{x} + C_2 e^{2x} - 5C_3 e^{-x} \end{cases}$

16. $\begin{cases} u' = u - 2v - w \\ v' = v - u + w \\ w' = u - w \end{cases}$.

Solution: $\lambda = -1, 0, 2$, $\begin{cases} u = C_1 + 3C_2 e^{2x} \\ v = -2C_2 e^{2x} + C_3 e^{-x} \\ w = C_1 + C_2 e^{2x} - 2C_3 e^{-x} \end{cases}$

17. $\begin{cases} u' = u - v - w \\ v' = u + v \\ w' = 3u + w \end{cases}$.

Solution: $\lambda = 1, 1 \pm 2i$, $\begin{cases} u = (2C_2 \sin 2x + 2C_3 \cos 2x)e^{x} \\ v = (C_1 - C_2 \cos 2x + C_3 \sin 2x)e^{x} \\ w = (-C_1 - 3C_2 \cos 2x + 3C_3 \sin 2x)e^{x} \end{cases}$

18. $\begin{cases} u' = 2u + v \\ v' = u + 3v - w \\ w' = 2v + 3w - u \end{cases}$.

Solution: $\lambda = 2, 3 \pm i$, $\begin{cases} u = C_1 e^{2x} + (C_2 \cos x + C_3 \sin x)e^{3x} \\ v = ((C_2 + C_3) \cos x + (C_3 - C_2) \sin x)e^{3x} \\ w = C_1 e^{2x} + ((2C_2 - C_3) \cos x + (2C_3 + C_2) \sin x)e^{3x} \end{cases}$

19. $\begin{cases} u' = 4u - v - w \\ v' = u + 2v - w \\ w' = u - v + 2w \end{cases}$. Solution: $\lambda = 2, 3, 3$, $\begin{cases} u = C_1 e^{2x} + (C_2 + C_3)e^{3x} \\ v = C_1 e^{2x} + C_2 e^{3x} \\ w = C_1 e^{2x} + C_3 e^{3x} \end{cases}$

20. $\begin{cases} u' = 2u - v - w \\ v' = 3u - 2v - 3w \\ w' = 2w - u + v \end{cases}$. Solution: $\lambda = 0, 1, 1$, $\begin{cases} u = C_1 + C_2 e^{x} \\ v = 3C_1 + C_3 e^{x} \\ w = -C_1 + (C_2 - C_3)e^{x} \end{cases}$

21. $\begin{cases} u' = 3u - 2v - w \\ v' = 3u - 4v - 3w \\ w' = 2u - 4v \end{cases}$.

Solution: $\lambda = -5, 2, 2,$ $\begin{cases} u = C_1 e^{2x} + C_3 e^{-5x} \\ v = C_2 e^{2x} + 3C_3 e^{-5x} \\ w = (C_1 - 2C_2)e^{2x} + 2C_3 e^{-5x} \end{cases}$

22. $\begin{cases} u' = u - v + w \\ v' = u + v - w \\ w' = 2w - v \end{cases}$.

Solution: $\lambda = 1, 1, 2,$ $\begin{cases} u = (C_1 + C_2 x)e^x + C_3 e^{2x} \\ v = (C_1 - 2C_2 + C_2 x)e^x \\ w = (C_1 - C_2 + C_2 x)e^x + C_3 e^{2x} \end{cases}$

23. $\begin{cases} u' = v - 2w - u \\ v' = 4u + v \\ w' = 2u + v - w \end{cases}$.

Solution: $\lambda = -1, -1, 1,$ $\begin{cases} u = (C_2 + C_3 x)e^{-x} \\ v = 2C_1 e^x - (2C_2 + C_3 + 2C_3 x)e^{-x} \\ w = C_1 e^x - (C_2 + C_3 + C_3 x)e^{-x} \end{cases}$

24. $\begin{cases} u' = 2u - v - w \\ v' = 2u - v - 2w \\ w' = 2w - u + v \end{cases}$.

Solution: $\lambda = 1, 1, 1,$ $\begin{cases} u = (C_1 + C_3 x)e^x \\ v = (C_2 + 2C_3 x)e^x \\ w = (C_1 - C_2 - C_3 - C_3 x)e^x \end{cases}$

25. $\begin{cases} u' = 7u + v + 2w \\ v' = 2u + 3v + w \\ w' = -8u - 2v - w \end{cases}$.

Solution: $\lambda = 3, 3, 3,$ $\begin{cases} u = (C_1 + C_2 x + C_3 \frac{x^2}{2})e^{3x} \\ v = (C_2 + C_3(x - 2))e^{3x} \\ w = (-2C_1 - 2xC_2 + C_3(1 - x^2))e^{3x} \end{cases}$

26. $\begin{cases} u' = -2u - v - w \\ v' = -3v + w \\ w' = -v - w \end{cases}$.

$$\text{Solution: } \lambda = -2, -2, -2, \begin{cases} u = (C_1 + 2C_2x + C_3(x^2 + x))e^{-2x} \\ v = (-C_2 - C_3x)e^{-2x} \\ w = (-C_2 - C_3(x+1))e^{-2x} \end{cases}$$

Section 2 Nonhomogeneous systems with constant coefficients

Section 2.2 Euler method

Solve using the method of reduction and Euler method and compare the solutions:

1. $\begin{cases} u' = v + 1 \\ v' = u + 1 \end{cases}$.

Solution: $\lambda = -1, 1,$ $\begin{cases} u = C_1e^x + C_2e^{-x} - 1 \\ v = C_1e^x - C_2e^{-x} - 1 \end{cases}$

2. $\begin{cases} u' = v + x \\ v' = u - t \end{cases}$.

Solution: $\lambda = -1, 1,$ $\begin{cases} u = C_1e^x - C_2e^{-x} + x - 1 \\ v = C_1e^x + C_2e^{-x} - x + 1 \end{cases}$

3. $\begin{cases} u' = v + 2e^x \\ v' = u + x^2 \end{cases}$.

Solution: $\lambda = -1, 1,$ $\begin{cases} u = C_1e^x + C_2e^{-x} + xe^x - x^2 - 2 \\ v = C_1e^x - C_2e^{-x} + (x-1)e^x - 2x \end{cases}$

4. $\begin{cases} u' = v - 5\cos x \\ v' = 2u + v \end{cases}$.

Solution: $\lambda = -1, 2,$ $\begin{cases} u = C_1e^{2x} + C_2e^{-x} - 2\sin x - \cos x \\ v = 2C_1e^{2x} - C_2e^{-x} + \sin x + 3\cos x \end{cases}$

5. $\begin{cases} u' = 2u - 4v + 4e^{-2x} \\ v' = 2u - 2v \end{cases}$.

Solution: $\lambda = \pm 2i,$ $\begin{cases} u = C_1(\cos 2x - \sin 2x) + C_2(\cos 2x + \sin 2x) \\ v = C_1\cos 2x + C_2\sin 2x + e^{-2x} \end{cases}$

6. $\begin{cases} u' = 4u + v - e^{2x} \\ v' = v - 2u \end{cases}$.

Solution: $\lambda = 2, 3,$ $\begin{cases} u = C_1e^{2x} + C_2e^{3x} + (x+1)e^{2x} \\ v = -2C_1e^{2x} - C_2e^{3x} - 2xe^{2x} \end{cases}$

7. $\begin{cases} u' = 2v - u + 1 \\ v' = 3v - 2u \end{cases}$.

Solution: $\lambda = 1, 1,$ $\begin{cases} u = (C_1 + 2C_2 x)e^x - 3 \\ v = (C_1 + C_2 + 2C_2 x)e^x - 2 \end{cases}$

8. $\begin{cases} u' = 2u + v + e^x \\ v' = -2u + 2x \end{cases}$.

Solution: $\lambda = 1 \pm i,$

$\begin{cases} u = (C_1 \cos x + C_2 \sin x)e^x + e^x + x + 1 \\ v = (-C_1(\cos x + \sin x) + C_2(\cos x - \sin x))e^x - 2e^x - 2x - 1 \end{cases}$

9. $\begin{cases} u' = 2u - v \\ v' = v - 2u + 18x \end{cases}$.

Solution: $\lambda = 0, 3,$ $\begin{cases} u = C_1 e^{3x} + C_2 + 3x^2 + 2x \\ v = -C_1 e^{3x} + 2C_2 + 6x^2 - 2x - 2 \end{cases}$

10. $\begin{cases} u' = u - v + 2\sin x \\ v' = 2u - v \end{cases}$.

Solution: $\lambda = \pm i,$

$\begin{cases} u = C_1 \cos x + C_2 \sin x + x \sin x - x \cos x \\ v = C_1(\sin x + \cos x) + C_2(\sin x - \cos x) - 2x \cos x + \sin x + \cos x \end{cases}$

11. $\begin{cases} u' = 2u - v \\ v' = u + 2e^x \end{cases}$.

Solution: $\lambda = 1, 1,$ $\begin{cases} u = (C_1 + C_2 x - x^2)e^x \\ v = (C_1 - C_2 + (C_2 + 2)x - x^2)e^x \end{cases}$

12. $\begin{cases} u' = 2u + v + 2e^x \\ v' = u + 2v - 3e^{4x} \end{cases}$.

Solution: $\lambda = 1, 3,$ $\begin{cases} u = C_1 e^x + C_2 e^{3x} + xe^x - e^{4x} \\ v = -C_1 e^x + C_2 e^{4x} - (x+1)e^x - 2e^{4x} \end{cases}$

13. $\begin{cases} u' = 2u - v \\ v' = 2v - u - 5e^x \sin x \end{cases}$.

Solution: $\lambda = 1, 3,$ $\begin{cases} u = C_1 e^x + C_2 e^{3x} + (2\cos x - \sin x)e^x \\ v = C_1 e^x - C_2 e^{3x} + (3\cos x + \sin x)e^x \end{cases}$

14. $\begin{cases} u' = -2u + 3v + 4w - 3x \\ v' = -6u + 7v + 6w + 1 - 7x \\ w' = u - v + w + x \end{cases}$.

$$\text{Solution: } \lambda = 1, 2, 3, \begin{cases} u = C_1 e^x + C_2 e^{2x} + C_3 e^{3x} \\ v = C_1 e^x + 3C_3 e^{3x} + x \\ w = C_2 e^{2x} - C_3 e^{3x} \end{cases}$$

15. $\begin{cases} u' = 4u + 3v - 3w \\ v' = -3u - 2v + 3w \\ w' = 3u + 3v - 2w + 2e^{-x} \end{cases}$.

$$\text{Solution: } \lambda = -2, 1, 1, \begin{cases} u = C_1 e^{-2x} + C_2 e^x + 3e^{-x} \\ v = -C_1 e^{-2x} + C_3 e^x - 3e^{-x} \\ w = C_1 e^{-2x} + C_2 e^x + C_3 e^x + 2e^{-x} \end{cases}$$

16. $\begin{cases} u' = u - 2v - w - 2e^x \\ v' = -u + v + w + 2e^x \\ w' = u - w - e^x \end{cases}$.

$$\text{Solution: } \lambda = -1, 0, 2, \begin{cases} u = C_2 + 3C_3 e^{2x} + 3e^x \\ v = C_1 e^{-x} - 2C_3 e^{2x} - \frac{3}{2} e^x \\ w = -2C_1 e^{-x} + C_2 + C_3 e^{2x} + e^x \end{cases}$$

Section 2.3 Method of variation of parameters (Lagrange method)

Solve applying the Lagrange method:

1. $\begin{cases} u' = -4u - 2v + \frac{2}{e^x - 1} \\ v' = 6u + 3v - \frac{3}{e^x - 1} \end{cases}$.

$$\text{Solution: } \lambda = -1, 0, \begin{cases} u = C_1 + 2C_2 e^{-x} + 2e^{-x} \ln |e^x - 1| \\ v = -2C_1 - 3C_2 e^{-x} - 3e^{-x} \ln |e^x - 1| \end{cases}$$

2. $\begin{cases} u' = 3u - 2v \\ v' = 2u - v + 15e^x \sqrt{x} \end{cases}$.

$$\text{Solution: } \lambda = 1, 1, \begin{cases} u = (C_1 + 2C_2 x - 8x^{5/2}) e^x \\ v = (C_1 + 2C_2 x - C_2 - 8x^{5/2} + 10x^{3/2}) e^x \end{cases}$$

3. $\begin{cases} u' = u - 2v \\ v' = u - v + \frac{1}{2 \sin x} \end{cases}$.

Solution: $\lambda = \pm i$,

$$\begin{cases} u = (x + C_1) \cos x + (-\ln |\sin x| + C_2) \sin x \\ v = \frac{1}{2}(x + C_1 + \ln |\sin x| - C_2) \cos x + (x + C_1 - \ln |\sin x| + C_2) \sin x \end{cases}$$

4. $\begin{cases} u' = 3u - 4v + \frac{e^x}{\sin 2x} \\ v' = 2u - v \end{cases}$.

Solution: $\lambda = 1 \pm 2i$,
$$\begin{cases} u = (C_1 + C_2 - x + \frac{1}{2}\ln|\sin 2x|)e^x \cos 2x \\ \quad + (C_1 - C_2 + x + \frac{1}{2}\ln|\sin 2x|)e^x \sin 2x \\ v = (C_2 - x)e^x \cos 2x + (C_1 + \frac{1}{2}\ln|\sin 2x|)e^x \sin 2x \end{cases}$$

5. $\begin{cases} u' = 3u + v \\ v' = -4u - v + \frac{e^x}{2\sqrt{x}} \end{cases}$.

Solution: $\lambda = 1 \pm 2i$, $\begin{cases} u = (-C_1(x+1) - C_2 + \frac{2}{3}x\sqrt{x})e^x \\ v = (C_1(2x+1) + 2C_2 + \sqrt{x} - \frac{4}{3}x\sqrt{x})e^x \end{cases}$

6. $\begin{cases} u' = 3u - 2v + \frac{e^{3x}}{e^x + 1} \\ v' = u - \frac{e^{3x}}{e^x + 1} \end{cases}$.

Solution: $\lambda = 1, 2$, $\begin{cases} u = C_1 e^x + 2C_2 e^{2x} - 3e^{2x} + (3e^x + 4e^{2x})\ln(e^x + 1) \\ v = C_1 e^x + C_2 e^{2x} - 3e^{2x} + (3e^x + 2e^{2x})\ln(e^x + 1) \end{cases}$

7. $\begin{cases} u' = -u - 2v + 2e^{-x} \\ v' = 3u + 4v + e^{-x} \end{cases}$.

Solution: $\lambda = 1, 2$, $\begin{cases} u = C_1 e^x + 2C_2 e^{2x} - 2e^{-x} \\ v = -C_1 e^x - 3C_2 e^{2x} + e^{-x} \end{cases}$

8. $\begin{cases} u' = -u - v + 4\cos 2x \\ v' = 3u - 2v + 8\cos 2x + 5\sin 2x \end{cases}$.

Solution: $\lambda = -1/2 \pm i\sqrt{3}/2$,
$$\begin{cases} u = 2e^{-x/2}\left(C_1 \cos \frac{\sqrt{3}}{2}x + C_2 \sin \frac{\sqrt{3}}{2}x\right) + 2\cos 2x + 3\sin 2x \\ v = e^{-x/2}\left((3C_1 - \sqrt{3}C_2)\cos \frac{\sqrt{3}}{2}x + (\sqrt{3}C_1 + 3C_2)\sin \frac{\sqrt{3}}{2}x\right) + 7\sin 2x \end{cases}$$

9. $\begin{cases} u' = v \\ v' = 4v - 5u + \frac{e^{2x}}{\cos x} \end{cases}$.

Solution: $\lambda = 2 \pm i$,
$$\begin{cases} u = e^{2x}(C_1 \cos x + C_2 \sin x) + e^{2x}(\cos x \ln|\cos x| + x \sin x) \\ v = e^{2x}((2C_1 + C_2)\cos x + (2C_2 - C_1)\sin x) \\ \quad + e^{2x}((2\cos x - \sin x)\ln|\cos x| + x(2\sin x + \cos x)) \end{cases}$$

10. $\begin{cases} u' = u + 2v - 9x \\ v' = 2u + v + 4e^x \end{cases}$.

Solution: $\lambda = -1, 3$, $\begin{cases} u = C_1 e^{-x} + C_2 e^{3x} + 5 - 3x - 2e^x \\ v = -C_1 e^{-x} + C_2 e^{3x} + 6x - 4 \end{cases}$

11. $\begin{cases} u' = -5u + v - 2w + e^{-x} + e^{-2x} \\ v' = -u - v + 3e^{-x} + 2e^{-2x} \\ w' = 6u - 2v + 2w - 2e^{-x} - 3e^{-2x} \end{cases}$.

Solution: $\lambda = -2, -1 \pm i,$

$$\begin{cases} u = C_1 e^{-2x} + C_2 \cos x e^{-x} + C_3 \sin x e^{-x} + 3e^{-x} - xe^{-2x} \\ v = C_1 e^{-2x} - C_2 \sin x e^{-x} + C_3 \cos x e^{-x} - e^{-x} - (x+3)e^{-2x} \\ w = -C_1 e^{-2x} - 2C_2 \cos x e^{-x} - 2C_3 \sin x e^{-x} - 6e^{-x} + (x - \tfrac{1}{2})e^{-2x} \end{cases}$$

12. $\begin{cases} u' = v \\ v' = w - \frac{\sin^3 x}{\cos^6 x} \\ w' = -u - v - w \end{cases}$.

Solution: $\lambda = -1, \pm i,$

$$\begin{cases} u = C_1 e^{-x} + (\tfrac{1}{5}\tan^5 x - C_2)\cos x - (\tfrac{1}{4}\tan^4 x + C_3)\sin x \\ v = -C_1 e^{-x} + (C_2 - \tfrac{1}{5}\tan^5 x)\sin x - (\tfrac{1}{4}\tan^4 x + C_3)\cos x \\ w = C_1 e^{-x} + (C_2 - \tfrac{1}{5}\tan^5 x)\cos x + (\tfrac{1}{4}\tan^4 x + C_3)\sin x \end{cases}$$

13. $\begin{cases} u' = 2u + 4v + w + e^{2x}\ln x \\ v' = 2v + w \\ w' = 4v - w \end{cases}$.

Solution: $\lambda = -2, 2, 3,$ $\begin{cases} u = (x\ln|x| - x + C_2)e^{2x} + 5C_3 e^{3x} \\ v = C_1 e^{-2x} + C_3 e^{3x} \\ w = -4C_1 e^{-2x} + C_3 e^{3x} \end{cases}$

14. $\begin{cases} u' = 2u + v - 3w + \tan x + \tan^2 x \\ v' = 3u - 2v - 3w + 1 \\ w' = u + v - 2w + \tan x + \tan^2 x \end{cases}$.

Solution: $\lambda = -2, -1, 1,$

$$\begin{cases} u = (C_1 - \tfrac{1}{2}e^{2x})e^{-2x} + (C_2 + \tan x \cdot e^x)e^{-x} + 2C_3 e^x \\ v = (-C_1 + \tfrac{1}{2}e^{2x})e^{-2x} + C_3 e^x \\ w = (C_1 - \tfrac{1}{2}e^{2x})e^{-2x} + (C_2 + \tan x \cdot e^x)e^{-x} + C_3 e^x \end{cases}$$

Exercises for section 2

Solve the following systems; if possible, apply different methods and compare the obtained solutions:

1. $\begin{cases} u' = v + \cos 2x - 2\sin 2x \\ v' = -u + 2v + 2\sin 2x + 3\cos 2x \end{cases}$.

Solution: $\lambda = 1, 1,$ $\begin{cases} u = C_1 e^x + C_2 x e^x + \cos 2x \\ v = C_1 e^x + C_2(x+1)e^x - \cos 2x \end{cases}$

2. $\begin{cases} u' = u - 2v - 2xe^x \\ v' = 5u - v - (2x+6)e^x \end{cases}$.

Solution: $\lambda = \pm 3i,$

$$\begin{cases} u = 2C_1\cos 3x + 2C_2\sin 3x + e^x \\ v = C_1(\cos 3x + 3\sin 3x) + C_2(\sin 3x - 3\cos 3x) - xe^x \end{cases}$$

3. $\begin{cases} u' = u - v + \cos xe^x \\ v' = u + v + \sin xe^x \end{cases}$.

Solution: $\lambda = 1 \pm i,$ $\begin{cases} u = C_1\cos xe^x + C_2\sin xe^x + x\cos xe^x \\ v = C_1\sin xe^x - C_2\cos xe^x + x\sin xe^x \end{cases}$

4. $\begin{cases} u' = 4u - 2v + \frac{1}{x^3} \\ v' = 8u - 4v - \frac{1}{x^2} \end{cases}$.

Solution: $\lambda = 0, 0,$ $\begin{cases} u = C_1 + C_2 x - \frac{1}{2x^2} + \frac{2}{x} - 2\ln x \\ v = 2C_1 + C_2(2x - \frac{1}{2}) + \frac{5}{x} - 4\ln x \end{cases}$

5. $\begin{cases} u' = -2u + 4v + \frac{1}{1+e^x} \\ v' = -2u + 4v - \frac{1}{1+e^x} \end{cases}$.

Solution: $\lambda = 0, 2,$

$$\begin{cases} u = 2C_1 + C_2 e^{2x} + (3e^{2x} - 4)\ln(1 + e^{-x}) - 3e^x + \frac{3}{2} \\ v = -C_1 + C_2 e^{2x} + (3e^{2x} - 2)\ln(1 + e^{-x}) - 3e^x + \frac{3}{2} \end{cases}$$

6. $\begin{cases} u' = 4u - 8v + \tan 4x \\ v' = 4u - 4v \end{cases}$.

Solution: $\lambda = \pm 4i,$

$$\begin{cases} u = C_1(\cos 4x + \sin 4x) + C_2(\cos 4x - \sin 4x) \\ \quad + \frac{1}{8}\left((\sin 4x - \cos 4x)\ln\left|\frac{1+\sin 4x}{1-\sin 4x}\right| - 2\right) \\ v = C_1\sin 4x + C_2\cos 4x - \frac{1}{8}\cos 4x\ln\left|\frac{1+\sin 4x}{1-\sin 4x}\right| \end{cases}$$

7. $\begin{cases} u' = 3u - 6v + \frac{1}{\cos^3 3x} \\ v' = 3u - 3v \end{cases}$.

Solution: $\lambda = \pm 3i,$

$$\begin{cases} u = (C_1 + C_2)\cos 3x + (C_1 - C_2)\sin 3x \\ \quad + \frac{1}{6\cos^2 3x}(2\sin 3x + \cos 6x(\sin 3x - \cos 3x)) \\ v = C_1\sin 3x + C_2\cos 3x - \frac{\cos 6x}{6\cos 3x} \end{cases}$$

8. $\begin{cases} u' = -3u + v \\ v' = -4u + v + \frac{1}{xe^x} \end{cases}$. Solution: $\lambda = -1, -1,$

$$\begin{cases} u = (C_1 + C_2 x)e^{-x} + xe^{-x}(\ln|x| - 1) \\ v = C_1(2x + 1)e^{-x} + 2C_2 e^{-x} + e^{-x}(2x\ln|x| + \ln|x| - 2x) \end{cases}$$

9. $\begin{cases} u' = 2u + v - \ln x \\ v' = -4u - 2v + \ln x \end{cases}$.

Solution: $\lambda = 0, 0$, $\begin{cases} u = -C_1(x+1) - C_2 + x + \frac{3}{4}x^2 - (x + \frac{1}{2}x^2)\ln x \\ v = C_1(2x+1) + 2C_2 - x - \frac{3}{2}x^2 + (x + x^2)\ln x \end{cases}$

10. $\begin{cases} u' = -3u - 2v + \frac{e^{2x}}{1+e^x} \\ v' = 10u + 6v \end{cases}$.

Solution: $\lambda = 1, 2$,
$\begin{cases} u = -2C_1 e^{2x} - C_2 e^x + 5e^x \ln(1 + e^x) + 4e^{2x}\ln(1 + e^{-x}) \\ v = 5C_1 e^{2x} + 2C_2 e^x - 10e^x \ln(1 + e^x) - 10e^{2x}\ln(1 + e^{-x}) \end{cases}$

11. $\begin{cases} u' = 5u - 6v + \frac{3e^{2x}}{\cos^3 3x} \\ v' = 3u - v \end{cases}$.

Solution: $\lambda = 2 \pm 3i$,
$\begin{cases} u = (C_1 + C_2)e^{2x}\cos 3x + (C_1 - C_2)e^{2x}\sin 3x \\ \quad + e^{2x}\tan 3x(\sin 3x + \cos 3x) + \frac{e^{2x}}{2\cos^2 3x}(\sin 3x - \cos 3x) \\ v = C_1 e^{2x}\sin 3x + C_2 e^{2x}\cos 3x - \frac{e^{2x}\cos 6x}{\cos 3x} \end{cases}$

12. $\begin{cases} u' = -2u + v + x\ln x \\ v' = -4u + 2v + 2x\ln x \end{cases}$.

Solution: $\lambda = 0, 0$, $\begin{cases} u = C_1 + C_2 x + \frac{1}{4}x^2(2\ln x - 1) \\ v = 2C_1 + C_2(2x+1) + \frac{1}{2}x^2(2\ln x - 1) \end{cases}$

13. $\begin{cases} u' = 4u - 2v \\ v' = 8u - 4v + \sqrt{x} \end{cases}$.

Solution: $\lambda = 0, 0$, $\begin{cases} u = C_1 + C_2 x - \frac{8}{15}x^{5/2} \\ v = 2C_1 + C_2(x - \frac{1}{2}) - \frac{16}{15}x^{5/2} + \frac{2}{3}x^{3/2} \end{cases}$

14. $\begin{cases} u' = 3u + 2v - \frac{1}{1+e^{-x}} \\ v' = -3u - 2v - \frac{1}{1+e^{-x}} \end{cases}$. Solution: $\lambda = 0, 1$,

$\begin{cases} u = -2C_1 - C_2 e^x + 4\ln(1 + e^x) + 5e^x \ln(1 + e^{-x}) \\ v = 3C_1 + C_2 e^x - 6\ln(1 + e^x) - 5e^x \ln(1 + e^{-x}) \end{cases}$

15. $\begin{cases} u' = 2u - v + w + \cos x \\ v' = 5u - 4v + 3w + \sin x \\ w' = 4u - 4v + 3w + 2\sin x - 2\cos x \end{cases}$.

Solution: $\lambda = -1, 1, 1$, $\begin{cases} u = C_2 e^x + C_3 x e^x + \sin x \\ v = C_1 e^{-x} + C_2 e^x + C_3(x+1)e^x \\ w = C_1 e^{-x} + 2C_3 e^x - 2\sin x \end{cases}$

16. $\begin{cases} u' = 2u + v - 3w + 2e^{2x} \\ v' = 3u - 2v - 3w - 2e^{2x} \\ w' = u + v - 2w \end{cases}$.

Solution: $\lambda = -2, -1, 1,$ $\begin{cases} u = C_1 e^{-2x} + C_2 e^{-x} + 2C_3 e^x + 3e^{2x} \\ v = -C_1 e^{-2x} + C_3 e^x + e^{2x} \\ w = C_1 e^{-2x} + C_2 e^{-x} + C_3 e^x + e^{2x} \end{cases}$

17. $\begin{cases} u' = 2u + v - 2w - x + 2 \\ v' = -u + 1 \\ w' = u + v - w - x + 1 \end{cases}$.

Solution: $\lambda = 1, \pm i,$ $\begin{cases} u = C_1 e^x + C_2 \sin x + C_3 \cos x \\ v = -C_1 e^x + C_2 \cos x - C_3 \sin x + x \\ w = C_2 \sin x + C_3 \cos x + 1 \end{cases}$

18. $\begin{cases} u' = -u + v + w + e^x \\ v' = u - v + w + e^{3x} \\ w' = u + v + w + 4 \end{cases}$.

Solution: $\lambda = -2, -1, 2,$
$\begin{cases} u = C_1 e^{-2x} + C_2 e^{-x} + C_3 e^{2x} + \frac{1}{6} e^x + \frac{3}{20} e^{3x} - 2 \\ v = -C_1 e^{-2x} + C_2 e^{-x} + C_3 e^{2x} - \frac{1}{6} e^x + \frac{7}{20} e^{3x} - 2 \\ w = -C_2 e^{-x} + 2C_3 e^{2x} - \frac{1}{2} e^x + \frac{1}{4} e^{3x} \end{cases}$

19. $\begin{cases} u' = 2u - v + w - 2e^{-x} \\ v' = u + 2v - w - e^{-x} \\ w' = u - v + 2w - 3e^{-x} \end{cases}$.

Solution: $\lambda = 1, 2, 3,$ $\begin{cases} u = C_2 e^{2x} + C_3 e^{3x} + \frac{1}{2} e^{-x} \\ v = C_1 e^x + C_2 e^{2x} + \frac{1}{2} e^{-x} \\ w = C_1 e^x + C_2 e^{2x} + c_3 e^{3x} + e^{-x} \end{cases}$

20. $\begin{cases} u' = 2u - 3v + x \\ v' = u - 2w - 3x^2 \\ w' = -v + 2w + 3x - 2 \end{cases}$.

Solution: $\lambda = 1, 1, 2,$
$\begin{cases} u = 2C_1 e^{2x} + 3C_2 e^x + 3C_3 x e^x + 9x^2 - 20x - 79 \\ v = C_2 e^x + C_3(x - 1)e^x + 6x^2 - 19x - 46 \\ w = C_1 e^{2x} + C_2 e^x + C_3 x e^x + 3x^2 - 16x - 30 \end{cases}$

Section 3 Solution of the Cauchy problem (initial value prob-

lem)

Solve the following Cauchy problems:

1. $\begin{cases} u' = 3u + 8v \\ v' = -u - 3v \end{cases}$, $\begin{cases} u(0) = 6 \\ v(0) = -2 \end{cases}$.

Solution: $\lambda = -1, 1$, $\begin{cases} u = -4e^x + 2e^{-x} \\ v = -e^x - e^{-x} \end{cases}$

2. $\begin{cases} u' + 3u + 4v = 0 \\ v' + 2u + 5v = 0 \end{cases}$, $\begin{cases} u(0) = 1 \\ v(0) = 4 \end{cases}$.

Solution: $\lambda = -1, -7$, $\begin{cases} u = -2e^{-x} + 3e^{-7x} \\ v = e^{-x} + 3e^{-7x} \end{cases}$

3. $\begin{cases} u' = u + v \\ v' = 4v - 2u \end{cases}$, $\begin{cases} u(0) = 0 \\ v(0) = -1 \end{cases}$.

Solution: $\lambda = 2, 3$, $\begin{cases} u = e^{2x} - e^{3x} \\ v = e^{2x} - 2e^{3x} \end{cases}$

4. $\begin{cases} u' = 4u - 5v \\ v' = u \end{cases}$, $\begin{cases} u(0) = 0 \\ v(0) = 1 \end{cases}$.

Solution: $\lambda = 2 \pm i$, $\begin{cases} u = -5e^{2x} \sin x \\ v = e^{2x}(\cos x - 2\sin x) \end{cases}$

5. $\begin{cases} u' = 2u - v + w \\ v' = u + w \\ w' = v - 2w - 3u \end{cases}$, $\begin{cases} u(0) = 0 \\ v(0) = 0 \\ w(0) = 1 \end{cases}$.

Solution: $\lambda = 0, \pm 1$, $\begin{cases} u = 1 - e^{-x} \\ v = 1 - e^{-x} \\ w = 2e^{-x} - 1 \end{cases}$

6. $\begin{cases} u' = v - w \\ v' = -v + w \\ w' = u - w \end{cases}$, $\begin{cases} u(0) = 0 \\ v(0) = 0 \\ w(0) = 1 \end{cases}$.

Solution: $\lambda = 0, -1 \pm i$, $\begin{cases} u = -e^{-x} \sin x \\ v = e^{-x} \sin x \\ w = e^{-x} \cos x \end{cases}$

7. $\begin{cases} u' = u - w \\ v' = v + w \\ w' = -u - v - w \end{cases}$, $\begin{cases} u(0) = 1 \\ v(0) = 1 \\ w(0) = -1 \end{cases}$.

Solution: $\lambda = -1, 1, 1,$ $\begin{cases} u = (1+x)e^x \\ v = (1-x)e^x \\ w = -e^x \end{cases}$

8. $\begin{cases} u' = 2u - v + w + 1 + e^{-x} \\ v' = 2u - v - 2w + 1 \\ w' = -u + v + 2w - 1 + e^{-x} \end{cases}$, $\begin{cases} u(0) = 0 \\ v(0) = 0 \\ w(0) = 0 \end{cases}$

Solution: $\lambda = 0, 1, 2,$ $\begin{cases} u = 0 \\ v = 0 \\ w = 1 - e^{-x} \end{cases}$.

9. $\begin{cases} u' = u + 2v - 9x \\ v' = 2u + v + 4e^x \end{cases}$, $\begin{cases} u(0) = 1 \\ v(0) = 2 \end{cases}$.

Solution: $\lambda = -1, 3,$ $\begin{cases} u = 2e^{3x} - 4e^{-x} + 5 - 3x - 2e^x \\ v = 2e^{3x} + 4e^{-x} + 6x - 4 \end{cases}$

10. $\begin{cases} u' = v + \tan^2 x - 1 \\ v' = -u + \tan x \end{cases}$, $\begin{cases} u(0) = 1 \\ v(0) = 3 \end{cases}$.

Solution: $\lambda = \pm i,$ $\begin{cases} u = \cos x + \sin x + \tan x \\ v = -\sin x + \cos x + 2 \end{cases}$

11. $\begin{cases} u' = 4u - 5v + 4x + 1 \\ v' = u - 2v + x \end{cases}$, $\begin{cases} u(0) = 1 \\ v(0) = 2 \end{cases}$.

Solution: $\lambda = -1, 3,$ $\begin{cases} u = \frac{11}{4}e^{-x} - \frac{5}{12}e^{3x} + \frac{1}{4}(x-2) - \frac{5}{12}(3x+2) \\ v = \frac{11}{4}e^{-x} - \frac{1}{12}e^{3x} + \frac{1}{4}(x-2) - \frac{1}{12}(3x+2) \end{cases}$

12. $\begin{cases} u' = 3u - v + w + e^x \\ v' = u + v + w - x \\ w' = 4u - v + 4w \end{cases}$, $\begin{cases} u(0) = \frac{41}{100} \\ v(0) = \frac{166}{100} \\ w(0) = -\frac{2}{100} \end{cases}$.

Solution: $\lambda = 1, 2, 5,$ $\begin{cases} u = \frac{1}{4}xe^x + \frac{3}{10}x + \frac{41}{100} \\ v = \frac{1}{4}xe^x + e^x + \frac{4}{5}x + \frac{33}{50} \\ w = -\frac{1}{4}xe^x + \frac{1}{4}e^x - \frac{1}{10}x - \frac{27}{100} \end{cases}$

Index

www.ingramcontent.com/pod-product-compliance
Lightning Source LLC
Chambersburg PA
CBHW031620210326
41599CB00021B/3242